W0106382

Springer Japan KK

Y. Maruyama, M. Hori, J.S. Janicki (Eds.)

Cardiac–Vascular Remodeling and Functional Interaction

With 286 Figures

Springer

Editors

Yukio Maruyama, M.D., Ph.D.
Professor and Chairman, First Department of Internal Medicine
Fukushima Medical College, Hikariga-oka 1, Fukushima, 960-12 Japan

Masatsugu Hori, M.D., Ph.D.
Chief, First Department of Medicine, Osaka University School of Medicine
2-2 Yamadaoka, Suita, Osaka, 565 Japan

Joseph S. Janicki, Ph.D.
Professor, Associate Dean of Research and Graduate Studies
College of Veterinary Medicine, Auburn University
106 Greene Hall, Auburn, AL 36849-5517, USA

Associate Editor

Kazuhira Maehara, M.D., Ph.D.
Associate Professor, First Department of Internal Medicine
Fukushima Medical College, Hikariga-oka 1, Fukushima, 960-12 Japan

ISBN 978-4-431-67043-8

Library of Congress Cataloging-in-Publication Data

Cardiac-vascular remodeling and functional interaction / Y. Maruyama, M. Hori, J.S. Janicki (eds.).
p. cm.
Includes bibliographical references and index.
ISBN 978-4-431-67043-8 ISBN 978-4-431-67041-4 (eBook)
DOI 10.1007/978-4-431-67041-4
1. Cardiovascular system—Pathophysiology—Congresses. 2. Cardiovascular system—Physiology—Congresses. 3. Adaptation (Physiology)—Congresses. I. Maruyama, Yukio, 1941- . II. Hori, M. (Masatsugu), 1945- . III. Janicki, Joseph S.
[DNLM: 1. Heart Diseases—physiopathology—congresses. 2. Ventricular Function, Left—physiology—congresses. 3. Myocardial Contraction—physiology—congresses. 4. Myocardium—metabolism—congresses. WG 210 C2673 1997]
RC669.9.C29 1997
616.1'07—dc20
DNLM/DLC
for Library of Congress 96-38651

Printed on acid-free paper

Preface

Cardiac performance is regulated not only by cardiac muscle properties but also by several other factors, including those associated with the neurohumoral system and the mechanical characteristics of the peripheral circulation. New information concerning these regulatory factors has furthered our understanding of the pathophysiology of cardiac dysfunction. However, controversy remains, along with a need to integrate these multidisciplinary findings. It was with this in mind, together with my continuing interest in the response of the normal and diseased heart to variations in loading conditions, that the satellite symposium entitled "Interactions Between Cardiac Function and Vascular Dynamics" was organized and dedicated to my mentor, Dr. T. Takishima.

The symposium was held in Fukushima, Japan, in 1992 following the Tenth International Conference of the Cardiovascular Systems Dynamics Society in Kobe, Japan, which was organized by the then president of the society, Dr. Masatsugu Hori. The Fukushima symposium and the Kobe conference were stimulating and informative. To commemorate these events, Dr. Hori, Dr. Janicki, and I decided to publish this book. It covers topics that were presented then as well as pertinent new material. As a result, the book includes not only updated reviews but also up-to-date findings that were not considered at the two scientific sessions. The high level attained in this book is due to the outstanding contributions from internationally renowned scientists. This final product of their efforts should prove to be a valuable source of information to the reader.

I am deeply indebted to all of the authors for their time and effort despite their busy schedules. I would also like to thank my colleagues at the Fukushima Medical College and the editorial staff of Springer-Verlag Tokyo for their continued generous support in the publication of this book.

Yukio Maruyama
April 1996

Contents

Part 3 Ventricular–Vascular Interaction

List of Contributors

Smits, J.F.M. 5
Spaan, J.A.E. 351
Starling, M.R. 229
Stergiopulos, N. 181
Suga, H. 381
Sugimachi, M. 189
Sunagawa, K. 189, 283
Suzuki, S. 381
Sys, S.U. 163
Tada, M. 67
Takaki, M. 381
Takeshita, A. 267
ter Keurs, H.E.D.J. 115
Tyberg, J.V. 257
van den Bos, G.C. 181
Wang, S.Y. 257
Westerhof, N. 181
Yada, T. 309
Yamashita, N. 67
Yano, M. 203
Yokoyama, M. 367

Introduction

Joseph S. Janicki[1], Masatsugu Hori[2], and Yukio Maruyama[3]

To meet the metabolic requirements of the body over a wide range of physical activity, the heart must be capable of increasing its cardiac output seven-to eightfold. This is accomplished by acute increases in heart rate and stroke volume. The elevations in stroke volume are produced through coordinated adjustments to the cardiac, respiratory, and peripheral circulatory systems which result in 1) enhanced myocardial contractility, 2) decreased systemic vascular impedance (i.e., reduced afterload), and/or 3) increased venous return leading to larger ventricular diastolic volumes (i.e., increased preload).

Over the past 20 years a great deal of research into extrinsic, physiologic factors which influence these direct determinants of stroke volume has been performed. The results of these investigative efforts in normal and diseased hearts are reviewed in the subsections entitled "Systolic–Diastolic Interaction" and "Ventricular–Vascular Interaction." These subsections contain chapters devoted to the interplay between the capacitance and conductance of the venous system and ventricular preload, to the extrinsic factors (i.e., ventricular interaction and pericardium) which affect the left-ventricular diastolic pressure–volume relation and restoring force, and to the mechanical characteristics of the systemic and pulmonary circulatory systems which influence the afterload of the left and right ventricles, respectively.

The manner by which these factors and their interaction with the heart are adversely altered in heart failure, hypertension, mitral valve regurgitation, and aging is also considered by several of the contributing authors. Evidence is presented which indicates that 1) an impaired ability of the ventricle to generate restoring forces may play a role in determining the level of exercise tolerance in patients with systolic ventricular dysfunction; 2) ventricular-arterial coupling in dysfunctional hearts is not optimal; 3) patients with reduced exercise capacity secondary to left-ventricular dysfunction have an inadequate decrease in afterload during exercise; 4) left-ventricular pump efficiency for performing forward stroke work is impaired in patients with long-term mitral regurgitation despite the outward appearance of normal left-ventricular performance; 5) there exists an age-related disturbance in ventriculo-vascular coupling; and 6) a defective endothelial function in patients with heart failure may contribute to abnormal control of regional blood flow.

---

[1] Department of Physiology and Pharmacology, Auburn University, Auburn, AL 36849, USA.
[2] First Department of Medicine, Osaka University Medical School, Suita, Osaka, 565 Japan.
[3] First Department of Internal Medicine, Fukushima Medical Collage, Fukushima, 960-12 Japan.

Another important factor in determining ventricular function is the maintenance of an adequate delivery of oxygen to the myocardium. The interaction between the coronary circulation and ventricular function is the focus of the last subsection, entitled "Coronary Circulation–Ventricular Interaction." In one chapter, the concept of mechanoenergetics is considered using the unifying myocardial oxygen consumption–pressure volume area–maximum elastance framework. Some of the findings using this approach appear to be similar to what has been described as the Gregg phenomenon. Other chapters in this subsection are concerned with the mechanism for this phenomenon as well as with the existence of a "venous Gregg effect." The importance of the dynamic interplay between the interstitial and intravascular compartments of the heart and the relation between microvascular resistance and heart rate, contractility, and systolic and diastolic blood pressures are also stressed. Finally, the functional role of the "coronary slosh phenomenon" and the ability of nitric oxide to regulate coronary flow as well as ventricular function are considered.

In addition to adverse alterations in the factors which regulate the responses of stroke volume and coronary perfusion to physical activity, the heart and vasculature undergo remodeling to compensate for chronic elevations in preload and/or afterload which in turn affects systolic–diastolic, ventricular–vascular and coronary circulation and ventricular interactions. This remodeling is considered in the first subsection, entitled "Ventricular–Vascular Adaptation and Remodeling." The nature and possible mechanisms of alterations in the coronary and systemic vascular systems during cardiac hypertrophy, ischemia, and heart failure are discussed in this subsection. Also the alteration (i.e., collagen synthesis and degradation) of the extracellular matrix of the heart in response to chronic increases in preload and afterload and the effect of this on ventricular systolic and diastolic function are presented. Recently the myocardium has been shown to contain a local renin–angiotensin system. The role that this system may play in the hypertrophic response to pressure and volume overload is also reviewed.

From this overview, it can be seen that there are many factors involved in the regulation of ventricular function. These factors are intricately interrelated so that an abnormality in one will impact negatively on the other factors as well as on ventricular function. As a consequence vicious cycles develop which lead to ventricular decompensation and heart failure. Thus, it is only through an understanding of these factors and their interrelations with the heart that the mechanisms responsible for decompensation and pathogenesis of heart failure will be delineated. We hope that the material to follow will provide a firm foundation for the attainment of this goal.

Part 1
Ventricular–Vascular Adaptation and Remodeling

Vascular Remodeling During Heart Failure

Sylvia Heeneman, Jos F.M. Smits, and Mat J.A.P. Daemen

Summary. This chapter describes the nature and possible mechanisms of alterations in the vascular system during cardiac hypertrophy and heart failure. Next to changes in cardiac muscle function, heart failure is associated with alterations in the vasculature. Functional and structural alterations in the coronary circulation, including a failure of adaptive growth of the coronary circulation, may contribute to the loss of muscle function of the heart and may be one of the factors that cause the transition from compensated cardiac hypertrophy to cardiac decompensation. In the last decade it has become clear that cardiac failure is also associated with changes in the peripheral circulation. Peripheral vascular alterations such as an increased aortic impedance, excessive vasoconstriction, and endothelium dysfunction may contribute to the deterioration of cardiac function and to the clinical symptoms of heart failure patients such as exercise intolerance and peripheral muscle fatigue. Potential mechanisms of peripheral vascular alterations are desensitization to neurohormonal activation and vascular remodeling.

Key words. Heart failure—Vascular remodeling—Coronary circulation—Peripheral circulation—Renin-angiotensin system—Atrial natriuretic factor—Nitric oxide

Introduction

The clinical syndrome of heart failure evolves from a mismatch between left-ventricular pump function and vascular impedance [1], usually resulting from a disturbed left-ventricular function. A loss of cardiac function induces a series of compensatory mechanisms to maintain blood pressure and sustain perfusion of vital organs. Compensatory mechanisms include the development of myocardial hypertrophy, activation of a number of neurohormonal responses, and coronary and peripheral vascular alterations. Prolonged or abnormal activation of these compensatory mechanisms may result in a vicious circle, with further deterioration of cardiac function. The final result is cardiac failure (Fig. 1).

Departments of Pathology and Pharmacology, Cardiovascular Research Institute Maastricht, University of Limburg, 6200MD Maastricht, The Netherlands.

LEFT VENTRICULAR DYSFUNCTION

⇓

COMPENSATING MECHANISMS

⇓

NEUROHUMORAL DESENSITIZATION
NEUROHUMORAL DYSBALANCE

⇓

LEFT VENTRICULAR FAILURE

Fig. 1. Mechanisms involved in the progression of left-ventricular dysfunction to heart failure

In the past, much research has focused on compensating mechanisms of the heart muscle itself, but in recent years it has become clear that changes in the periphery and especially in the peripheral skeletal muscle and circulation are also important in the development of heart failure. This fact is illustrated by the observation that the degree of ventricular dysfunction does not correlate with the clinical severity of heart failure [2], and that patients with a normal systolic function display a number of symptoms of heart failure [3,4]. Thus, factors other than ventricular dysfunction per se determine the clinical status of the patient with heart failure. Possible candidates are (1) alterations in the peripheral skeletal structure and function and (2) alterations in the peripheral vascular structure and function.

Skeletal muscle abnormalities, which may contribute to reduced exercise capacity during heart failure, have been studied with a variety of techniques. Metabolic studies have been performed with phosphorus-31 nuclear magnetic resonance (^{31}P-NMR) using the ratio of inorganic phosphate and phosphocreatinine (P_i/PCr). During exercise, adenosine diphosphate (ADP) determines mitochondrial oxidative phosphorylation in the skeletal muscles. The inorganic phosphate/phosphocreatinine ratio (P_i/PCr) correlates with ADP concentration. Thus, by measuring this ratio at different work levels by ^{31}P-NMR, alterations in the control of oxidative phosphorylation can be identified.

Glycolysis increases intracellular lactate and decreases muscle pH. Changes in muscle pH correlate with glycolytic activity during exercise [5]. Patients with heart failure show a pronounced increase in the P_i/PCr ratio and a drop in muscle pH [6–8], indicating muscle ischemia and stimulation of anaerobic glycolysis during heart failure. These studies also show that metabolic abnormalities do not result from muscle underperfusion per se, as forearm blood flow does not change during the experiments. Therefore, structural skeletal muscle abnormalities were considered as a possible cause for these alterations.

Muscle biopsy studies indeed show marked changes in muscle structure of patients with heart failure, including a smaller percentage of slow-twitch type I fibers and a higher percentage of fast-twitch, glycolytic type II fibers [9–11]. Enzyme studies showed a maintenance of glycolytic muscle metabolism in patients with heart failure [9], but decreases in mitochondrial oxidative enzymes [10]. Also, reductions in mitochondrial volume density and surface density of mitochondrial cristae were observed in patients with heart failure [11]. The result is a muscle with low aerobic capacity and excess glycolytic metabolism.

The biochemical and structural changes in the skeletal muscle of patients with heart failure are comparable to those found during deconditioning in healthy subjects

[12,13]. In addition, exercise training improves the impaired oxidative capacity and reverses biochemical and structural changes of the skeletal muscle of patients with heart failure [14–16]. It is not clear whether decreased blood flow contributes to the changes in skeletal muscle metabolism and structure. Although several studies have shown no changes in blood flow during exercise [6–8], at least one study did show impaired flow to muscle in the same conditions [17]. Also, blood flow redistribution to the working muscle must be considered because it has been shown that less flow is directed toward red "oxidative" fibers despite an adequate increase in systemic blood flow [18].

Changes in the vascular system include alterations in both the coronary and the peripheral circulation. A maladaptive growth of the coronary circulation has been observed in several models of cardiac hypertrophy. Alterations in the peripheral circulation include changes in peripheral vascular tone, decreased arterial compliance, excessive vasoconstriction, and endothelium dysfunction. These changes, as well as activation of neurohumoral mechanisms and skeletal muscle alterations, contribute to the development of clinical symptoms of exercise intolerance, dyspnea, and fatigue (Table 1).

There are few data available on the potential mechanisms leading to peripheral vascular alterations in heart failure. Possible factors described in the literature include desensitization to the neurohormonal activation, physical deconditioning, and vascular remodeling. In this chapter, the nature of peripheral alterations in heart failure is described. Possible control mechanisms of the structure and function of the peripheral vasculature are outlined with special emphasis on the role of the renin-angiotensin system.

Vascular Alterations During Heart Failure

Vascular Alterations in the Heart

Expansion of the cardiac mass during cardiac hypertrophy results mostly from an increase in cardiac myocyte volume and only to a small extent from an increase in myocyte number [19–21]. Although changes in the myocyte compartment are important for the maintenance of cardiac function, changes in the nonmyocyte compartment, which represents most of the numbers of cells in the heart, also contribute to

TABLE 1. Compensatory mechanisms induced after a loss of cardiac function.

Cardiac alterations
Myocyte
Vascular
Extracellular matrix
Peripheral alterations
Skeletal muscle
Vasculature
Increase in arterial compliance
Vasoconstriction
Endothelial dysfunction
Structural changes

(the loss of) cardiac function in hypertrophy and failure. Nonmyocytes include interstitial fibroblasts and endothelial cells.

Cardiac hypertrophy after rat myocardial infarction is associated with increases in interstitial DNA synthesis [22] and interstitial collagen [23]. Both the increased interstitial DNA synthesis and collagen could be blocked by administration of the angiotensin-converting enzyme (ACE) inhibitor captopril [22], indicating a role for the renin–angiotensin system in the control of interstitial cell growth and interstitial collagen deposition in cardiac hypertrophy.

Endothelial cells constitute another important cell population of the cardiac interstitium. Although cardiac hypertrophy is associated with an increase in the number and length of capillaries, the capillary-to-fiber ratio decreases, resulting in a relative energy starvation. This decrease in capillary-to-fiber ratio is present in pressure- and infarct-induced cardiac hypertrophy, but not in volume overload-induced hypertrophy. The role of angiotensin II (AII) in this "angiogenic" response is not clarified. In the rat infarct model, both the AT1 antagonist losartan [24] and the AT2 antagonist PD123.319 partly inhibit interstitial cell DNA synthesis [25]. This suggests a stimulatory role for AII on interstitial (endothelial and fibroblast) DNA synthesis. Because AT1, but not AT2, receptor blockade reduces collagen increase after myocardial infarct (MI), we postulate that AT2 receptors mediate DNA synthesis in the endothelial cells [24,25].

Peripheral Vascular Alterations

Changes in Aortic Compliance

Afterload, which is an important determinant of cardiac output, consists of a static resistance component (determined by the resistance arteries) and a pulsatile component (determined by the elastic properties of the large-conduit arteries) [26]. Arterial compliance is a contributing factor to aortic impedance and depends on the elastic properties of the large conduit arteries. Changes in compliance may contribute to the disruption of optimal ventriculoarterial coupling that is seen in heart failure [27]. It has been shown that aortic impedance is increased during ventricular failure in animal models of heart failure [28,29] and in human heart failure [30–33]. Also, these changes may precede alterations in peripheral vascular resistance [28].

Peripheral Vasoconstriction

Next to possible changes in the pulsatile component of afterload, changes in the resistance component during heart failure are well known. The early observations of Zelis et al. showed that patients with heart failure exhibit reduced responses to vasodilator stimuli [34] and abnormal blood flow responses to exercise [35]. Also, attenuated increases in cardiac output and stroke volume following exercise were found in patients with heart failure [36], all observations indicating excessive peripheral vasoconstriction during heart failure. Vasoconstriction and a reduced capacity to maximal vasodilation [34] (see following) may prevent the development of severe hypotension in situations of prominent vasodilation such as in exercise [37].

Endothelial Dysfunction

Attention has recently been focused on an impairment of endothelium-dependent vasodilation during heart failure. The vascular endothelium modulates vascular smooth muscle relaxation via endothelium-derived relaxing factor (EDRF). Palmer et al. [38] showed that nitric oxide (NO) is responsible for this EDRF activity. In a canine model of heart failure, endothelium-dependent responses to acetylcholine in the femoral artery were found to be depressed [39], but endothelium-independent relaxation to nitroglycerine remained unchanged. This observation was later confirmed in various other studies, both in humans [40,41] and in animal models for heart failure [42,43]. All these studies suggest that the basal release of NO is preserved in heart failure while the stimulated release of NO in the peripheral resistance vessels is impaired. Of interest is the observation that in the rat myocardial infarct model endothelium dysfunction even precedes the onset of heart failure [44].

Possible Mechanisms of Vascular Alterations in Heart Failure

Several mechanisms may be responsible for the vascular changes just described during heart failure, including (abnormal) neurohumoral activation and vascular remodeling (mediated by chronic reductions in flow) or (abnormal) neurohumoral activation.

Abnormal Neurohumoral Activation

Ventricular systolic dysfunction stimulates the activity of a variety of neurohumoral systems, such as the sympathetic nervous system. Increased release of norepinephrine enhances myocardial contractility and results in vasoconstriction, which both contribute to short-term blood pressure maintenance. Increases in plasma norepinephrine concentrations have frequently been observed in patients and animal models for heart failure [45,46], and seem to result from increased release and spillover in combination with decreased clearance [45]. Although the beneficial effects of this sympathetic activation may prevail in the acute phase of systolic dysfunction, the increases in afterload may ultimately cause a further impairment of ventricular function.

Elevated plasma concentrations of norepinephrine and epinephrine indeed correlate with poor prognosis [47,48]. In addition, it has been shown that the failing heart becomes less responsive to the inotropic effects of catecholamines, caused by a downregulation of β-adrenergic receptors and decoupling of these receptors from their signal transduction mechanisms [49–51]. Concomitant with the reduced responsiveness of the heart to sympathetic activation, baroreflex dysfunctions have been reported in heart failure [52]. Interestingly, baroreflex dysfunction can also be observed before the onset of clinical symptoms of heart failure [52].

The depressed cardiac function and the increased sympathetic activity may also be responsible for the activation of another humoral system: the renin–angiotensin system (RAS). As for the catecholamines, increases in plasma concentrations of renin and AII have been described during heart failure [53–55]. AII is a potent vasoconstrictor and induces vasoconstriction in various vascular beds. It also induces vascular smooth muscle cell growth, an effect that is partly mediated via the α_1-adrenergic receptor [56–58], as well as an increase in the synthesis of extracellular matrix in the vessel wall [59]. Next to a circulating RAS, it has been shown that components of the RAS can be synthesized within various organs such as the heart, vasculature, and kidney, which points to the existence of local renin–angiotensin systems [60,61]. The cardiac RAS may play an important role in cardiac remodeling during heart failure [62], while the vascular RAS might be important in the local regulation of vascular tone and vascular growth [63,64]. Thus, induction of a myocardial infarction in the rat is associated with an increased ACE mRNA expression and ACE activity in the infarcted area [65]; similarly, increased ACE mRNA levels have also been found in human failing hearts [62].

An important aspect of the potential role of the circulating and tissue RAS is the timing of the activation. Most studies show that the plasma concentration of AII, plasma renin activity, and ACE activity are increased during the acute phase of heart failure [66,67]; these stabilize during compensation but increase again when cardiac function further diminishes [68–70]. Several studies show a discrepancy between the circulating and tissue RAS in this time frame [68–70] with activated tissue systems with normal plasma concentrations, indicating that the tissue RAS and not the systemic may be important in the long-term adaptation of cardiovascular tissues in heart failure.

Another humoral factor that is activated during heart failure is atrial natriuretic peptide (ANP). ANP is released by the atria in response to atrial stretch, and its plasma concentrations are increased during heart failure [71,72]. Next to ANP, plasma concentrations of a second natriuretic peptide, brain natriuretic peptide (BNP), are also increased during heart failure [73]. BNP, released by the ventricles, is a 32-amino-acid peptide and shares with ANP a high degree of structural homology and diuretic, natriuretic, and vasodilating properties [74]. Plasma levels of ANP and BNP correlate with the clinical severity of heart failure [75]. However, it should be noted that the effects of ANP on the total venous compliance and vasodilating capacities of the forearm are attenuated during heart failure [76,77]. Thus, despite its potential vasodilating and natriuretic capacities, ANP does not seem to be able to counterregulate the peripheral vasoconstriction and volume retention during heart failure. It is thought that downregulation of the guanylate cyclase-coupled ANP receptor causes the blunted affect of ANP during heart failure [78,79].

It has become evident that heart failure also induces changes in endothelium-derived neurohumoral factors. Several studies reported increases in the plasma concentrations of endothelin, a potent vasoconstrictor produced by endothelial cells, and of NO, an endothelium-derived vasodilating factor, during heart failure [80,81]. Both endothelin and nitrate plasma levels have been shown to correlate with poor prognosis and severity of heart failure [81–84].

Thus, there is ample evidence that the plasma levels of several potent vasoactive substances are increased in heart failure (Table 2). Factors such as norepinephrine, AII, and NO are involved in the short-term regulation of vessel wall tone. However, pharmacological interventions that block the effect of these agents do not restore normal skeletal muscle function or vascular tone. Therefore, increased plasma levels of several vasoconstrictor and vasodilator agents cannot solely explain the peripheral vascular changes seen during heart failure. For instance, impaired metabolic vasodilation cannot be restored with α-adrenergic blockade [34]. In fact, sympathetic vasoconstrictor effects are attenuated in heart failure. Plasma norepinephrine levels are lower in patients with heart failure as compared to normal subjects during comparable percentages of peak exercise O_2 consumption [85]. Other stimuli of the activity of the sympathetic nervous system, such as the cold pressor test [86] and the

TABLE 2. Neurohormones activated in heart failure.

Vasoconstrictors (SMC growth stimulators)
Angiotensin II
Catecholamines
Endothelin
Vasodilators (SMC growth inhibitors)
Atrial natriuretic peptide
Brain natriuretic peptide
Nitric oxide

SMC, Smooth muscle cell.

head-up tilt or electrical stimulation [87–89], do not result in the expected sympathetic peripheral vasoconstriction.

A dissociation between increased activity and end-organ effect during heart failure is also observed for the RAS. Although angiotensin I-converting enzyme (ACE) inhibition therapy during heart failure improves cardiac function [90,91], ACE inhibition does not restore the impaired metabolic vasodilation during exercise in patients with heart failure [92]. Recent observations from our laboratory (S. Heeneman, unpublished observations, 1995) showed that the pressor effects of a continuous infusion of AII given directly after induction of a rat myocardial infarction were suppressed. In contrast to sham-operated rats, rats with a myocardial infarction did not develop hypertension (Fig. 2A).

Structural Changes

We are thus left with the observation that the effects of several vasoactive agents, which concentrations are elevated in heart failure, cannot be blocked with the appropriate antagonists. Structural changes in the end-organs may explain this apparent paradox. The observation that several peripheral compensatory mechanisms such as the RAS have different short- and long-term effects during heart failure favors the existence of slowly developing structural changes. A role for structural vascular alterations is also suggested by the delayed reversal of impaired maximal vasodilation after cardiac transplantation [93,94].

Vascular structural changes have received much attention in hypertension research. In the classical vascular remodeling hypothesis of Folkow [95], wall thickening in resistance arteries of patients with hypertension takes place at the expense of the lumen. Histological data suggest that vascular remodeling leads to a rearrangement of the same amount of material around the smaller lumen. Thus, cross-sectional areas of normotensive and hypertensive vessels may be the same, concomitant with a decrease in both internal and external diameters [96].

On the basis of the early work of Zelis et al. [97], in which an increase in the arterial vascular sodium content was shown during experimental heart failure, a "vascular stiffness" component was suggested to explain the reduced maximal vasodilatory response. Subsequently, it was shown that diuretic therapy in patients with heart failure could enhance metabolic vasodilation during exercise. Approximately one-third of the reduced vasodilation could be attributed to increased sodium and water content [98].

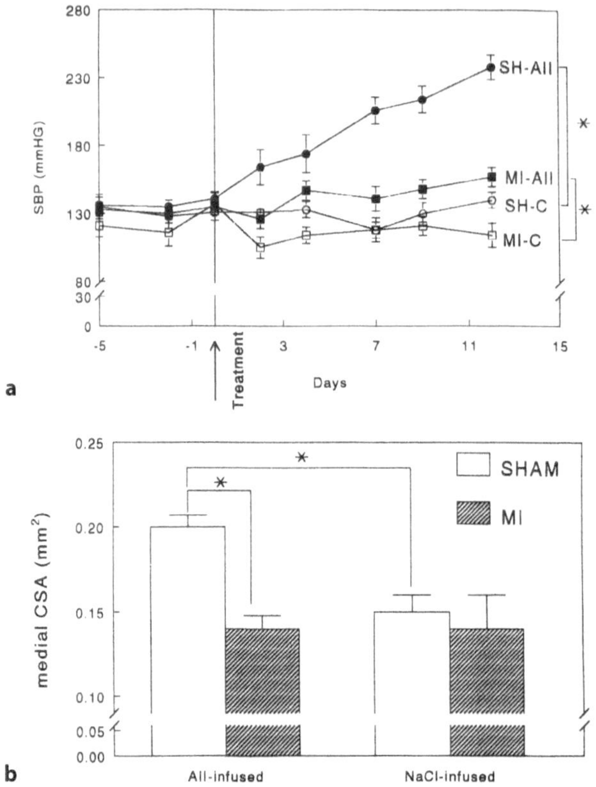

FIG. 2. **a**. Effects of a 2-week NaCl or angiotensin II (AII) infusion (250 ng/kg per min) in sham and myocardial infarction rats on systolic blood pressures (mmHg, measured by tail-cuff plethysmography). *SH-NaCl*, sham rats infused with NaCl (*open circles*); *SH-AII*, sham rats infused with AII (*solid circles*); *MI-NaCl*, myocardial infarction rats infused with NaCl (*open squares*); *MI-AII*, myocardial infarction rat infused with AII (*solid squares*). *, $P < .05$, Mann–Whitney. **b** Effects of a 2-week NaCl or AII infusion (250 ng/kg per min) on medial cross-sectional area (mm^2) of the superior mesenteric artery of sham and myocardial infarction animals.*, $P < .05$, Mann–Whitney

A role for vascular remodeling in the peripheral changes seen during heart failure has been postulated in view of the increases of plasma concentrations of AII, catecholamines, and endothelin (see Table 2). These neurohormones are not only vasocontrictors but are also potent stimulators of vascular growth [56,99–101]. Therefore, vascular hypertrophy, which is an increase in vessel wall mass, was postulated to contribute to the functional peripheral alterations during heart failure, such as increased resistance and reduced maximal vasodilation. Indeed, microangiopathic alterations resulting from hyalinosis of the basement membranes of terminal skin arterioles have been reported in patients with heart failure [102], as well as a decrease in arterial diameter and a decrease in medial thickness of large-conduit arteries [29,33]. In contrast, the medial thickness of resistance arteries of the skeletal muscle bed is increased in a rat model of heart failure [103].

A recent study from our laboratory also showed that while medial cross-sectional areas and internal and external diameters of large-conduit arteries decreased, internal and external diameters of at least two types of resistance arteries increased during experimental heart failure. Moreover, these changes were shown to develop slowly in time [104]. Thus, evidence is accumulating that structural vascular changes do exist in experimental heart failure and, at least for the large-conduit arteries, this cannot be characterized as remodeling or hypertrophy. A reduction of vessel mass is surprising in view of the increased plasma concentrations of various vascular growth-promoting factors.

We believe that increased levels of vasodilating smooth muscle cell (SMC) growth-inhibiting factors, such as ANP, balance the effects of increased levels of vasoconstrictor/SMC growth-promoting factors such as AII or endothelin. This neurohumoral balance may explain the lack of effect of a continuous AII infusion, when given in the first 2 weeks after the induction of a myocardial infarction in rats, on systemic blood pressure and vessel wall mass (S. Heeneman, unpublished observations, 1995) (Fig. 2a,b).

The later development of structural vascular alterations in heart failure may be caused by the development of a "neurohumoral dysbalance," because plasma ANP and BNP concentrations, for example, have been shown to remain at increased levels during long-term experimental heart failure, while plasma AII concentrations returned to normal [70].

Conclusions

Left-ventricular dysfunction initiates a number of compensatory responses to maintain cardiac output and peripheral perfusion. Although well-functioning, balanced, compensatory mechanisms prevent the development of heart failure, depleted compensatory mechanisms lead to heart failure. Peripheral skeletal muscle and vascular functional and structural alterations are profound in heart failure and can affect cardiac function. The reduction in compliance of the large-conduit arteries leads to an increased impedance to left-ventricular outflow while an increase in resistance artery diameter increases afterload. Peripheral vascular structural changes may explain the lack of effect of several short-term pharmacological interventions to normalize peripheral vascular function in heart failure. We would like to suggest that a neurohumoral dysbalance contributes to the development of heart failure. The result is a desensitized vascular system that does not respond adequately to growth stimulatory (or inhibitory) stimuli.

References

1. Elzinga G, Westerhof N (1991) Matching between ventricle and arterial load. Circ Res 68:1495–1500
2. Franciosa JA, Park M, Levine TB (1981) Lack of correlation between exercise capacity and indexes of resting left ventricular performance in heart failure. Am J Cardiol 47:33–39
3. Dougherty AH, Naccarelli GV, Gray EL, Hicks CH, Goldstein RA (1984) Congestive heart failure with normal systolic function. Am J Cardiol 54:778–782
4. Soufer R, Wohlgelernter D, Vita NA, Amuchestegui M, Sostman HD, Berger HJ, Zaret BL (1985) Intact systolic left ventricular function in clinical congestive heart failure. Am J Cardiol 55:1032–1036
5. Wilson JR, Mancini DM (1993) Skeletal muscle metabolic dysfunction—implications for exercise intolerance in heart failure. Circulation 87:104–109

6. Massie B, Conway M, Yonge R, Frostick S, Ledingham J, Sleight P, Radda G, Rajagopalan B (1987) Skeletal muscle metabolism in patients with congestive heart failure: relation to clinical severity and blood flow. Circulation 76:1009–1019
7. Massie BM, Conway M, Rajagopalan B, Yonge R, Frostick S, Ledingham J, Sleight P, Radda G (1988) Skeletal muscle metabolism during exercise under ischemic conditions in congestive heart failure. Evidence for abnormalities unrelated to blood flow. Circulation 78:320–326
8. Wiener DH, Fink LI, Maris J, Jones RA, Chance B, Wilson JR (1986) Abnormal skeletal muscle bioenergetics during exercise in patients with heart failure: role of reduced muscle blood flow. Circulation 73:1127–1136
9. Mancini DM, Coyle E, Coggan A, Beltz J, Ferraro N, Montain S, Wilson JR (1989) Contribution of intrinsic skeletal muscle changes to ^{31}P-NMR skeletal muscle metabolic abnormalities in patients with chronic heart failure. Circulation 80:1338–1346
10. Sullivan MJ, Green HJ, Cobb FR (1990) Skeletal muscle biochemistry and histology in ambulatory patients with long-term heart failure. Circulation 81:518–527
11. Drexler H, Riede U, Munzel T, Koning H, Funke E, Just H (1992) Alterations of skeletal muscle in chronic heart failure. Circ Res 85:1751–1759
12. Holloszy JO, Coyle EF (1984) Adaptations of skeletal muscle to endurance exercise and their metabolic consequences. J Appl Physiol 56:831–838
13. Holloszy JO, Booth FW (1976) Biochemical adaptations to endurance exercise in muscle. Annu Rev Physiol 38:273–291
14. Minotti JR, Johnson EC, Hudson TL, Zuroske G, Murata G, Fukushima E, Cagle TG, Chick TW, Massie BM, Icenogle MV (1990) Skeletal muscle response to exercise training in congestive heart failure. J Clin Invest 86:751–758
15. Adamopoulos S, Coats AJ, Brunotte F, Arnolda L, Meyer T, Thompson CH, Dunn JF, Stratton J, Kemp GJ, Radda GK, Rajagopalan, B (1993) Physical training improves skeletal muscle metabolism in patients with chronic heart failure. J Am Coll Cardiol 21:1101–1106
16. Sullivan MJ, Higginbotham MB, Cobb FR (1988) Exercise training in patients with severe left ventricular dysfunction. Hemodynamic and metabolic effects. Circulation 78:506–515
17. Wilson JR, Martin JL, Schwartz D, Ferraro N (1984) Exercise intolerance in patients with chronic heart failure: role of impaired nutritive flow to skeletal muscle. Circulation 69:1079–1087
18. Drexler H, Faude F, Hoing S, Just H (1987) Blood flow distribution within skeletal muscle during exercise in the presence of chronic heart failure: effect of milrinone. Circulation 76:1344–1352
19. Anversa P, Beghi C, Kikkawa Y, Olivetti G (1985) Myocardial response to infarction in the rat—morphometic measurement of infarct size and myocyte cellular hypertrophy. Am J Pathol 118:484–492
20. Capasso JM, Bruno S, Cheng W, Li P, Rodgers R, Darzynkiewicz Z, Anversa P (1992) Ventricular loading is coupled with DNA synthesis in adult cardiac myocytes after acute and chronic myocardial infarction in rats. Circ Res 71:1379–1389
21. Quaini F, Cigola E, Lagrasta C, Saccani G, Quaini E, Rossi C, Olivetti G, Anversa P (1994) End-stage cardiac failure in humans is coupled with the induction of proliferating cell nuclear antigen and nuclear mitotic division in ventricular myocytes. Circ Res 74:1050–1063
22. van Krimpen C, Smits JFM, Cleutjens JPM, Debets JJM, Schoemaker RG, Struyker Boudier HAJ, Bosman FT, Daemen MJAP (1991) DNA synthesis in the non-infarcted cardiac interstitium after left coronary artery ligation in the rat: effects of Captopril. J Mol Cell Cardiol 23:1245–1253
23. Cleutjens JPM, Verluyten MJA, Smits JFM, Daemen MJAP (1995) Collagen remodeling after myocardial infarction in the rat heart. Am J Pathol 147:325–338
24. Smits JF, van-Krimpen C, Schoemaker RG, Cleutjens JP, Daemen MJ (1992) Angiotensin II receptor blockade after myocardial infarction in rats: effects on hemodynamics, myocardial DNA synthesis, and interstitial collagen content. J Cardiovasc Pharmacol 20:772–778

25. Kuizinga MC, Cleutjens JPM, Smits JFM, Daemen MJAP (1992) Griffonia simplicifolia (GSI): A suitable marker rat cardiac microvascular marker on paraffin embedded tissue. J Mol Cell Cardiol 24(suppl V):S57
26. Milnor WR (1975) Arterial impedance as ventricular afterload. Circ Res 36:565–570
27. Sasayama S, Asanoi H (1991) Coupling between the heart and arterial system in heart failure. Am J Med 90:14S–18S
28. Eaton GE, Cody RJ, Binkley PF (1993) Increased aortic impedance precedes peripheral vasoconstriction at the early stage of ventricular failure in the paced canine model. Circulation 88:2714–2721
29. Gabella MA, Raya TE, Goldman S (1995) Large artery remodeling after myocardial infarction. Am J Physiol 268:H2092–H2103
30. Pepine CJ, Nichols WW, Conti CR (1978) Aortic input impedance in heart failure. Circulation 58:460–465
31. Stefanadis C, Stratos C, Boudoulas H, Kourouklis C, Toutouzas P (1990) Distensibility of the ascending aorta: comparison of invasive and non-invasive techniques in healthy men and in men with coronary artery disease. Eur Heart J 11:990–996
32. Lage SG, Kopel L, Monachini MC, Medeiros CJ, Pileggi F, Polak JF, Creager MA (1994) Carotid arterial compliance in patients with congestive heart failure secondary to idiopathic dilated cardiomyopathy. Am J Cardiol 74:691–695
33. Arnold JMO, Marchiori GE, Imrie JR, Burton GL, Pflugfelder PW, Kostuk WJ (1991) Large artery function in patients with chronic heart failure. Studies of brachial artery diameter and hemodynamics. Circulation 84:2418–2425
34. Zelis R, Mason DT, Braunwald E (1968) A comparison of the effects of vasodilator stimuli on peripheral resistance vessels in normal subjects and in patients with congestive heart failure. J Clin Invest 47:960–970
35. Zelis R, Mason DT, Braunwald E (1969) Partition of blood flow to the cutaneous and muscular bed of the forearm at rest and during leg exercise in normal subjects and in patients with heart failure. Circ Res 24:799–806.
36. Reddy HK, Weber KT, Janicki JS, McElroy PA (1988) Hemodynamic, ventilatory and metabolic effects of light isometric exercise in patients with chronic heart failure. J Am Coll Cardiol 12:353–358
37. Zelis R, Sinoway LI, Musch TI, Davis D, Just H (1988) Regional blood flow in congestive heart failure: concept of compensatory mechanisms with short and long time constants. Am J Cardiol 62:2E–8E
38. Palmer RM, Ferrige AG, Moncada S (1987) Nitric oxide release accounts for the biological activity of endothelium-derived relaxing factor. Nature 327:524–526
39. Kaiser L, Spickard RC, Olivier NB (1989) Heart failure depresses endothelium-dependent responses in canine femoral artery. Am J Physiol 256:H962–H967
40. Drexler H, Hayoz D, Münzel T, Hornig B, Just H, Brunner HR, Zelis R (1992) Endothelial function in chronic congestive heart failure. Am J Cardiol 69:1596–1601
41. Kubo SH, Rector TS, Bank AJ, Williams RE, Heifetz SM (1991) Endothelium-dependent vasodilation is attenuated in patients with heart failure. Circulation 84:1589–1596
42. Drexler H, Lu W (1992) Endothelial dysfunction of hindquarter resistance vessels in experimental heart failure. Am J Physiol 262:H1640–H1645
43. Ontkean M, Gay R, Greenberg B (1991) Diminished endothelium-derived relaxing factor activity in an experimental model of chronic heart failure. Circ Res 69:1088–1096
44. Teerlink JR, Clozel M, Fischli W, Clozel JP (1993) Temporal evolution of endothelial dysfunction in a rat model of chronic heart failure. J Am Coll Cardiol 22:615–620
45. Davis D, Baily R, Zelis R (1988) Abnormalities in systemic norepinephrine kinetics in human congestive heart failure. Am J Physiol 254:E760–E766
46. Davis D, Sinoway LI, Robison J, Minotti JR, Day FP, Baily R, Zelis R (1987) Norepinephrine kinetics during orthostatic stress in congestive heart failure. Circ Res 61:I87–I90

47. Swedberg K, Eneroth P, Kjekshus J, Wilhelmsen L (1990) Hormones regulating cardiovascular function in patients with severe congestive heart failure and their relation to mortality. Circulation 82:1730–1736
48. Francis GS, Cohn JN, Johnson G, Rector TS, Goldman S, Simon A (1993) Plasma norepinephrine, plasma renin activity and congestive heart failure. Relations to survival and the effects of therapy in V-HeFT II. Circulation 87:VI40–VI48
49. Bristow MR, Minobe WA, Raynolds MV, Port JD, Rasmussen R, Ray PE, Feldman AM (1993) Reduced beta-1 receptor messenger RNA abundance in the failing human heart. J Clin Invest 92:2737–2745
50. Bristow MR (1993) Changes in myocardial and vascular receptors in heart failure. J Am Coll Cardiol 22:61A–71A
51. Insel PA (1993) β-Adrenergic receptors in heart failure. J Clin Invest 92:2564
52. Higgins CB, Vatner SF, Eckberg DL, Braunwald E (1972) Alterations in the baroreceptor reflex in conscious dogs with heart failure. J Clin Invest 51:715–724
53. Francis GS, Benedict C, Johnstone DE, Kirlin PC, Nicklas J, Liang CS, Kubo SH, Rudin-Toretsky E, Yusuf S (1990) Comparison of neuroendocrine activation in patients with left ventricular dysfunction with and without cogestive heart failure. A substudy of the studies of left ventricular dysfunction (SOLVD). Circulation 82:1724–1729
54. Rouleau JL, de-Champlain J, Klein M, Bichet D, Moye L, Packer M, Dagenais GR, Sussex B, Arnold JM, Sestier F (1993) Activation of neurohumoral systems in postinfarction left ventricular dysfunction. J Am Coll Cardiol 22:390–398
55. McAlpine HM, Morton JJ, Leckie B, Rumley A, Gillen G, Dargie HJ (1988) Neuroendocrine activation after acute myocardial infarction. Br Heart J 60:117–124
56. Daemen MJ, Lombardi DM, Bosman FT, Schwartz SM (1991) Angiotensin II induces smooth muscle cell proliferation in the normal and injured rat arterial wall. Circ Res 68:450–456
57. Van-Kleef EM, Smits JF, De-Mey JG, Cleutjens JP, Lombardi DM, Schwartz SM, Daemen MJ (1992) Alpha-1 adrenoreceptor blockade reduces the angiotensin II-induced vascular smooth muscle cell DNA synthesis in the rat thoracic aorta and carotid artery. Circ Res 70:1122–1127
58. van Kleef EM, Smits JFM, Schwartz SM, Daemen MJAP (1996) Doxazosin blocks the angiotensin II-induced smooth muscle cell DNA synthesis in the media but not in the neointima of the rat carotid artery after balloon injury. Cardiovasc Res 31:324–330
59. Kato H, Suzuki H, Tajima S, Ogata Y, Tominaga T, Sato A, Saruta T (1991) AII stimulates collagen synthesis in cultured vascular smooth muscle cells. J Hypertens 9:17–22
60. Thie M, Harrach B, Schonherr E, Kresse H, Robenek H, Rauterberg J (1993) Responsiveness of aortic smooth muscle cells to soluble growth mediators is influenced by cell-matrix contact. Arterioscler Thromb 13:994–1004
61. Naftilan AJ, Zuo WM, Inglefinger J, Ryan TJ Jr, Pratt RE, Dzau VJ (1991) Localization and differential regulation of angiotensinogen mRNA expression in the vessel wall. J Clin Invest 87:1300–1311
62. Studer R, Reinecke H, Muller B, Holtz J, Just H, Drexler H (1994) Increased angiotensin I converting enzyme expression in the failing human heart. J Clin Invest 94:301–310
63. Oliver JA, Sciacca RR (1984) Local generation of angiotensin II as a mechanism of regulation of peripheral vascular tone in the rat. J Clin Invest 74:1247–1251
64. Dzau VJ (1993) Vascular renin-angiotensin system and vascular protection. J Cardiovasc Pharmacol 22(suppl 5):S1–S9
65. Passier RJ, Smits JFM, Verluyten MJA, Studer R, Drexler H, Daemen MJAP (1995) Activation of angiotensin converting enzyme expression in the infarct zone following myocardial infarction. Am J Physiol 269:H1268–H1276
66. Dzau VJ, Colucci WS, Hollenberg NK, Williams GH (1981) Relation of the renin-angiotensin-aldosteron system to clinical state in congestive heart failure. Circulation 63:645–651

67. Schunkert H, Tang S, Litwin SE, Diamant D, Riegger G, Dzau VJ, Ingelfinger JR (1993) Regulation of intrarenal and circulating renin-angiotensin systems in severe heart failure in the rat. Cardiovasc Res 27:731–735
68. Dzau VJ (1993) Tissue renin-angiotensin system in myocardial hypertrophy and failure. Arch Intern Med 153:937–942
69. Hirsch AT, Talsness CE, Schunkert H, Paul M, Dxau VJ (1991) Tissue-specific activation of cardiac angiotensin-converting enzyme in experimental heart failure. Circ Res 69:475–481
70. Huang H, Arnal J, Llorens-Cortes C, Challah M, Alhenc-Gelas F, Corvol P, Michel JB (1994) Discrepancy between plasma and lung angiotensin-converting enzyme activity in experimental heart failure. A novel aspect of endothelium dysfunction. Circ Res 75:454–461
71. Svanegaard J, Angelo-Nielsen K, Pindborg T (1992) Plasma concentration of atrial natriuretic peptide at admission and risk of cardiac death in patients with acute myocardial infarction. Br Heart J 68:38–42
72. Lerman A, Gibbons RJ, Rodeheffer RJ, Bailey KR, McKinley LJ, Heublein DM, Burnett JC (1993) Circulating N-terminal atrial natriuretic peptide as a marker for symptomless left-ventricular dysfunction. Lancet 341:1105–1109
73. Morita E, Yasue H, Yoshimura M, Ogawa H, Jougasaki M, Matsumura T, Mukoyama M, Nakao K (1993) Increased plasma levels of brain natriuretic peptide in patients with acute myocardial infarction. Circulation 88:82–91
74. Sudoh T, Kangawa K, Minamino N, Matsuo H (1988) A new natriuretic peptide in porcine brain. Nature 332:78–81
75. Johnston CI (1992) Renin angiotensin system—a dual tissue and hormonal system for cardiovascular control. J Hypertens 10:S13–S26
76. Chien Y, Pegram BL, Kardon MB, Frohlich ED (1992) ANF does not increase total body venous compliance in conscious rats with myocardial infarction. Am J Physiol 262:H432–H436
77. Hirooka Y, Takeshita A, Imaizumi T, Suzuki S, Yoshida M, Ando S, Nakamura M (1990) Attenuated forearm vasodilative response to intra-arterial atrial natriuretic peptide in patients with heart failure. Circulation 82:147–153
78. Tsutamoto T, Kanamori T, Morigami N, Sugimoto Y, Yamaoka O, Kinoshita M (1993) Possibility of downregulation of atrial natriuretic peptide receptor coupled to guanylate cyclase in peripheral vascular beds of patients with chronic severe heart failure. Circulation 87:70–75
79. Smits P, Jansen TLTA, Thien T (1993) Possibility of downregulation of atrial natriuretic peptide receptor coupled to guanylate cyclase in peripheral vascular beds of patients with chronic severe heart failure. Circulation 88:811–812
80. Tomoda H (1993) Plasma endothelin-1 in acute myocardial infarction with heart failure. Am Heart J 125:667–672
81. Pacher R, Berglerklein J, Globits S, Teufelsbauer H, Schuller M, Krauter A, Ogris E, Rodler S, Wutte M, Hartter E (1993) Plasma big endothelin-1 concentrations in congestive heart failure patients with or without systemic hypertension. Am J Cardiol 71:1293–1299
82. Omland T, Lie RT, Aakvaag A, Aarsland T, Dickstein K (1994) Plasma endothelin determination as a prognostic indicator of 1-year mortality after acute myocardial infarction. Circulation 89:1573–1579
83. Winlaw DS, Smythe GA, Keogh AM, Schyvens CG, Spratt PM, MacDonald PS (1994) Increased nitric oxide production in heart failure. Lancet 344:373–374
84. Winlaw DS, Smythe GA, Keogh AM, Schyvens CG, Spratt PM, Macdonald PS (1995) Nitric oxide production and heart failure. Lancet 345:390–391
85. Francis GS, Goldsmith SR, Ziesche S, Nakajima H, Cohn JN (1985) Relative attenuation of sympathetic drive during exercise in patients with congestive heart failure. J Am Coll Cardiol 5:832–839
86. Ferguson DW, Abboud FM, Mark AL (1984) Selective impairment of baroreflex-mediated vasoconstrictor responses in patients with ventricular dysfunction. Circulation 69:451–460

87. Goldsmith SR, Francis GS, Levine TB, Cohn JN (1983) Regional blood flow response to orthostatsis in patients with congestive heart failure. J Am Coll Cardiol 1:1391–1395
88. Kassis E, Jacobsen TN, Mogensen F, Amtorp O (1986) Sympathetic reflex control of skeletal muscle blood flow in patients with congestive heart failure: evidence for beta-adrenergic circulatory control. Circulation 74:929–938
89. Wilson JR, Matthai W, Lanoce V, Frey M, Ferraro N (1988) Effect of experimental heart failure on peripheral sympathetic vasoconstriction. Am J Physiol 254:H727–H733
90. Schoemaker RG, Debets JJM, Struyker-Boudier HA, Smits JFM (1991) Delayed but not immediate captopril therapy improves cardiac function in conscious rats, following myocardial infarction. J Mol Cell Cardiol 23:187–197
91. Pfeffer MA (1995) ACE inhibition in acute myocardial infarction. N Engl J Med 332:118–120
92. Wilson JR, Ferraro N (1985) Effect of the renin-angiotensin system on limb circulation and metabolism during exercise in patients with heart failure. J Am Coll Cardiol 6:556–563
93. Sinoway LI, Minotti JR, Davis D, Pennock JL, Burg JE, Musch TI, Zelis R (1988) Delayed reversal of impaired vasodilation in congestive heart failure after heart transplantation. Am J Cardiol 61:1076–1079
94. Kubo SH, Rector TS, Bank AJ, Tschumperlin LK, Raij L, Brunsvold N, Kraemer MD (1993) Effects of cardiac transplantation on endothelium-dependent dilation of the peripheral vasculature in congestive heart failure. Am J Cardiol 71:88–93
95. Folkow B (1982) Physiological aspects of primary hypertension. Physiol Rev 62:347–504
96. Mulvany M, Aalkjaer C (1990) Structure and function of small arteries. Physiol Rev 70:921–963
97. Zelis R, Delea CS, Coleman HN, Mason DT (1970) Arterial sodium content in experimental heart failure. Circ Res 41:213–216
98. Sinoway L, Minotti J, Musch T, Goldner D, Davis D, Leaman D, Zelis R (1987) Enhanced metabolic vasodilation secondary to diuretic therapy in decompensated congestive heart failure secondary to coronary artery disease. Am J Cardiol 60:107–111
99. Geisterfer AA, Peach MJ, Owens GK (1988) Angiotensin II induces hypertrophy, not hyperplasia, of cultured rat aortic smooth muscle cells. Circ Res 62:749–756
100. Yamori Y, Mano M, Nara Y, Horie R (1987) Catecholamine-induced polyploidization in vascular smooth muscle cells. Circulation 75:I92–I95
101. Chua BHL, Krebs CJ, Chua CC, Diglio CA (1992) Endothelin stimulates protein synthesis in smooth muscle cells. Am J Physiol 262:E412–E416
102. Wroblewski H, Kastrup J, Norgaard T, Mortensen S, Hauso S (1992) Evidence of increased microvascular resistance and arteriolar hyalinosis in skin in congestive heart failure secondary to idiopathic dilated cardiomyopathy. Am J Cardiol 69:769–774
103. Schieffer B, Wollert KC, Berchtold M, Saal K, Riede U, Drexler H (1994) Development and prevention of skeletal muscle structural alternations in experimental chronic heart failure. Circulation 90(pt 2):I–262
104. Heeneman S, Leenders PJA, Aarts PLJW, Smits JFM, Arends JW, Daemen MJAP (1995) Peripheral vascular alterations during experimental heart failure. Do they exist? Arterioscler Thromb Vasc Biol 15:1503–1511

Myocardial Interstitial Collagen Matrix Remodeling in Response to a Chronic Elevation in Ventricular Preload or Afterload

Joseph S. Janicki[1], Scott E. Campbell[2], Jeffrey R. Henegar[1], and Gregory L. Brower[1]

Summary. Fibrillar collagen is an essential component of the extracellular matrix of the heart that surrounds and interconnects the coronary microcirculation, individual myocytes, groups of cardiac myofibrils, muscle fibers, and muscle fiber bundles. Accordingly, it provides for cardiac myocyte and muscle fiber alignment, and in so doing it imparts a mechanical support to the myocardium that assists in the maintenance of ventricular shape, size, and function. Interstitial fibrillar collagen concentration is known to be increased from 40% to 150% in genetic and experimental left-ventricular hypertension and with pathological elevations of circulating angiotensin II or mineralocorticoids. A significant portion of this fibrosis may be related to multifocal myocyte necrosis and intramyocardial coronary artery damage, which occur in these models. As a consequence of this increase in fibrillar collagen, both the myocardium and the ventricle become stiffer. Systolic function is preserved by the accompanying hypertrophy and enhanced contractility unless the increase in myocardial collagen consists of a significant degree of endomyocardial reparative fibrosis. In contrast, chronic volume overload induces an increase in myocardial collagenase activity with subsequent collagen degradation that results in ventricular dilatation and a decrease in ventricular stiffness. The ventricular dilatation, change in ventricular shape, and increase in distensibility in dilated cardiomyopathy and end-stage heart failure are also mediated by collagenase activation and fibrillar collagen breakdown.

Key words. Ventricular function—Ventricular dilatation—Myocyte necrosis—Reparative fibrosis—Myocardial collagenase activity—Cardiomyopathy—Heart failure

Introduction

Fibrillar collagen is an essential component of the extracellular matrix of the heart that surrounds and interconnects the coronary microcirculation, individual myocytes, groups of cardiac myofibrils, muscle fibers, and muscle fiber bundles.

[1] Department of Physiology and Pharmacology, Auburn University, Auburn, AL 36849, USA.
[2] Department of Anatomy and Structural Biology, University of South Dakota, Vermillion, SD 57069, USA.

Accordingly, it provides for muscle fiber and cardiac myocyte alignment, and in so doing it imparts a mechanical support to the myocardium that assists in the maintenance of ventricular shape, size, and function. Alterations of this matrix can adversely affect ventricular function. For example, it is well known that a significant increase in the concentration of collagen results in diastolic dysfunction with an abnormal elevation in ventricular diastolic stiffness [1–11]. Conversely, disruption and degradation of collagen fibers have been shown to result in ventricular dilatation and sphericalization and a decrease in ventricular diastolic stiffness [1,12–14]. In this chapter, we review in detail myocardial collagen remodeling induced by chronic pathophysiological alterations in vascular impedance, circulating angiotensin II or aldosterone levels, and ventricular filling volume, as well as that seen in dilated cardiomyopathy and heart failure. Also, the relationship between the myocardial collagen matrix and ventricular function is considered. We begin with a brief description of the micro- and ultrastructural organization and mechanical properties of myocardial collagen.

Mechanical Properties of Fibrillar Collagen and Organization of the Myocardial Collagen Matrix

Once deposited in fibrillar form and subsequently cross-linked, extracellular collagen is extremely stable and resistant to degradation. It consists of three peptide chains that are intertwined to form a right-handed super-helix. This conformation renders it highly resistant to all proteases except collagenase. The concentration of collagen in the normal adult heart is relatively small; morphometric assessments indicate that only 2%–4% of the normal myocardium consists of collagen. However, fibrillar collagen exhibits a high tensile strength and low extensibility. Thus, it would be expected that small changes in the amount of myocardial collagen would strongly influence the passive mechanical properties of the myocardium. The proportion of collagen types could also influence the passive mechanical behavior of the myocardium. Of the five types of fibrillar collagen, only I, III, and V are found in the adult heart [3,15–19], with type I being the most abundant [3,19]. For example, in the nonhuman primate, 85% of myocardial collagen is type I, 15% is type III, and 5% is type V [3]. Although tissue with predominantly type I collagen (e.g., tendons) is stiffer than tissue with mostly type III collagen (e.g., the uterus), most of the fibrillar collagen in cardiac tissue appears to be a copolymerization of types I and III collagen molecules [16–18].

In addition to its concentration and proportion of types, other characteristics of collagen that may influence the diastolic function of the heart include the spatial alignment of collagen fibers, the crimp properties of the fiber, collagen fibril and fiber diameter, and the degree of cross-linking. Collagen fibers with an orientation parallel to locally generated stresses contribute significantly to the elasticity of the tissue, while fibers that are aligned perpendicularly contribute little.

The crimp property of fibrillar collagen is a measure of the degree of corrugation or coiling of a collagen fiber. The greater the amount of crimping, the longer a fiber can be extended without developing significant stress or tension. Thereafter, further stretch results in an exponential increase in fiber stress. Coiled perimysial fibers oriented parallel or obliquely to the long axes of myocytes have been observed to undergo focal straightening as the myocardium is stretched [18]. Tissue that is normally subjected to high tensile stresses is composed of collagen fibers of larger diameter with a greater percentage of covalent cross-links relative to a tissue exposed to lower stresses. Little is known with regard to the effect of changes in the proportions of collagen types, the degree of cross-linking, and the crimp properties on myocardial functional behavior.

The connective tissue network of the heart is depicted as having three levels of organization, the epimysium, the perimysium, and the endomysium. The epimysium is a sheath of connective tissue composed primarily of collagen with small amounts of elastin and microfibrils; it lies just beneath the epicardial and endocardial surfaces of the heart. The perimysium forms a connective tissue network surrounding groups of cardiac myocytes. It includes tendinous extensions of the epimysium that terminate in a collagen weave around groups of myocytes. Collagen fibers, referred to as strands, interconnect adjacent weaves. Coiled perimysial fibers are also included in the perimysium. These fibers are oriented parallel or obliquely to the long axis of muscle fibers and may play a role in limiting muscle distention [20]. The endomysial network consists of the collagen fibers that surround individual myocytes and collagen struts; the struts project to the lateral surfaces of adjacent myocytes as well as to neighboring capillaries. These struts insert into the basal lamina at the Z-band of the sarcomere and are believed to play an important role in the efficient transduction of generated force to the ventricular chamber [21] and in the prevention of myocyte slippage [22].

Collagen Matrix Remodeling Associated with Chronic Elevation in Afterload

Interstitial fibrillar collagen concentration is known to be increased from 40% to 150% in genetic [2,5,8,23] and experimental left-ventricular (3,6–8,17,24,25] hypertension and secondary to increases in circulating angiotensin II [26–28] or mineralocorticoids [24,26]. Results obtained in studies with excess angiotensin II and mineralocorticoids are included here because of the elevations in afterload induced by their administration. Generally, remodeling of the collagen matrix is observed at both the light and electron microscopic levels and consists of an increase in the concentration of collagen related to fibroblast proliferation [29–31]. Remodeling of the peri- and endomysial components includes thickening of collagen fibrils, fibers, and tendons; increase in the density of the weave and network surrounding myocytes; and newly formed fibers [6,25,32].

In an effort to further characterize myocardial fibrosis, Anderson et al. [33] suggested that three types be considered, namely, microscopic scars, interfibrosis, and perivascular fibrosis. Three types of fibrosis have been identified at the light microscopic level to be associated with chronic elevations in afterload, angiotensin II, or aldosterone: the collagenous connective tissue encircling and separating individual muscle fibers increases (i.e., interfibrosis or interstitial fibrosis), the area occupied by collagen in the adventitia of intramyocardial arteries expands and extends (i.e., perivascular fibrosis), and areas of focal, confluent fibrosis, which are thought to represent microscopic scars, appear [5,6,34].

Weber and co-workers [34–37] distinguished myocardial fibrosis as being either reactive (in the absence of myocyte necrosis) or reparative (secondary to myocyte necrosis) fibrosis according to its morphological presentation. Basically, they assumed that increases in interstitial (interfibrosis) and perivascular fibrosis represented a reactive process while microscopic scars (i.e., confluent areas of fibrosis) were the result of a reparative fibrosis. Such an assumption, while plausible, remains to be proven. This is particularly true for models of experimental hypertension [28–31,34,38,39] and excess angiotensin II [27,28,40–44] or mineralocorticoids [31,37] in which extensive multifocal myocyte necrosis and coronary vascular damage are known to occur in both ventricles. That is, it is possible that increased perivascular fibrosis may be the result of angiotensin II-induced coronary vascular damage or adjacent myocyte necrosis [28,40–44] and therefore be considered a reparative fibrosis. In fact, Kabour et al. [27] found many of the angiotensin II-induced necrotic foci to be concentric to damaged intramyocardial coronary arteries, and Rodrigues and co-workers [45] considered coronary vascular injury to be the basis of patchy multifocal myocyte necrosis secondary to renovascular hypertension.

Similarly, in situations in which extensive multifocal myocyte necrosis is occurring, one cannot discern with absolute certainty that increases in interstitial fibrosis are not

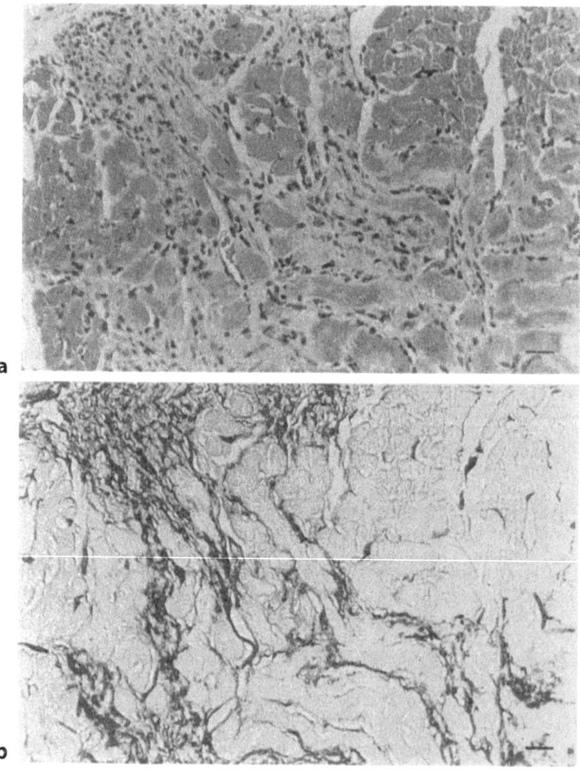

Fig. 1a,b. Photomicrographs from 5-μm-thick serial sections of myocardium illustrate a region of angiotensin II-induced myocyte necrosis (angiotensin II was infused subcutaneously for 9 days at 150 ng/min). **a** Myocardium is stained with hematoxylin and eosin to demonstrate the areas of myocyte necrosis and substantial nonmyocyte infiltration. **b** Myocardium is stained with collagen-specific sirius red F3BA to delineate the areas of reparative fibrosis. These areas of fibrosis could be misread as an increase in interstitial or reactive fibrosis unrelated to myocyte necrosis. (From [27], with permission)

Table 1. Overview of histological findings.

	WEEKS 1–2	WEEKS 3–4	WEEKS 5–6	WEEKS 7–8
ALDO	No change	Cell clusters — Fbs — MyoFbs — Macrophages — Lymphocytes — No PMNs — Early scars	Same as wk 3–4 except: — ↓ MyoFbs — More mature scars — Perivasc/interstit fibrosis	Same as wk 5–6 except: — All cell types decreased — More fibrosis
	DAY 1	DAY 2	DAY 4	DAY 7
AngII	No change	Cell clusters — Fbs — MyoFbs — Macrophages — Lymphocytes — PMNs — No fibrosis	Same as day 2 except: — More Fbs, MyoFbs, Macrophages — ↓ Lymphocytes, PMNs — Early scars	Same as day 4 except: — No PMNs — More mature scars — Perivasc fibrosis
	DAY 14	DAY 28	DAY 42	DAY 56
	Same as day 7 except: — More fibrosis	Same as day 14 except: — ↓ Macrophages — More fibrosis	Same as day 28 except: — ↓ all cell types — No lymphocytes	— Only Fbs — More fibrosis
	DAY 1	DAY 2	DAY 4	DAY 7
URI	No change	Cell clusters — Fbs — MyoFbs — No macrophages — Lymphocytes — No PMNs — No fibrosis	— ↑ Fbs, MyoFbs — Macrophages — Lymphocytes — PMNs — Minimal collagen changes	Same as day 4 except: — No macrophages, lymphocytes, PMNs — Early scars
	DAY 14	DAY 28	DAY 42	DAY 56
	Same as day 7 except: — ↓ MyoFbs — More mature scars — Perivasc/interstit fibrosis	Same as day 14 except: — No MyoFbs — More fibrosis	Same as day 28	Same as day 42 except: — ↓ Fbs — More fibrosis

Fbs, Fibroblasts; MyoFbs, myofibroblasts; PMNs, neutrophils; Perivasc, perivascular; Interstit, interstitial; ALDO, chronic aldosterone infusion; AngII, chronic angiotensin II infusion; URI, unilateral renovascular ischemia.
From [31], with permission.

the result of this necrosis and a subsequent reparative fibrosis. As can be seen in Fig. 1, many of the areas where myocyte necrosis has occurred (i.e., areas of inflammatory infiltrate; Fig. 1a) appear as increases in interstitial fibrosis (i.e., red-stained collagen; Fig. 1b). Thus, caution is warranted in attempting to identify or define a nonnecrotic mechanism for the fibrosis seen in experimental models of hypertension or circulating hormonal excess [46,47] in which myocyte necrosis and coronary vascular damage are known to occur.

The temporal patterns of cellular and myocardial fibrotic responses associated with renovascular hypertension and with chronic elevations in angiotensin II or aldosterone have recently been described [31] (Table 1). A necrotic/inflammatory response that eventually resulted in increased myocardial fibrosis was involved in all three of these experimental models. While inflammatory cells were seen as early as day 2 in the renovascular hypertension and angiotensin II infusion models, they were not seen in the one-kidney, high-salt, aldosterone model before 3–4 weeks. It is interesting to note that a significant increase in blood pressure did not occur until 3–4 weeks after elevation of the circulating levels of aldosterone [48]. The results of Campbell et al. [31] indicated that significant fibrosis occurs in both ventricles within 2 weeks of the acute inflammatory response in these three models. Also, the interstitial fibrosis does not appear to be progressive despite sustained hypertension and elevated levels of angiotensin II or aldosterone. The acute nature of the consequent myocardial damage indicates that the mechanism is receptor mediated and that the subsequent insensitivity of the myocardium to further damage is the result of receptor downregulation [27,28,42].

Functional Consequences of an Abnormal Increase in Myocardial Collagen

The relation between collagen concentration and ventricular diastolic function has recently been reviewed [1]. Both the myocardium [7,8] and the ventricle [2–6,9–12] become stiffer (Tables 2, 3) as a consequence of an increase in myocardial collagen concentration, such as that described in the previous section. In most of these studies, the elevation in collagen concentration was accompanied by a significant myocyte hypertrophy, and one could argue that myocyte enlargement, not fibrosis, is responsible for the observed diastolic dysfunction. Indeed, Schraeger et al. [49] concluded that regression of left-ventricular hypertrophy in spontaneously hypertensive rats (SHR) decreased chamber stiffness and that the increased collagen concentration does not necessarily influence chamber stiffness.

However, the results of others tend to refute this argument. Bing et al. [50], using β-amino proprionitrile, were able to prevent the increase in collagen content associated with chronic aortic constriction and concluded that the elevations in resting tension normally observed in hypertrophied papillary muscle are caused by the increase in

TABLE 2. Experimentally induced myocardial fibrosis, hypertrophy, and diastolic stiffness: papillary or trabecular muscle studies.

Reference	Species	Model	Stiffness (% Inc)	LVH (% Inc)	Hpro (% Inc)	Nonmyocyte space (% Inc)
Bing et al. [7]	Rat	AAC-4	50	40	31	—
Holubarsch et al. [8]	Rat	GBT-4	17*	53	13*	19*
		GBT-8	42	84	24	40
		SHR-40	6*	44	36	12*
		SHR-80	40	90	73	59

% Inc, percent increase relative to appropriate control; LVH, left-ventricular hypertrophy; Hpro, hydroxyproline; AAC-4, aorta arch constriction for 4 weeks; GBT, Goldblatt renovascular hypertension for 4 (GBT-4) and 8 (GBT-8) weeks; SHR, spontaneous hypertension for 40 (SHR-40), and 80 (SHR-80) weeks.

*, Not significantly different from control.

From [1], with permission.

TABLE 3. Experimentally induced myocardial fibrosis, hypertrophy, and diastolic stiffness: left ventricular studies.

Reference	Species	Model	Stiffness (% Inc)	LVH (% Inc)	Hpro (% Inc)	CVF (% Inc)
Weber et al. [3]	Primate	PN-4	−36	34	29	78
		PN-33	23	41	37	93
		PN-88	23	38	15	61
Doering et al. [6]	Rat	RHT-4	29	17	—	117
		RHT-8	42	21	—	72
Jalil et al. [10]	Rat	PN-10	27	24	—	140
		RHT-ISO-9	172	39	—	573
		PN-ISO-14	80	53	—	817
Narayan et al. [5]	Rat	SHR-36	24	27	—	174
Brilla et al. [2]	Rat	SHR-14	32	33	—	64
		SHR-26	41	41	—	150

% Inc., LVH and Hpro same as in Table 2. CVF, Collagen volume fraction; PN, perinephritic hypertension for 4 (PN-4), 10 (PN-10), 33 (PN-33), and 88 (PN-88) weeks; RHT, abdominal aorta constriction plus unilateral renal ischemia for 4 (RHT-4) and 8 (RHT-8) weeks; RHT-ISO-9, RHT for 8 weeks followed by isoproterenol (ISO) for 10 days; PN-ISO-14, PN for 10 weeks followed by ISO for 10 days and sacrifice 3 weeks later; SHR, spontaneous hypertension for 14 (SHR-14), 26 (SHR-26), and 36 (SHR-36) weeks.
From [1], with permission.

collagen content and not the hypertrophy. By administering hydralazine, Narayan et al. [5] were able to prevent myocyte hypertrophy in SHR but not the abnormal accumulation of collagen. The consequence of this fibrosis was an elevated passive myocardial stiffness which was similar to that measured in the untreated SHR with fibrosis and hypertrophy. In contrast to the findings of Schraeger, Brilla and colleagues [2] reported a return to normal passive stiffness in SHR following 12 weeks of treatment with the angiotensin-converting enzyme inhibitor, lisinopril, which resulted in complete regression of the elevated levels of collagen concentration to that found in control WKY rats. Finally, Gelpi and co-workers [51] reported left-ventricular diastolic chamber compliance to be normal in dogs with normal collagen concentration despite significant hypertrophy secondary to perinephritic hypertension.

While there is little doubt that excessive accumulations of myocardial collagen affect diastolic function, the influence of increased collagen on systolic function is difficult to discern because of the accompanying hypertrophy and a possible compensatory increase in contractility. Several studies have reported systolic function to be either normal or enhanced despite a significant increase in myocardial collagen concentration [3,10,11]. However, a decrease in systolic function has been noted when the hypertrophy is accompanied by significant endomyocardial reparative fibrosis [10,52–54]. Whether this is caused by (1) a significant loss of endomyocardial myocytes, (2) the encasement by inelastic collagen of groups of muscle in the endomyocardium so as to impose a physical restraint on muscle distension and curtail the length-dependent property of cardiac muscle, or (3) some other mechanism yet to be determined. Similarly, the hypothesis that diastolic dysfunction secondary to excessive collagen accumulation eventually leads to systolic dysfunction and decompensated heart failure has not been experimentally tested [54].

Collagen Matrix Remodeling and Altered Ventricular Function Associated with Chronic Elevation in Preload

Pathological conditions that lead to a chronic elevation in preload include aortic and mitral valve regurgitation, dilated cardiomyopathy, anemia, and ischemic heart disease. These conditions characteristically lead to chamber dilatation and hypertrophy. Most studies of remodeling of the collagen matrix leading to ventricular dilatation have focused primarily on the endpoint of what most certainly is a dynamic process. Thus, a biochemical, histological, or morphological assessment of interstitial collagen using cardiac tissue obtained from hearts subjected to several months of an experimentally induced increase in ventricular loading, from explanted hearts at the time of cardiac transplantation, from endomyocardial biopsies, or from hearts at autopsy provides a description of the collagen network "after the fact" with no insight into the temporal relation of the remodeling process that may have preceded. This process conceivably could include a breakdown and subsequent reconstruction of the collagen matrix.

As mentioned earlier, the conformation of collagen renders it highly resistant to all proteinases except collagenase. Latent collagenase is present throughout the myocardium [55], and most of it appears to be bound to collagen. Thus, there exists the potential for collagen degradation to exceed synthesis should there be significant activation of this latent collagenase system.

When an arrested heart was incubated in a solution of collagenase for 90 min, a significant shift to the right of the left-ventricular pressure–volume curve occurred [14]. Caulfield et al. [13] investigated the structural and functional consequences of collagen degradation using an assumed in vivo model of global collagenase activation. Their findings indicated left-ventricular dilatation and increased diastolic compliance following a nearly complete loss of ultrastructural collagen. Preliminary data from our laboratory indicated that a sustained elevation in ventricular volume secondary to the creation of an infrarenal abdominal aorta–vena cava fistula is associated with significant activation of the myocardial collagenase system; twelve hours after the creation of the fistula, collagenase activity was elevated by 50% relative to control activity. During the next 5 days of volume overload, it further increased to levels that were 80% greater than control levels. During this period, a significant decline (i.e., 40%) in collagen concentration also occurred. For the remainder of the study (i.e., 8 weeks), collagenase activity decreased to a level that was 33% greater than control collagenase activity, and there was a trend for collagen concentration to return to its normal value. These early alterations to the fibrillar collagen matrix appear to correlate with significant ventricular enlargement, increased compliance, and a decrease in contractility [56]. The fact that collagen concentration was nearly normal after 8 weeks of a sustained increase in volume in this study reinforces the earlier statement regarding the need to study the temporal response of the collagen matrix following the initiation of the volume overload condition. Thus, it is highly probable that a similar sequence of collagen degradation followed by increased synthesis occurred in the studies by others who found collagen concentration in the volume-overloaded heart to be normal despite marked ventricular dilatation [57–59].

Evidence is accumulating that collagenase activation and fibrillar collagen breakdown are responsible for the dilatation, the change in shape, and the increase in distensibility of the cardiomyopathic left ventricle [60]. For example, in the cardiomyopathic hamster, there is a continuing increase in collagenase activity with collagen degradation eventually exceeding collagen synthesis around 240 days of age [61]. It is around 240 days of age that the cardiomyopathic Syrian hamster develops significant left-ventricular dilatation and wall thinning, and shortly thereafter congestive heart failure is clinically present [62]. Microscopic evidence of fibrillar collagen degradation found in and around necrotic areas [63] is also found in remote areas [60], resulting in an inadequate interstitial collagen matrix. The fact that collagen degradation becomes significant around the age that is associated with the development of marked ventricular dilatation and wall thinning indicates a strong cause-and-effect relation between collagenase activity and ventricular remodeling in the dilated cardiomyopathic heart.

In the cardiomyopathic hamster, myocardial collagen concentration also is significantly greater than the age-invariant control values as early as 150 days of age; most of the excess collagen is the result of multifocal myocyte necrosis and reparative fibrosis. Collagen concentration continues to increase up to an age of 210 days and thereafter, as mentioned, begins to decline. Thus, in view of the marked increase in collagenase activity, it can be surmised that significant collagen degradation is occurring even though total myocardial collagen concentration remains significantly greater in the diseased heart compared to the age-matched control heart.

Human findings are similar. The interstitial collagen remodeling that occurs with human cardiac disease and which is associated with myocyte necrosis includes reparative or replacement fibrosis [64–68]. Left-ventricular collagen concentration was found to be 30% and 100% greater in explanted hearts from patients with coronary heart disease and dilated cardiomyopathy, respectively, than in normal postmortem hearts. Reddy et al. [60] assessed collagenase activity in endomyocardial biopsy tissue obtained from patients with end-stage dilated cardiomyopathy and from post-transplant patients with presumably normal hearts. They found collagenase activity in the cardiomyopathic tissue to be increased eightfold above that measured in normal hearts. Whether the scar tissue undergoes subsequent degradation, however, is unimportant with regard to the shape and size of the ventricle. Instead, it is degradation of the collagen matrix components responsible for the maintenance of myocyte alignment that results in significant ventricular dilatation.

Finally, in end-stage heart failure one would predict that an inadequate collagen matrix is responsible for progressive left-ventricular dilatation, wall thinning, and sphericalization. Experimentally, chronic supraventricular tachycardia is widely used to create a model of heart failure. Within 3 weeks of rapid pacing, both ventricles are significantly dilated and wall thickness is reduced. Histological changes characteristic of collagen degradation, including reduced collagen concentration, significant myocardial edema, and disrupted fibrillar collagen tethers, have been reported to occur within 6h of rapid ventricular pacing [59]. With continued rapid pacing, these changes persisted for weeks [59,69]. After 3 weeks of rapid pacing, Spinale et al. [69] found hydroxyproline concentration to be significantly reduced.

Thus, a breakdown of myocardial fibrillar collagen appears to play a major role in the pathophysiology of heart failure. However, in heart failure secondary to dilated cardiomyopathy or other causes, it is not possible to discern interstitial collagen remodeling by simply measuring the concentration of myocardial collagen that may be elevated because of extensive reparative fibrosis. Instead, one needs to directly assess the status of the collagen matrix that surrounds and interconnects myocytes and groups of myocytes together with myocardial collagenase activity.

Acknowledgments. This work was supported in part by a grant from the National Heart, Lung, and Blood Institute (RO1-HL-46461) and the American Heart Association, Missouri Affiliate (93-GS-016). Gregory Brower is the recipient of a National Research Service Award (F32 HL09111).

References

1. Janicki JS, Matsubara BB (1993) Myocardial collagen and left ventricular diastolic dysfunction. In: Gaasch WH, LeWinter MM (eds) Left ventricular diastolic dysfunction. Lea and Febiger, Philadelphia, pp 125–140
2. Brilla CG, Janicki JS, Weber KT (1991) Cardioprotective effects of lisinopril in rats with genetic hypertension and left ventricular hypertrophy. Circulation 83:1771–1779
3. Weber KT, Janicki JS, Shroff SG, Pick R, Chen RM, Bashey RI (1988) Collagen remodeling of the pressure-overloaded, hypertrophied nonhuman primate myocardium. Circ Res 62:757–765
4. Thiedemann KU, Holubarsch C, Medugorac I, Jacob R (1983) Connective tissue content and myocardial stiffness in pressure overload hypertrophy: a combined study of morphological, morphometric, biochemical, and mechanical properties. Basic Res Cardiol 78:140–155
5. Narayan S, Janicki JS, Shroff SG, Pick R, Weber KT (1989) Myocardial collagen and mechanics after preventing hypertrophy in hypertensive rats. Am J Hypertens 2:675–682
6. Doering CW, Jalil JE, Janicki JS, Pick R, Aghili S, Abrahams C, Weber KT (1988) Collagen network remodeling and diastolic stiffness of the rat left ventricle with pressure overload hypertrophy. Cardiovasc Res 22:686–695
7. Bing OHL, Matsushita S, Fanburg BL, Levine HJ (1971) Mechanical properties of rat cardiac muscle during experimental hypertrophy. Circ Res 28:234–245
8. Holubarsch C, Holubarsch T, Jacob R, Medugorac I, Thiedemann KU (1983) Passive elastic properties of myocardium in different models and stages of hypertrophy: a study comparing mechanical, chemical, and morphometric parameters. Perspect Cardiovasc Res 7:323–336
9. Borg TK, Ranson WF, Moshlehy FA, Caulfield JB (1981) Structural basis of ventricular stiffness. Lab Invest 44:49–54
10. Jalil JE, Doering CW, Janicki JS, Pick R, Shroff SG, Weber KT (1989) Fibrillar collagen and myocardial stiffness in the intact hypertrophied rat left ventricle. Circ Res 64:1041–1050
11. Carroll EP, Janicki JS, Pick R, Weber KT (1989) Myocardial stiffness and reparative fibrosis following coronary embolization in the rat. Cardiovasc Res 23:655-661
12. Janicki JS, Brower GL, Henegar JR (1995) Interstitial collagen remodeling in chronic heart failure. Basic Appl Myol 5:339–348
13. Caulfield JB, Norton P, Weaver RD (1992) Cardiac dilatation associated with collagen alterations. Mol Cell Biochem 118:171–179
14. O'Brien LJ, Moore CM (1966) Connective tissue degradation and distensibility characteristics of the non-living heart. Experientia (Basel) 22:845–847
15. Eghbali M, Blumenfeld OO, Seifter S, Buttrick PM, Leinwand LA, Robinson TF, Zern MA, Giambrone MA (1989) Localization of types I, III, IV collagen mRNAs in rat heart cells by in situ hybridization. J Mol Cell Cardiol 21:103–113
16. Shekhonin BV, Domogatsky SP, Idelson GL, Koteliansky VE (1988) Participance of fibronectin and various collagen types in the formation of fibrous extracellular matrix in cardiosclerosis. J Mol Cell Cardiol 20:501–508

17. Contard F, Koteliansky V, Marotte F, Dubus I, Rappaport L, Samuel JL (1991) Specific alterations in the distribution of extracellular matrix components within rat myocardium during the development of pressure overload. Lab Invest 64:65–75
18. Robinson TF, Cohen-Gould L, Factor SM, Eghbali M, Blumenfeld OO (1988) Structure and function of connective tissue in cardiac muscle: collagen types I and III in endomysial struts and pericellular fibers. Scanning Microsc 2:1005–1015
19. Medugorac I (1982) Characterization of intramuscular collagen in the mammalian left ventricle. Basic Res Cardiol 77:589–598
20. Robinson TF, Geraci MA, Sonnenblick EH, Factor SM (1988) Coiled perimysial fibers of papillary muscle in rat heart: morphology, distribution, and changes in configuration. Circ Res 63:577–592
21. Robinson TF, Factor SM, Capasso JM, Wittenberg BA, Blumenfeld OO, Seifter S (1987) Morphology, composition and function of struts between cardiac myocytes of rat and hamster. Cell Tissue Res 249:247–255
22. Caulfield JB, Borg TK (1979) The collagen network of the heart. Lab Invest 40:364–372
23. Pearlman ES, Weber KT, Janicki JS, Pietra G, Fishman AP (1982) Muscle fiber orientation and connective tissue content in the hypertrophied human heart. Lab Invest 46:158–164
24. Weber KT, Brilla CG (1991) Pathologic hypertrophy and cardiac interstitium: fibrosis and renin-angiotensin-aldosterone system. Circulation 83:1849–1865
25. Abrahams C, Janicki JS, Weber KT (1987) Myocardial hypertrophy in the macaque fascicularis: structural remodeling of the collagen matrix. Lab Invest 56:676–683
26. Brilla CG, Pick R, Tan LB, Janicki JS, Weber KT (1990) Remodeling of the rat right and left ventricles in experimental hypertension. Circ Res 67:1355–1364
27. Kabour A, Henegar JR, Janicki JS (1994) Angiotensin II induced myocyte necrosis: role of the angiotensin II receptor. J Cardiovasc Pharmacol 23:547–553
28. Kabour A, Henegar JR, Devineni VR, Janicki JS (1995) Prevention of angiotensin II induced myocyte necrosis and coronary vascular damage by lisinopril and losartan in rats. Cardiovasc Res 29:543–548
29. Morkin E, Ashford TP (1968) Myocardial DNA synthesis in experimental cardiac hypertrophy. Am J Physiol 215:1409–1413
30. Grove D, Zak R, Nair KG, Aschenbrenner V (1969) Biochemical correlates of cardiac hypertrophy IV. Observations on the cellular organization of growth during myocardial hypertrophy in the rat. Circ Res 25:473–485
31. Campbell SE, Janicki JS, Weber KT (1995) Temporal differences in fibroblast proliferation and phenotype expression in response to chronic administration of angiotensin II or aldosterone. J Mol Cell Cardiol 27:1545–1560
32. Caulfield JB (1983) Alterations in cardiac collagen with hypertrophy. Perspect Cardiovasc Res 8:49–57
33. Anderson KR, Sutton MG St J, Lie JT (1979) Histopathological types of cardiac fibrosis in myocardial disease. J Pathol 128:79–85
34. Pick R, Janicki JS, Weber KT (1989) Myocardial fibrosis in nonhuman primate with pressure overload hypertrophy. Am J Pathol 135:771–781
35. Weber KT, Pick R, Jalil JE, Janicki JS, Carroll EP (1989) Patterns of myocardial fibrosis. J Mol Cell Cardiol 21(Suppl V):121–131
36. Silver MA, Pick R, Brilla CG, Jalil JE, Janicki JS, Weber KT (1990) Reactive and reparative fibrillar collagen remodelling in the hypertrophied rat left ventricle: two experimental models of myocardial fibrosis. Cardiovasc Res 24:741–747
37. Brilla CG, Weber KT (1992) Reactive and reparative myocardial fibrosis in arterial hypertension in the rat. Cardiovasc Res 26:671–677
38. Bishop SP, Melsen LR (1976) Myocardial necrosis, fibrosis, and DNA synthesis in experimental cardiac hypertrophy induced by sudden pressure overload. Circ Res 39:238–245
39. Gavras H, Brown JJ, Lever AF, MacAdam RF, Robertson JIS (1971) Acute renal failure, tubular necrosis, and myocardial infarction induced in the rabbit by intravenous AngII. Lancet 2:19–22

40. Gavras H, Kremer D, Brown JJ, Gray B, Lever AF, MacAdam RF, Medina A, Morton JJ, Robertson JIS (1975) Angiotensin and norepinephrine induced myocardial lesions: experimental and clinical studies in rabbits and man. Am Heart J 89:321–332
41. Tan LB, Jalil JE, Pick R, Janicki JS, Weber KT (1991) Cardiac myocyte necrosis induced by angiotensin II. Circ Res 69:1185–1195
42. Henegar JR, Brower GL, Kabour A, Janicki JS (1995) Catecholamine response to chonic ANG II infusion and its role in myocyte and coronary vascular damages. Am J Physiol 269:H1564–H1569
43. Giacomelli F, Anversa P, Wiener J (1976) Effect of angiotensin-induced hypertension on rat coronary arteries and myocardium. Am J Pathol 84:111–138
44. Bhan RJ, Giacomelli F, Wiener J (1982) Adrenoreceptor blockade in angiotensin-induced hypertension: effect on rat coronary arteries and myocardium. Am J Pathol 108:60–71
45. Rodrigues MAM, Bregagnollo EA, Montenegro MR, Tucci PJF (1992) Coronary vascular and myocardial lesions due to experimental constriction of the abdominal aorta. Int J Cardiol 35:253–257
46. Brilla CG, Maisch B, Weber KT (1993) Renin-angiotensin system and myocardial collagen matrix remodeling in hypertensive heart disease: in vivo and in vitro studies on collagen matrix regulation. Clin Invest 71:S35–S41
47. Weber KT, Brilla CG (1992) Myocardial fibrosis and the renin-angiotensin-aldosterone system. J Cardiovasc Pharmacol 20:S48–S54
48. Brilla CG, Weber KT (1992) Mineralocorticoid excess, dietary sodium, and myocardial fibrosis. J Lab Clin Med 120:893–901
49. Schraeger JA, Canby CA, Rongish BJ, Kawai M, Tomanek RJ (1994) Normal left ventricular diastolic compliance after regression of hypertrophy. J Cardiovasc Pharmacol 23:349–357
50. Bing OHL, Fanburg BL, Brooks WW, Matsushita S (1978) The effect of the lathyrogen β-amino proprionitrile (BAPN) on the mechanical properties of experimentally hypertrophied rat cardiac muscle. Circ Res 43:632–637
51. Gelpi RJ, Pasipoularides A, Lader AS, Patrick TA, Chase N, Hittinger L, Shannon RP, Bishop SP, Vatner SF (1991) Changes in diastolic cardiac function in developing and stable perinephritic hypertension in conscious dogs. Circ Res 68:555–567
52. Jalil JE, Janicki JS, Pick R, Abrahams C, Weber KT (1989) Fibrosis-induced reduction of endomyocardium in the rat after isoproterenol treatment. Circ Res 65:258–264
53. Hittinger L, Shannon RP, Bishop SP, Gelpi RJ, Vatner SF (1989) Subendocardial exhaustion of blood flow reserve and increased fibrosis in conscious dogs with heart failure. Circ Res 65:971–980
54. Weber KT, Jalil JE, Janicki JS, Pick R (1989) Myocardial collagen remodeling in pressure overload hypertrophy. A case for interstitial heart disease. Am J Hypertens 2:931–940
55. Montfort I, Perez-Tamayo R (1975) The distribution of collagenase in normal rat tissues. J Histochem Cytochem 23:910–920
56. Brower GL, Henegar JR, Janicki JS (1996) Temporal evaluation of left ventricular remodeling and function in rats with chronic volume overload. Am J Physiol (in press)
57. Bartosova D, Chvapil M, Korecky B, Poupa O, Rakusan K, Turek Z, Vizek M (1969) The growth of the muscular and collagenous parts of the rat heart in various forms of cardiomegaly. J Physiol 200:285–295
58. Michel JB, Salzmann JL, Ossondo Nlom M, Bruneval P, Barres D, Camilleri JP (1986) Morphometric analysis of collagen network and plasma perfused capillary bed in the myocardium of rats during evolution of cardiac hypertrophy. Basic Res Cardiol 81:142–154
59. Weber KT, Pick R, Silver MA, Moe GW, Janicki JS, Zucker IH, Armstrong PW (1990) Fibrillar collagen and the remodeling of the dilated canine left ventricle. Circulation 82:1387–1401
60. Reddy HK, Tyagi SC, Tjahja IE, Voelker DJ, Campbell SE, Weber KT (1993) Enhanced endomyocardial collagenase activity in dilated cardiomyopathy: a marker of dilatation and architectural remodeling. Circulation 88:I–407

61. Janicki JS, Tyagi SC, Henegar JR, Campbell SE (1993) Myocardial collagenase activity and ventricular dilatation in cardiomyopathic hamsters. Circulation 88:I-381
62. Gertz EW (1972) Cardiomyopathic Syrian hamster: a possible model of human disease. Prog Exp Tumor Res 16:242–260
63. Cohen-Gould L, Robinson TF, Factor SM (1987) Intrinsic connective tissue abnormalities in the heart muscle of cardiomyopathic Syrian hamsters. Am J Pathol 127:327–334
64. Bishop JE, Greenbaum R, Gibson DG, Yacoub M, Laurent GJ (1990) Enhanced deposition of predominantly type I collagen in myocardial disease. J Mol Cell Cardiol 22:1157–1165
65. Heneghan MA, Malone D, Dervan PA (1991) Myocardial collagen network in dilated cardiomyopathy. Morphometry and scanning electron microscopy study. Ir J Med Sci 160:399–401
66. Rossi MA (1991) Patterns of myocardial fibrosis in idiopathic cardiomyopathies and chronic Chagasic cardiomyopathy. Can J Cardiol 7:287–294
67. Schaper J, Froede R, Hein S, Buck A, Hashizume H, Speiser B, Friedl A, Bleese N (1991) Impairment of the myocardial ultrastructure and changes of the cytoskeleton in dilated cardiomyopathy. Circulation 83:504–514
68. Yoshikane H, Honda M, Goto Y, Morioka S, Ooshima A, Moriyama K (1992) Collagen in dilated cardiomyopathy: scanning electron microscopic and immunohistochemical observations. Jpn Circ J 56:899–910
69. Spinale FG, Tomita M, Zellner JL, Cook JC, Crawford FA (1991) Collagen remodeling and changes in LV function during development and recovery from supraventricular tachycardia. Am J Physiol 261:H308–H318

Myocardial Hypertrophy and Coronary Circulation Before and After Relief of Pressure Overload

SHOGEN ISOYAMA

Summary. Cardiac hypertrophy is one of the major risk factors for cardiac events. Antihypertensive treatment reduces the incidence of such events. This chapter focuses on the following two issues: the relationship between physiological (decreased coronary dilator reserve and impaired autoregulation) and morphological changes (remodeling of arterial microvessels) in the coronary arterial system of pressure-overloaded hearts, and changes in this relationship after the relief of pressure overload. Vascularity and vascular hypertrophy are considered limiting factors for dilator reserve in cardiac hypertrophy. The age at which hemodynamic overload or hypertrophy begins is an important factor for capillary or arteriolar angiogenesis. After maturation, cardiac hypertrophy is accompanied by a decrease in dilator reserve and impairment of coronary autoregulation, which can probably be attributed to hypertrophy of the coronary arterial microvessels (medial thickening and deposition of perivascular collagen) and to endothelial and smooth muscle cell dysfunction. In principle, decreased dilator reserve or impaired coronary autoregulation in cardiac hypertrophy is reversible after relief of the overload. The duration of pressure overload before its relief affects the reversibility of decreased dilator reserve and impaired autoregulation, because the duration determines the regression of excess deposition of collagen. At present, there are no data concerning the reversibility of these changes in pressure-overloaded human hearts. Further studies are needed to clarify how antihypertensive treatment reduces the incidence of cardiac events in patients.

Key words. Cardiac hypertrophy—Coronary reserve—Autoregulation—Regression—Angiogenesis—Vascular hypertrophy—Arterial microvessels

Introduction

Epidemiological studies indicate that cardiac hypertrophy is one of the major risk factors for cardiac events including congestive heart failure, myocardial ischemia, and sudden death [1–7]. On the other hand, antihypertensive treatment reduces the inci-

First Department of Internal Medicine, Tohoku University School of Medicine, Sendai, 980-77 Japan.

dence of such events [8]. In the clinical situation, systemic hypertension is the most common underlying cause of left-ventricular hypertrophy [1]. In hearts hypertrophied by pressure overload, alterations in the coronary circulation and remodeling of coronary arterial microvessels have been observed by a number of investigators [9–23]. At present, it is not known whether this process, once it has begun, can be either prevented or reversed by antihypertensive treatment. This review focuses on the following two issues: (i) the relationship between physiological (decreased coronary dilator reserve and impaired autoregulation) and morphological (remodeling of arterial microvessels) changes in the coronary arterial system of pressure-overloaded hearts; and (ii) changes in this relationship after the relief of pressure overload.

Remodeling of Coronary Arterial Microvessels in Hypertensive Cardiac Hypertrophy

Vascularity

Vascularity and vascular hypertrophy are considered limiting factors for coronary dilator reserve in hypertensive cardiac hypertrophy. The factors that determine whether angiogenesis or rarefaction occurs at the capillary or arteriolar level remain controversial, as both are observed in hearts with renal hypertension [15,19,24] or mechanical aortic constriction [25] and in hearts of spontaneously hypertensive rats [11].

Several studies have discussed some important modulating factors for angiogenesis in hypertrophied hearts. First, the capacity for capillary or arteriolar growth decreases with age [26,27]. The age at which hemodynamic overload or myocardial hypertrophy begins appears to be critical for angiogenesis in humans [28] and animals [20,27,29]. Angiogenesis at the capillary or arteriolar level does occur in parallel with an increase in cardiac mass during the developmental phase and results in normal coronary dilator reserve [20,28,29]. In pressure-overloaded hearts of lambs, treatment with protamine compromised the increase in capillary number or density, resulting in decreased dilator capacity compared with nontreated hearts [30]. Therefore, pressure overload imposed on hearts during the developmental phase promotes angiogenesis in parallel with an increase in muscle mass or may prolong the period in which angiogenesis can occur.

Second, the duration of pressure overload has been suggested to be important for angiogenesis. Even in hearts of adult animals, long-term pressure overload by 6-month renovascular hypertension [15,31] or pulmonary artery banding [17] caused an increase in arteriolar number or capillary density [15,17,31]. Furthermore, cardiac failure induced by long-term hypertension promoted capillary proliferation in the heart [24]. The type of pressure overload, i.e., the genetic model of hypertension (spontaneously hypertensive rats), renal hypertension, or the constriction model of the thoracic aorta, also may be important, because circulating factors or factors in the myocardial tissue itself that modulate proliferation of endothelial and smooth muscle cells differ among those models.

Molecular Basis for Angiogenesis

The family of angiogenesis factors includes polypeptides and nonpolypeptides. At present, at least eight purified and sequenced polypeptide growth factors have been shown to be responsible for physiological and pathological angiogenesis [26,32,33]. The molecular basis for angiogenesis in hearts has been studied in the case of myocardial tissue hypoxia induced by ischemia. In response to ischemia [34] or hypoxia [34–36], monocytes and platelets are known to produce growth factors [platelet-derived growth factor (PDGF), transforming growth factor-alpha (TGF-α), TGF-β, tumor necrosis factor-α, basic and acidic fibroblast growth factor (FGF), etc.] that may act as mitogens for endothelial and smooth muscle cells, in addition to the mitogens (vascular endothelial growth factor [34,35,37], basic [37] and acidic FGF [38]) which are produced by ischemic cardiomyocytes, by smooth muscle cells, and by endothelial cells themselves. It has been reported that intra- or extraluminal (periadventitial) administration of basic FGF protected the ischemic myocardium from necrosis and improved myocardial function, suggesting that angiogenesis had occurred [39,40]. Thus, basic FGF seems to promote angiogenesis in the ischemic myocardium, although it is not clear that endogenous or exogenous basic FGF is essential for angiogenesis in the ischemic myocardium.

Several initiation factors for angiogenesis in cardiac hypertrophy are suggested: high blood velocity, increased tangential wall stress, or tissue hypoxia [33,41–43]. A few studies have reported the possible participation of mast cells in angiogenesis during physiological growth or during the development of hypertensive hypertrophy of the heart [17,44]. However, the molecular basis for angiogenesis has not yet been determined: Which is the primary stimulus, tissue hypoxia or high coronary blood flow? Which type of cell is stimulated? How does (and which of) the growth-promoting or -inhibitory substances [26] modulate the proliferation of endothelial or smooth muscle cells?

Vascular Hypertrophy in Hearts with Coronary Hypertension

Systemic hypertension has two combined effects on the heart: pressure overload to the left-ventricular myocardium and coronary hypertension (Fig. 1). Hypertensive changes in the coronary arterial tree, i.e., vascular wall thickening and perivascular fibrosis, are basically the same in the pulmonary [45,46] and systemic arterial trees [47]. Mechanical stretch, tension, or pressure directly stimulates smooth muscle cells

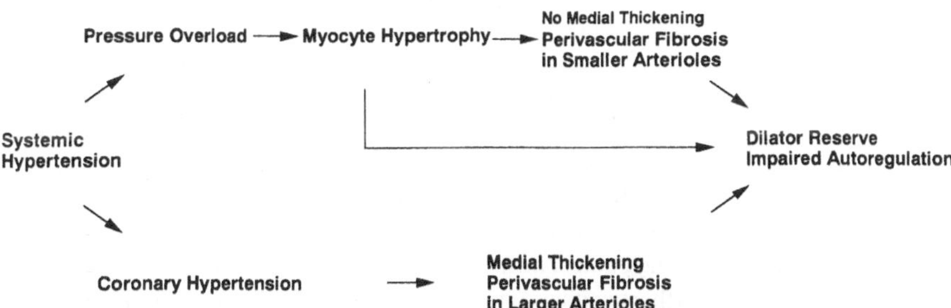

Fig. 1. In systemic hypertension, remodeling of coronary arterial microvessels is produced through two different mechanisms: hypertensive vascular hypertrophy (medial thickening and perivascular fibrosis in arteries and larger arterioles) and remodeling that results from the activation of the extravascular tissue (perivascular fibrosis without medial thickening in smaller arterioles). These vascular changes would contribute to decreased dilator reserve and the impairment of autoregulation

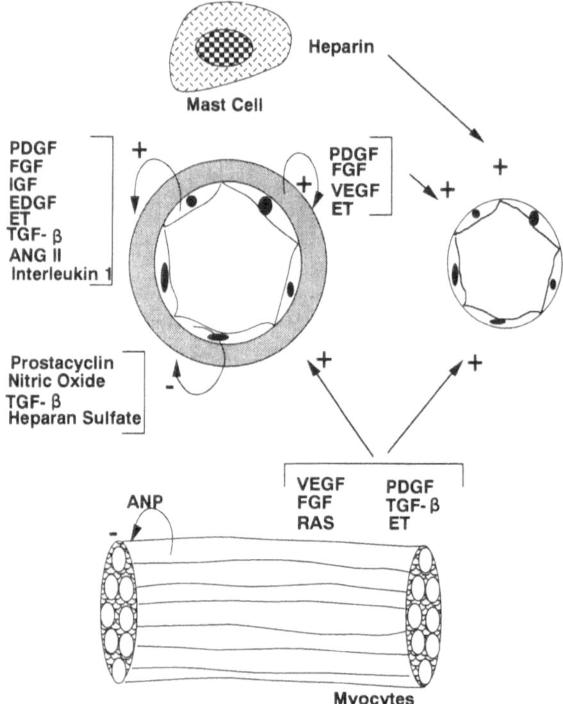

FIG. 2. Possible molecular basis for the two mechanisms of coronary vascular remodeling in hypertensive cardiac hypertrophy. *PDGF*, Platelet-derived growth factor; *FGF*, fibroblast growth factor; *IGF*, insulin-like growth factor; *EDGF*, endothelium-derived growth factor; *ET*, endothelin; *TGF-β*, transforming growth factor-β; *Ang II*, angiotensin II; *VEGF*, vascular endothelial growth factor; *ANP*, atrial natriuretic peptide; *RAS*, renin-angiotensin system; +, growth-promoting effect; –, growth-inhibitory effect

to become hyperplastic [48,49] and fibroblasts to express collagen [50]. These mechanical factors simultaneously stimulate smooth muscle cells, endothelial cells, and fibroblasts to release a large number of growth factors, cytokines, and vasoregulatory molecules [14,51–56] (Fig. 2). These mediators have been reported to participate in the pathogenesis of hypertensive remodeling through an autocrine or paracrine mechanism within the peripheral vascular wall [54,57]. Although the evidence showing that the mechanism operates in vivo was obtained in the peripheral vasculature, this mechanism would operate in the coronary vascular wall as well.

Medial hypertrophy (smooth muscle hypertrophy or hyperplasia) and excess deposition of perivascular collagen in coronary arterial microvessels were reported in hypertrophied hearts with coronary hypertension [10,12,13,23,58–61]. The medial thickening occurred in the arteries and relatively larger arterioles but not in the smaller arterioles in nonhypertrophied hearts with coronary hypertension [10]. The decrease in the ratio of luminal area to vascular wall area produced by the medical thickening would be linked to an increase in minimal coronary vascular resistance and a decrease in dilator reserve [10,12,22,60]. The physiological observation concerning intracoronary pressure or resistance distribution measured at the epicardial arterial trees [62] is correlated with those size-related structural changes in the arterial microvessels. That is, a greater pressure drop was observed in the larger arterial trees, and the smaller arterial microvessels are thereby protected from an elevation of intravascular pressure.

In rat hearts with aortic banding, increases in mRNAs of fetal forms of cellular fibronectin and collagen were preceded by the induction of TGF-β_1 mRNA [63,64]. The accmulation of fibronectin mRNA was observed within the walls of the right and left coronary arteries [64], but was not observed in the wall of any vein. In addition, cyclic stretch induced type III collagen mRNA in in vitro cardiac fibroblasts [50]. Therefore, the trigger for the upregulation of the fibronectin gene in nonmuscle and smooth muscle cells is vascular stretch alone or in combination with growth factors such as TGF-β_1 or other hormonal peptides.

This type of coronary vascular remodeling is apparently not always associated with the presence of myocardial hypertrophy, even in pressure-overloaded left ventricles with coronary hypertension. For example, the superimposition of pressure overload on aged hearts may accelerate the underlying age-related vascular changes [11,16] and decrease the dilator reserve further, in spite of an insignificant degree of myocardial hypertrophy [65,66].

Vascular Changes in Hypertrophied Hearts Without Coronary Hypertension

The effects of myocardial pressure overload alone on the remodeling of coronary arterial microvessels were examined in the right-ventricular free wall with pulmonary artery banding [10]. The remodeling pattern in hypertrophied hearts without coronary hypertension is characterized by collagen accumulation in the perivascular tissues of smaller arterioles [10]. No medial hypertrophy was observed over the whole range of arterial microvessels [10] (see Fig. 1). This pattern of remodeling in coronary arterial microvessels is also observed in myocardial tissue samples obtained from patients with valvular aortic stenosis [13] or idiopathic cardiomyopathy [23]. Of interest, in rat hearts treated with continuous infusion of angiotension II, fibronectin mRNA was induced particularly in the perivascular tissue of smaller arterioles [67] through its direct action and independently of arterial pressure. This localization of fibronectin mRNA is compatible with the fact of collagen accumulation in hypertrophied hearts without coronary hypertension [10,13].

Substances released from the extravascular tissue or cells activated by pressure overload may play important roles in the arterial microvessel remodeling of hypertrophied hearts without coronary hypertension (see Fig. 2). The cardiac renin–angiotensin system is upregulated in hypertrophied hearts [68] (induction of angiotensin-converting enzyme, ACE [69], renin, or angiotensinogen [70] at the protein and mRNA level). In addition, cardiomyocytes can be sources of PDGF-A chain [71], PDGF-B chain [72], acidic FGF [38,71,72], atrial natriuretic peptide [73], endothelin-1 [74], and TGF-β_1 [72] under some pathological conditions. The vascular remodeling that resulted from the surrounding tissue activation would be specific for the hearts of subjects with systemic hypertension. Therefore, in such hearts, the remodeling pattern of coronary arterial microvessels is the summation of two patterns resulting from two different mechanisms: hypertensive vascular remodeling and remodeling caused by the activation of the extravascular tissues (see Fig. 2).

Coronary Arterial Microvessels After Normalization of Systemic Hypertension

Reversibility of Circulation Abnormalities

In principle, the coronary circulation abnormalities in hypertensive cardiac hypertrophy are reversible, as is the myocyte hypertrophy. Both the decreases in dilator reserve and the impairment of coronary autoregulation can be reversed after relief of pressure overload. The restoration of normal dilator reserve was observed in rodent hearts with genetic hypertension [59,61,75,76], renal hypertension [77], or aortic banding [78–81], and in canine hearts with aortic banding [82]. The rectification of impaired autoregulation was also observed in rodent hearts with aortic banding [78,80]. In addition, both decreased dilator reserve and impaired autoregulation were observed during banding of the ascending aorta, and both were reversed after debanding [80]. These observations may suggest that the sites of arterial microvessels or the principal mechanisms responsible for regulation of maximal dilation and autoregulation are the same.

Several factors affect the time required for the reversal of coronary circulation abnormalities. Six months after the relief of 3-month pressure overload in a

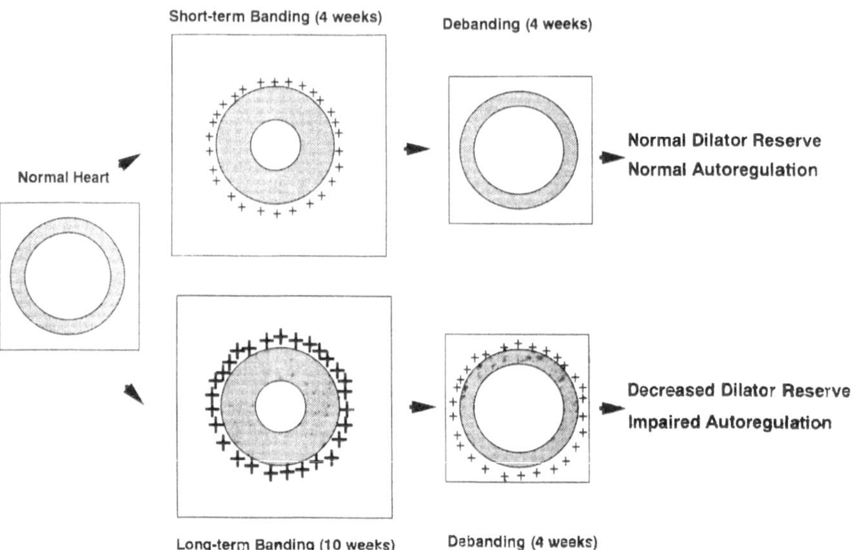

FIG. 3. Regression of myocardial and vascular hypertrophy with perivascular fibrosis after relief of pressure overload. Short-term (4 weeks) and long-term (10 weeks) banding of the ascending aorta produced the same degree of myocardial hypertrophy and medial thickening in rat hearts. The two different durations of banding also produced the same degree of decreases in coronary dilator reserve measured in the Langendorff preparation. Four weeks after debanding, myocardial hypertrophy and medial thickening regressed completely, and this regression did not depend on the duration of pressure overload before its relief. The remaining perivascular fibrosis in hearts debanded after long-term banding seemed to be the determining factor of whether decreased dilator reserve and impaired autoregulation could be reversed after debanding (see [8,9,11,12] for details). Size of box indicates muscle mass

canine model of ascending aortic banding, the dilator reserve measured after the administration of adenosine was normal, but myocardial hypertrophy had regressed only partially [82]. On the other hand, 4-week debanding after 4-week banding of the ascending aorta produced complete reversal of both myocardial hypertrophy and coronary circulation abnormalities in rat hearts [80,81]. The time required for restoration in a large animal model seems to be longer than that in a rodent model.

The duration of pressure overload before its relief also affects the time required for restoration. Coronary circulation abnormalities remained in long-term (10 weeks) banded rat hearts [78,79] after 4-week debanding, whereas both the circulation abnormalities and myocardial hypertrophy in short-term (4 weeks) banded hearts were completely reversed [80,81] (Fig. 3). In addition, antihypertensive treatment required a longer period (12–20 weeks) to restore the dilator reserve in spontaneously hypertensive rats [59,61,75,76,83]. Although the duration of pressure overload before its relief did not alter the regression of myocyte hypertrophy [79], a longer duration does seem to prolong the time needed for reversal of coronary circulation abnormalities. This longer period can be accounted for by the difficulties in the repair of structural changes in the systemic hypertensive arteries [47].

Structural Restoration of Coronary Arterial Microvessels

In the systemic [47] and pulmonary arterial trees [46], normalization of hypertension regresses medial thickening and excess deposition of extracellular matrix protein such as collagen and elastin. In deoxycorticosterone acetate (DOCA) salt hypertensive rats, inhibition of collagen deposition reduced arterial pressure [84]. In addition, in a two-kidney, one-clip renovascular hypertension model, nephrectomy completely normalized the arterial pressure after a short-term duration of hypertension, but did not do so after long-term hypertension [85]. These observations suggest that the excess deposition plays a key role in maintaining or normalizing hypertension [45,84,85].

Because it is unlikely that relief of pressure overload induces angiogenesis, the reversal of the coronary circulation abnormalities would be produced mainly by the regression of vascular hypertrophy. Medial thickening and perivascular fibrosis of the larger coronary arteries caused by elevated intravascular pressure can regress after normalization of the pressure. The degradation rate of collagen is one of the factors responsible for the difference in the reversibility of coronary circulation abnormalities after short- or long-term pressure overload [58,60] (see Fig. 3). Judging from the turnover of collagen, the deposition of collagen is decreased in a relatively early stage after normalization of hypertension [46,58,60]. It has been observed that collagen turnover of the peripheral arterial wall is shorter in hypertensive compared to normotensive rodent animals [86], suggesting that the collagenolytic system is activated. Therefore, the activation might persist after the normalization of hypertension and reduce the excess deposition of collagen in the early stage.

Reversal of Coronary Circulation Abnormalities Versus Myocyte Hypertrophy

In hypertensive hearts, the regression of myocyte hypertrophy is not always accompanied by the reversal of coronary circulation abnormalities [79]. Conversely, the regression of abnormalities in the coronary circulation was also observed despite the presence of remaining myocyte hypertrophy [61,75,82]. The independent regression of myocyte hypertrophy and coronary circulation abnormalities would be derived from the fact that hypertension can develop myocyte hypertrophy and result in remodeling of the coronary vasculature independently (see Fig. 1).

Pharmacological Normalization of Hypertension in Animal Studies

In the systemic circulation, the time at which antihypertensive treatment is begun is one of the factors that determine the level of arterial pressure after discontinuation of the treatment [87–89]. In spontaneously hypertensive rats, treatment with an ACE inhibitor that began before birth or the establishment of hypertension reduced arterial pressure in the long term and prevented morphological changes in the peripheral arteries after its discontinuation. However, treatment started after the establishment of hypertension did not have the same effect [87–89]. This is probably because structural alterations of arteries and arterioles had already taken place by the time of initiation of the treatment. These observations suggest that it is difficult to normalize vascular structure in patients with established hypertension, and that normalization of structural vascular changes requires a longer period of antihypertensive treatment.

The pharmacological normalization of systemic hypertension and consequent alterations in the coronary arterial microvessels or of myocardial hypertrophy has been studied in spontaneously hypertensive rats. Lowering the systemic arterial blood pressure regressed medial hypertrophy of the coronary arterial trees [59,61,83], increased the distensibility of the vascular wall [59,61], and improved the dilator reserve [59,61,75,83]. The regression of coronary vascular hypertrophy can be brought about by various antihypertensive drugs such as hydralazine [61,75], calcium channel blockers [83], and ACE inhibitors [59,75], as an improvement in the decreased dilator reserve can be observed even without regression of myocardial hypertrophy [75]. However, whether a pharmacological reduction of blood pressure regresses myocardial hypertrophy does not simply depend on the induced change in arterial pressure, but also on the mechanism of the drug. Calcium channel blockers [83] and ACE inhibitors [59,75,76] regressed myocardial hypertrophy, but hydralazine did not [61,75]. Thus, pharmacological normalization of hypertension does not always regress both myocardial and vascular hypertrophy, although normalization of pressure overload by ascending aortic debanding regresses both [58,60,80,81].

In pressure-overloaded hearts, elevation of systolic pressure (primary stimulus) or activation of the second step such as the cardiac renin–angiotensin system by the primary stimulus is important for the induction of myocardial hypertrophy [90–92]. However, it is not clear whether the mechanism for the maintenance of cardiac hypertrophy is the same as for its induction in pressure-overloaded hearts. Of interest, an angiotensin II receptor blocker accelerated the regression of myocardial hypertrophy induced by isoproterenol administration, suggesting that the renin–angiotensin system is important for the maintenance of hypertrophy [93]. The differing effects of the reduction of arterial pressure on the regression of myocardial hypertrophy among drugs [94] might result from different actions on the maintenance mechanism.

Future Directions in the Clinical Situation from Observations in Animal Studies

In the clinical setting, myocardial hypertrophy induced by hypertension has been observed to regress after nonmedical or medical correction. In patients with essential hypertension, the reduction of blood pressure by sodium restriction regressed myocardial hypertrophy [95]. Regression of myocardial hypertrophy became evident after 8–12 weeks of antihypertensive treatment [96,97], and remained decreased 4 weeks after cessation of the treatment [96] in hypertensive humans. Regression of myocardial hypertrophy depended on the type of drugs in humans as in animals [1,97,98].

To my knowledge, there are no data concerning the reversibility of decreased dilator reserve and impaired coronary autoregulation in pressure-overloaded human hearts. If the abnormalities are reversible, how rapidly do the abnormalities resolve? How does the age at which antihypertensive treatment begins or the duration of pressure overload before the treatment begins affect the reversibility? What kind of antihypertensive drugs can reverse the abnormalities? Do the abnormalities regress independently of the presence or absence of myocardial hypertrophy in hypertensive patients? Further studies are needed to answer these questions.

Acknowledgment. This study was partly supported by a Grant-in-Aid for Scientific Research (06670688) from the Ministry of Education, Science, and Culture of Japan.

References

1. Frohlich ED, Apstein C, Chobanian AV, Devereux RB, Dustan HP, Dzau V, Fauad-Tarazi F, Horan MJ, Marcus M, Massie B, Pfeffer MA, Re RN, Roccellar EJ, Savage D, Shub C (1992) The heart in hypertension. N Engl J Med 327:998–1008
2. Devereux RB, de Simone G, Koren MJ, Roman MJ, Laragh JH (1991) Left ventricular mass as a predictor of development of hypertension. Am J Hypertens 4:603S–607S
3. Koren MJ, Devereux RB, Casale PN, Savage DD, Laragh JH (1991) Relation of left ventricular mass and geometry to morbidity and mortality in uncomplicated essential hypertension. Ann Intern Med 114:345–352
4. Casale PN, Devereux RB, Milner M, Zullo G, Harshfield GA, Pickering TG, Laragh JH (1986) Value of echocardiographic measurement of left ventricular mass in predicting cardiovascular morbid events in hypertensive man. Ann Intern Med 105:173–178
5. Kannel WB, Abbott RD (1986) A prognostic comparison of symptomatic left ventricular hypertrophy and unrecognized myocardial infarction: the Framingham study. Am Heart J 111:391–397
6. Kannel WB (1983) Prevalence and natural history of electrocardiographic left ventricular hypertrophy. Am J Med 75(suppl 3A):4–11
7. Kannel WB, Gordon T, Offutt D (1969) Left ventricular hypertrophy by electrocardiogram: prevalence, incidence, and mortality in the Framingham study. Ann Intern Med 71:89–105
8. Alderman MH, Ooi WL, Madhavan S, Cohen H (1989) Treatment-induced blood pressure reduction and the risk of myocardial infarction. JAMA 262:920–924
9. Isoyama S (1994) Interplay of hypertrophy and myocardial ischemia. In: Lorell BH, Grossman W (eds) Diastolic relaxation of the heart. Kluwer, Boston, pp 203–211
10. Ito N, Nitta Y, Ohtani H, Ooshima A, Isoyama S (1994) Remodeling of microvessels by coronary hypertension or cardiac hypertrophy in rats. J Mol Cell Cardiol 26: 49–59

11. Vitullo JC, Penn MS, Rakusan K, Wicker P (1993) Effects of hypertension and aging on coronary arteriolar density. Hypertension (Dallas) 21:406–414
12. Schwartzkopff B, Motz W, Frenzel H, Vogt M, Kanuer S, Strauer BE (1993) Structural and functional alterations of the intramyocardial coronary arterioles in patients with arterial hypertension. Circulation 88:993–1003
13. Schwarzkopff B, Frenzel H, Dieckerhoff J, Betz P, Flasshove M, Schulte HD, Mundhenke M, Motz W, Strauer BE (1992) Morphometric investigation of human myocardium in arterial hypertension and valvular aortic stenosis. Eur Heart J 13(suppl D):17–23
14. Harrison DG, Marcus ML, Dellsperger KC, Lamping KG, Tomanek RJ (1991) Pathophysiology of myocardial perfusion in hypertension. Circulation 83(suppl III):III-14–III-18
15. Tomanek RJ, Wessel TJ, Harrison DG (1991) Capillary growth and geometry during long-term hypertension and myocardial hypertrophy in dogs. Am J Physiol (Heart Circ Physiol 30) 261:H1011–H1018
16. Tomanek RJ, Aydelotte MR, Butters CA (1990) Late-onset hypertension in old rats alters myocardial microvessels. Am J Physiol (Heart Circ Physiol 28) 259:H1681–H1687
17. Olivetti G, Lagrasta C, Ricci R, Sonnenblick EH, Capasso JM, Anversa P (1989) Long-term pressure-induced cardiac hypertrophy: capillary and mast cell proliferation. Am J Physiol (Heart Circ Physiol 26) 257:H1766–H1772
18. Jeremy RW, Fletcher PJ, Thompson J (1989) Coronary pressure-flow relations in hypertensive left ventricular hypertrophy: comparison of intact autoregulation with physiological and pharmacological vasodilation in the dog. Circ Res 65:224–236
19. Cimini C, Weiss HR (1988) Microvascular morphometry and perfusion in renal hypertension-induced cardiac hypertrophy. Am J Physiol (Heart Circ Physiol 24) 255:H1384–H1390
20. Bache RJ (1988) Effects of hypertrophy on the coronary circulation. Prog Cardiovasc Dis 31:403–440
21. Harrison DG, Florentine MS, Brooks LA, Cooper SM, Marcus ML (1988) The effect of hypertension and left ventricular hypertrophy on the lower range of coronary autoregulation. Circulation 77:1108–1115
22. Marcus ML, Harrison DG, Chilian WM, Koyanagi S, Inou T, Tomanek RJ, Martins JB, Eastham CL, Hiratzka LF (1987) Alterations in the coronary circulation in hypertrophied ventricles. Circulation 75(suppl I):I-19–I-25
23. Tanaka M, Fujiwara H, Onodera T, Wu DJ, Matsuda M, Hamashima Y, Kawai C (1987) Quantitative analysis of narrowings of intramyocardial small arteries in normal hearts, hypertensive hearts, and hearts with hypertrophic cardiomyopathy. Circulation 75:1130–1139
24. Anversa P, Capasso JM (1991) Loss of intermediate-sized coronary arteries and capillary proliferation after left ventricular failure in rats. Am J Physiol (Heart Circ Physiol 29) 260:H1552–H1560
25. Breisch EA, White FC, Nimmo LE, Bloor CM (1986) Cardiac vasculature and flow during pressure-overload hypertrophy. Am J Physiol (Heart Circ Physiol 20) 251:H1031–H1037
26. Klagsbrun M, D'Amore PA (1991) Regulators of angiogenesis. Annu Rev Physiol 53:217–239
27. Rakusan K, de Rochemont WDM, Braasch W, Tschopp H, Bing RJ (1967) Capacity of the terminal vascular bed during normal growth, in cardiomegaly, and in cardiac atrophy. Circ Res 21:209–215
28. Rakusan K, Flanagan MF, Geva T, Southern J, Praagh RV (1992) Morphometry of human coronary capillaries during normal growth and the effect of age in left ventricular pressure-overload hypertrophy. Circulation 86:38–46
29. Bache RJ, Alyono D, Sublett E, Dai X-Z (1986) Myocardial blood flow in left ventricular hypertrophy developing in young and adult dogs. Am J Physiol (Heart Circ Physiol 20) 251:H949–H956

30. Flanagan MF, Fujii AM, Colan SD, Flanagan RG, Lock JE (1991) Myocardial angiogenesis and coronary perfusion in left ventricular pressure-overload hypertrophy in the young lamb. Evidence for inhibition with chronic protamine administration. Circ Res 68:1458–1470
31. Tomanek RJ, Schalk KA, Marcus ML, Harrison DG (1989) Coronary angiogenesis during long-term hypertension and left ventricular hypertrophy in dogs. Circ Res 65:352–359
32. Folkman J, Shing Y (1992) Angiogenesis. J Biol Chem 267:10931–10934
33. Shaper W (1991) Angiogenesis in the adult heart. Basic Res Cardiol 86(suppl 2): 51–56
34. Hashimoto E, Ogita T, Nakaoka T, Matsuoka R, Takao A, Kira Y (1994) Rapid induction of vascular endothelial growth factor expression by transient ischemia in rat heart. Am J Physiol (Heart Circ Physiol 36) 267:H1948–H1954
35. Banai S, Shweiki D, Pinson A, Chandra M, Lazarovici G, Keshet E (1994) Upregulation of vascular endothelial growth factor expression induced by myocardial ischaemia: implications for coronary angiogenesis. Cardiovasc Res 28:1176–1179
36. Ladoux A, Frelin C (1993) Hypoxia is a strong inducer of vascular endothelial growth factor mRNA expression in the heart. Biochem Biophys Res Commun 195:1005–1010
37. Brogi E, Wu T, Namiki A, Isner JM (1994) Indirect angiogenic cytokines upregulate VEGF and bFGF gene expression in vascular smooth muscle cells, whereas hypoxia upregulates VEGF expression only. Circulation 90:649–652
38. Bernotat-Danielowski S, Sharma HS, Schott RJ, Schaper W (1993) Generation and localisation of monoclonal antibodies against fibroblast growth factors in ischemic collateralised porcine myocardium. Cardiovasc Res 27:1220–1228
39. Harada K, Grossman W, Friedman M, Edelman ER, Prasad PV, Keighley CS, Manning WJ, Sellke F, Simons M (1994) Basic fibroblast growth factor improves myocardial function in chronically ischemic porcine hearts. J Clin Invest 94:623–630
40. Yanagisawa-Miwa A, Uchida Y, Nakamura F, Tomaru T, Kido H, Kamijo T, Sugimoto T, Kaji K, Utsuyama M, Kurashima C, Ito H (1992) Salvage of infarct myocardium by angiogenic action of basic fibroblast growth factor. Science 257:1401–1403
41. Torry RJ, O'Brien DM, Connell PM, Tomanek RJ (1992) Dipyridamole-induced capillary growth in normal and hypertrophic hearts. Am J Physiol (Heart Circ Physiol 31) 262:H980–H986
42. Rakusan K, Cicutti N, Kazda S, Turek Z (1994) Effect of nifedipine on coronary capillary geometry in normotensive and hypertensive rats. Hypertension (Dallas) 24:205–211
43. Tajima M, Katayose D, Bessho M, Isoyama S (1994) Acute ischaemic preconditioning and chronic hypoxia independently increase myocardial tolerance to ischaemia. Cardiovasc Res 28:312–319
44. Rakusan K, Sarkar K, Turek Z, Wicker P (1990) Mast cells in the rat heart during normal growth and in cardiac hypertrophy. Circ Res 66:511–516
45. Poiani GJ, Wilson FJ, Fox JD, Sumka JM, Peng BW, Liao W-C, Tozzi CA, Riley DJ (1992) Liposome-entrapped antifibrotic agent prevents collagen accumulation in hypertensive pulmonary arteries of rats. Circ Res 70:912–922
46. Poiani GJ, Tozzi CA, Yohn SE, Pierce RA, Belsky SA, Berg RA, Yu SY, Deak SB, Riley DJ (1990) Collagen and elastin metabolism in hypertensive pulmonary arteries of rats. Circ Res 66:968–978
47. Heagerty AM, Aalkjaer, Bund SJ, Korsgaard N, Mulvany MJ (1993) Small artery structure in hypertension. Dual processes of remodeling and growth. Hypertension (Dallas) 21:391–397
48. Hishikawa K, Nakaki T, Marumo T, Hayashi M, Suzuki H, Kato R, Saruta T (1994) Pressure promotes DNA synthesis in rat cultured vascular smooth muscle cells. J Clin Invest 93:1975–1980
49. Yang Z, Noll G, Lüscher TF (1993) Calcium antagonists differently inhibit proliferation of human coronary smooth muscle cells in response to pulsatile stretch and platelet-derived growth factor. Circulation 88:832–836

50. Carver W, Nagpal ML, Nachtigal M, Borg TK, Terracio L (1991) Collagen expression in mechanically stimulated cardiac fibroblasts. Circ Res 69:116–122
51. Morishita R, Higaki J, Miyazaki M, Ogihara T (1992) Possible role of the vascular renin-angiotensin system in hypertension and vascular hypertrophy. Hypertension (Dallas) 19(suppl II):II-62–II-67
52. Lever AF (1992) Angiotensin II, angiotensin-converting enzyme inhibitors, and blood vessel structure. Am J Med 92(suppl 4B):4B-35S–4B-38S
53. Shiota N, Miyazaki M, Okunishi H (1992) Increase of angiotensin-converting enzyme gene expression in the hypertensive aorta. Hypertension (Dallas) 20:168–174
54. Dzau VJ, Gibbons GH (1991) Endothelium and growth factors in vascular remodeling of hypertension. Hypertension (Dallas) 18(suppl III):III-115–III-121
55. Tozzi CA, Poiani GJ, Harangozo AM, Boyd CD, Riley DJ (1989) Pressure-induced connective tissue synthesis in pulmonary artery segements is dependent on intact endothelium. J Clin Invest 84:1005–1012
56. Okamura T, Miyazaki M, Inagami T, Toda N (1986) Vascular renin-angiotensin system in two-kidney, one-clip hypertensive rats. Hypertension (Dallas) 8:560–565
57. Morishita R, Gibbons GH, Ellison KE, Lee W, Zhang L, Yu H, Kaneda Y, Ogihara T, Dzau VJ (1994) Evidence for direct local effect of angiotensin in vascular hypertrophy. In vivo gene transfer of angiotensin-converting enzyme. J Clin Invest 94:978–984
58. Ito N, Isoyama S, Takahashi T, Takishima T (1993) Coronary dilator reserve and morphological changes after relief of pressure-overload in rats. J Mol Cell Cardiol 25:3–14
59. Brilla CG, Janicki JS, Weber KT (1991) Cardioreparative effects of lisinopril in rats with genetic hypertension and left ventricular hypertrophy. Circulation 84:1771–1779
60. Isoyama S, Ito N, Satoh K, Takishima T (1992) Collagen deposition and the reversal of coronary reserve in cardiac hypertrophy. Hypertension (Dallas) 20:491–500
61. Anderson PG, Bishop SP, Digerness SB (1989) Vascular remodeling and improvement of coronary reserve after hydralazine treatment in spontaneously hypertensive rats. Circ Res 64:1127–1136
62. Kanatsuka H, Lamping KG, Eastham CL, Marcus ML, Dellsperger KC (1991) Coronary microvascular resistance in hypertensive cats. Circ Res 68:726–733
63. Villarreal FJ, Dillmann WH (1992) Cardiac hypertrophy-induced changes in mRNA levels for TGF-β_1, fibronectin and collagen. Am J Physiol (Heart Circ Physiol 31) 262:H1861–H1866
64. Samuel JL, Barrieux A, Dufour S, Dubus I, Contard F, Koteliansky V, Farhadian F, Marotte F, Tiery J-P, Rappaport L (1991) Accumulation of fetal fibronectin mRNAs during the development of rat cardiac hypertrophy induced by pressure overload. J Clin Invest 88:1737–1746
65. Isoyama S (1994) Hypertension and age-related changes in the heart. Drugs & Aging 5:102–115
66. Isoyama S, Sato F, Takishima T (1991) Effect of age on coronary circulation after imposition of pressure-overload in rats. Hypertension (Dallas) 17:369–377
67. Crawford DC, Chobanian AV, Brecher P (1994) Angiotensin II induces fibronectin expression associated with cardiac fibrosis in the rat. Circ Res 74:727–739
68. Haber HL, Powers ER, Gimple LW, Wu CC, Subbiah K, Johnson WH, Feldman MD (1994) Intracoronary angiotensin-converting enzyme inhibition improves diastolic function in patients with hypertensive left ventricular hypertrophy. Circulation 89:2616–2625
69. Schunkert H, Dzau VJ, Tang SS, Hirsch AT, Apstein CS, Lorell BH (1990) Increased rat cardiac angiotensin-converting enzyme activity and mRNA expression in pressure overload left ventricular hypertrophy: effects on coronary resistance, contractility, and relaxation. J Clin Invest 86:1913–1920
70. Sawa H, Tokuchi F, Mochizuki N, Endo Y, Furuta Y, Shinohara T, Takada A, Kawaguchi H, Yasuda H, Nagashima K (1992) Expression of angiotensinogen gene and localization of its protein in the human heart. Circulation 86:138–146

71. Zhao X-M, Yeoh T-K, Frist WH, Porterfield DL, Miller GG (1994) Induction of acidic fibroblast growth factor and full-length platelet-derived growth factor expression in human cardiac allografts. Analysis by PCR, in situ hybridization, and immunohistochemistry. Circulation 90:677–685
72. Sarzani R, Ardaldi G, Takasaki I, Brecher P, Chobanian AV (1991) Effects of hypertension and aging on platelet-derived growth factor and platelet-derived growth factor receptor expression in rat aorta and heart. Hypertension (Dallas) 18(suppl III):III-93–III-99
73. Dzau VJ (1993) Local contractile and growth modulators in the myocardium. Clin Cardiol 16(suppl II):II-5–II-9
74. Ito H, Hirata Y, Adachi S, Tanaka M, Tsujino M, Koike A, Nogami A, Marumo F, Hiroe M (1993) Endothelin-1 is an autocrine/paracrine factor in the mechanism of angiotensin II-induced hypertrophy in cultured rat cardiomyocytes. J Clin Invest 92:398–403
75. Canby CA, Tomanek RJ (1989) Role of lowering arterial pressure on maximal coronary flow with and without regression of cardiac hypertrophy. Am J Physiol (Heart Circ Physiol 26) 257:H1110–H1118
76. Clozel J-P, Kuhn H, Hefti F (1989) Effects of chronic ACE inhibition on cardiac hypertrophy and coronary vascular reserve in spontaneously hypertensive rats with developed hypertension. J Hypertens 7:267–75
77. Wicker P, Tarazi RC, Kobayashi K (1983) Coronary blood flow during the development and regression of left ventricular hypertrophy in renovascular hypertensive rats. Am J Cardiol 51:1744–1749
78. Sato F, Isoyama S, Takishima T (1992) Effects of duration of pressure overload on the reversibilities of impaired coronary autoregulation in rats. Int J Cardiol 37:131–143
79. Ito N, Isoyama S, Kuroha M, Takishima T (1990) Duration of pressure overload alters regression of coronary circulation abnormalities. Am J Physiol (Heart Circ Physiol 27) 258:H1753–H1760
80. Sato F, Isoyama S, Takishima T (1990) Normalization of impaired coronary circulation in hypertrophied rat hearts. Hypertension (Dallas) 16:26–34
81. Isoyama S, Ito N, Kuroha M, Takishima T (1989) Complete reversibility of physiological coronary vascular abnormalities in hypertrophied hearts produced by pressure-overload in the rat. J Clin Invest 84:288–294
82. Ishihara K, Zile MR, Nagatsu M, Nakano K, Tomita M, Kanazawa S, Clamp L, DeFreyte G, Carabello BA (1992) Coronary blood flow after the regression of pressure-overload left ventricular hypertrophy. Circ Res 71:1472–1481
83. Strauer BE (1990) Significance of coronary circulation in hypertensive heart disease for development and prevention of heart failure. Am J Cardiol 65:34G–41G
84. Iwatsuki K, Cardinale GJ, Spector S, Udenfriend S (1977) Reduction of blood pressure and vascular collagen in hypertensive rats by beta-aminopropionitrile. Proc Natl Acad Sci USA 74:360–362
85. Gilligan JP, Spector S (1984) Synthesis of collagen in cardiac and vascular walls. Hypertension (Dallas) 6(suppl III):III-44–III-49
86. Nissen R, Cardinale GJ, Udenfriend S (1978) Increased turnover of arterial collagen in hypertensive rats. Proc Natl Acad Sci USA 75:451–453
87. Wu J-N, Berecek KH (1993) Prevention of genetic hypertension by early treatment of spontaneously hypertensive rats with the angiotensin-converting enzyme inhibitor captopril. Hypertension (Dallas) 22:139–146
88. Lee RMKW, Berecek KH, Tsoporis J, McKenzie R, Triggle CR (1991) Prevention of hypertension and vascular changes by captopril treatment. Hypertension (Dallas) 17:141–150
89. Harrap SB, Van der Merwe WM, Griffin SA, Macpherson F, Lever AF (1990) Brief angiotensin-converting enzyme inhibitor treatment in young spontaneously hypertensive rats reduces blood pressure long-term. Hypertension (Dallas) 16:603–614

90. Kojima M, Shiojima I, Yamazaki T, Komuro I, Yunzeng Z, Ying W, Mizuno T, Ueki K, Tobe K, Kadowaki T, Nagai R, Yazaki Y (1994) Angiotensin II receptor antagonist TCV-116 induces regression of hypertensive left ventricular hypertrophy in vivo and inhibits the intracellular signaling pathway of stretch-mediated cardiomyocyte hypertrophy in vitro. Circulation 89:2204–2211
91 Zierhut W, Zimmer H-G, Gerdes AM (1991) Effect of angiotensin-converting enzyme inhibition on pressure-induced left ventricular hypertrophy in rats. Circ Res 69:609–617
92. Baker KM, Chernin MI, Wixson SK, Aceto JF (1990) Renin-angiotensin system involvement in pressure-overload cardiac hypertrophy in rats. Am J Physiol (Heart Circ Physiol 28) 259:H324–H332
93. Golomb E, Abassi ZA, Cuda G, Stylianou M, Panchal VR, Trachewsky D, Keiser HR (1994) Angiotensin II maintains, but does not mediate, isoproterenol-induced cardiac hypertrophy in rats. Am J Physiol (Heart Circ Physiol 36) 267:H1496–H1506
94. Levy D, Garrison RJ, Savage DD, Kannel WB, Castelli WP (1990) Prognostic implications of echocardiographically determined left ventricular mass in the Framingham heart study. N Engl J Med 332:1561–1566
95. Jula AM, Karanko HM (1994) Effects on left ventricular hypertrophy of long-term nonpharmacological treatment with sodium restriction in mild-to-moderate essential hypertension. Circulation 89:1023–1031
96. Asmar RG, Pannier B, Santoni JPH, Laurent ST, London GM, Levy BI, Safar ME (1988) Reversion of cardiac hypertrophy and reduced arterial compliance after converting enzyme inhibition in essential hypertension. Circulation 78:941–950
97. Fouad-Tarazi FM, Liebson PR (1987) Echocardiographic studies of regression of left ventricular hypertrophy in hypertension. Hypertension (Dallas) 9(suppl II):II-65–II-68
98. Motz WH, Scheler S, Strauer BE (1992) Medical repair of hypertensive left ventricular remodeling. J Cardiovasc Pharmacol 20(suppl 1):S-32–S-36

Cardiac Hypertrophy and the Renin–Angiotensin System

Louis J. Dell'Italia

Summary. In hearts with myocardial damage secondary to myocardial infarction, chronic ischemia, inflammation, or pressure or volume overload, there is a complex sequence of compensatory events that ultimately result in an adversely remodeled myocardium and a dilated, thin-walled, spherical ventricle. For a period of time, a preclinical heart failure state may exist in which there is ventricular dysfunction caused by myocardial damage but no clinical evidence of cardiac insufficiency, circulatory congestion, or edema. However, in an attempt to maintain this state, there are ongoing myocardial adaptations resulting in a continual state of remodeling with progressive ventricular dilatation mediated by changes in myocyte morphology, intracellular calcium regulation, and extracellular matrix production. Although a number of growth-promoting factors have been implicated in cardiac hypertrophy and remodeling, angiotensin II (ANG II) is assumed to play a major role in this process because it is a potent growth factor for myocytes and fibroblasts in the heart [1,2]. It has been postulated that the cardiac renin–angiotensin system (RAS) is activated during compensated heart failure and that the continued presence of ANG II could have important local pathological functions in the transition to heart failure (Fig. 1). Alternatively, the heart is a target organ for ANG II, which produces a positive inotropic and chronotropic effect on myocytes [4–6], stimulates the release of norepinephrine from cardiac sympathetic nerves [7,8], and acts as a growth factor for myocytes [1,2]. Thus, the failing heart may be dependent on local ANG II production to provide inotropic support and to promote myocyte hypertrophy to minimize wall stress. However, the identification of alternative ANG II-forming mechanisms in the heart and the role of angiotensin-converting enzyme (ACE) on bradykinin degradation have raised questions regarding the mechanism of ANG II production in the heart and the role of ANG II per se in cardiac hypertrophy and remodeling. This chapter reviews these issues and the results of animal models investigating the cardiac renin–angiotensin system in cardiac hypertrophy caused by pressure and volume overload.

Key words. Hypertrophy—Angiotensin II—Heart failure—Angiotensin-converting enzyme—Kinin

---

Birmingham Veteran Affairs Medical Center, University of Alabama at Birmingham, Department of Medicine, Division of Cardiovascular Disease, University Station, Birmingham, AL 35294, USA

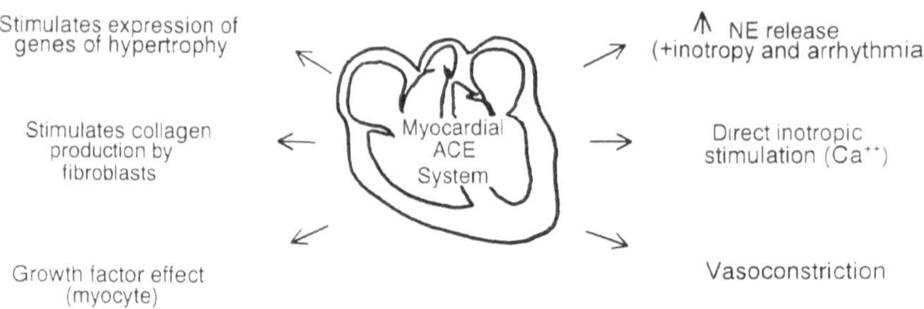

FIG. 1. Diagram of the pathophysiological consequences of intracardiac renin–angiotensin and angiotensin II activated by increases in regional wall stress. *ACE*, Angiotensin-converting enzyme. (From [3], with permission)

Left-Ventricular Hypertrophy Caused by Pressure Versus Volume Overload

Pressure overload results in the development of concentric left-ventricular (LV) hypertrophy, in which parallel sarcomere replication produces increased wall thickness and a concentric pattern of chamber hypertrophy. It is hypothesized that increased systolic wall stress (stress = pressure × radius/2 thickness) causes this pattern of myocyte and chamber hypertrophy whereby increased wall thickness offsets increased pressure, normalizing stress [9,10]. In contrast, volume overload increases diastolic wall stress and triggers a series replication of sarcomeres, resulting in an eccentric pattern of chamber hypertrophy. Wall thickness (left-ventricular mass) increases in proportion to the volume overload so that adequate mass is present to pump the extra volume while wall stress remains normal.

There is a distinction, however, between the low-pressure volume overload of mitral regurgitation and the high-pressure volume overload of aortic regurgitation and aortocaval fistula. Mitral regurgitation presents a unique hemodynamic stress to the left ventricle in that a large part of the excess volume is ejected into the low-pressure left atrium. This form of chronic hemodynamic stress is very different from pressure overload, in which shortening load is increased throughout systole, and differs from the high-pressure volume overload of aortocaval fistula and aortic regurgitation in which the excess volume is ejected into the high-pressure aorta. Consequently, mitral regurgitation places a much smaller systolic load on the left ventricle and results in far less hypertrophy than pressure overload and high-pressure volume overload [11,12].

Cardiac myocyte growth is the common denominator in myocardial hypertrophy. The pattern of myocyte remodeling and cardiac chamber geometry that develops in LV hypertrophy is determined by the hypertrophic stimulus. Furthermore, the hypertrophic remodeling of the myocardium may or may not include the growth of nonmyocyte cells (endothelial cells, macrophages, fibroblasts, smooth muscle cells), which include two-thirds of the cell population of the heart. Myocardial remodeling is mediated by locally and systemically generated trophic factors that are produced in response to hemodynamic stress. The extent of or lack of accumulation of perivascular and fibrillar collagen combined with the changes in myocyte geometry determine the pattern of remodeling at the chamber level in response to the hemodynamic load.

Pressure overload produces an increase in myocyte cross-sectional area relative to length as the result of the in-parallel addition of myofibrillar units [13]. In most animal models of pressure overload, there is pathological accumulation of collagen in the left ventricle. Collagen accumulates in perivascular areas surrounding intramyocardial arteries and arterioles and extends from this perivascular location into the interstitial space [14]. Recent work in the spontaneously hypertensive rat has demonstrated that levels of fibronectin and pro-alpha 1(I) and pro-alpha 1(III) collagen mRNA were significantly elevated (fourfold) in failing versus nonfailing, hypertrophied hearts [15]. Whether the increase in collagen production progresses over the time-course of a particular hemodynamic stress and whether it is a cause or a consequence of heart failure is unresolved.

In contrast, the high-pressure volume overload of aortocaval fistula produces myocyte hypertrophy that is characterized by increases in both length and cross-sectional area. Gerdes and co-workers have demonstrated a one-third increase in length and a two-thirds increase in cross-sectional area at 1 week and 1 month after creation of aortocaval fistula in the rat [16,17]. In dogs with aortocaval fistula, total increases in LV mass have ranged between 43% and 100%, depending on both shunt size and duration of the volume overload [18]. Weber and co-workers reported no increase in myocardial interstitial collagen content by the picrosirius polarization technique at 10 weeks following placement of aortocaval fistulae [19]. However, Covell and co-workers demonstrated a qualitative change in collagen content manifested by a greater degree of cross-linking between types I and III collagen in dogs with aortocaval fistula who had increased LV end-diastolic pressure and chamber stiffness at the time they were killed [20]. Thus, it appears that a qualitative as well as quantitative examination of structural collagen in the heart may be necessary to define changes in this extracellular supporting latticework that may affect chamber distensibility and stiffness.

ISOLATED MYOCYTES

NORMAL **MITRAL REGURGITATION**

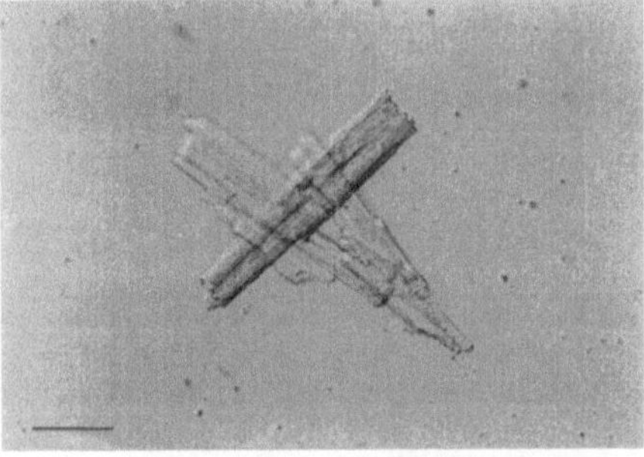

Fig. 2. Representative isolated myocytes from a normal dog (*left*) and a dog with mitral regurgitation (*right*). Myocyte length increases as cross-sectional area remains unchanged. (From [22], with permission)

In the rat model of experimentally induced aortocaval fistula, Ruzicka and co-workers demonstrated decreased LV collagen accumulation in spite of increased renin activity, while collagen content increased in the pressure-overloaded right ventricle [21]. Taken together, these results suggested that, in contrast to the stimulatory effect of pressure overload and systolic wall stress on collagen accumulation, volume overload and diastolic wall stress do not promote collagen accumulation in the extracellular matrix.

The canine model of mitral regurgitation presents a lesser systolic stress to the left ventricle than either high-pressure volume overload or pressure overload because a large part of the excess volume is ejected into the low-pressure left atrium. We, and Carabello and co-workers, have found that percutaneous rupture of the mitral valve in the dog produces a 20%–30% increase in myocyte length associated with an unchanged or reduced cross-sectional area (Fig. 2) [22,23]. There is a significant de-

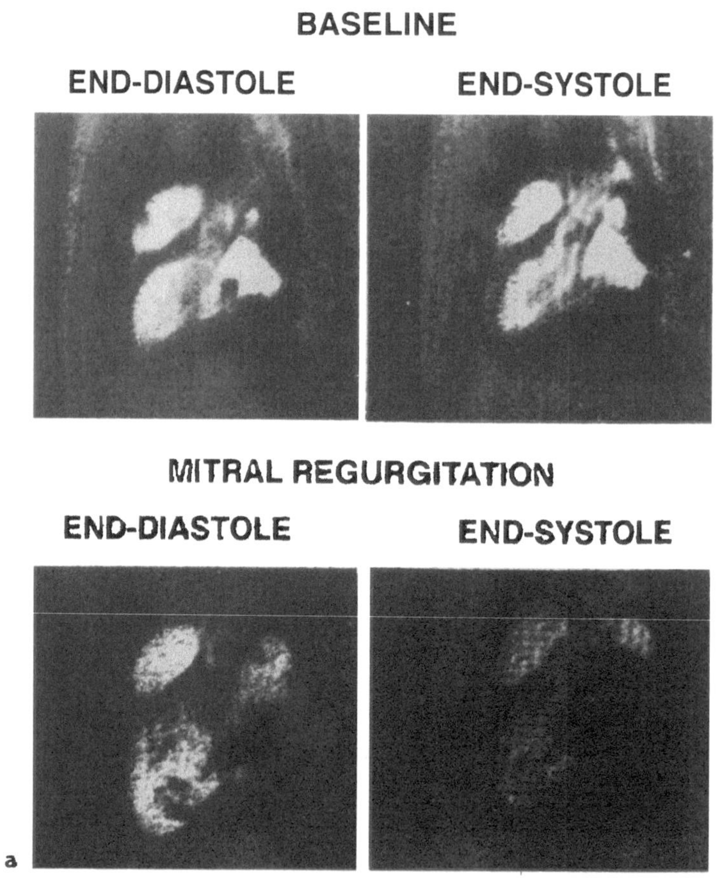

FIG. 3a,b. Long-axis (a) and short-axis (b) end-diastolic and end-systolic magnetic resonance images (MRI) of the heart at baseline and 5 months after induction of mitral regurgitation demonstrate the increase in end-diastolic volume and decrease in wall thickness that occur as a result of a decrease in the left-ventricular (LV) mass/volume ratio. (From [25], with permission)

BASELINE

END-DIASTOLE END-SYSTOLE

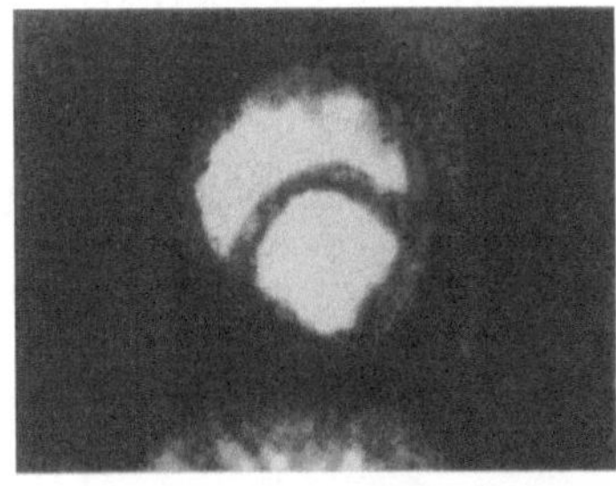

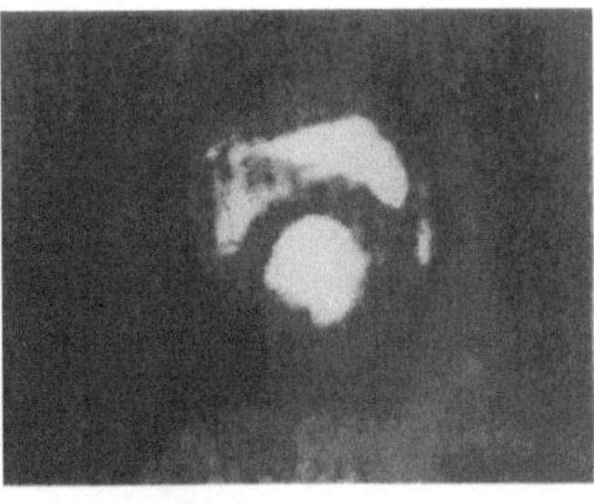

MITRAL REGUGITATION

END-DIASTOLE END-SYSTOLE

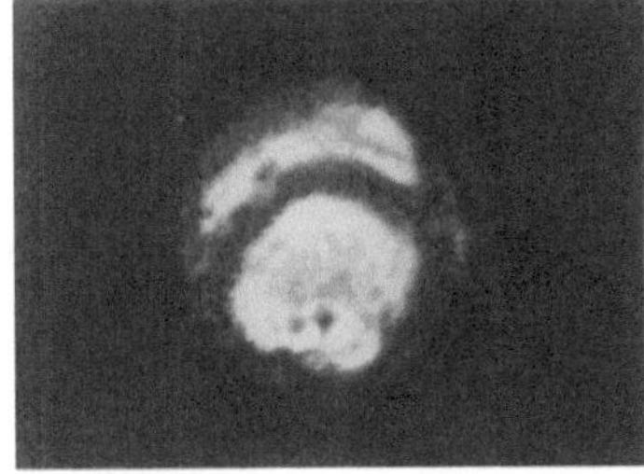

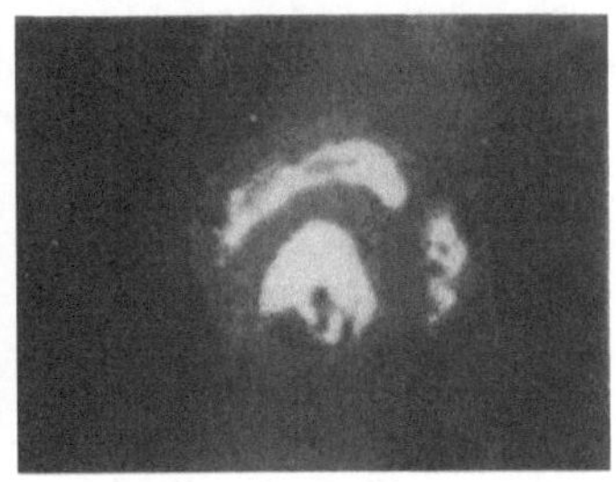

b

FIG. 3a,b. *Continued*

crease in the LV mass-to-volume ratio resulting from a 50%–100% increase in LV end-diastolic volume and a maximum 30% increase in LV mass 3–5 months after chordal rupture [25] (Fig. 3). In those animals exhibiting frank congestive heart failure, there is a striking loss of myofibrils [23; Sanford P. Bishop, from our work in the canine mitral regurgitation model]. It is of interest that Carabello and co-workers found a greater degree of cardiocyte hypertrophy in the failing, dilated hearts with thin walls, suggesting actual cardiocyte loss from these ventricles with hypertrophy of the remaining cardiocytes. From this work, the two most noteworthy findings of the failing left ventricle in this model are a lack of appropriate growth response to the high diastolic wall stress and a loss of myofibrils from the cardiocyres of ventriclcs exhibiting the poorest contractile function.

In the canine model of chronic mitral regurgitation, we found a focal loss of the collagen fiber weave surrounding individual myocytes and myocyte bundles using scanning electron microscopy [22] (Fig. 4). The fine collagen weave connects cardiac myocytes laterally by collagen struts and inserts into the sarcolemma at the level of the Z-lines [26]. This remodeling of the pattern of collagen structure in the heart was not accompanied by an alteration in total intracardiac collagen content by hydroxyproline analysis. This finding has previously been reported in a dilated cardiomyopathy in the dog induced by rapid right-ventricular pacing [27,28]. The global decrease in

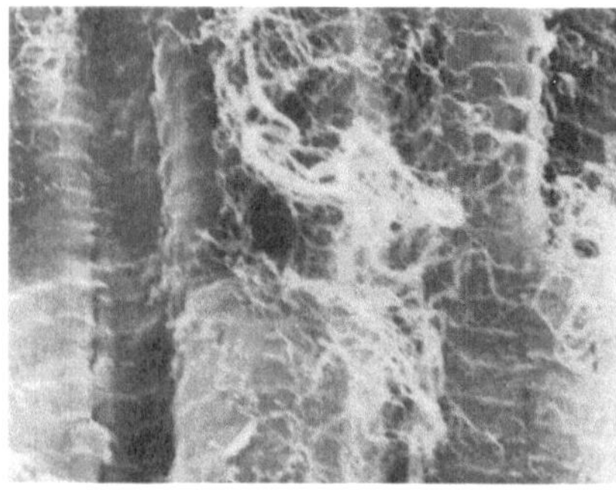

FIG. 4. Scanning electron micrographs of the collagen weave in a normal dog (*left*) and in a dog with mitral regurgitation (*right*) demonstrate the characteristic loss of the collagen struts in the mitral regurgitation heart. (From [24], with permission)

LV mass/volume ratio in the mitral regurgitation and pacing tachycardia hearts was characterized by an increase in myocyte length without an increase in cross-sectional area combined with the focal loss of collagen weave. Thus, elongation of myocytes coupled with the structural changes in the LV extracellular matrix represent compensatory remodeling, which permits the increase in preload that is critically important in maintaining hemodynamic compensation in volume overload hypertrophy. The inability to increase preload and stroke volume appropriately in response to increasing volume is a hallmark of decompensation. As an illustration of this concept, Komamura demonstrated that the failing ventricle subjected to chronic pacing tachycardia was unable to further increase preload in response to volume expansion, suggesting that at this stage it was operating at the limit of this chronic adaptation and that the Frank–Starling mechanism had been exhausted [27].

In summary, animal models clearly demonstrate characteristic patterns of myocyte remodeling in response to the pressure versus volume overload. There also appears to be a trend toward greater interstitial collagen accumulation in pressure versus the volume overload animal models that may be related ultimately to the marked differences in shortening load. In support of this hypothesis, recent studies in the hearts of humans undergoing valve replacement for volume overload demonstrated significantly greater interstitial collagen in the high-pressure volume overload of aortic regurgitation than in the low-pressure volume overload of mitral regurgitation hearts [29].

To better define the remodeling process at the chamber level, we have used magnetic resonance imaging (MRI) to acquire tomographic sectioning of the entire heart, thereby providing a three-dimensional anatomy to accurately measure masses and volumes as well as wall thickness and curvature [30]. We have applied a finite-element modeling approach to the estimation of in vivo ventricular geometry from stacked cine-MRI slices of the heart of dogs subjected to 5–6 months of mitral regurgitation [30]. This analysis demonstrated an asymmetrical decrease in LV surface endocardial curvatures corresponding to the greater radial expansion of the endocardial surface in the lateroseptal direction rather than in the anteroposterior dimension. This resulted in encroachment of the interventricular septum (IVS) into the right ventricle in diastole (Fig. 5). This rearrangement in septal position combined with no change in right-ventricular (RV) free wall geometry produced a decrease in RV end-diastolic volume in dogs who did not develop significant pulmonary hypertension. During systole, the excursion of the IVS toward the left ventricle increased fourfold, thereby contributing to the twofold increase in LV stroke volume in mitral regurgitation

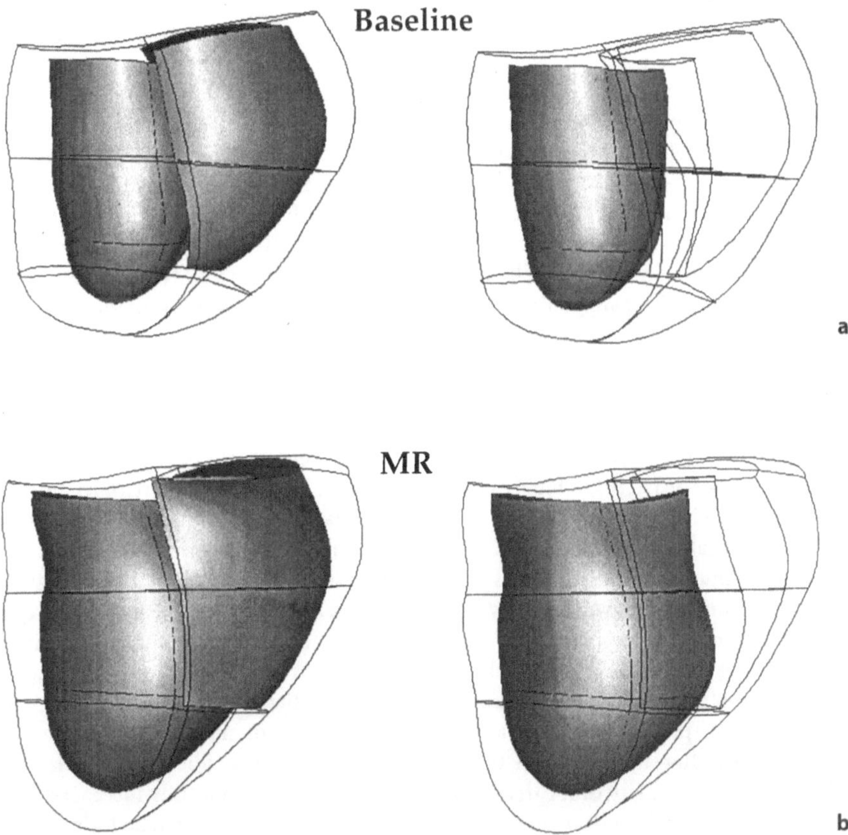

Fig. 5a,b. Shaded surface display of the left and right ventricles at end-diastole at baseline (**a**) and 5 months after induction of mitral regurgitation (*MR*) (**b**) from three-dimensional reconstruction of stacked magnetic resonance short-axis images. This display demonstrates the asymmetrical dilatation of the left ventricle with expansion of the interventricular septum into the right ventricle, decreasing right-ventricular diastolic volume. (From [24], with permission)

compared to baseline studies. The contribution of the IVS to LV stroke output and the maintenance of RV stroke volume in spite of a decrease in RV preload suggest a mechanical coupling of the left and right ventricles. Alternatively, in the absence of favorable loading conditions in the right ventricle, contractile performance of the RV chamber could be enhanced by increased circulating or tissue humoral substances (i.e., catecholamines, angiotensin II) that act directly on the right ventricle as well as the left ventricle.

Renin–Angiotensin System and Myocyte Hypertrophy

Both the circulating and the local cardiac renin–angiotensin–aldosterone systems (RAAS) may contribute to the development of LV hypertrophy and interstitial fibrosis in pressure overload via the angiotensin II type 1 (AT_1) receptor signaling [1]. Angiotensin II (ANG II) stimulates protein synthesis in isolated embryonic chick myocytes [2] and c-*fos*, c-*jun*, and c-*myc* proto-oncogene induction in isolated neonatal rat myocytes [31], thus demonstrating load-independent effects on myocyte growth. In these cell cultures, ANG II is a potent stimulus for hypertrophy, suggesting that it may act as a growth factor via activation of these proto-oncogenes. Stretch of neonatal cardiomyocytes has been reported to cause acute release of ANG II and an increase in protein synthesis [32]. An increase in steady-state levels of angiotensinogen mRNA occurred as early at 6h after stretch, reaching 4.8-fold increases over control in 24h. However, pretreatment with saralasin (an antagonist of AT_1 and AT_2 receptors) and CV-11974 (an AT_1 antagonist) inhibits the increase in myocyte protein synthesis by approximately 70% [33]. Thus, ANG II-induced stimulation of protein synthesis in myocyte cell cultures in vitro is blocked by ANG II type 1 (AT_1) receptor antagonists, suggesting that ANG II-activated signal transduction involves AT_1 receptors.

Recent work in the isolated, buffer-perfused adult rat heart demonstrated that ANG II perfusion activated protein kinase C and stimulated cardiac protein synthesis without induction of the c-*fos* and c-*jun* proto-oncogenes [34]. Further, Knowlton and co-workers have demonstrated divergent gene regulatory pathways for atrial natriuretic factor (ANF), a highly conserved genetic marker of hypertrophy both in vitro and in vivo, in neonatal and adult ventricular myocardium [35]. The authors concluded that because neonatal rat ventricular myocytes may be more permissive than adult muscle myocardial cells for upregulation of the embryonic gene program, additional molecular switches may be required for the activation of hypertrophy in the adult versus the neonatal heart, suggesting the need for in vivo studies to confirm the findings in neonatal rat cell cultures.

Data from numerous in vivo animal models suggest that ANG II, acting through the AT_1 receptor, plays a critical role in the development or maintenance of cardiac hypertrophy. Recent studies in the congenital cardiomyopathic hamster demonstrated that the density of AT_1 receptors in the heart was significantly increased at 25 days of life, before the development of histologically detectable cardiac lesions [36]. Constriction of the abdominal aorta in the rat resulted in significant increases in intracardiac AT_1 receptor mRNA levels at 72h, and cardiac hypertrophy was prevented by losartan (an AT_1 receptor antagonist) but not by treatment with PD 123319 (an AT_2 receptor antagonist) [37]. Thus, the AT_1 receptor mediates the well-known pressor and trophic effects of ANG II, but the signaling mechanism and physiological role of AT_2 receptors have not been established. Transgenic mice lacking a functional AT_2 receptor had an exaggerated pressor response to exogenous ANG II [38,39]. Thus, AT_1 and AT_2 appear to mediate opposing hemodynamic effects: AT_2 antagonizes the AT_1-mediated pressor action of ANG II.

In Vivo Studies of Pressure and Volume Overload

In vitro studies have demonstrated that angiotensin II (ANG II) acts as a growth factor for cardiac myocytes and fibroblasts [1,2,31–35]. All the components of the renin–angiotensin system (RAS) have been identified in the heart at both the mRNA and protein levels [40]. Increased gene transcript levels of angiotensinogen [41], angiotensin-converting enzyme (ACE) [42,43], and angiotensin receptors [37] have been identified in the pressure-overloaded rat heart, suggesting that increased local production of ANG II and subsequent ANG II receptor signaling may mediate concentric myocyte hypertrophy and interstitial fibrosis in this model. Increased expression of components of the RAS in the heart has also been associated with a different pattern of myocardial remodeling, i.e., eccentric myocyte hypertrophy with little to no interstitial fibrosis, in the volume overload model induced by placement of an aortocaval fistula in the rat [44–48].

As described in the previous section, chronic pressure overload in the rat left ventricle results in increases in myocyte cross-sectional area relative to length, interstitial collagen, and gene transcript levels of the components of the RAS [37,40–43]. In contrast, the high-pressure volume overload of aortocaval fistula in the rat produces an eccentric LV remodeling characterized by increases in myocyte length and cross-sectional area with little to no increase in interstitial collagen, in spite of increases in cardiac RAS components [44–48]. A threefold increase in steady-state ACE mRNA levels correlated with LV weight/body weight ($r = .696$, $P = .0002$) [44] and severity of heart failure [45] in rats 40 days after fistula placement. Thus, both pressure overload and high-pressure volume overload produce upregulation of the cardiac RAS but result in different patterns of remodeling of the myocyte and extracellular matrix in the rat heart.

Experimentally induced mitral regurgitation in the dog produces a limited amount of LV hypertrophy over 3–5 months compared to either the pressure overload of aortic constriction or the volume overload of aortocaval fistula [9,22,25]. We demonstrated that LV mass and myocyte length increased 25% as myocyte cross-sectional area remained unchanged in the mitral regurgitation hearts. Membrane-bound ACE activity and ANG II levels were increased in the left ventricle but not in the right ventricle of mitral regurgitation hearts compared to normal dog hearts, suggesting that ACE levels were regulated at the myocardial level in response to volume overload. The nature of the stimulus for the increase in ACE activity is a matter of controversy, but as it did not occur in the right ventricle and was negatively related to LV mass/end-diastolic volume (EDV) ratio and end-diastolic wall stress in the mitral regurgitation hearts, stretch is a likely candidate (Figs. 6, 7).

Thus, in vitro and in vivo studies demonstrate that stretch or an increase in shortening load upregulate the cardiac RAS but result in distinctly different phenotypic myocardial remodeling. A recent study comparing pressure and volume overload hypertrophy in the rat demonstrated that transforming growth factor-β_3 and insulin-

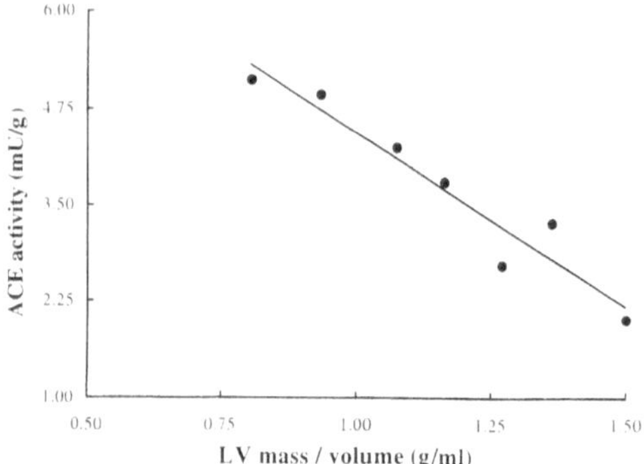

FIG. 6. Linear regression plot of ACE activity versus left-ventricular (*LV*) mass/volume in mitral regurgitation hearts. $P < .001$, $r = .93$ (From [25], with permission)

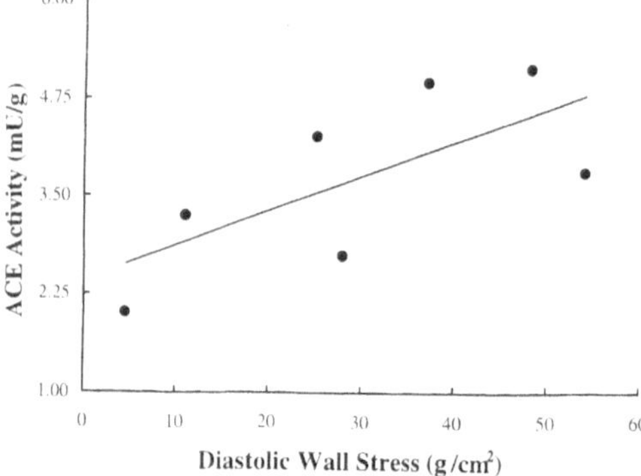

FIG. 7. Linear regression plot of ACE activity versus LV diastolic wall stress in mitral regurgitation hearts. $P < .05$, $r = .71$ (From [25], with permission)

like growth factor-1 mRNA levels were increased with pressure but not volume overload, but acidic fibroblast growth factor mRNA was unchanged with pressure but decreased with volume overload [49]. Taken together, stimulus-specific peptide growth factors and signaling events coupled to gene expression may work in combination with the RAS to produce morphologically and functionally distinct forms of myocardial hypertrophy.

ACE Inhibitor and AT_1 Receptor Blockade Therapy

ACE inhibitor treatment has been reported to attenuate LV hypertrophy in response to both pressure [50–52] and volume overload [47,48,53,54] in the rat, presumably by inhibiting tissue ACE and decreasing ANG II generation, thus blunting the trophic effect of ANG II on LV myocytes and fibroblasts. Furthermore, doses of ACE inhibitor that do not affect systemic arterial pressure can reverse cardiac hypertrophy induced by banding of the aorta in a rat model of pressure overload [55]. Baker and co-workers demonstrated that abdominal aortic constriction in the rat produced cardiac hypertrophy that was associated with significant upregulation of angiotensinogen mRNA in the left ventricle at 7 and 15 days after surgery [41]. Subpressor doses of enalapril completely prevented the increase in LV mass even though carotid artery pressures were not different in conscious, awake, aortic-constricted animals receiving or not receiving enalapril. These data suggested a direct growth effect of ANG II on the left ventricle as a result of increased local production because angiotensinogen mRNA was increased while circulating plasma activity was not, compared to sham-operated rats at the time they were killed.

Studies of experimentally induced pressure overload hypertrophy in the rat have evaluated the effects of ACE inhibitor therapy initiated 6 weeks after aortic banding using m-mode echocardiography [51,52]. These studies demonstrated regression of LV hypertrophy, improvement in systolic and diastolic chamber function, and a significant survival benefit despite a persistently elevated LV systolic pressure (190–210 mmHg) that did not differ from the untreated rats subjected to ascending aortic constriction. These studies suggested that the beneficial effects of ACE inhibitor therapy on cardiac function and cardiovascular morbidity and mortality occur in spite of a blunting of the hypertrophic response to normalize wall stress. However, there was improved contractile response to increasing doses of calcium in hypertrophied hearts receiving ACE inhibitor compared to untreated hypertrophied hearts, which appeared to compensate for the regression on hypertrophy [51]. Whether the improved function is related to an effect of ACE inhibitor on calcium responsiveness at the level of the myocyte or myofibrillar calcium sensitivity requires further investigation.

In addition to increasing local synthesis of ANG II from ANG I, ACE or kininase II catalyzes the degradation of endogenous kinins, e.g., bradykinin, to inactive metabolites. Thus, ACE inhibitors not only inhibit the conversion of ANG I to ANG II but also prevent the breakdown of bradykinin to inactive peptides. Bradykinin is considered a physiological antagonist to ANG II in that bradykinin inhibits ANG II-stimulated protein synthesis in vascular smooth muscle tissue and blocks contraction to agonists. Bradykinin exerts its effect through endothelial bradykinin-2 receptors in part by activating endothelial nitric oxide synthase (e-NOS). Nitric oxide has been shown to have a number of important physiological and homeostatic effects in vascular smooth muscle cells, including inhibition of contraction and growth [56].

In the rat model of aortic constriction, addition of a bradykinin-2 receptor antagonist (HOE 140) to ACE inhibitor therapy negated the regression of LV hypertrophy [57]. In a dog model of transmural myocardial necrosis, therapy with an AT_1 receptor-blocking drug was ineffective in attenuating the increase in LV mass when compared to ACE inhibitor therapy with ramipril [58]. Concomitant treatment with ramipril and HOE 140 prevented regression of LV hypertrophy achieved by ramipril alone independent of changes in systolic blood pressure [59]. These studies demonstrate the potential importance of bradykinin to the mechanism by which ACE inhibitors may promote hypertrophy regression in the myocardium. However, an alternative mechanism may be mediated by a bradykinin effect on vascular compliance resulting in diminution of a late systolic pressure peak at the root of the aorta from the reflected wave, thus decreasing load not detected in a simple arterial pressure measurement.

The basis for impaired contractile performance of hearts in moderate to severe stages of hypertrophy remains uncertain. There is evidence that the enhanced accumulation and remodeling of interstitial collagen in the hypertrophied hearts impairs myocardial stiffness, thereby causing diastolic dysfunction and altered pumping activity of the heart [60]. On the other hand, in almost all models of hypertrophy caused by mechanical overload it also appears that the mechanisms that govern Ca^{2+} movements are depressed [61]. As a consequence, it has been postulated that abnormal Ca^{2+} homeostasis or myofilament sensitivity to Ca^{2+} can be responsible for the decrease in contractile strength and the impaired relaxation observed in the hypertrophied myocardium.

ANG II, bradykinin, and nitric oxide affect contraction and relaxation of muscle. ANG II has been shown to potentiate the impaired relaxation of isolated cardiac myocytes from rats with pressure overload hypertrophy [62]. Recent work in isolated ejecting guinea pig hearts demonstrated that both substance P and bradykinin induced an enhancement of LV relaxation, which was attributable to the paracrine release of nitric oxide [63]. Because myocardial ACE activity is increased in the pressure-overloaded rat heart, blunted degradation of bradykinin by ACE inhibitor therapy could improve diastolic function in vivo either by direct coronary vasodilatation or by modulation of endothelial-derived nitric oxide and subsequent effects on myocyte function via endothelial–myocardial coupling. Indeed, captopril caused a progressive acceleration of LV relaxation without significantly affecting systolic parameters or coronary artery flow in the isolated ejecting guinea pig heart [64] and in the hypertrophied rat heart [42,43]. These findings are in keeping with recent studies in humans demonstrating that infusion of enalaprilat into the left anterior descending artery of patients with LV hypertrophy improved regional diastolic function of the anterior wall in the absence of systemic hemodynamic and neurohormonal effects [65].

Thus, in addition to structural changes in the extracellular matrix, endothelial dysfunction resulting in decreased nitric oxide could contribute to the reduction in diastrolic distensibility in cardiac hypertrophy. Taken together, the combined ACE inhibitor effect of decreasing ANG II and increasing bradykinin may be a mechanism of improved structure and mechanical function in the hypertrophied heart.

Mechanisms of ANG II Production

The effect of ACE inhibitors on myocardial hypertrophy and remodeling has conventionally been attributed to inhibition of tissue ACE resulting in decreased ANG II generation and a blunted trophic effect on LV myocytes and fibroblasts. However, interpretation of the origins of ANG II in the heart and of the mechanisms of intracardiac action of ACE inhibitors is complicated by the finding of multiple ANG II-forming pathways in cardiac tissue [66]. In particular, a serine protease with an extremely high affinity for ANG I, "chymostatin-sensitive angiotensin-generating enzyme" (heart chymase), has been found in high concentrations in the human heart [67]. It is insensitive to ACE inhibition and has higher specificity and catalytic activity for conversion of ANG I to ANG II than ACE [68,69]. This enzyme represents approximately 90% of the ANG II-forming capacity in tissue extracts from human myocardium, suggesting that ACE is not the major ANG II-forming enzyme in the human left ventricle in vitro [67]. Thus, heart chymase provides a highly active alternative ANG II-forming mechanism that may become increasingly important in the setting of ACE inhibitor therapy, which results in increased ANG I levels, preferentially shunting this substrate to heart chymase. The identification of this enzyme in the human heart has provided the impetus for use of AT_1 receptor-blocking drugs in LV hypertrophy and heart failure.

The findings of high chymase activity in human heart tissue extracts in vitro, however, are difficult to reconcile with the important clinical role of ACE inhibitors in preventing cardiac hypertrophy and the transition to heart failure. Several recent studies in patients have suggested that ACE-mediated conversion of ANG I to ANG II across the coronary vascular bed represents the predominant pathway of ANG II generation in the human heart in vivo. In patients with orthotopic transplanted hearts, the fractional conversion of radiolabeled ANG I across the myocardial circulation decreased 89.5% during intracoronary enalaprilat infusion [70]. In another study, infusion of enalaprilat into the left anterior descending artery of patients with LV hypertrophy from aortic stenosis improved regional diastolic function of the anterior wall in the absence of systemic hemodynamic and neurohormonal effects [65]. The results of the two studies in human subjects suggest an important role for ACE in ANG II formation across the coronary vascular bed in spite of the much greater ANG II-forming capacity of heart chymase compared to ACE in in vitro studies in the human heart. The unavailability of an approved specific chymase inhibitor precludes evaluation in humans to determine the functional significance of ANG II formation by chymase across the myocardial vascular bed.

Most of the animal studies evaluating the role of the cardiac RAS in the pathogenesis of myocardial hypertrophy and heart failure and its sensitivity to ACE inhibition have been performed in the rat. However, conversion of ANG I to ANG II from chymase has been reported not to exist in the rat heart [71,72]. We have recently reported that 80% of ANG II-forming enzyme activity was from chymase-like activity while only 6% was from ACE in tissue extracts of normal dog heart and of heart from dogs subjected to chronic volume overload hypertrophy [25]. However, in vivo studies demonstrated that intracoronary infusion of ANG I plus captopril decreased the arteriovenous difference of ANG II across the myocardium by 60% in the normal dog heart [73]. Furthermore, the extent of elevation of diastolic wall stress in volume overload hypertrophy caused by mitral regurgitation was related to intracardiac ACE activity but not to chymase activity [25]. The results of these in vitro and in vivo studies in the dog are similar to the results in the human heart and suggest a physiologically important role for ACE in intracardiac ANG II formation in vivo in spite of the much greater ANG II-forming capacity of chymase compared to ACE in vitro.

The difference between in vivo and in vitro ANG II formation from ACE and chymase may be related to differences in the distribution and compartmentalization of chymase and ACE in the heart. In situ hybridization and electron microscopic immunocytochemical studies have demonstrated that mast cells, endothelial cells, and other cell types within the interstitial extracellular matrix are the sites of synthesis and storage of chymase [74]. In contrast, ACE is bound to the cell membranes of endothelial cells and fibroblasts, with its catalytic site exposed to the extracellular surface [75]. This situation could make exogenously administered ANG I more accessible to ACE in vivo in the intravascular space, while studies of tissue extracts disrupt the physical compartmentalization of ACE and chymase and allow more access between ANG I substrate and chymase in vitro.

Chymase-like activity is widely distributed in a number of tissues in the baboon, including those of the heart [76]. Systemic infusion of [Pro^{11}D-Ala^{12}] ANG I, a selective and specific substrate for human chymase but not for ACE, produced systemic arterial vasoconstriction in the conscious baboon that was reversed by an ANG II receptor blocker but not by captopril [76]. Future studies in animal models having chymase activity in the heart are required to assess the relative roles of ACE versus chymase in intracardiac ANG II formation. Additional studies with direct sampling of cardiac interstitial fluid space are needed to assess the relative roles of the intravascular versus the interstitial ANG II formation in the heart and to determine the relative roles of ACE versus chymase-like enzyme in the process.

Other Neurohormonal Systems and Cardiac Hypertrophy

Endothelin (ET) was originally described as a strong vasoconstrictor but recent studies have demonstrated that ET also has growth factor-like properties in a wide variety of mammalian cells, including myocytes [77]. Cultured rat myocytes express abundant preproET-1 (ppET-1) transcripts and release mature ET-1 into medium. Cardiomyocyte hypertrophy stimulated by ANG II is partially blocked by a receptor antagonist selective for ET_A subtype [78]. Fujisaki and co-workers have recently demonstrated that ppET-1 expressed by rat cardiac fibroblasts was upregulated by ANG II in the same manner as in cardiomyocytes [79]. In the aortic-constricted rat, LV ET-1 levels increased by day 8 post constriction and showed a significant positive correlation with the degree of cardiac hypertrophy [80]. The density of ET-1 binding sites was also coordinately increased with ET-1 levels. In situ hybridization studies have demonstrated that ventricular cardiocytes expressed ppET-1 mRNA in hypertrophied hearts. As in vitro studies have demonstrated that ANG II can induce ET-1 synthesis and release, ACE inhibitors could indirectly alter intracardiac ET-1 levels by reducing intracardiac ANG II. Further work will elucidate the interaction of ANG II and ET-1 in the heart in the pathogenesis of hypertrophy.

Another mechanism of action of ACE inhibitor therapy may protect the heart from the toxic effects of ANG II. Myocyte necrosis has been demonstrated in normal rat hearts during ANG II infusion and in rat hearts subjected to a suprarenal but not infrarenal aortic band [81]. Captopril was effective in preventing myocyte injury in renovascular hypertension [81]. However, it is not clear whether ANG II injures myocytes through a direct effect or via indirect effects involving other cell types or mediators. Henegar and co-workers recently demonstrated that myocyte necrosis and coronary vascular damage in the normal rat heart appeared to be caused by a local release of catecholamines from nerve endings early in the course of ANG II infusion and that this effect could be prevented by both losartan and propranolol [82]. Furthermore, catecholamines have been shown to decrease protein synthesis in adult cardiocytes [83]. Thus, it can be postulated that ACE inhibitors protect the myocardium by attenuating local ANG II-mediated norepinephrine release from presynaptic nerve terminals in the heart [8], thus reducing local concentrations of the cytotoxic catecholamine. In support of this potential mechanism, Carabello and co-workers demonstrated that dogs with experimentally induced mitral regurgitation (MR) treated with β-adrenergic blockade for 3 months had significantly greater LV mass/body weight and myocyte myofibrillar density and normalization of myocyte contractile function compared to unblocked MR dogs [84].

Other growth factors, such as aldosterone and transforming growth factor-β_1 (TGF-β_1), contribute to myocardial fibrosis independent of hemodynamic load [85]. Recent studies in the Goldblatt model of two-kidney, one-clip hypertension in the rat have shown that losartan and spironolactone significantly decreased collagen production

[86]. Other studies in the rat heart suggest that ANG II-induced fibrosis may be modulated in part by nitric oxide. Inhibition of nitric oxide synthase with N^G-nitro-L-arginine-methyl ester (L-NAME) in the rat heart resulted in a greater susceptibility to myocardial fibrosis during administration of subpressor doses of ANG II [87]. This effect was not related to systemic blood pressure changes. These results suggest a synergistic effect of increased ANG II and chronic inhibition of nitric oxide production in the development of cardiac hypertrophy and fibrosis.

Evidence of a Functional RAS in the Human Heart

The mRNA for ACE [88,89], heart chymase [74,88], angiotensinogen [90], mineralocorticoid (aldosterone) receptor [91], and ANG II receptors has been demonstrated in the human heart [92]. In contrast, the presence of renin mRNA has been difficult to establish in the heart. ACE mRNA is significantly upregulated in the failing human heart compared to normals [88,89]. Studer and co-workers reported twofold higher ACE mRNA levels in hearts from patients with end-stage heart failure than in presumably normal hearts derived from potential transplant donors whose hearts were not used for transplantation [88]. However, the variability of ACE mRNA expression in donor and failing hearts was very large, perhaps related to problems in sampling, the intrinsic limitations of the quantitative polymerase chain reaction (PCR) technique, and variability in state of compensation of the hearts. In cardiomyopathy patients, both ACE activity and steady-state ACE mRNA expression were higher in the failing versus the nonfailing ventricles [88]. Studer and co-workers recently demonstrated that steady-state ACE mRNA levels in hearts hypertrophied by the pressure overload of aortic stenosis were significantly higher than in hearts with dilated cardiomyopathy [93].

Conclusion

The progression to heart failure in pressure and volume overload is a multifactorial process. Multiple changes in the pattern of myocardial gene expression result in a remodeling of the myocardium that over time is insufficient to fully compensate for increased cardiac load. The RAAS is only one of many molecular mechanisms involved in regulating myocyte hypertrophy, contractile function, and the extracellular matrix, paving the way for possible combination drug strategies. Studies from various experimental animal models of hypertrophy and heart failure now suggest that a specific hemodynamic stress selectively turns on genes and trophic factors that, in concert with ANG II, produce phenotypically different remodeling of the myocyte and the extracellular matrix. Future work should include investigation of the relative roles of chymase and ACE in ANG II production in the heart, as well as the role of other growth factors, in myocardial remodeling in response to various acute and chronic hemodynamic stresses.

References

1. Sadoshima J, Izumo S (1993) Molecular characterization of angiotensin II-induced hypertrophy of cardiac myocytes and hyperplasia of cardiac fibroblasts. Critical role of the AT1 receptor subtype. Circ Res 73:413–423

2. Baker KM, Aceto JF (1990) Angiotensin II stimulation of protein synthesis and cell growth in chick heart cells. Am J Physiol 259:H610–H618
3. Oparil S, Meng QC, Sun S, Chen YF, Dell'Italia LC (1996) Tissue angiotensin converting enzyme. In: Catravas J, Callow A, Ryan U (eds) Vascular endothelium: responses to injury. Plenum, New York
4. Freer RJ, Pappano AJ, Peach MJ, Bing KT, McLean MJ, Vogel S, Spereleakis N (1976) Mechanism for the positive inotropic effect of angiotensin II on isolated cardiac muscle. Circ Res 39:178–183
5. Meulemans AL, Andries LJ, Brutsaert DL (1990) Does endocardial endothelium mediate positive inotropic response to angiotensin I and angiotensin II? Circ Res 66:1591–1601
6. Moravec CS, Schluchter MD, Parnandi L, Czerska B, Stewart RW, Rosenkranz E, Bond M (1990) Inotropic effects of angiotensin II on human cardiac muscle in vitro. Circulation 82:1973–1984
7. Blumberg AL, Ackerly JA, Peach MJ (1975) Differentiation of neurogenic and myocardial angiotensin II receptors in isolated rabbit atria. Circ Res 36:719–726
8. Brasch, Sieroslawski HL, Dominiak P (1993) Angiotensin II increases norepinephrine release from atria by acting on angiotensin subtype 1 receptors. Hypertension (Dallas) 22(5):699–704
9. Grossman W (1980) Cardiac hypertrophy: useful adaptation or pathological process? Am J Med 69:576–590
10. Grossman W, Jones D, McLaurin LP (1974) Wall stress and patterns of hypertrophy in the human ventricle. J Clin Invest 56:56–64
11. Carabello BA, Zile MR, Tanaka R, Cooper IV G (1992) Left ventricular hypertrophy due to volume overload versus pressure overload. Am J Physiol (Heart Circ Physiol 32) 263:H1137–H1144
12. Carabello BA, Nakano B, Corin W, Biederman R, Spann JF Jr (1989) Left ventricular function in experimental volume overload hypertrophy. Am J Physiol (Heart Circ Physiol 25) 256:H974–H981
13. Anversa P, Ricci R, Olivetti G (1986) Quantitative structural analysis of the myocardium during physiologic growth and induced cardiac hypertrophy: a review. J Am Coll Cardiol 7:1140–1149
14. Weber KT, Sun Y, Guarda E (1994) Structural remodeling of the hypertensive heart and the role of hormones. Hypertension (Dallas) 23(2):869–877
15. Boluyt MO, O'Neill L, Meredith AL, Bing OHL, Brooks WW, Conrad CH, Crow MT, Lakatta EG (1994) Alterations in cardiac gene expression during the transition from stable hypertrophy to heart failure. Marked upregulation of genes encoding extracellular matrix proteins. Circ Res 75:23–32
16. Liu Z, Hilbelink DR, Gerdes AM (1991) Regional changes in hemodynamics and cardiac myoyte size in rats with aortocaval fistulas 2. Long-term effects. Circ Res 69:59–65
17. Liu Z, Hilbelink DR, Crockett WB, Gerdes AM (1991) Regional changes in hemodynamics and cardiac myocyte size in rats with aortocaval fistulas: 1. Developing and established hypertrophy. Circ Res 69:52–58
18. Lerault F, Rouleau JL, Juneau C, Rose C, Rakusan K (1990) Functional and morphological characteristics of compensated and decompensated cardiac hypertrophy in dogs with chronic infrarenal aorto-caval fistulas. Circ Res 66:846–849
19. Weber KT, Pick R, Silver MA, Moe GW, Janicki JS, Zucker JH, Armstrong PW (1990) Fibrillar collagen and remodeling of the canine left ventricle. Circulation 82:1387–1401
20. Imoto DS, Covell JW, Harper E (1988) Increase in cross-linking of type I and type III collagens associated with volume-overload hypertrophy. Circ Res 63:399–408
21. Ruzicka M, Keeley FW, Leenen FHH (1994) The renin-angiotensin system and volume overload-induced changes in cardiac collagen and elastin. Circulation 90:198–196
22. Dell'Italia LJ, Balcells E, Meng QC, Bishop SP, Straeter-Knowlen IM, Hankes GH, Dillon R, Elton T, Oparil S (1995) Effect of ramipsil on cardiac ultrastructure and

intracardiac ACE activity and angiotensin II levels in chronic mitral regurgitation in the dog. Circulation 92(8):I-669

23. Urabe Y, Mann DL, Kent RL, Nakano K, Tomanek RJ, Carabello BA, Cooper G III (1992) Cellular and ventricular contractile dysfunction in experimental canine mitral regurgitation. Circ Res 70:131–147
24. Dell'Italia L (1995) The canine model of mitral regurgitation. Heart Failure 11(5):208–218
25. Dell'Italia LJ, Meng QC, Balcells E, Straeter-Knowlen IM, Hankes GH, Dillon R, Cartee RE, Orr R, Bishop SP, Oparil S, Elton T (1995) Increased ACE and chymase-like activity in cardiac tissue of dogs with chronic mitral regurgitation. Am J Physiol (Heart Circ Physiol 38) 269:H2065–H2073
26. Caulfield JB, Borg TK (1979) The collagen network of the heart. Lab Invest 40:364–372
27 Komamura K, Shannon RP, Ihara T, Shen Y, Mirsky I, Bishop SF, Vatner SF (1994) Exhaustion of Frank–Starling mechanism in conscious dogs with heart failure. Am J Physiol (Heart Circ Physiol) 265:H1119–H1131
28. Zellner JI, Spinale FG, Eble DM, Hewett KW, Crawford FA (1991) Alterations in myocyte shape and basement membrane attachment with tachycardia-induced heart failure. Circ Res 69:590–600
29. Villari B, Vassalli G, Campbell SE, Hess OM (1995) Differences in left ventricular adaptation to chronic volume overload. Circulation 92(8):I-791
30. Young A, Orr R, Smaill B, Dell'Italia LJ (1996) Three-dimensional changes in left and right ventricular geometry in chronic mitral regurgitation. Am J Physiol (Heart Circ Physiol) (in press)
31. Sadoshima J, Izumo S (1993) Signal transduction pathways of angiotensin II-induced *c-fos* gene expression in cardiac myocytes in vitro: roles of phospholipid-derived second messengers. Circ Res 73:424–438
32. Sadoshima J, Xu Y, Slayter HS, Izumo S (1993) Autocrine release of angiotensin II mediates stretch-induced hypertrophy of cardiac myocytes in vitro. Cell 75:977–984
33. Yamazaki T, Komuro I, Kudoh S, Zou Y, Shiojima I, Mizuno T, Takano H, Hiroi Y, Ueki K, Tobe K, Kadowaki T, Nagai R, Yazaki Y (1995) Angiotensin II partly mediates mechanical stress-induced cardiac hypertrophy. Circ Res 77:258–265
34. Schunkert H, Sadoshima J, Cornelius T, Kagaya Y, Weinberg EO, Izumo S, Riegger G, Lorell BH (1995) Angiotensin II-induced growth responses in isolated adult rat hearts. Evidence for load-independent induction of cardiac protein synthesis by angiotensin II. Circ Res 76:489–497
35. Knowlton KU, Rockman HA, Itani M, Vovan A, Seidman CE, Chien KR (1995) Divergent pathways mediate the induction of ANF transgenes in neonatal and hypertrophic ventricular myocardium. J Clin Invest 96:1311–1318
36. Lambert C, Massillon Y, Meloche S (1995) Upregulation of cardiac angiotensin II AT_1 receptors in congenital cardiomyopathic hamsters. Circ Res 77:1001–1007
37. Everett AD, Tufro-McReddie A, Fisher A, Gomez RA (1994) Angiotensin receptor regulates cardiac hypertrophy and transforming growth factor-β_1 expression. Hypertension (Dallas) 23:587–592
38. Hein L, Barsh GS, Pratt RE, Dzau VJ, Kobilka BK (1995) Behavioural and cardiovascular effects of disrupting the angiotensin II type-2 receptor gene in mice. Nature 377:744–747
39. Ichiki T, Labosky PA, Shiota C, Okuyama S, Imagawa Y, Fogo A, Niimura F, Ichikawa I, Hogan BLM, Inagami T (1995) Effects on blood pressure and exploratory behaviour of mice lacking angiotensin II type-2 receptor. Nature 377:748–750
40. Lindpaintner K, Ganten D (1991) The cardiac renin-angiotensin system: an appraisal of present experimental and clinical evidence. Circ Res 68:905–921
41. Baker KM, Chernin M, Wixon SK, Aceto JF (1990) Renin-angiotensin system involvement in pressure-overload cardiac hypertrophy in rats. Am J Physiol (Heart Circ Physiol) 259:H324–H332
42. Schunkert H, Dzau VJ, Tang SS, Hirsch AT, Apstein CS, Lorell BH (1990) Increased rat cardiac angiotensin-converting enzyme activity and mRNA expression in pressure

overload left ventricular hypertrophy: effects on coronary resistance, contractility and relaxation. J Clin Invest 86:1913–1920
43. Schunkert H, Jackson B, Tang SS, Schoen FJ, Smits JFM, Apstein CS, Lorell BH (1993) Distribution and functional significance of cardiac ACE in hypertrophied rat hearts. Circulation 87:1328–1339
44. Iwai N, Shimoike H, Kinoshita M (1995) Cardiac renin-angiotensin system in the hypertrophied heart. Circulation 92:2690–2696
45. Pieruzzi F, Abassi ZA, Keiser HR (1995) Expression of renin-angiotensin system components in the heart, kidneys, and lungs of rats with experimental heart failure. Circulation 92:3105–3112
46. Boer PH, Ruzicka M, Lear W, Harmsen E, Rosenthal J, Leenen FHH (1994) Stretch-mediated activation of the cardiac renin gene. Am J Physiol (Heart Circ Physiol 36) 267:H1630–H1636
47. Ruzicka M, Skarda V, Leenen FHH (1995) Effects of ACE inhibitors on circulating versus cardiac angiotensin II in volume overload-induced cardiac hypertrophy in rats. Circulation 92:3568–3573
48. Ruzicka M, Leenen FHH (1995) Relevance of blockade of cardiac and circulatory angiotensin-converting enzyme for the prevention of volume overload-induced cardiac hypertrophy. Circulation 91:16–19
49. Calderone A, Takahashi N, Izzo NJ, Thaik CM, Colucci WS (1995) Pressure- and volume-induced left ventricular hypertrophies are associated with distinct myocyte phenotypes and different induction of peptide growth factor mRNAs. Circulation 92:2385–2390
50. Pahor M, Bernabei R, Sgadari AN, Gambassi G, Giudice PL, Pacifici L, Ramacci MT, Lagrasta C, Olivette G, Carbonin P (1991) Enalapril prevents cardiac fibrosis and arrhythmias in hypertensive rats. Hypertension (Dallas) 18:148–157
51. Weinberg EO, Schoen FJ, Gearge D, Kagaya Y, Douglas PS, Litwin SE, Schunkert H, Benedict CR, Lorell BH (1994) Angiotensin-converting enzyme inhibition prolongs survival and modifies the transition to heart failure in rats with pressure overload hypertrophy due to ascending aortic stenosis. Circulation 90:1410–1422
52. Litwin SE, Katz SE, Weinberg EO, Lorell BH, Aurigemma GP, Douglas PS (1995) Serial echocardiographic-doppler assessment of left ventricular geometry and function in rats with pressure-overload hypertrophy. Chronic angiotensin-converting enzyme inhibition attenuates the transition to heart failure. Circulation 91:2642–2654
53. Qing G, Garcia R (1992) Chronic captopril and losartan (Dup 753) administration in rats with high output heart failure. Am J Physiol (Heart Circ Physiol 32) 263:H833–H840
54. Gay RG (1990) Captopril reduces left ventricular enlargement induced by chronic volume overload. Am J Physiol (Heart Circ Physiol 28) 259:H796–H803
55. Linz W, Scholkens BA, Ganten D (1989) Converting enzyme inhibition specifically prevents the development and induces regression of cardiac hypertrophy in rats. Clin Exp Hypertens Part A Theory Pract 11(7):1325–1350
56. Garg VC, Hassid A (1989) Nitric oxide-generating vasodilators and 8-bromo-cyclic guanosine monophosphate inhibit mitogenesis and proliferation of cultured rat vascular smooth muscle cells. J Clin Invest 83:1774–1777
57. Linz W, Scholkens BA (1992) A specific β_2-bradykinin receptor antagonist HOE 140 abolishes the antihypertrophic effect of ramipril. Br J Pharmacol 105: 771–772
58. McDonald KM, Garr MD, Carlyle PF, Francis GS, Hauer K, Hunter DW, Parrish T, Stillman A, Cohn JN (1994) The relative effects of $alpha_1$ adrenoreceptor blockade, converting enzyme inhibitor therapy and angiotensin II subtype I receptor blockade on ventricular remodeling in the dog. Circulation 90:3034–3036
59. McDonald KM, Mock J, D'Aloia A, Parrish T, Hauer K, Francis G, Stillman A, Cohn JN (1995) Bradykinin antagonism inhibits the antigrowth effect of converting enzyme inhibition in the dog myocardium after discrete transmural myocardial necrosis. Circulation 91:2043–2048

60. Janicki JS, Matsubara BB (1994) Myocardial collagen and left ventricular diastolic function. In: Gaasch WH, LeWinter MM (eds) Left ventricular diastolic function and heart failure. Lea and Febiger, Philadelphia, p 125–140
61. Masashi A, Matsui H, Periasamy M (1994) Sarcoplasmic reticulum gene expression in cardiac hypertrophy and heart failure. Circ Res 74:555–564
62. Neyses L, Vetter H (1990) Impaired relaxation of hypertrophied myocardium is potentiated by angiotensin II. J Hypertens 81(suppl III):III-123–III-129
63. Grocott-Mason R, Anning P, Evans H, Lewis MJ, Shah AM (1994) Modulation of left ventricular relaxation in the isolated ejecting heart by endogenous nitric oxide. Am J Physiol 267:H1804–H1813
64. Anning PB, Grocott-Mason RM, Lewis MJ, Shah AM (1995) Enhancement of left ventricular relaxation in the isolated heart by an angiotensin-converting enzyme inhibitor. Circulation 92:2660–2665
65. Friedrich SP, Lorell BH, Rousseau MF, Hyashida W, Hess OM, Douglas PS, Gordon S, Keighley CS, Benedict C, Krayenbuehl HP, Grossman W, Pouleur H (1994) Increased angiotensin-converting enzyme inhibition improves diastolic function in patients with left ventricular hypertrophy due to aortic stenosis. Circulation 90:2761–2771
66. Dzau VJ (1989) Multiple pathways of angiotensin production in the blood vessel wall: evidence, possibilities and hypotheses. J Hypertens 7:933–936
67. Urata H, Healy B, Stewart RW, Bumpus FM, Husain A (1990) Angiotensin II-forming pathways in normal and failing human hearts. Circ Res 66:883–890
68. Urata H, Kinoshita A, Misono KS, Bumpus FM, Husain A (1990) Identification of a highly specific chymase as the major angiotensin II-forming enzyme in the human heart. J Biol Chem 265(36):22348–22357
69. Husain A, Kinoshita A, Sung SS, Urata H, Bumpus FM (1994) The cardiac renin-angiotensin system. Futura, Armonk, NY, pp 309–332
70. Zisman LS, Abraham WT, Meizell GE, Vamvakias BN, Quaife RA, Lowes BD, Roden RL, Peacock SJ, Groves BM, Bristow MR, Perryman MB (1995) Angiotensin II formation in the intact human heart: predominance of the angiotensin converting enzyme pathway. J Clin Invest 95:1490–1498
71. Okunishi H, Oka Y, Shiota N, Kawamoto T, Song K, Miyazaki M (1993) Marked species-differences in the vascular angiotensin II-forming pathways: humans vs. rodents. Jpn J Pharmacol 62:207–210
72. Chandrasekharan UM, Sanker S, Glynias MJ, Karnik SS, Husain A (1996) Angiotensin II-forming activity in a reconstructed ancestral chymase. Science 271:502–505
73. Balcells E, Meng QC, Hageman G, Palmer RW, Durand J, Dell'Italia LJ (1996) Angiotensin II formation in dog heart is mediated by different pathways in vivo and in vitro. Am J Physiol (in press)
74. Urata H, Boehm KD, Philip A, Kinoshita A, Gabrovsek J, Bumpus FM, Husain A (1993) Cellular localization and regional distribution of an angiotensin II-forming chymase in the heart. J Clin Invest 91:1269–1281
75. Johnston CI (1994) Tissue angiotensin converting enzyme in cardiac and vascular hypertrophy, repair, and remodeling. Hypertension (Dallas) 23:258–268
76. Hoit B, Shao Y, Kinoshita A, Gabel M, Hussain A, Walsh RA (1995) Effects of angiotensin II generated by an angiotensin converting enzyme—independent pathway on left ventricular performance in the conscious baboon. J Clin Invest 95:1519–1527
77. Kramer BK, Nishida M, Kelly RA, Smith TV (1992) Endothelins. Myocardial actions of a new class of cytokines. Circulation 85:350–356
78. Ito H, Hirata Y, Adachi S, Tanaka M, Tsujino M, Koike A, Nogami A, Marumo F, Hiroe M (1993) Endothelin-1 is an autocrine/paracrine factor in the mechanism of angiotensin II-induced hypertrophy in cultured rat cardiomyocytes. J Clin Invest 92:398–403
79. Fujisaki H, Ito H, Hirata Y, Tanaka M, Hata M, Lin M, Adachi S, Akimoto H, Marumo F, Hiroe M (1995) Natriuretic peptides inhibit angiotensin II-induced proliferation of rat cardiac fibroblasts by blocking endothelin-1 gene expression. J Clin Invest 96:1059–1065

80. Arai M, Yoguchi A, Iso T, Takahashi T, Imai S, Murata K, Suzuki T (1995) Endothelin-1 and its binding sites are upregulated in pressure overload cardiac hypertrophy. Am J Physiol 268:H2084–2091
81. Tan LB, Jalil JE, Pick R, Janicki JS, Weber KT (1991) Cardiac myocyte necrosis induced by angiotensin II. Circ Res 69:1185–1195
82. Henegar JR, Brower GL, Kabour A, Janicki JS (1995) Catecholamine response to chronic ANG II infusion and its role in myocyte and coronary vascular damage. Am J Physiol (Heart Circ Physiol 38) 269:H1564–H1569
83. Mann DL, Kent RL, Parsons B, Cooper G IV (1992) Adrenergic effects on the biology of the adult mammalian cardiocyte. Circulation 85:790–804
84. Tsutsui H, Spinale FG, Nagatsu M, Schmid PG, Ishihara K, DeFreyte G, Cooper G IV, Carabello BA (1994) Effects of chronic beta-adrenergic blockade on the left ventricular and cardiac myocyte abnormalities of chronic mitral regurgitation. J Clin Invest 93:2639–2648
85. Young M, Fullerton M, Dilley R, Funder J (1994) Mineralocorticoids, hypertension, and cardiac fibrosis. J Clin Invest 93:2578–2583
86. Nicoletti A, Heudes D, Hinglais N, Appay M, Philippe M, Sassy-Prigent C, Bariety J, Michel J (1995) Left ventricular fibrosis in renovascular hypertensive rats. Hypertension (Dallas) 26:101–111
87. Hou J, Kato H, Cohen RA, Chobanian AV, Breche P (1995) Angiotensin II-induced cardiac fibrosis in the rat is increased by chronic inhibition of nitric oxide synthase. J Clin Invest 96:2469–2477
88. Studer R, Reinecke H, Muller B, Holtz J, Hanjorg J, Drexler H (1994) Increased angiotensin-I converting enzyme gene expression in the failing human heart. J Clin Invest 94:301–310
89. Zisman L, Bush EW, Taft CS, Bristow MR, Perryman MB, Raynolds MV (1994) Increase in angiotensin-converting gene expression and activity in the failing human ventricle. Circulation 90(4):I–578
90. Sawa S, Kawaguchi H, Mochizuki N, Endo Y, Kudo T, Tokuchi F, Fijioka Y, Nagashima K, Kitabatake A (1994) Distribution of angiotensinogen in diseased human hearts. Mol Cell Biochem 132:15–23
91. Lombes M, Alfaidy N, Eugene E, Lessana A, Farman N, Bonvalet JP (1995) Prerequisite for cardiac aldosterone action. Mineralocorticoid receptor and 11-beta-hydroxysteroid dehydrogenase in the human heart. Circulation 92:175–182
92. Urata H, Healy B, Stewart RW, Bumpus FM, Husain A (1989) Angiotensin II receptors in normal and failing human hearts. J Clin Endocrinol Metab 69:54–66
93. Studer R, Susch G, Muller B, Oechslin E, Hess OM, Drexler H (1994) Role of pressure overload and wall stress for cardiac gene expression of angiotensin-converting enzyme in humans. Circulation 90(4):I–451

The Role of Manganese Superoxide Dismutase in the Acquisition of Tolerance of the Heart to Ischemia: Molecular Adaptation to Ischemia

Masashi Nishida[1,2], Tsunehiko Kuzuya[1,2], Shiro Hoshida[2], Nobushige Yamashita[2], Masatsugu Hori[2], and Michihiko Tada[1,2]

Summary. The heart acquires tolerance to ischemic stress after exposure to brief, nonlethal ischemia (ischemic preconditioning). We examined the mechanism of tolerance acquisition from the aspect of oxygen radical metabolism in the heart. Manganese superoxide dismutase (Mn-SOD), which scavenges the initial reductive metabolite of molecular oxygen on ischemia–reperfusion, was induced after various external stimuli including heat shock, α_1-adrenergic stimulation, and ischemia (hypoxia). The induction was blocked by a protein kinase C (PKC) inhibitor, staurosporine, and the specific inhibition of the Mn-SOD induction by antisense oligodeoxyribonucleotides abolished acquisition of tolerance after the stimuli. Therefore, we propose that the de novo synthesis of Mn-SOD could be one of the molecular mechanisms by which the heart adapts to ischemic stress.

Key words. Preconditioning—Mn-SOD—α-Adrenergic agonist—Heat shock—Second window—Ischemia

Introduction

Increasing evidence suggests that the myocardium undergoes adaptation after sublethal cellular stresses, such as ischemia [1], heat stress [2], endotoxin [3], and cytokines [4]. The preconditioning phenomenon, in which brief, nonlethal ischemia increases the tolerance of the heart to subsequent lethal ischemia, has been observed in experimental models [1] and in clinic patients [5–7]. The mechanism of the preconditioning phenomenon has been examined in regard to adrenergic stimulation [8], protein kinase c activation [9], or adenosine metabolism [10]. However, the signal transduction system and the final effector of the protection are still a matter of discussion.

Recently, we [11] and another group [12] found that the protective effect disappeared 3h after the "classical" preconditioning and reappeared 24h after the initial nonlethal ischemia. The late-phase effect of ischemic preconditioning was named the second window of preconditioning. Although de novo protein synthesis is not in-

[1] Department of Medicine and Pathophysiology.
[2] First Department of Medicine, Osaka University Medical School, Suita, Osaka, 565 Japan.

volved in the mechanism of classical preconditioning [13], the time-course of the second window suggests that the synthesis of intrinsic proteins after initial ischemic stress is involved in the mechanism of the late-phase effect. Among them, the induction of heat shock proteins was reported to have a parallel with the acquisition of tolerance to ischemia–reperfusion [14].

In an in vivo model of ischemia-reperfusion, we found that an intrinsic radical scavenger, manganese superoxide dismutase (Mn-SOD), was induced in the ischemic myocardium [15] coincident with the acquisition of tolerance to ischemia–reperfusion [11]. Oxygen radicals are produced in ischemia-reperfused heart tissue and cause myocardial injury by means of lipid peroxidation of the sarcoplasmic membrane [16]. Therefore, the induction of Mn-SOD on ischemic preconditioning has a rationale in terms of site protection against oxidative stress at ischemia–reperfusion. In this review, we focus on the aspect of oxygen radical metabolism in the ischemia-reperfused heart and discuss the role of intrinsic antioxidants in the heart, especially that of manganese superoxide dismutase in the acquisition of tolerance to ischemic stress.

Oxygen Radicals in the Ischemia-Reperfused Heart

Oxygen radicals, produced at reflow of ischemic myocardium, were proposed to be one of the culprits that cause myocardial injury at ischemia–reperfusion [17]. In the late 1980s, several groups successfully detected the presence of oxygen radical species in ischemia-reperfused myocardium by electron paramagnetic resonance (EPR). Zweier's [18] and Baker's [19] groups directly measured an oxygen-centered radical signal in frozen myocardial samples, and Garlick's group [20] detected radical derivatives from oxygen radicals in coronary effluent by spin trap technique soon after reperfusion of ischemic myocardium. Oxygen radicals not only were produced at the time of reperfusion but also induced a chain reaction of lipid peroxidation following the course of reperfusion.

We also detected the same oxygen-derived free radicals at the late phase of reperfusion [21]. After a 90-min occlusion of the canine left anterior descending coronary artery (LAD), cardiac tissue was freeze-clamped at 1 h and at 3 h after reperfusion, and 5,5-dimethyl-1-pyrroline *N*-oxide (DMPO) adducts of the tissue sample were detected by EPR. A DMPO-OH signal, which represents the hydroxyl radical species, was observed at both 1 h and 3 h after reperfusion in much larger quantities than the signal from normal cardiac tissue (Fig. 1). These results suggest that oxygen radicals were produced in myocardial tissue after reperfusion of the ischemic lesion and that the radical reaction proceeded during the course of extension of cardiac injury after reperfusion.

A number of studies therefore were designed to scavenge radical species to protect the myocardium from reperfusion injury; however, the results were controversial [22]. Exogenous radical scavengers (e.g., SOD or catalase) successfully reduced reperfusion injury in in vivo models [23–26]. On the other hand, in other reports the same radical scavengers failed to salvage reperfused myocardium [27–19]. The experimental protocols of these reports differed regarding duration of ischemia and reperfusion, timing of the addition of radical scavengers, and consciousness of the animals.

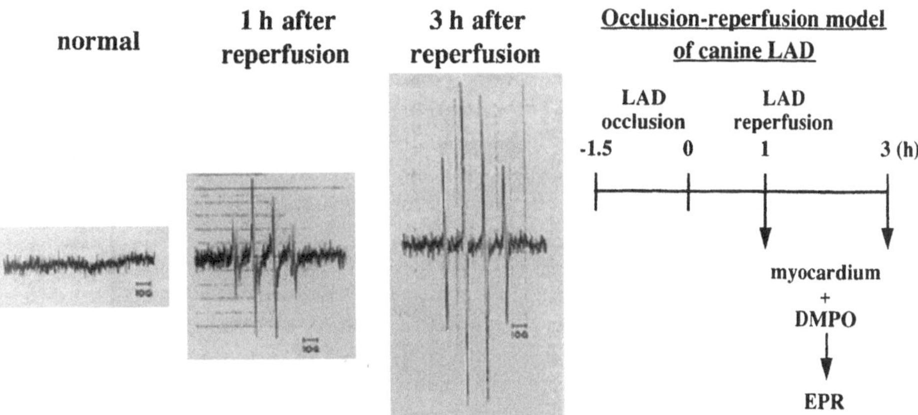

FIG. 1. Electron paramagnetic resonance (*EPR*) spectra of myocardial samples. EPR spin-trapping spectra were recorded from normal myocardium and the ischemic myocardium at 1 h and 3 h of reperfusion. *LAD*, Left anterior descending artery; *DMPO*, 5,5-dimethyl-1-pyrroline N-oxide

One possible reason for the controversy is the difference between the site of radical production and the site of distribution of exogenous radical scavengers. In the ischemia-reperfused heart, oxygen radicals were generated mainly from three sources: the xanthine–xanthine oxidase system of vascular endothelial cells, reduced nicotinamide adenine dinucleotide (NADH) oxidase of activated neutrophils in the bloodstream or in myocardial tissue, and the mitochondrial electron transport system of cardiac myocytes. These systems could produce superoxide anion, the initial product of oxygen reduction, in the coronary circulation, in the extracellular space of cardiac tissue, and inside the mitochondria of myocytes. Exogenous radical scavengers were mostly distributed in the bloodstream or extracellular space.

Therefore, oxygen radicals produced by endothelial cells and neutrophils could be scavenged by the scavengers. However, they reacted with difficulty with superoxide generated by the mitochondrial electron transport system, although a conjugated, long-lived SOD was developed and exhibited better results in reaching and scavenging cardiac superoxide [30,31]. Recent studies showed, however, that the heart is not merely threatened by oxygen radicals, but also has its own intrinsic radical scavenger system, such as Mn-SOD, Zn-SOD, catalase, and glutathione redox system. These enzymes were revealed to be reactively induced by exogenous stimuli, such as endotoxin [3], cytokines [4], and hyperthermia [32]. Among them, Mn-SOD is located in cardiac mitochondria and is supposed to play a major role in scavenging superoxide generated by the electron transport system in the front line.

Mn-SOD Induction on Ischemia–Reperfusion: In Vivo/In Vitro

The induction and activation of Mn-SOD thus could be the mechanism producing the protective effect of the second window of preconditioning. To examine this hypothesis, we measured the Mn-SOD content in the ischemic myocardium of the dog heart after four 5-min coronary occlusions [15]. Mn-SOD protein was measured by enzyme-linked immunosorbent assay soon after and at 3h, 12h, and 24h after repeated ischemia (Fig. 2a). Mn-SOD content in the subendocardium increased gradually with a peak (60% increase) observed 24h after sublethal ischemia. At this peak point, myocardial Mn-SOD activity, measured simultaneously by the nitroblue tetrazolium method, was also increased by about 80% of normal control. We could not see any differences in other antioxidant enzyme activities, including Cu-Zn SOD, catalase, and glutathione peroxidase, in this experiment.

We have also demonstrated that such an ischemic preconditioning protocol results in a delayed protective response against myocardial necrosis after subsequent prolonged ischemia in the dog [11]. We examined the effect of repeated brief ischemia on the limitation of infarct size (Fig. 2b). Immediately after, or at 3, 12, and 24h after four 5-min occlusions of LAD, dogs were subjected to 90 min of occlusion followed by 5h of reperfusion. When the second ischemia was applied immediately after the first sublethal ischemia, the percentage of risk area infarcted was markedly decreased to 14% compared with 42% in the control. When the time interval between sublethal and sustained ischemia was 3 or 12h, no statistical differences were seen in infarct size between preconditioned and sham-operated groups. However, the infarct-limiting effect was observed again 24h after ischemic preconditioning. The time-courses of induction of Mn-SOD and reappearance of tolerance to ischemia–reperfusion in the myocardium were identical, suggesting that the Mn-SOD protects the myocardium from injury caused by ischemia–reperfusion.

To investigate the role of enhanced SOD activity in protection against ischemia-reperfusion injury directly, we examined whether the preconditioning phenomenon

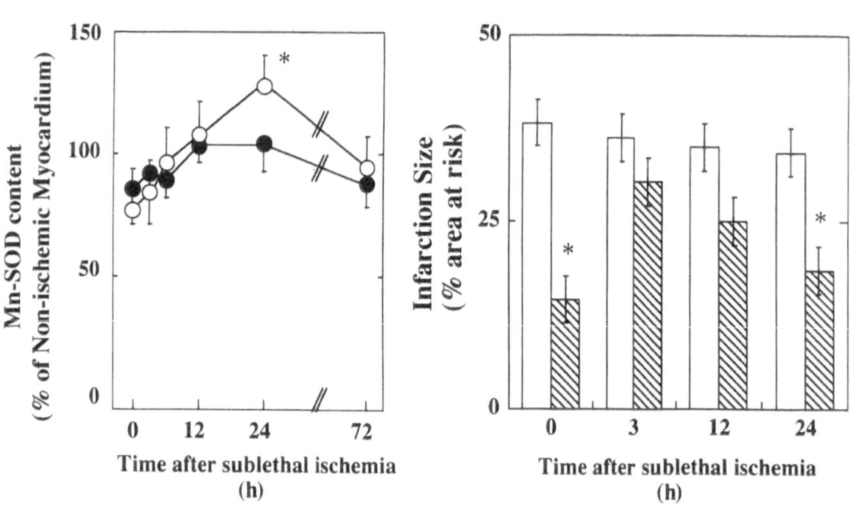

Fig. 2a,b. Manganese superoxide dismutase (*Mn-SOD*) content and the size of myocardial infarction after ischemic preconditioning. **a** Serial changes in level of Mn-SOD protein in subepicardial (*open circles*) and subendocardial (*closed circles*) sides of ischemic myocardium were measured. Each point shows the percent change in Mn-SOD level over that in each side of the nonischemic myocardium at the respective time. **b** The infarcted area (expressed as a percentage of the area at risk) is compared in preconditioned (*hatched bars*) and sham-operated (*open bars*) dogs

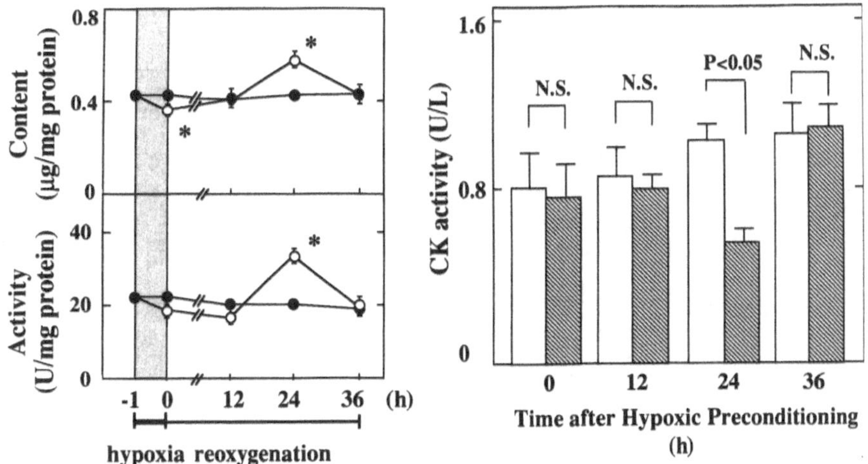

FIG. 3a,b. Mn-SOD induction and the injury of cardiac myocytes after hypoxic preconditioning. **a** The SOD activity in cultured myocytes with (*open circles*) or without (*closed circles*) hypoxic preconditioning was determined by the nitroblue tetrazolium (NBT) method. The Mn-SOD content in the supernatant was determined by enzyme-linked immunosorbent assay (ELISA). **b** Creatine kinase (*CK*) release after second reoxygenation for 1 h following sustained hypoxia for 3 h. Control cells were exposed to normoxic conditions instead of hypoxic preconditioning for 1 h. *Open columns*, control myocytes; *hatched columns*, preconditioned myocytes. *N.S.*, Not significant

could be mimicked in cultured rat myocytes by exposing them to hypoxia (7 mmHg)–reoxygenation (143 mmHg) before exposure to sustained hypoxia–reoxygenation [33]. Figure 3a shows the change in cardiac Mn-SOD content and activity after exposure to hypoxia: the upper panel shows Mn-SOD content and the lower panel shows the activity of Mn-SOD in cultured cardiac myocytes. In control cells, which were subjected to normoxia instead of hypoxia for the first hour, Mn-SOD content and activity either did not change or showed a slight decrease during the following 36 h. On the other hand, in the cells exposed to hypoxia, both the activity and the content of Mn-SOD increased markedly at 24 h after reoxygenation from hypoxia. Mn-SOD induction was regulated at the transcriptional level because pro Mn-SOD messenger RNA after hypoxia–reoxygenation increased after reoxygenation and reached its peak at 30 min after reoxygenation.

We next examined creatine kinase (CK) release from myocyte cultures after exposure to prolonged hypoxia for 3 h followed by reoxygenation (Fig. 3b). In the control experiments in which cells were exposed to normoxia instead of hypoxia for the first hour, CK release after the second hypoxia remained constant up to 36 h. However, in the cells exposed to 1 h of hypoxia, CK release from myocyte cultures was markedly reduced when the second hypoxia was applied 24 h after the first hypoxia compared with the cells without simulated preconditioning. The time-course of this increase in myocardial tolerance to hypoxia of these myocytes after exposure to a brief preceding hypoxia was apparently similar to that of Mn-SOD induction.

To examine the cause-and-effect relationship between Mn-SOD induction and tolerance to hypoxia, we attempted to block Mn-SOD induction after hypoxia using antisense oligodeoxyribonucleotides to pro Mn-SOD mRNA. After confirming that

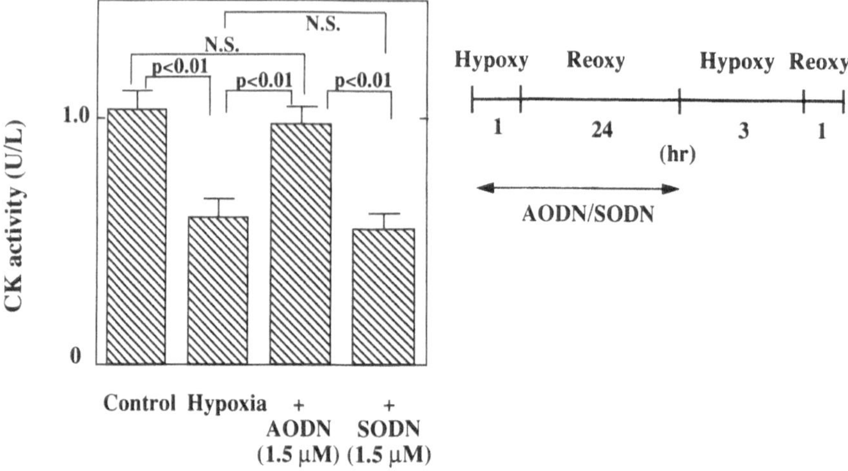

FIG. 4. The effect of antisense oligodeoxyribonucleotides (*AODN*) to Mn-SOD on CK release from cultured cardiac myocytes. AODN and sense ODN (*SODN*) were applied to preconditioned myocyte cultures at 1.5 μ*M* from 18 h before first hypoxia (*Hypoxy*) to 24 h after reoxygenation (*Reoxy*). In the control group, cultured cells were exposed to normoxic condition instead of hypoxic preconditioning. Creatine kinase (*CK*) activity in the culture medium was measured 1 h after reoxygenation after sustained hypoxia

antisense oligodeoxyribonucleotides to pro Mn-SOD mRNA inhibited the induction of Mn-SOD protein, we examined the effect of the oligonucleotides on CK release from myocytes after exposure to prolonged hypoxia-reoxygenation (Fig. 4). Cardiac myocytes were exposed to 1 h of hypoxia with or without antisense oligonucleotides to Mn-SOD at 24 h before exposure to prolonged hypoxia. CK release from myocyte cultures was attenuated by 51% when the cells were exposed to a preceding brief hypoxia compared to that from cells without such exposure. However, antisense oligonucleotides to Mn-SOD abolished the expected decrease in CK release from myocytes with hypoxic preconditioning, and sense oligonucleotides did not attenuate CK release. These results indicate that Mn-SOD induction in cardiac myocytes after exposure to brief hypoxia is the mechanism for acquisition of tolerance to lethal hypoxia in cardiac myocytes.

Induction of Mn-SOD by Heat Shock or Adrenergic Stimulation

The induction of Mn-SOD could be one of the common mechanisms by which cardiac myocytes respond to external stresses and acquire tolerance toward such stresses. Hyperthermia is a typical intervention that causes cardiac injury but is also known to induce resistance to various stimuli within myocytes through the induction of rescue proteins such as heat shock protein (HSP) [34]. Therefore, we applied heat shock to cultured cardiac myocytes and examined the induction of Mn-SOD protein and mRNA together with HSP72, a protein known to be induced after heat stress. Pro Mn-SOD mRNA exhibited transient expression that peaked at 30 min after heat shock for

1h at 42°C (Fig. 5a). Heat shock protein 72 mRNA expression probed with a polymerase chain reaction (PCR) fragment of mouse cDNA was also increased dramatically at 30 min after heat shock (Fig. 5b).

As we had confirmed that heat shock induces both Mn-SOD and HSP72 in myocardial cells, we examined whether Mn-SOD and HSP72, thus induced, play roles in the acquisition of tolerance to hypoxia after heat stress. Heat stress for 1h at 42°C increased the tolerance of myocytes to hypoxia, and CK release from the myocytes after 3h of hypoxia followed by 1h of reoxygenation decreased by 50% compared with control cells without heat stress. Antisense oligonucleotides to Mn-SOD that inhibited Mn-SOD induction after heat stress attenuated significantly this decrease in CK release. As the antisense oligonucleotide specifically inhibited Mn-SOD induction, HSP72 level after heat shock was not reduced by the oligonucleotides.

These results indicate that Mn-SOD is induced after heat stress together with HSP72 and plays a pivotal role in the acquisition of tolerance to ischemia after heat shock. The role of Mn-SOD in cardioprotection might be independent from the induction of HSP72 because inhibition of Mn-SOD alone by antisense oligonucleotides abolished the tolerance to hypoxia–reoxygenation. Because we could not design a proper oligonucleotide to inhibit HSP72 induction after heat shock, the experiment to suppress HSP72 alone was not performed.

Sympathetic stimulation is another stimulus that induces adaptation of myocytes to external stresses. α_1-Adrenergic stimulation is known to induce adaptive cardiac hypertrophy by augmentation of protein synthesis at the transcription level [35]. In the classical preconditioning phenomenon of the heart, the α_1-adrenergic stimulation pathway is also reported to conduct a protective effect of preconditioning through adenosine [36], the K channel [37], or protein kinase C [38]. Therefore, we hypothesized that the induction of Mn-SOD in cardiac myocytes in the second-window phenomenon of ischemic preconditioning could be mediated through the same α_1-adrenergic receptor-mediated mechanism as the classical preconditioning.

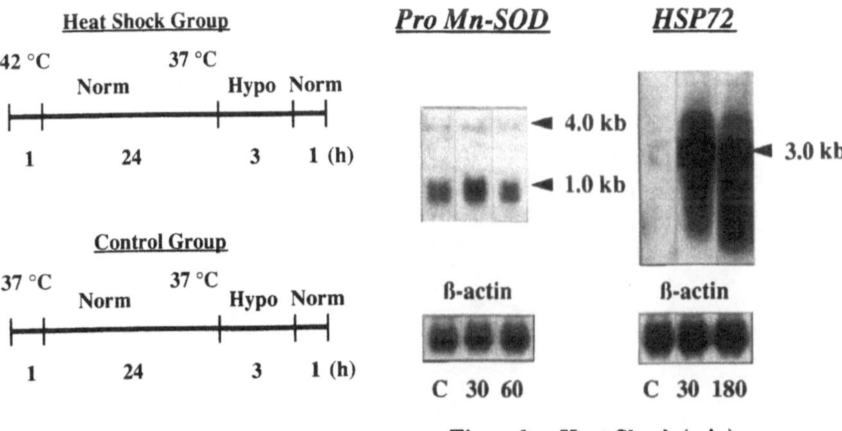

Fig. 5. The expression of pro Mn-SOD and HSP72 mRNA after heat shock. Cultured cardiac myocytes were exposed to heat shock at 42°C. The expression of pro Mn-SOD mRNA and HSP72 mRNA after the exposure was examined by Northern analysis

When norepinephrine was added to myocyte cultures, Mn-SOD activity in the culture dishes 24h after the stimulation increased dose dependently to 0.2 μ*M* and declined thereafter (Fig. 6, top). Although total protein content in the cell was also increased dose dependently up to 20 μ*M* of norepinephrine as reported previously [35] (Fig. 6, middle), the increase in Mn-SOD was shown to be specifically larger than the nonspecific increase in proteins by dividing Mn-SOD activity with the total protein amount (Fig. 6, bottom). Pro Mn-SOD mRNA expression after norepinephrine addition examined by Northern blotting showed pro Mn-SOD transcription was augmented at 30 min after the stimulation.

To confirm the receptor specificity of this phenomenon, we examined the effect of adrenergic receptor antagonists and an agonist on Mn-SOD induction. The Mn-SOD activity in myocytes augmented by the addition of norepinephrine (0.2 μ*M*) was not attenuated by the α_2-adrenergic blocker yohimbine (2 μ*M*) and the β-adrenergic blocker propranolol (2 μ*M*). However, an α_1-adrenergic blocker, prazosin (2 μ*M*), abolished the increase in Mn-SOD activity induced by norepinephrine. The α_1-adrenergic agnoist methoxamine (20 μ*M*) also increased Mn-SOD activity in the myocytes 24 h after the addition. These results indicate the induction of Mn-SOD after norepinephrine was conducted through α_1 -adrenergic receptors. Mn-SOD activity increased by α_1-adrenergic stimulation was attenuated by the addition of antisense oligonucleotides to pro Mn-SOD mRNA. The protein kinase C inhibitor staurosporine (100 n*M*) also attenuated the increase in Mn-SOD activity.

The tolerance of myocytes to hypoxia–reoxygenation was also examined under these conditions. Cardiac myocytes were exposed to α_1-adrenergic stimulation for 24h, and then hypoxia (3h) followed by reoxygenation (1h) was applied to the cells. CK release from the myocytes was significantly attenuated by α_1-adrenergic stimulation. However, antisense oligonucleotides to Mn-SOD, which inhibited the induction of Mn-SOD by norepinephrine, abolished the expected decrease in CK release from hypoxia-reoxygenated myocytes.

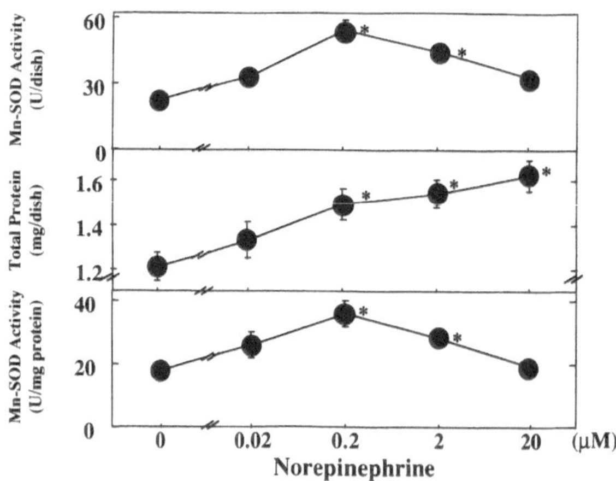

Fig. 6. Dose response of norepinephrine (NE) stimulation on Mn-SOD activity in cultured cardiac myocytes. Mn-SOD activity in cell homogenates (*top*) and the total amount of cell protein (*middle*) were measured after NE stimulation for 24h (0–20 μ*M*). *Bottom:* The ratio of Mn-SOD activity to total protein was calculated for each concentration of NE

Conclusion

In this chapter, we focused on the role of Mn-SOD, an intrinsic radical scavenger in the heart, in the heart's acquisition of tolerance to ischemia–reperfusion.

1. Oxygen radicals produced on ischemia–reperfusion lead to the chain reaction of radical production. Although exogenous radical scavengers can scavenge radicals produced in the extracellular space, radicals initially produced in mitochondria are not easily reached by exogenous scavengers.
2. Mn-SOD, which is located within the mitochondria, and scavenge superoxide, was induced by various stresses, such as brief ischemia (preconditioning), heat shock, or α_1-adrenergic stimulation.
3. The induction of Mn-SOD was correlated with acquisition of tolerance to ischemia (hypoxia). Antisense oligodeoxyribonucleotides to pro Mn-SOD mRNA, which inhibited the induction of Mn-SOD, abolished the expected tolerance to ischemia.
4. Although HSP72 was also induced in cardiac myocytes, inhibition of Mn-SOD induction specifically abolished the tolerance. The protein kinase C blocker, staurosporine, attenuated Mn-SOD induction.

These results suggest that the induction of Mn-SOD is the mechanism by which the heart acquires tolerance to ischemia (hypoxia). The induction of rescue proteins in the myocytes [39] could be a novel mechanism of the heart to adapt against external stresses. Although the inhibition of Mn-SOD alone specifically abolished the tolerance in our cellular model, the induction of other rescue proteins such as HSP, which acts as a molecular chaperone [40], might be involved in the adaptive mechanism. Further studies should be done regarding the interaction or the sequence of action of these proteins and the signal transduction mechanism by which external stresses lead to the induction of the rescue proteins.

References

1. Murry CE, Jennings RB, Reimer KA (1986) Preconditioning with ischemia: a delay of lethal cell injury in ischemic myocardium. Circulation 74:1124–1136
2. Currie RW, Tanguay RM, Kingma JGJ (1993) Heat-shock response and limitation of tissue necrosis during occlusion/reperfusion in rabbit hearts [see comments]. Circulation 87:963–971
3. Brown JM, Grosso MA, Terada LS, Whitman GJR, Banerjee A, White CW, Harken AH, Repine JE (1989) Endotoxin pretreatment increases endogenous myocardial catalase activity and decreases ischemia-reperfusion injury of isolated rat hearts. Proc Natl Acad Sci USA 86:2516–2520
4. Brown JM, White CW, Terada LS, Grosso MA, Shanley PF, Mulvin DW, Banerjee A, Whitman GJ, Harken AH, Repine JE (1990) Interleukin-1 pretreatment decreases ischemia/reperfusion injury. Proc Natl Acad Sci USA 87:5026–5030
5. Komamura K, Kitakaze M, Nishida K, Naka M, Tamai J, Uematsu M, Korestune Y, Nanto S, Hori M, Inoue M, Kamada T, Kodama K (1994) Progressive decreases in coronary vein flow during reperfusion in acute myocardial infarction: clinical documentation of the no-reflow phenomenon after successful thrombolysis. J Am Coll Cardiol 24:370–377
6. Nakagawa Y, Ito H, Kitakaze M, Kusuoka H, Hori M, Kuzuya T, Higashino Y, Fujii K, Minamino T (1995) Effect of angina pectoris on myocardial protection in patients with reperfused anterior wall myocardial infarction: retrospective clinical evidence of "preconditioning." J Am Coll Cardiol 25:1076–1083

7. Ottani F, Galvani M, Ferrini D, Sorbello F, Limonetti P, Pantoli D, Rusticali F (1995) Prodromal angina limits infarct size. A role for ischemic preconditioning. Circulation 91:291–297
8. Benerjee A, Locke-Winter C, Rogers KB, Mitchell MB, Brew EC, Cairns CB, Bensard DD, Harken AH (1993) Preconditioning against myocardial dysfunction after ischemia and reperfusion by an α_1-adrenergic mechanism. Circ Res 73:656–670
9. Downey JM, Cohen MV, Ytrehus K, Liu Y (1994) Cellular mechanisms in ischemic preconditioning: the role of adenosine and protein kinase C. Ann NY Acad Sci 723:82–98
10. Liu GS, Thornton J, Van Winkle DM, Stanley AW, Olsson RA, Downey JM (1991) Protection against infarction afforded by preconditioning is mediated by A1 adenosine receptors in rabbit heart. Circulation 84:350–356
11. Kuzuya T, Hoshida S, Yamashita N, Fuji H, Oe H, Hori M, Kamada T, Tada M (1993) Delayed effects of sublethal ischemia on the acquisition of tolerance to ischemia. Circ Res 72:1293–1299
12. Marber MS, Latchman DS, Walker JM, Yellon DM (1993) Cardiac stress protein elevation 24 hours after brief ischemia or heat stress is associated with resistance to myocardial infarction. Circulation 88:1264–1272
13. Thornton J, Stripin S, Liu GS, Swafford A, Stanley AW, Van Winkle DM, Downey JM (1990) Inhibition of protein synthesis does not block myocardial protection afforded by preconditioning. Am J Physiol 259:H1822–H1825
14. Yellon DM, Pasini E, Cargnoni A, Marber MS, Latchman DS, Ferrari R (1992) The protective role of heart stress in the ischaemic and reperfused rabbit myocardium. J Mol Cell Cardiol 24:895–907
15. Hoshida S, Kuzuya T, Fuji H, Yamashita N, Oe H, Hori M, Suzuki K, Taniguchi N, Tada M (1993) Sublethal ischemia alters myocardial antioxidant activity in canine heart. Am J Physiol 264:H33–H39
16. Ambrosio G, Flaherty JT, Duilio C, Tritto I, Santoro G, Elia PP, Condorelli M, Chiariello M (1991) Oxygen radicals generated at reflow induce peroxidation of membrane lipids in reperfused hearts. J Clin Invest 87:2056–2066
17. Guarnieri C, Flamigni F, Caldarera CM (1980) Role of oxygen in the cellular damage induced by re-oxygenation of hypoxic heart. J Mol Cell Cardiol 12:797–808
18. Zweier JL, Flaherty JT, Weisfeldt ML (1987) Direct measurement of free radical generation following reperfusion of ischemic myocardium. Proc Natl Acad Sci USA 84:1404–1407
19. Baker JE, Felix CC, Olinger GN, Kalyanaramn B (1988) Myocardial ischemia and reperfusion: direct evidence for free radical generation by electron spin resonance spectroscopy. Proc Natl Acad Sci USA 85:2786–2789
20. Garlick PB, Davies MJ, Hearse DJ, Slater TF (1987) Direct detection of free radicals in the reperfused rat heart using electron spin resonance. Circ Res 61:757–760
21. Kuzuya T, Hoshida S, Kim Y, Nishida M, Fuji H, Kitabatake A, Tade M, Kamada T (1990) Detection of oxygen-derived free radical generation in the canine postischemic heart during late phase of reperfusion. Circ Res 66:1160–1165
22. Opie LH (1989) Reperfusion injury and its pharmacologic modification. Circulation 80:1049–1062
23. Werns SW, Shea MJ, Driscoll EM, Cohen C, Abrams GD, Pitt B, Lucchesi BR (1985) The independent effects of oxygen radical scavengers on canine infarct size. Reduction by superoxide dismutase but not catalase. Circ Res 56:895–898
24. Ambrosio G, Becker LC, Hutchins GM, Weisman HF, Weisfeldt ML (1986) Reduction in experimental infarct size by recombinant human superoxide dismutase: insights into the pathophysiology of reperfusion injury. Circulation 74:1424–1433
25. Przyklenk K, Kloner RA (1989) "Reperfusion injury" by oxygen-derived free radicals? Effect of superoxide dismutase plus catalase, given at the time of reperfusion, on myocardial infarct size, contractile function, coronary microvasculature, and regional myocardial blood flow. Circ Res 64:86–96

26. Jolly SR, Kane WJ, Bailie MB, Abrams GD, Lucchesi BR (1984) Canine myocardial reperfusion injury. Its reduction by the combined administration of superoxide dismutase and catalase. Circ Res 54:277–285
27. Gallagher KP, Buda AJ, Pace D, Gerren RA, Shlafer M (1986) Failure of superoxide dismutase and catalase to alter size of infarction in conscious dogs after 3 hours of occlusion followed by reperfusion. Circulation 73:1065–1076
28. Richard VJ, Murry CE, Jennings RB, Reimer KA (1988) Therapy to reduce free radicals during early reperfusion does not limit the size of myocardial infarcts caused by 90 minutes of ischemia in dogs. Circulation 78:473–480
29. Uraizee A, Reimer KA, Murry CE, Jennings RB (1987) Failure of superoxide dismutase to limit size of myocardial infarction after 40 minutes of ischemia and 4 days of reperfusion in dogs. Circulation 75:1237–1248
30. Watanabe N, Inoue M, Morino Y (1989) Inhibition of postischemic reperfusion arrhythmias by an SOD derivative that circulates bound to albumin with prolonged in vivo half-life. Biochem Pharmacol 38:3477–3483
31. Inoue M, Ebashi I, Watanabe N, Morino Y (1989) Synthesis of a superoxide dismutase derivative that circulates bound to albumin and accumulates in tissues whose pH is decreased. Biochemistry 28:6619–6624
32. Knowlton AA, Brecher P, Apstein CS (1991) Rapid expression of heat shock protein in the rabbit after brief cardiac ischemia. J Clin Invest 87:139–147
33. Yamashita N, Nishida M, Hoshida S, Kuzuya T, Hori M, Taniguchi N, Kamada T, Tada M (1994) Induction of manganese superoxide dismutase in rat cardiac myocytes increases tolerance to hypoxia 24 hours after preconditioning. J Clin Invest 94:2193–2199
34. Currie RW, Karmazyn M, Kloc M, Mailer K (1988) Heat-shock response is associated with enhanced postischemic ventricular recovery. Circ Res 63:543–549
35. Waspe LE, Ordahl CP, Simpson PC (1990) The cardiac beta-myosin heavy chain isogene is induced selectively in alpha$_1$-adrenergic receptor-stimulated hypertrophy of cultured rat heart myocytes. J Clin Invest 85:1206–1214
36. Kitakaze M, Hori M, Morioka T, Minamino T, Takashima S, Sato H, Shinozaki Y, Chujo M, Mori H, Inoue M, Kamada T (1994) Alpha$_1$-adrenoceptor activation mediates the infarct size-limiting effect of ischemic preconditioning through augmentation of 5′-nucleotidase activity. J Clin Invest 93:2197–2205
37. Gross GJ (1995) ATP-sensitive potassium channels and myocardial preconditioning. Basic Res Cardiol 90:85–88
38. Tsuchida A, Liu Y, Liu GS, Cohen MV, Downey JM (1994) Alpha$_1$-adrenergic agonists precondition rabbit ischemic myocardium independent of adenosine by direct activation of protein kinase C. Circ Res 75:576–585
39. Takuwa Y, Yanagisawa M, Takuwa N, Masaki T (1989) Endothelin, its diverse biological activities and mechanisms of action. Prog Growth Factor Res 1:195–206
40. Menon V, Yang J, Ku Z, Thomason DB (1995) Decrease in heart peptide initiation during head-down tilt may be modulated by HSP-70. Am J Physiol 268:C1375–C1380

Remodeling: How Vessels Narrow

Stephen M. Schwartz

Introduction

Vascular narrowing is the fundamental defect characterizing atherosclerosis and hypertension. This loss of physiological caliber occurs despite the remarkable ability of normal vessels to maintain the caliber appropriate for the blood flow needed for the subservient organ. Indeed, the ability to adjust caliber must be very primitive. The vessel tree, like any closed circuit, can only continue to conduct blood if each branch has an appropriate resistance to flow. Flow, moreover, is proportional to the fourth power of the radius. Thus, the normal vessel wall must and does have an exquisite mechanism for controlling lumen size.

This maintenance of a normal caliber in the presence of changes in wall mass has been called remodeling by vascular biologists studying normal blood flow and the adaptation to hypertension. The terminology has become confusing, because "remodeling" has also been used by cardiologists to explain the loss of lumen following angioplasty, that is, restenosis despite lack of evidence that the loss can be accounted for by an increase in intimal mass [1,2]. As we discuss here, the problem of restenosis may well turn out to be a problem of remodeling or perhaps a result of healing of the wounds made in the vessel wall by overdilation during angioplasty [3].

Remodeling may also be a fundamental issue in atherosclerotic progression. As atherosclerotic lesions enlarge, the wall accomodates. The result is the normalization of wall tension and, especially, the maintenance of lumen size, as described by Glagov et al. [4] and by Clarkson et al. [5]. Glagov suggested that at some point the ability to remodel is overwhelmed and further lesion growth narrows the lumen. Clarkson, however, found that there was no limit, i.e., that lumen size is not a simple function of lesion size. The same concepts may apply to restenosis. On the right side of Fig. 1 we see three possible results following angioplasty of the narrowed vessel. The gain in lumen size may be lost because the cracks produced by the angioplasty heal, or the gain might be lost because of formation of neointima, or the wall itself may remodel. Evidence for the neointimal hypothesis, that is, loss of mass, is surprisingly sparse. Ultrasound studies in human restenotic vessels failed to show a correlation of wall mass with loss of lumen size [2]. The absence of evidence that intimal hyperplasia

Department of Pathology, University of Washington, Seattle, WA 98195-7335, USA

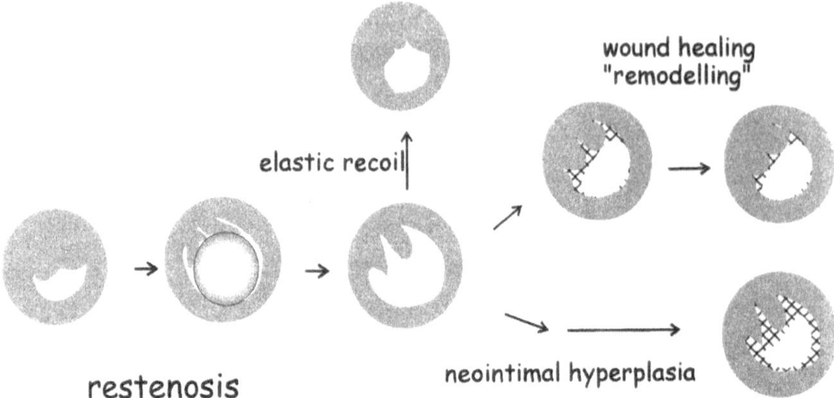

FIG. 1. Restenosis can, in theory, occur in three ways. Early on, the vessel may recoil from dilation; this is called "elastic recoil." Later, narrowing can occur by one of two mechanisms. Most of the discussion, until recently, has focused on loss of lumen because of formation of new intimal mass, i.e., "neointimal hyperplasia." Cell kinetic studies and intravascular ultrasound (IVUS) data, however, suggest that this is not the usual case. The alternative, as shown here, is that the dilation results from fissures formed by the balloon and that loss of gain may simply occur by wound healing. This is often called "remodeling"; however, as we discussed, the term remodeling should be used for the normal physiological property of adapting to demands for flow and response to changes in intraluminal pressure. An important issue in the wound healing mechanism is that it represents a loss of the initial gain produced by the balloon, not a further stenosis beyond that existing at the time of angioplasty

contributes to a lumen of size in restenosis after conventional angioplasty contrasts with studies showing true loss of intima as the result of formation of new mass inside a stent [6] (Fig. 2).

Origins of Blood Vessels

To understand remodeling, we need to begin by understanding vascular development. In the developing embryo, aggregates of endothelial cells line up as primitive vascular tubes form in positions ready to be connected to create a primitive vascular circuit [7,8]. The primitive tubes join each other and extend new vessels by a process that is very similar to "angiogenesis" seen in adult animals as a response to tumors or inflammation or in wound healing. Thus, vascularization of new organs occurs by formation of new branches that extend from the existing vascular network. The branding pattern, however, is itself genetic. This is apparent from genetic studies identifying genes associated with specific abnormalities in the layout of the vascular circuit [9] as well as from studies of animals manipulated by adding a transgene that alters organ size by means of a growth hormone [10]. In this experiment, the number of branches of the mesenteric tree, down to the level of vessels penetrating the gut wall itself, was preserved despite an approximately twofold increase in size of the perfused organs and, presumably, in perfusion volume. The numbers of layers of medial

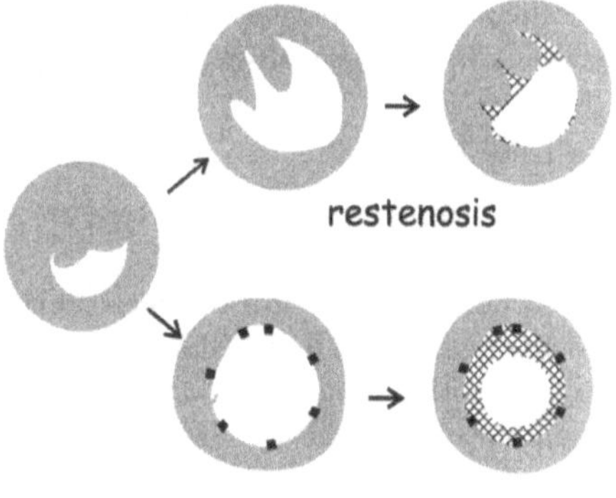

FIG. 2. Unlike "restenosis" after angioplasty, stents make a permanent new lumen. This lumen can be lost only by formation of new tissue inside the stent, a process we suggest should be called "stent stenosis"

smooth muscle cells was also normal, but lumen size and wall thickness were increased, presumably to accommodate the increase in flow and the increase in wall tension. Thus, endothelium sets the vascular pattern while smooth muscle may control the adaptation resulting from the increased diameter of the vessel to rearrangement in flow.

Smooth muscle cells arise in a quite different manner than endothelial cells: the primitive endothelial tubes somehow recruit smooth muscle coats to differentiate from the surrounding mesenchyme shortly after the dorsal aortas appear as recognizable tubes. The origins of the cells making up these coats of vascular smooth muscle remain mysterious. Because primitive endothelial cell precursors migrate widely over the embryo, an individual endothelial cell in a caudad vessel or a craniad vessel may have its origin at the opposite end of the embryo. Thus, smooth muscle cells appear to arise locally; indeed, some vascular smooth muscle cells in the head and thoracic regions have been described as coming from "mesectoderm," that is, from the mesoderm associated with the neural crest [11; Topouzis S, Smith-Anderson M, Majesky MW (1995) Smooth muscle cells of neural crest versus mesoderm lineage differ in growth and receptor-mediated signaling responses to TGF-β (in manuscript)]. Moreover, studies of vessels developing inside organs imply that endothelial cells are derived from invading vessels, while the smooth muscle cells may be derived from within the organ primordium. Thus, embryologically, smooth muscle cells from different vascular beds may well be quite different, while endothelial cells may all share a common origin [12].

This variation in smooth muscle cell origins is of special interest in relationship to the origin of elastin. Layers of elastin delineate the atherosclerosis-prone intima from the rest of the vessel wall. Studies in avian embryos as well as in adult rats responding to balloon catheter injury indicate that the vessel wall may contain a unique subset of "elastogenic" smooth muscle cells. In the bird, these cells appear to be derived from mesectoderm, and in balloon injury, similar cells selectively proliferate to form the neointima seen after injury [11,13,14] (Topouzis et al., in manuscript). It is intriguing to consider the possibility that cells which comprise the intima have different origins and different properties from the smooth muscle cells of the underlying media. Interestingly, the cells of cardiac cushions include a subset derived from the overlying endothelium [15], and there is limited evidence that the normal intima also contains a subset of dendritic cells with unique properties [16].

Role of Vascular Layers in Smooth Muscle Responses to Injury

We know nothing of the relative contributions of intimal versus medial cells to vascular narrowing. Redistribution of smooth muscle cells to form a "neointima" may occur by proliferation of smooth muscle cells already trapped in the intima after formation of the internal elastic lamina or by migration of cells from the media. However, convincing evidence suggests that medial mass is a function of endothelial products produced in response to changes in blood flow [12]. While there is clear evidence that hypertensive arteries are thickened, stereologic studies show that this can occur without an increase in mass or cell number. Apparently, the components of the wall are somehow redistributed. This poorly understood phenomenon, by changing the wall/lumen ratio, may elevate blood pressure because contractile stimuli now operate at a greater mechanical advantage [18]. This morphological change, the redistribution of vessel wall components to adapt to pathological or physiological changes in blood flow, is called remodeling [18–21].

Much remains to be learned about the relative role of intimal and medial changes in remodeling. To date we only know that the media remodels. We do not know if the presence of an intima enhances or inhibits this response, nor do we know whether the intima itself can remodel.

A Model for Neointimal Formation

Formation of new intima, that is, neointimal hyperplasia, is an extremely common pathological response to many stimuli, including normal aging and hypertension as well as a wide variety of chemical, physical, and biological forms of tissue injury [22–26]. As already discussed, however, it may be incorrect to assume that neointimal formation after injury occurs via the same process as occurs during development of the normal intima seen in atherosclerosis-prone sites before formation of frank lesions [27–29]. Neointimal formation after injury has been extensively studied in vessels like the rat carotid artery that normally have little or no spontaneously formed intima. This has become a standard model for understanding vascular responses to injury. Because there is no intima to begin with, it is relatively easy to analyze formation of the neointima as a new layer, i.e., the formation of a layer of connective tissue cells between the endothelium and the intimal elastic laminae. The injury to rat carotid artery produced by a balloon catheter includes complete removal of the endothelium, adhesion of a monolayer of platelets, and activation of thrombin [30], as well as damage to inner smooth muscle layers of the tunica media [31]. Of most importance, migration of smooth muscle cells from the underlying intact media to form a neointima, and proliferation of the newly arrived "neointimal" cells following angioplasty in experimental animals, have been proposed as the initiating series of events in the pathogenesis of atherosclerosis [32–34].

A number of growth factors have been shown to stimulate smooth muscle cell growth in vitro. It is important to note that these include three categories of molecules: molecules produced by the vessel wall cells themselves, molecules derived from plasma proteins, and molecules only present in the vessel wall following platelet or leukocyte adhesion. At one time there was a consensus that the initial replication of medial smooth muscle cells was caused by factors released from platelets [33]; this now seems unlikely. Platelets adhere to the vessel wall immediately after injury, but are mostly gone by 24h after injury [35]. Similarly, thrombin generation is greatly decreased after 48h [30]. After that time, the surface is relatively nonthrombogenic and smooth muscle replication falls, although remaining above normal levels. However, when animals are made thrombocytopenic by pretreatment with antiplatelet antiserum, this initial smooth muscle replication still occurs [36]. Finally, infused platelet-derived growth factor (PDGF) is at best a weak mitogen and anti-PDGF antibodies do not alter the proliferative response [37,38]. There is also little evidence, at least in the rat, of interaction of the acutely injured wall with leukocytes. Thus, it is likely that the first wave of smooth muscle proliferation is the result of growth factors derived not from platelets or leukocytes but directly from the injured vessel wall [36,39].

The most important of vessel wall mitogens may be molecules produced by the vessel wall cells themselves. Lindner and Reidy have used antibodies to fibroblast growth factor (FGF) and recombinant FGF to study this early phase of the lesion. FGF is mitogenic in vivo and is present in the normal wall, and the initial replicative response is blocked by the antibody. It seems likely, at least in the carotid model, that the "first wave" of DNA synthesis is the result of the release of endogenous FGF from traumatized smooth muscle cells [40]. Thus, FGF seems to be a sufficient but not necessarily exclusive explanation for medial smooth muscle replication after balloon injury. We specify that it is not exclusive because there is compelling evidence that angiotensin is also an endogenous mitogen. Antagonists to the AT1 angiotensin receptor block the first wave [41], and infused angiotensin is mitogenic for the intact medial smooth muscle cells (Su, Reidy, and Schwartz, unpublished data).

This first wave of replication in the media is important for formation of neointima because antisense growth inhibitors directed at the first wave also inhibit the extent of intimal thickening [42]. However, we do not know why medial cell proliferation should alter the extent of intimal accumulation. Somehow, effective inhibitors of medial replication also block the cell migration needed to move medial cells into the intima. This is surprising because the process of smooth muscle cell migration appears to be separable from cell proliferation per se. We found that cells which have not undergone DNA replication are found among those that have migrated into the intima [43].

Although platelets are not responsible for the initial proliferative response, they may play a critical role in later stages of neointimal formation. A number of molecules in the platelet, including PDGF, are chemotactic for smooth muscle cells [44]. As already noted, thrombocytopenic animals show a normal initial proliferation of smooth muscle cells in the media. However, if the thrombocytopenia is maintained, there is a decrease in the final extent of intimal thickening [36]. The obvious caveats include the possibility of side effects of maintaining injections of antiplatelet antiserum for several days, as well as the side effects of thrombocytopenia itself. Nonetheless, the possibility that platelets are critical to the migration stage of neointima formation is an intriguing modification of the existing platelet hypothesis [33,36].

Once formed, the neointima continues to grow in size by cell replication. A relatively high percentage of intimal smooth muscle cell nuclei continue to replicate for weeks, as indicated by pulse labeling with [^{3}H]thymidine. The ongoing replication is particularly evident at the luminal surface [45]. The mechanism controlling this chronic proliferation is not known. It also appears that the neointima may show an exaggerated proliferative response to physiological stimuli. When animals were infused chronically with angiotensin II, the neointima produced a three- to fivefold increase in replication as compared with the underlying vessel wall [46]. This is particularly intriguing in view of the use of converting enzyme inhibitors to treat hypertensive patients undergoing angioplasty and in view of a report that converting enzyme inhibitors decrease the formation of neointima after balloon injury [47]. The chronic enhanced proliferation of neointima may reflect endogenous synthesis of mitogens. This concept is supported by in vitro studies showing that smooth muscle cells from the neointima have unique properties also seen in smooth muscle cells from the tunica media of vessels in the newborn rat.

Relationship of Neointimal Formation to Formation of Normal Intima Atherosclerotic Lesions

We know that human atherosclerotic lesions are monoclonal [48]. This makes it unlikely that formation of neointima is a model for normal growth of the intima. The various explanations of monoclonality, i.e., existence of a proliferative subset or benign transformation of plaque smooth muscle cells, both imply unique properties of the plaque smooth muscle cell that would become prime targets in understanding the ontogeny of this most important vascular disease. Lesions used to study atherosclerosis in animals, however, do not appear to be polyclonal, although some mosaic hares did show a clonal drift following fat feeding [49–51]. Interpretation of this sort of data is confusing because the predominant experimental lesions are macrophage rich.

Monoclonality of atherosclerotic lesions adds evidence to the suggestion that the intima is composed of a unique population of cells. Velican and Velican suggested that sites of early intimal formation correlate highly with sites of atherosclerotic plaque development [27,28]. As the intima may originate from small numbers of smooth muscle cells trapped during development, such preexisting clones, if they exist, could explain the clonality of plaques without requiring transformation to a benign neoplasm or expansion of a proliferogenic subset. The implication for the current discussion, however, is that intimal cells may well have unique properties which may alter the way the vessel responds, especially in regard to remodeling.

Unique Properties of Neointima

Perhaps the most critical property of the neointima may be an as-yet-unexplained propensity of the intima, as compared with the media, to stimulate thrombosis. Jorgensen et al. [35] found that balloon catheter injury to normal artery results in rapid deposition of a platelet monolayer and leukocyte adherence to the damaged surface, but little penetration of leukocytes into the wall and a general absence of fibrin formation. A second mechanical injury produces a markedly different response. Platelet aggregates and fibrin-rich thrombi are readily found, and adherent leukocytes frequently penetrate into the damaged neointima. The molecular basis for this change in response is not known. It is tempting, however, to consider the possibility that the procoagulant, prothrombic properties of the neointima are related to the vascular occlusive processes seen at sites of atherosclerosis.

Taubman [52], Fuster et al. [53–55], and others [56,57] have suggested this procoagulant activity is caused by tissue factor, a hypothesis upheld by our observation that plaque smooth muscle cells, as well as other types of intimal smooth muscle cells, express tissue factor [58]. Until recently, most investigators had assumed that the normal wall contained tissue factor, and that simple denudation of the endothelium was sufficient to initiate coagulation. Because this activating factor does not have a pro-form, this view was apparently incorrect. First, numerous studies of balloon-injured vessels have failed to show fibrin formation unless the vessel was being injured a second time or multiple times [35,45,59]. Second, when antibodies and cDNAs to tissue factor became available, both immunocytochemistry and in situ hybridization of normal arteries failed to show tissue factor except in the adventitia. In contrast, the atherosclerotic plaque contains tissue factor as both an RNA and a protein [58,60]. The procoagulant is associated with two cell types: intimal smooth muscle cells and plaque macrophages. Whether similar induction of tissue factor occurs following angioplasty is not known.

Neointimal tissue factor could be important for remodeling or response to injury in two ways. First, intimal fibrin might be important for wound healing and contracture. Jay Degen of the University of Cincinnati has found that epithelial regeneration in fibrogen-knockout animals occurs but the cells move in disorderly patterns (personal communication). Similar contributions of intramural fibrin in wound healing following vascular injury could have profound effects on the remodeling required to maintain the vessel lumen or to produce a loss of the gain caused when the vessel wall is fractured during angioplasty [61]. Second, the apo(a)-overexpressing mouse surprisingly accumulates lipid in its intima [62]. Because apo(a) is a plasminogen inhibitor, these data suggest that intimal coagulation is somehow related to lipid accumulation factor.

Properties of Neointimal Cells In Vitro

As already noted, restenosis after angioplasty cannot be attributed to an increase in new mass. Even if the *amount* of neointima does not explain the size of the lumen, it is possible that the *qualities* of the intima may be important. The cells of the neointima may have special properties that may contribute to atherosclerosis: one such property is production of tissue.

There are many reasons to wonder about the unique properties of smooth muscle subsets. As we have already discussed, atherosclerotic plaques are monoclonal—does this reflect an origin from a cell with special properties? Surprisingly little has been written about cultured intimal cells and almost nothing about nonatherosclerotic intimal cells in vivo [63]. We have, however, made an analogy between neointima cultured from the rat and cells revived from arteries of 2-week-old rat pups. Smooth muscle cells from 2-week-old rats in culture are quite distinct from cells from 3-month-old animals [14,64–66]. Walker et al. [67] moreover found that neointimal smooth muscle cells in culture grew as a monolayer of epithelioid-shaped cells. Both the newborn "pup" cells and the neointimal cells could be grown in medium prepared without platelet release, implying a distinct mechanism of growth control as compared with the usual adult smooth muscle cells.

Differences characteristic of pup cells include secretion of PDGF, synthesis of PDGF-B mRNA, synthesis of a cytochrome P-450, and lack of expression of the PDGF-α receptor [63,65,68]. Constitutive expression of elastin and osteopontin is also characteristic of the pup cells in vitro, consistent with a high level of expression of these genes in the neointima in vivo. We have recently shown that a subset of neointimal smooth muscle cells in the rat aortic injury model express PDGF-B. It is intriguing to posit that these PDGF-B-expressing cells in vivo are the predecessors of the PDGF-positive cells seen when neointimal cells are placed in culture [69]. It is important to point out again that differences in morphology, growth factor production, and gene expression in different smooth muscle cell isolates described here are stable over many cell generations (>20) in vitro. Thus, at least for the rat, it is likely that the normal artery wall contains at least two subsets of smooth muscle cells.

Limited evidence also supports the idea of unique subsets in human intimal smooth muscle. We have known for 20 years that human smooth muscle cells have a short replicative life span in vitro [70]. Recently we showed that plaque smooth muscle cells also have a very high rate of spontaneous apoptosis [71], implying that the short replicative lifespan may be caused by loss of generations in vitro to cell death. Because apoptosis is genetically controlled [71], we assume that these data imply an as-yet-unidentified genetic difference between plaque and normal medial cells.

Atherosclerotic Plaque Progression and Restenosis

While a lot has been written about restenosis after angioplasty, surprisingly, one can dilate a stenotic atherosclerotic vessel with a balloon angioplasty catheter and the vessel will remain dilated 60%–70% of the time. As already reviewed, pathologists have long known that atherosclerotic vessels dilate. As people grow older, their coronary arteries have larger lumens—a process that continues even as atherosclerotic lesions progress [4]. We might consider this dilatation as normal remodeling. In the same sense, loss of lumen as a plaque grows is a failure of normal remodeling. Glagov suggested that this failure resulted when the plaque mass exceeded some critical value [4]. Alternatively, failure to remodel could reflect the special properties of the plaque or its smooth muscle cells.

This should be regarded as a phenomenal success, especially if neointimal formation is occurring. We suggest here that restenosis is actually a combination of two normal processes: remodeling and wound healing. Given the known formation of neointima after injury of normal vessels, the simplest explanation for the loss of gain seen after angioplasty in about one-third of the time is neointimal hyperplasia. This hypothesis presents two problems. First, as just discussed, under normal circumstances a vessel will dilate as intimal mass increases. Indeed, Jamal et al. [72] showed a 100% increase in wall mass following angioplasty in rabbit vessels with no loss of lumen. Thus, changes in mass without a failure of normal remodeling need not narrow a vessel. Second, and perhaps most importantly, our cell kinetic studies failed to show a replicative wave to atherectomy specimens taken at a wide range of times after an initial angioplasty [73]. Analysis of a series of autopsy specimens seems to confirm these cell kinetic data. In collaboration with Nobuyushi, we found no correlation between the intimal mass on total wall mass and the extent of lumen size (O'Brien et al., unpublished data). Similarly, preliminary studies using intravascular ultrasound showed no correlation of wall mass with restenosis [74]. These data at first seem to be at variance with the behavior of the rat vessel subjected to balloon injury. However, the difference may simply point out that the rat system is an example of how a neointima is formed, while the human data represent a loss of lumen size. There is, to our knowledge, no evidence that angioplasty causes a decrease in lumen size from that existing before injury.

While restenosis may not be simply the result of the mass of neointima, we need to consider how else the intima might cause vessels to lose their normal ability to maintain a large lumen despite growth of an atherosclerotic plaque [4]. It is intriguing to consider the possibility that intima-specific genes could be responsible for an abnormal remodeling to produce a small lumen. Similarly, there is ample evidence that advanced atherosclerotic lesions are characterized by plaque breakdown and fissures [75,76]. It is conceivable that wound healing, an effort to heal these injuries, produces narrowing as a result of loss of the normal ability of the atherosclerotic vessel to remodel.

References

1. O'Brien ER, Alpers CE, Stewart DK, Ferguson M, Tran N, Gordon D, Benditt EP, Hinohara T, Simpson JB, Schwartz SM (1993) Proliferation in primary and restenotic coronary atherectomy tissue: implications for antiproliferative therapy. Circ Res 73:223–231
2. Mintz GS, Douek PC, Bonner RF, Kent KM, Pichard AD, Satier LF, Leon MB (1993) Intravascular ultrasound comparison of de novo and restenotic coronary artery lesions (abstract). J Am Coll Cardiol 21:118A
3. Schwartz SM, Murry CD, O'Brien ER (1996) Vessel wall response to injury. Sci Am Sci Med Mar/Apr: 12–21
4. Glagov S, Weisenberg E, Zarins CK, Stankunavicius R, Kolettis GJ (1987) Compensatory enlargement of human atherosclerotic coronary arteries. N Engl J Med 316:1371–1375
5. Clarkson TB, Prichard RW, Morgan TM, Petrick GS, Klein KP (1994) Remodeling of coronary arteries in human and nonhuman primates. JAMA 271(4):289–294
6. Dussaillant GR, Mintz GS, Pichard AD, Kent KM, Satler LF, Popma JJ, Wong SC, Leon MB (1995) Small stent size and intimal hyperplasia contribute to restenosis: a volumetric intravascular ultrasound analysis. J Am Coll Cardiol 26(3):720–724
7. Noden DM (1989) Embryonic origins and assembly of blood vessels. Am Rev Respir Dis 140:1097–1103
8. Coffin JD, Harrison J, Schwartz SM, Heimark R (1991) Angioblast differentiation and morphogenesis of the vascular endothelium in the mouse embryo. Dev Biol 148: 51–62
9. Fishman MC, Stainier DY (1994) Cardiovascular development. Prospects for a genetic approach. Circ Res 74(5):757–763
10. Dilley RJ, Schwartz SM (1989) Vascular remodeling in the growth hormone transgenic mouse. Circ Res 65:1223–1240
11. Majesky MW, Topouzis S (1995) Smooth muscle lineage diversity and atherosclerosis. Atherosclerosis 10:56
12. Pardanaud L, Yassine F, Dieterlen-Lievre F (1989) Relationship between vasculogenesis, angiogenesis and haemopoiesis during avian ontogeny. Development (Camb) 105:473–485

13. Rosenquist TH, McCoy JR, Waldo KL, Kirby ML (1988) Origin and propagation of elastogenesis in the developing cardiovascular system. Anat Rec 221:860–871
14. Majesky MW, Giachelli CM, Schwartz SM (1992) Rat carotid neointimal smooth muscle cells re-express a developmentally regulated phenotype during repair of arterial injury. Circ Res 71:759–768
15. Wunsch AM, Little CD, Markwald RR (1994) Cardiac endothelial heterogeneity defines valvular development as demonstrated by the diverse expression of JB3, an antigen of the endocardial cushion tissue. Dev Biol 165:585–601
16. Bobryshev YV, Lord RS (1995) Ultrastructural recognition of cells with dendritic cell morphology in human aortic intima. Contacting interactions of vascular dendritic cells in athero-resistant and athero-prone areas of the normal aorta. Arch Histol Cytol 58(3):307–322
17. Langille BL, O'Donnell F (1986) Reductions in arterial diameter produced by chronic diseases in blood flow are endothelium-dependent. Science 231:405–407
18. Folkow B (1982) Physiological aspects of primary hypertension. Physiol Rev 62: 347–504
19. Heagerty AM, Aalkjaer C, Bund SJ, Korsgaard N, Mulvany MJ (1993) Small artery structure in hypertension. Dual processes of remodeling and growth. Hypertension 21:391–397
20. Alkjaer C, Heagerty AM, Mulvany MJ (1987) In vitro characteristics of vessels from patients with essential hypertension. J Clin Hypertens 3:317–322
21. Korsgaard N, Aalkjaer C, Heagerty AM, Izzard AS, Mulvany MJ (1993) Histology of subcutaneous small arteries from patients with essential hypertension. Hypertension (Dallas) 22:523–526
22. Albelda SM, Oliver PD, Romer LH, Buck CA, Endo CAM (1990) A novel endothelial cell-cell adhesion molecule. J Cell Biol 110:1227–1237
23. Gage AA, Fazekas C, Riley EE (1967) Freezing injury to large blood vessels in dogs. Surgery (St Louis) 61:748–754
24. Bjorkerud S, Bondjers G (1971) Arterial repair and atherosclerosis after mechanical injury. Part 2. Tissue response after induction of a large superficial necrosis (deep longitudinal injury). Atherosclerosis 14:259–276
25. Poole JCF, Cromwell SB, Benditt EP (1971) Behavior of smooth muscle cells and formation of extracellular structures in the reaction of arterial walls to injury. Am J Pathol 62:391–414
26. Haudenschild CC, Prescott MF, Chobanian AV (1980) Effects of hypertension and its reversal on aortic lesions of the rat. Hypertension (Dallas) 2:33–44
27. Velican D, Velican C (1976) Intimal thickening in developing coronary arteries and its relevance to atherosclerotic involvement. Atherosclerosis 23:345
28. Velican C, Velican D (1980) The precursors of coronary atherosclerotic plaques in subjects up to 40 years old. Atherosclerosis 37:33–46
29. Sims FH, Gavin JB, Vanderwee MA (1989) The intima of human coronary arteries. Am Heart J 118:32–38
30. Hatton MW, Moar SL, Richardson M (1989) De-endothelialization in vivo initiates a thrombogenic reaction at the rabbit aorta surface. Am J Pathol 135:499–508
31. Clowes AW, Clowes MM, Reidy MA (1983) Kinetics of cellular proliferation after arterial injury. I. Smooth muscle growth in the absence of endothelium. Lab Invest 49:327–333
32. Schwartz SM, Heimark RL, Majesky MW (1990) Developmental mechanisms underlying pathology of arteries. Physiol Rev 70:1177–1209
33. Ross R (1986) The pathogenesis of atherosclerosis—an update. N Engl J Med 314:488–500
34. Ross R (1993) The pathogenesis of atherosclerosis: a perspective for the 1990s. Nature 362:801–809
35. Jorgensen L, Grothe AG, Groves HM, Kinlough-Rathbone RL, Richardson M, Mustard JF (1988) Sequence of cellular responses in rabbit aortas following one and two injuries with a balloon catheter. Br J Exp Pathol 69:473–486

36. Fingerle J, Johnson R, Clowes AW, Majesky MW, Reidy MA (1989) Role of platelets in smooth muscle cell proliferation and migration after vascular injury in rat carotid artery. Proc Natl Acad Sci USA 86:8412–8416
37. Jackson CL, Raines EW, Ross R, Reidy MA (1993) Role of endogenous platelet-derived growth factor in arterial smooth muscle cell migration after balloon catheter injury. Arterioscler Thromb 13:1218–1226
38. Jawien A, Bowen-Pope DF, Lindner V, Schwartz SM, Clowes AW (1992) Platelet-derived growth factor promotes smooth muscle migration and intimal thickening in a rat model of balloon angioplasty. J Clin Invest 89:507–511
39. Majesky MW, Reidy MA, Bowen-Pope DF, Wilcox JN, Schwartz SM (1990) Platelet-derived growth factor (PDGF) ligand and receptor gene expression during repair of arterial injury. J Cell Biol 111:2149–2158
40. Lindner V, Reidy MA (1991) Proliferation of smooth muscle cells after vascular injury is inhibited by an antibody against basic fibroblast growth factor. Proc Natl Acad Sci USA 88:3739–3743
41. Prescott M, Webb R, Reidy MA (1991) Angiotensin-converting enzyme inhibitors versus angiotensin II, AT1 receptor antagonist: effects on smooth muscle cell migration and proliferation after balloon catheter injury. Am J Pathol 139:1291–1296
42. Bennett MR, Schwartz SM (1995) Antisense therapy for angioplasty restenosis—some critical considerations. Circulation 92:1981–1993
43. Clowes AW, Schwartz SM (1985) Significance of quiescent smooth muscle migration in the injured rat carotid artery. Circ Res 56:139–145
44. Grotendorst GR, Seppa HEJ, Kleinman HK, Martin GR (1981) Attachment of smooth muscle cells to collagen and their migration toward platelet-derived growth factor. Proc Natl Acad Sci USA 78:3669–3672
45. Schwartz SM, Stemerman MB, Benditt EP (1975) The aortic intima. II. Repair of the aortic lining after mechanical denudation. Am J Pathol 81:15–42
46. Daemen MJAP, Lombardi DM, Bosman FT, Schwartz SM (1991) Angiotensin II induces smooth muscle cell proliferation in the normal and injured arterial wall. Circ Res 68:450–456
47. Powell J, Clozel J, Muller R (1989) Inhibitors of angiotensin-converting enzyme prevent myointimal proliferation after vascular injury. Science 245:186–188
48. Benditt EP, Benditt JM (1973) Evidence for a monoclonal origin of human atherosclerotic plaques. Proc Natl Acad Sci USA 70:1753–1756
49. Lee KT, Thomas WA, Florentin RA, Reiner JM, Lee WM (1976) Evidence for a polyclonal origin and proliferation heterogeneity of atherosclerotic lesions induced by dietary cholesterol in swine. Ann NY Acad Sci 275:336–347
50. Thomas WA, Kim DN (1983) Biology of disease: atherosclerosis as a hyperplastic and/or neoplastic process. Lab Invest 48:245–255
51. Janakidevi K, Lee KT, Kroms M, Imai H, Thomas WA (1984) Mosaicism in female hybrid hares heterozygous for glucose-6-phosphate-dehydrogenase. VI. Production of monotypism in the arotas of four of 10 mosaic hares fed cholesterol oxidation products. Exp Mol Pathol 41:354–362
52. Taubman MB (1993) tissue factor regulation in vascular smooth muscle: a summary of studies performed using in vivo and in vitro models. Am J Cardiol 72:55C–60C
53. Fuster V, Badimon JJ, Badimon L (1992) Clinical-pathological correlations of coronary disease progression and regression. Circulation 86:III1–III11
54. Fuster V, Badimon L, Badimon J, Chesebro JH (1992) The pathogenesis of coronary artery disease and the acute coronary syndromes (1). N Engl J Med 326: 242–250
55. Fuster V, Badimon L, Badimon J, Chesebro JH (1992) The pathogenesis of coronary artery disease and the acute coronary syndromes (2). N Engl J Med 326: 310–318
56. Landers SC, Gupta M, Lewis JC (1994) Ultrastructural localization of tissue factor on monocyte-derived macrophages and macrophage foam cells associated with atherosclerotic lesions. Virchows Arch 425:49–54

57. Brand K, Banka CL, Mackman N, Terkeltaub RA, Fan ST, Curtiss LK (1994) Oxidized LDL enhances lipopolysaccharide-induced tissue factor expression in human adherent monocytes. Arterioscler Thromb 14:790–797
58. Wilcox JN, Smith KM, Schwartz SM, Gordon D (1989) Localization of tissue factor in the normal vessel wall and in the atherosclerotic plaque. Proc Natl Acad Sci USA 86:2839–2843
59. Reidy MA (1985) A reassessment of endothelial injury and arterial lesion formation. Lab Invest 53:513–520
60. Drake TA, Morrissey JH, Edgington TS (1989) Selective cellular expression of tissue factor in human tissues. Am J Pathol 134:1087–1097
61. Kuntz RE, Gibson M, Nobuyoshi M, Baim DS (1993) Generalized model of restenosis after conventional balloon angioplasty stenting and directional atherectomy. J Am Coll Cardiol 21:15–25
62. Lawn RM, Wade DP, Hammer RE, Chiesa G, Verstuyft JG, Rubin EM (1992) Atherogenesis in transgenic mice expressing human apolipoprotein(a). Nature 360: 670–672
63. Schwartz SM, deBlois D, O'Brien ER (1995) The intima: soil for atherosclerosis and restenosis. Circ Res 77:445–465
64. Majesky MW, Benditt EP, Schwartz SM (1988) Expression and developmental control of platelet-derived growth factor A-chain and B-chain/sis genes in rat aortic smooth muscle cells. Proc Natl Acad Sci USA 85:1524–1528
65. Seifert RA, Schwartz SM, Bowen-Pope DF (1984) Developmentally regulated production of platelet-derived growth factor-like molecules. Nature 311:669–671
66. Lemire JM, Covin CW, White S, Giachelli CM, Schwartz SM (1994) Characterization of cloned aortic smooth muscle cells from young rats. Am J Pathol 144:1068–1081
67. Walker L, Bowen-Pope DF, Ross R, Reidy MA (1986) Production of PDGF-like molecules by cultured arterial smooth muscle cells accompanies proliferation after arterial injury. Proc Natl Acad Sci USA 83:7311–7315
68. Sjolund M, Hedin U, Sejersen R, Heldin CH, Thyberg J (1988) Arterial smooth muscle cells express PDGF-A chain mRNA, secrete a PDGF-like mitogen, and bind exogenous PDGF in a phenotype- and growth state-dependent manner. J Cell Biol 106:403–413
69. Lindner V, Reidy MA (1995) Platelet-derived growth factor ligand and receptor expression by large vessel endothelium in vivo. Am J Pathol 146:1488–1497
70. Moss NS, Benditt EP (1975) Human atherosclerotic plaque cells and leiomyoma cells: comparison of in vitro growth characteristics. Am J Pathol 78:175–190
71. Bennett MR, Evan GI, Schwartz SM (1995) Apoptosis of human vascular smooth muscle cells derived from normal vessels and coronary atherosclerotic plaques. J Clin Invest 95:2266–2274
72. Jamal A, Bendeck M, Langille BL (1992) Structural changes and recovery of function after arterial injury. Arterioscler Thromb 12:307–317
73. O'Brien ER, Alpers CE, Stewart DK, Ferguson M, Tran N, Gordon D, Benditt EP, Hinohara T, Simpson JB, Schwartz SM (1993) Proliferation in primary and restenotic coronary atherectomy tissue: implications for antiproliferative therapy. Circ Res 73:223–231
74. Mintz GS, Douek PC, Bonner RF (1993) Intravascular ultrasound comparison of de novo and restenotic coronary artery lesions. J Am Coll Cardiol 21:118A
75. Arbustini E, Grasso M, Diegoli M, Pucci A, Bramerio M, Ardissino D, Angoli L, de Servi S, Bramucci E, Mussini A (1991) Coronary atherosclerotic plaques with and without thrombus in ischemic heart syndromes: a morphologic, immunohistochemical, and biochemical study. Am J Cardiol 68:36B–50B
76. Davies MJ (1992) Anatomic features in victims of sudden coronary death. Coronary artery pathology. Circulation 85(suppl 1):119–124

Part 2
Systolic–Diastolic Interaction

Interaction Between Left-Ventricular Function and Loading Condition in Cardiovascular Disease

Yukio Maruyama, Kazuhira Maehara, Tomiyoshi Saito, Shuichi Saitoh, Masahiko Sato, and Minoru Mitsugi

Summary. In this chapter, interactions between ventricular performance, loading condition, and coronary circulation are described from the standpoint of pathogenesis of the progression or amelioration of cardiac dysfunction. A decrease in peripheral resistance increases stroke volume until a certain critical level of afterload reduction; thus, afterload-reducing therapy is frequently performed in patients with heart failure. When afterload reducing is performed in cardiac dysfunction with regional ischemia, increased stroke volume is also found, as observed in global ischemia. However, the meaning of augmented ejected volume from the heart differs between regional and global ischemia. In regional ischemia, cardiac output continues to increase after progressive afterload reduction despite the appearance of functional aggravation in the ischemic region. Thus, the afterload, at which the maximal cardiac output is obtained, may not necessarily indicate optimal loading level in this type of cardiac dysfunction. This must be considered clinically when adopting afterload reducing therapy in regional ischemia. Moreover, when intraaortic balloon pumping (IABP) is used for patients with heart failure from coronary artery disease, the aortic pressure level of the pre-IABP stage must be also considered to attain suitable cardiac pump function. Decrease in arterial compliance leads to systolic hypertension and then finally to heart failure via increased ventricular afterload during systole. Conversely, in heart failure arterial compliance seems to decrease as a result of changes in vessel wall characteristics, and cardiac dysfunction might then further decline. When decreased arterial compliance is introduced to a depressed heart, ventricular ejection is impaired significantly, compared to the heart with normal contractility; moreover, the rate of decrease of stroke volume may be different, depending on the cause of cardiac dysfunction. An increase in diastolic filling pressure is frequently found in heart failure. Elevation of left-ventricular filling pressure increases zero-flow pressure of the left coronary pressure–flow relationship without showing significant change in the slope of the coronary pressure–flow relationship. The presence of the pericardium further increases the zero-flow pressure value of coronary pressure–flow relationship, which is obtained at the same filling pressure without the pericardium. In addition, increased filling pressure decreases diastolic compliance in both the related ventricle and the ventricle of the opposite side. Accordingly, when diastolic filling pressure

---

First Department of Internal Medicine, Fukushima Medical College, Fukushima, 960-12 Japan.

increases, cardiac chamber interactions are found in chamber stiffness and also in the coronary pressure–flow relationship. As supportive therapy for myocardial ischemic injury and subsequent ischemia-induced cardiac dysfunction, synchronized retroperfusion is used as well as IABP. Synchronized retroperfusion delivers arterial blood flow during diastole and increases downstream venous pressure in the pathway of the coronary sinus; this reduces coronary driving pressure in diastole, and the coronary pressure–flow relationship is significantly shifted to the right. However, impairment of coronary pressure–flow relationship following diastolic downstream pressure elevation is much less than that in the systolic change when compared at the same mean downstream pressure elevation, indicating that the phase of downstream pressure elevation is also important in coronary inflow mechanics. From the same experimental series, it becomes apparent that coronary sinus pressure elevation via synchronized retroperfusion promotes collateral flow to the ischemic region, when concomitant with cardiac contraction. Finally there are many unclear but important interactions including these issues. These interactions play an important role in regulating cardiac function, especially in disease, and such information is therefore necessary for understanding pathophysiology or performing treatment in cardiovascular diseases.

Key words. Left-ventricular function—Afterload impedance—Preload—Coronary circulation—Intraaortic balloon pumping

Introduction

In the cardiovascular system, left-ventricular performance, loading condition, and coronary hemodynamics are coupled with each other, and interactions among those variables greatly affect cardiovascular function. In chronic heart failure especially, sympathetic stimulation continuously occurs and vasoconstrictive substances from neurohumoral compensatory adaptation are released, so that peripheral resistance is increased; this is accompanied frequently by a decreased arterial compliance and an elevation of diastolic filling pressure which is simultaneously induced. In addition, the coronary vascular bed is embedded in the myocardium, so that the extravascular systolic compressive force following the elevated loading state as well as that in diastole modifies coronary vascular characteristics. Moreover, it is conceivable that an elevation of coronary venous back pressure via coronary sinus or Thebesian vessels caused by an increased filling pressure interferes with coronary circulation.

Thus, although there are many studies regarding various interactions, three topics are discussed in this chapter: (1) interaction between afterload resistance or preload changes and cardiac performance in normal and depressed contractile states; (2) interaction between arterial compliance changes and cardiac performance in normal and depressed contractile states; and (3) interaction between coronary hemodynamics and cardiac overload. Throughout this discussion, myocardial ischemia is used as the basic model of the depressed contractile state. Each issue is interrelated, as was described, and has many controversial problems; we discuss these, mainly on the basis of our data from the standpoint of mechanical interaction.

Interaction Between Afterload Resistance or Preload Changes and Cardiac Performance in Normal and Depressed Contractile States

Peripheral vascular resistance greatly affects ventricular pumping action. Because arterial blood pressure is determined by variables such as contractility, preload, and arterial vascular resistance, it is suitable to consider afterload peripheral resistance as an afterload to the heart. When aortic pressure is elevated, myocardial oxygen consumption increases via a rise in cardiac work, while augmenting coronary blood supply; i.e., myocardial oxygen demand and supply are balanced in the normal coronary circulation, even if the loading condition is greatly altered. In contrast, in various pathological states such as ventricular hypertrophy, especially when accompanied by heart failure or coronary stenosis where the relation between coronary blood supply and demand becomes easily unbalanced, it is of interest how arterial peripheral resistance affects cardiac performance. That is, effects of afterload modulation on cardiac performance may differ in different severity and also in different pathological states of a disease [1].

There are many patients with coronary stenosis, and afterload-reducing therapy for ischemic heart disease frequently has been done in the clinical setting. However, it is probable that afterload reduction may lead to a decrease of coronary blood supply in a perfused area of the constricted coronary artery. This chapter thus focuses on afterload-induced modification of cardiac function in the presence of coronary stenosis. In the in situ condition, afterload change alters preload level and the two variables intimately influence each other. For example, elevation of arterial peripheral resistance is accompanied by increase of end-diastolic ventricular pressure and volume; left-ventricular function thereby increases via the Frank–Starling mechanism and end-diastolic pressure then slowly decreases to baseline while maintaining augmented stroke volume (Anrep effect). The Anrep effect is reported to be a synergistic action of the Frank–Starling mechanism and the gardenhose effect, in which increased intravascular pressure distends the surrounding myocardial tissue and then increases myofibrillar preload [2]. Thus, afterload changes induce preload alteration, and this preload-related hemodynamic change further modifies them. Therefore, it is necessary to see how afterload change itself modifies ischemia-induced hemodynamics.

We have developed an afterload hydraulic model that has four components: peripheral resistance, arterial compliance, characteristic resistance, and inertia; this afterload system is connected to an excised heart perfused with arterial blood from a support dog. In this experimental model, afterload changes are decoupled from preload changes, which are adjusted by the height of the preload reservoir. Accordingly, this model enables us to clarify the sole effect of the loading component, e.g., peripheral resistance or preload change on cardiac performance [3–7]. In our previous study, arterial resistance was changed widely by using this hydraulic model with and without coronary artery constriction, while keeping heart rate, left-ventricular end-diastolic pressure, and other afterload components constant. As a result, reduction in peripheral resistance increased stroke volume, accompanied with a fall of aortic pressure, but its further extreme reduction decreased ventricular ejection; below approximately 65 mmHg mean aortic pressure, stroke volume started to decrease. However, its critical value of aortic pressure became higher when there was significant impairment of the coronary circulation such as left main coronary artery stenosis.

This principle can also be extended to stenosis of a single coronary arterial branch, although the response of the stroke volume differs from systolic shortening in the ischemic region; that is, stroke volume expressed by systolic shortening of both nonischemic and ischemic regions does not show a decrease and continues to augment in the range of afterload reduction tested, although systolic shortening deteriorates in the ischemic region. The reason is that systolic shortening of the nonischemic region during ischemia shows augmentation compared with the extent of an increase of systolic shortening in hearts without ischemia in the same region at the same aortic pressure [7]. This increase in systolic shortening of the nonischemic region may be the result of interaction between nonischemic and ischemic regions; the ischemic region, by failing to shorten, may reduce afterload to the nonischemic region. By this increased systolic shortening in the nonischemic region, cardiac output continues to increase depending on the reduction of peripheral resistance.

In contrast, afterload elevation via an increase of peripheral resistance decreases cardiac output in a depressed contractile state with stenosis of a coronary arterial branch. However, it should be noted that systolic shortening of the ischemic myocardium decreases only slightly with the elevation of afterload; the load dependency of the ischemic myocardium is less than that of the nonischemic myocardium. The small dependency on afterload is a result of alleviation of ischemia produced by an increase in blood flow of the constricted coronary artery at high aortic pressure [8]. Because afterload elevation induces preload elevation in the in situ condition, as previously described, effects of afterload elevation combined with preload elevation were investigated in the same experimental preparation [8]. The preload elevation counterbalances the beneficial effects of afterload elevation on alleviating ischemic myocardium.

According to the concept of afterload mismatch, when preload reserve is consumed in the severely depressed contractile state, any afterload augmentation leads to a marked reduction of cardiac pumping action, and therefore afterload reduction will improve ventricular function as a whole, even if the preload-reducing effect observed following this situation might sometimes decrease cardiac output.

Preload-reducing therapy has also been commonly performed to reduce the symptoms of pulmonary congestion in acute heart failure, to improve prognosis, and also to induce ventricular remodeling in chronic heart failure. The beneficial effect of preload reducing itself on cardiac dysfunction with stenosis of a coronary arterial branch has also been confirmed by our experimental series [9], and therefore there seems to be no problem in reducing preload, if significantly elevated. However, up to a certain level preload is an important compensatory factor to augment cardiac pump function, especially in cardiac dysfunction. According to a recent report obtained in a closed circulation model, preload reduction results in a greater coronary reserve, implying higher efficiency of ventricular pressure generation, while afterload reduction may not be beneficial for improving myocardial oxygen balance [10]. This study was performed by using the simulation model, and moreover the results are predicted only within the range of intact coronary autoregulation [10].

Therefore, although we showed the directional change following afterload alteration in our previous reports [7,11], it is also necessary to investigate in the in vivo condition how beneficially unloading therapy affects ventricular performance in the pathological state of cardiac dysfunction and to determine whether an optimal afterload level can obtain the maximum effect of afterload reduction. These problems have been examined by using intraaortic balloon pumping in which the phasic and constant displacement of blood is accomplished via balloon deflation in systole [12–16].

Intraaortic Balloon Pumping

Intraaortic balloon pumping has been used in the clinical setting to treat acute myocardial infarction, unstable angina, and heart failure [12,13]. Intraaortic balloon pumping induces systolic unloading and enhances coronary blood flow by diastolic augmentation. These combined effects provide beneficial conditions to alleviate myocardial ischemia and assist the pumping action in the failing heart. However, there are conflicting data regarding increasing oxygen availability in the presence of severe coronary stenosis [12], and the mechanism of intraaortic balloon pumping benefit in such patients occurs predominantly through afterload reduction rather than augmented coronary blood flow.

According to Rosenbaum et al. [14], hemodynamics improved in both nonischemic and ischemic cardiomyopathy patients following intraaortic balloon counterpulsation. However, the mechanism of sustained benefit in these two groups seems to be different; afterload reduction appears to be more important in patients with nonischemic cardiomyopathy but reduction in filling pressures may be the main mechanism in patients with ischemic cardiomyopathy. It is not yet confirmed whether their observation is true. As for optimal loading for the heart in unloading therapy, we hypothesize that the beneficial effect of intraaortic balloon pumping on cardiac performance depends on the level of aortic pressure in preintraaortic balloon pumping.

Our preliminary data tested in a heart with severe coronary stenosis (flow reduction, 40%–70%), by varying mean aortic pressure to three different levels using nitroprusside or methoxamine infusion, i.e., 80, 100, and 120 mmHg, are shown in Fig. 1. Increase of ventricular ejection was modest, as expected from previous data [15] (i.e., 20%–30% increase), but the augmentation is not the same among these afterload pressures. At a mean aortic pressure of 100 mmHg, an increase of cardiac output (stroke volume) was the largest [16]. Since left-ventricular filling pressure similarly decreases in each level of aortic pressure, and coronary inflow augmentation is minimal although significant, effects of intraaortic balloon pumping are considered to be produced mainly from different unloading effects of afterload. It is of interest that there seems to be an optimal range of afterload to attain maximal effects on global and regional cardiac performance in the heart with severe coronary stenosis and that this is observed by the same intervention of intraaortic balloon pumping. However, the optimal value of mean aortic pressure at 100 mmHg may vary with different cardiovascular diseases and their severity [17]. Accordingly, further study is necessary for a general conclusion in various cardiovascular diseases regarding the optimal level of aortic pressure, as well as preload level.

Interaction Between Arterial Compliance Changes and Cardiac Performance in Normal and Depressed Contractile States

Interaction between cardiac function and arterial compliance has been intensively reviewed [18–26]. Briefly, the component of arterial compliance cushions flow pulsation and reduces left-ventricular load in systole by storing about 50% of ejected

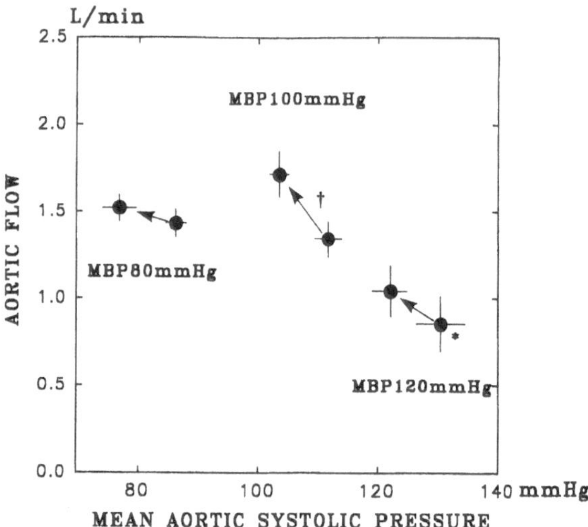

FIG. 1. Relationship between decrease in mean aortic systolic pressure (systolic unloading) and increase in aortic flow by intraaortic balloon pumping (IABP) at different levels of mean aortic pressure (*MBP*). Data are shown as mean ± SEM. *Arrows* indicate the effect of IABP. Aortic flow before IABP was significantly reduced at MBP 120 mmHg compared with that at MBP 80 and at 100 mmHg (*, $P < .05$). The increase in aortic flow by IABP at MBP 100 mmHg was significantly larger than at MBP 80 mmHg (†, $P < .05$)

volume from the left ventricle [18]. Accordingly, in this increased stiffness an increase of left-ventricular systolic pressure development is introduced. Moreover, arterial stiffness leads to decreased diastolic blood pressure. Enhancement of systolic pressure stimulates the development of cardiac hypertrophy and deteriorates left-ventricular ejection, thus predisposing to left-ventricular failure. Wave reflection must also be considered in the relation between arterial compliance and ventricular performance. In a compliant arterial system, the major boost to aortic pressure from wave reflection is found after the aortic valve is closed, so that systolic left-ventricular load is not influenced by wave reflection, but coronary perfusion via an increased diastolic pressure is augmented.

In contrast, in a stiffened arterial tree, arterial wave velocity increases and wave reflection returns earlier, adding a boost to systolic pressure [19]. As a result, left-ventricular systolic load further increases. Thus, as arterial pressure and flow waves are the composites of forward- and backward-traveling waves, arterial properties are very important for determining cardiac pumping action. That is, the aortic pressure and flow waves during systole are constructed not only by ventricular action, but also by vascular properties and wave reflection. Arterial elastance increases with aging, with an increase in blood pressure, and in pathological changes such as atherosclerosis. Therefore, the interaction between arterial stiffening and ventricular action is becoming an important issue.

Although the concept previously mentioned for vascular and ventricular interaction following vascular stiffening is generally accepted, it is difficult to see how arterial compliance change alone affects ventricular performance in situ because a change of one afterload component will inevitably affect the others in the peripheral circulation and evoke regulatory (reflex) mechanisms. Accordingly, cardiac performance has been investigated when only arterial compliance is changed [4,5,21]. In increased arterial compliance, enhancement of late systolic flow contributes to an increase in stroke volume without showing much pressure change. Thus, in capacitance changes, there is no relation between stroke volume and mean left-ventricular pressure [4]. These findings are related to the continuation of ventricular ejection after end-systole and the increased time between end-systole and end-ejection with increasing arterial compliance [5]. It is thus clear that stroke volume decreases because of increased pulsatile load following decreased arterial compliance itself.

It is conceivable that responses of nonpulsatile power, pulsatile power, and ventricular ejection to changes in afterload may not be the same in various contractile states. When cardiac function is already compromised, decreased arterial compliance may further worsen cardiac function. However, these aspects of changes in the pulsatile component of the afterload are not fully resolved in previous studies that were mainly performed in normal in situ ventricles. To solve this problem, the effects of changes in arterial compliance on left-ventricular ejection have been investigated using four different experimental situations, i.e., a normal control state, an augmented contractile state, and globally and regionally depressed contractile states, in isolated canine hearts [21]. The results suggest that decreased arterial distensibility has a further deleterious effect on cardiac output, when cardiac dysfunction is already present, especially as the result of regional ischemia [21].

In this examination a Windkessel hydraulic model is used as an afterload system in which wave travel does not occur, and thus effects of decreased arterial distensibility are considered to be underestimated because less-attenuated reflected waves during systole, found in increased vascular elasticity in a pathological state such as heart failure, further reduce stroke volume. In addition, although our data were obtained during constant perfusion, a decreased coronary blood supply particularly in the subendocardial layer resulting from decreased perfusion pressure following reduced arterial compliance in the in vivo condition [27] might be also deleterious, especially to the heart with depressed contractility.

Interaction Between Coronary Hemodynamics and Cardiac Overlaod

As previously mentioned, the coronary artery is embedded in the myocardium, so that the extravascular systolic compressive force as well as that in diastole greatly affects coronary vascular resistance [28–37]. Conversely, pressure increase in not only the arterial tree but also the venous one reduces ventricular compliance [2,38,39]. In congestive heart failure, diastolic filling pressure generally increases and aortic pressure tends to decrease, and therefore the driving pressure of coronary perfusion [diastolic aortic pressure minus right-atrial pressure (i.e., downstream pressure)] seems to fall. In fact, it has been shown that elevation of filling pressure impedes coronary inflow and alters the coronary pressure–flow relationship, i.e., elevated downstream pressure causes increased zero-flow pressure of the coronary pressure–flow relationship without changing the slope of its relationship [40–44]. Thus, impairment of coronary circulation by increased filling pressure might further aggravate cardiac performance in the failing heart.

In addition, effects of diastolic filling pressure elevation on coronary pressure-flow relation are enhanced [31] when the pericardium is acutely stretched. This pericardium-related action following increased filling pressure will be also demonstrated as reduced diastolic ventricular compliance [45]. Accordingly, if diastolic filling pressure increases rapidly, as is found in acute heart failure, pericardial stretch-induced impairment of coronary inflow must be taken into account.

Coronary Flow Dynamics

Hitherto, effects of filling pressure on coronary flow dynamics, which has been mainly investigated in nonbeating, arrested hearts, imply impediment of coronary inflow supply. It should be noted, however, that squeezing action in systole partly promotes coronary inflow [28,29]. Thus, it remains not fully resolved whether data obtained in an arrested state are applicable for the beating heart; that is, there are few data comparing the effects of diastolic filling pressure or coronary sinus pressure between arrested and beating conditions, using the same preparation.

In our study, an elevation of coronary sinus pressure to 30 mmHg significantly shifted the coronary pressure–flow relationship to the right in an arrested state compared to a beating condition, indicating that impairment of coronary inflow following an increase of coronary sinus pressure is slight, in contrast to the expectation from flow impediment in systole in the beating state [41].

Synchronized Retroperfusion

In synchronized retroperfusion, which is used in the clinical setting as supportive therapy for ischemic myocardium [6–9] by delivering arterial blood in a retrograde fashion during diastole to the ischemic area through the coronary sinus, coronary sinus pressure increases during diastole as expected, and as a result coronary driving pressure seems to decrease and may impair oxygen availability. On the other hand, coronary venous drainage basically occurs during systole; therefore, it is plausible that back-pressure elevation in systole has some effect on coronary inflow. However, it remains unclear how the back-pressure elevation during systole or diastole impairs coronary inflow dynamics. In our study, using the same experimental preparation of an excised perfused heart, we investigated how synchronized retroperfusion affects coronary arterial inflow and how synchronized retroperfusion differs from systolic coronary sinus pressure elevation.

When mean coronary sinus pressure in each group of controls, lower retrograde perfusion during diastole or coronary sinus partial occlusion, and higher retrograde perfusion flow during diastole or coronary sinus complete occlusion were adjusted to 14, 30, and 50 mmHg, respectively, the mean values of the slopes in the coronary pressure–flow relations did not differ among these interventions (Fig. 2a). The relation between zero-flow pressure and mean coronary sinus pressure in each condition is shown in Fig. 2b. From these results, it becomes apparent that synchronized retroperfusion does not greatly impair coronary arterial inflow supply, compared with the state causing systolic coronary sinus pressure elevation [46]. These data strongly suggest that the blood volume contained in the heart during systole is an important determinant factor in coronary inflow mechanics [47].

Synchronized retroperfusion preserves the squeezing effect of systole, and therefore the impeding effect on coronary inflow following downstream pressure elevation

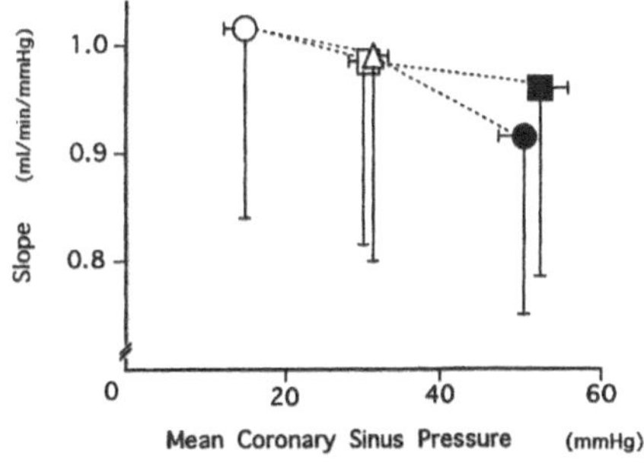

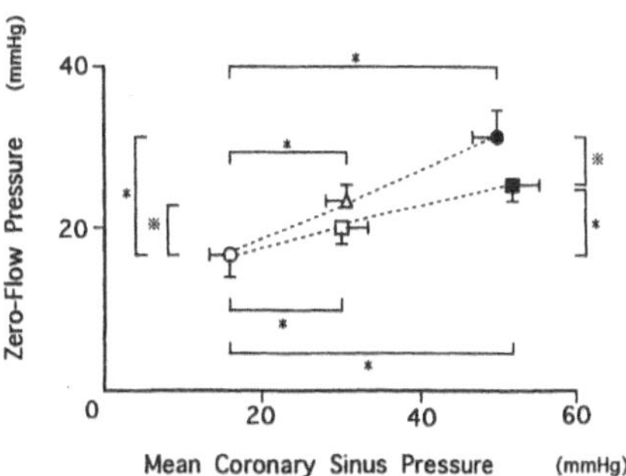

FIG. 2a,b. Slope (**a**) and zero-flow pressure (**b**) of the coronary pressure–flow relationship plotted against mean coronary sinus prssure in each group of control (*open circles*), lower retrograde perfusion during diastole (*open squares*), coronary sinus partial occlusion (*open triangles*), higher retrograde perfusion flow during diastole (*solid squares*), and coronary sinus complete occlusion (*solid circles*). Data are shown as mean ± SEM ($n = 7$)

is relatively smaller than the effect of systolic coronary sinus pressure elevation caused by coronary sinus occlusion. However, it still remains to be clarified from the standpoint of coronary circulation why synchronized retroperfusion is beneficial for reducing myocardial infarct size and improving cardiac function, although increased washout of noxious metabolites that accumulate during myocardial ischemia has been assumed to be one possible mechanism of coronary sinus pressure elevation [48]. Collateral flow is known to play an important role in salvaging the ischemic myocardium. However, it is not yet established that coronary sinus pressure elevation may affect coronary collateral circulation in the presence of regional ischemia. It is probable that a prominent reduction of contractile force caused by ischemia may give rise to the so-called sink effect, and this effect could cause increased collateral shunting to the nonischemic bed as well as in the ischemic bed [49].

There are few data, however, to explain how cardiac contraction itself with and without coronary sinus pressure elevation affects coronary collateral flow. By using an isolated maximally coronary vasodilated canine heart in which collateral channels are abundant, this problem has been investigated. Because inflow to the left anterior descending coronary artery was intercepted in the beating and nonbeating states with and without coronary sinus pressure elevation of 30 mmHg, while keeping the right and left circumflex coronary artery perfusion pressure constant, the measured regional myocardial blood flows in the left anterior descending coronary perfusion area were considered to be collateral flow supplied from the surrounding vessels. As a result, cardiac contraction and coronary sinus pressure elevation impeded collateral flow independently, but the coexistence of these two factors, that is, coronary sinus pressure elevation concurrent with cardiac contraction, augments collateral flow [50].

According to Manor et al. [51], increased right-atrial pressure results in a reduction of flow to the collateral-dependent myocardium and reduces perfusion of the unoccluded coronary vessel. One important methodological difference between these two experiments is the phase of back-pressure elevation used, i.e., diastolic back-pressure elevation in ours compared to back-pressure elevation not limited to diastole in theirs, suggesting that the phase of back-pressure elevation is also an important determinant of coronary collateral circulation as well as that of coronary inflow mechanics.

References

1. Opie LH (1987) Drug for the heart, 2nd edn. Grune and Stratton, New York, p 131
2. Schipke JD, Stocks I, Sunderdie KU, Arnold G (1993) Effect of changes in aortic pressure and in coronary arterial pressure on left ventricular geometry and function. Anrep vs. gardenhose effect. Basic Res Cardiol 88:621–367
3. Maruyama Y, Nunokawa T, Koiwa Y, Isoyama S, Ikeda K, Ino-oka E, Takishima T (1983) Mechanical interaction between the ventricles. Basic Res Cardiol 78:544–599
4. Ishide N, Shimizu Y, Maruyama Y, Koiwa Y, Nunokawa T, Isoyama S, Kitaoka S, Tamaki K, Ino-oka E, Takishima T (1980) Effects of changes in the aortic input impedance on systolic pressure-ejected volume relationships in the isolated supported canine left ventricle. Cardiovasc Res 14:229–243
5. Nishioka O, Maruyama Y, Ashikawa K, Isoyama S, Satoh S, Suzuki H, Watanabe J, Watanabe H, Shimizu Y, Ino-oka E, Takishima T (1987) Effects of changes in afterload impedence on left ventricular ejection in isolated canine heart: dissociation of end ejection from end systole. Cardiovasc Res 21:107–118
6. Isoyama S, Maruyama Y, Koiwa Y, Ishide N, Kitaoka S, tamaki K, Sato S, Shimizu Y, Ino-oka E, Takishima T (1981) Experimental study of afterload reducing therapy. The effect of the reduction of systemic vascular resistance on cardiac output, aortic pressure and coronary circulation in isolated, ejecting canine heart. Circulation 64: 490–498
7. Isoyama S, Maruyama T, Ashikawa K, Sato S, Suzuki H, Watanabe J, Shimizu Y, Ino-oka E, Takishima T (1983) Effects of afterload reduction on global left ventricular and regional myocardial functions in the isolated canine heart with stenosis of a coronary arterial branch. Circulation 67:139–146
8. Maruyama Y, Isoyama S, Ashikawa K, Satoh S, Suzuki H, Nishioka O, Watanabe J, Takishima T (1989) Effects of afterload elevation on the ischemic myocardium in isolated, paced canine heart with partial coronary stenosis. Am J Cardiol 63:40E–44E
9. Satoh S, Maruyama Y, Ashikawa K, Isoyama S, Suziki H, Watanabe J, Nishioka O, Ino-oka E, Takishima T (1986) Effects of preload alteration on the degree of ischemia and function on the degree of ischemia and function of ischemic myocardium under constant mean aortic pressure, coronary perfusion pressure and heart rate in isolated perfused canine heart. Jpn Circ J 50:1100–1112
10. Barnea O (1994) Mathematical analysis of coronary autoregulation and vascular reserve in closed-loop circulation. Comput Biomed Res 27:263–275
11. Maruyama Y, Ashikawa K, Isoyama S, Satoh S, Suzuki H, Watanabe J, Shimizu Y, Ino-oka E, Takishima T (1984) Pressure-length loop in the ischemic segment during left circumflex coronary artery stenosis and its modification by afterload reducing in excised perfused canine hearts. Basic Res Cardiol 79:155–163
12. Kern MJ, Aguirre F, Bach R, Donohue T, Siegle R, Segal J (1993) Augmentation of coronary blood flow by intra-aortic balloon pumping in patients after coronary angioplasty. Circulation 87:500–511
13. Lazar HL, Yang XM, Rivers S, Treanor P, Bernard S, Shemin RJ (1992) Retroperfusion and balloon support to improve coronary revascularization. J Cardiovasc Surg 33:538–544
14. Rosenbaum AM, Murali S, Uretsky BF (1994) Intra-aortic balloon counterpulsation as a "bridge" to cardiac transplantation effects in nonischemic and ischemic cardiomyopathy. Chest 106:1683–1688
15. Aguirre FV, Kern MJ, Bach R, Donohue T, Caracciolo E, Flynn MS, Wolford T (1994) Intraaortic balloon pump support during high-risk coronary angioplasty. Cardiology 84:175–186

16. Maehara K, Saito T, Saito S, Niitsuma T, Maruyama Y (1995) Afterload dependency of the effects of intra-aortic balloon pumping in dogs with coronary artery stenosis. In: Fifth Antwerp–La Jolla–Kyoto research conference on cardiac function, program and abstracts 90:9-7-7
17. Samuelsson OG, Wilhelmsen LW, Pennert KM, Wedel H, Berglund GL (1990) The J-shaped relationship between coronary heart disease and achieved blood pressure level in treated hypertension: further analysis of 12 years of follow-up of treated hypertensives in the primary prevention trial in Gothenburg, Sweden. J Hypertens 8:547–555
18. Belz GG (1995) Elastic properties and Windkessel function of the human aorta. Cardiovasc Drugs Ther 9:73–83
19. O'Rourke M (1994) Arterial stiffening and vascular/ventricular interaction. J Hum Hypertens 8(suppl 1):S9–S15
20. Westerhof N, O'Rourke MF (1995) Haemodynamic basis for the development of left ventricular failure in systolic hypertension and for its logical therapy. J Hypertens 13:943–952
21. Maruyama Y, Nishioka O, Nozaki E, Kinohita H, Kyono H, Koiwa Y, Takishima T (1993) Effects of arterial distensibility on left ventricular ejection in the depressed contractile state. Cardiovasc Res 27:182–187
22. Elzinga G, Westerhof N (1973) Pressure and flow generated by the left ventricle against different impedance. Circ Res 32:178–186
23. O'Rourke MF (1967) Steady and pulsatile energy losses in the systemic circulation under normal conditions and in simulated arterial disease. Cardiovasc Res 1:313–326
24. Urschel COW, Cavil JAW, Sounenblick EH, Ross J, Braunwald E (1968) Effects of decreased aortic compliance on performance of the left ventricle. Am J Physiol 214:298–304
25. Sunagawa K, Maughan WL, Sagawa K (1985) Stroke volume effect of changing arterial input impedance over selected frequency ranges. Am J Physiol 248:H477–H484
26. Weber KT, Janicki JS, Hunter WC, Shroff S, Pearlman ES, Fishiman AP (1982) The contractile behavior of the heart and its functional coupling to the circulation. Prog Cardiovasc Dis 24:375–400
27. Watanabe H, Ohtsuka S, Kakihana M, Sugishita Y (1993) Coronary circulation in dogs with an experimental decrease in aortic compliance. J Am Coll Cardiol 21:1497–1506
28. Hoffman JIE, Spaan JAE (1990) Pressure-flow relations in coronary circulation. Physiol Rev 70(2):331–390
29. Maruyama Y, Takishima T (1990) Effect of ventricular and extraventricular pressureon the coronary artery pressure-flow relationship. In: Kajiya F, Klassen GA, Spaan JAE, Hoffman JIE (eds) Coronary circulation, basic mechanism and clinical relevance. Springer-Verlag, Berlin, Heidelberg, New York, Tokyo, pp 127–138
30. Satoh S, Watanabe J, Keitoku M, Itoh N, Maruyama Y, Takishima T (1988) Influences of pressure surrounding the heart and intracardiac pressure on the diastolic coronary pressure-flow relation in the excised canine heart. Circ Res 63:788–797
31. Watanabe J, Maruyama Y, Satoh S, Keitoku M, Takishima T (1987) Effect of pericardium on the diastolic left coronary pressure-flow relationship in isolated dog heart. Circulation 75:670–675
32. Satoh S, Maruyama Y, Watanabe J, Keitoku M, Hangai K, Takishima T (1990) Coronary zero flow pressure and intracardial pressure in transiently attested heart. Cardiovasc Res 24:358–363
33. Aversano T, Klocke FJ, Mates RM, Canty JM Jr (1984) Preload induced alterations in capacitance-free diastolic pressure-flow relationship. Am J Physiol 246:H410–H417
34. Ellis AK, Klocke FJ (1979) Effects of preload on the transmural distribution of perfusion and pressure-flow relationships in the canine coronary vascular bed. Circ Res 46:68–77
35. Krams R, Sipkema P, Westerhof N (1989) Varying elastance concept may explain coronary systolic flow impediment. Am J Physiol 257:H1471–H1479
36. Rabbany SY, Kresh JY, Noordergraaf A (1989) Intramyocardial pressure: interaction of myocardial fluid pressure and fiber stress. Am J Physiol 257:H357–H364

37. Watanabe J, Levine MJ, Bellotto F, Johnson RG, Grossman W (1993) Left ventricular diastolic chamber stiffness and intramyocardial corronary capacitance in isolated dog hearts. Circulation 88:2929–2940
38. Watanabe J, Levine MJ, Bellotto F, Johnson RG, Grossman W (1990) Effects of coronary venous pressure on left ventriclar diastolic distensibility. Circ Res 67:923–932
39. Hangai K, Satoh S, Sato F, Watanabe J, Maruyama Y, Takishima T (1993) Continuous measurement of canine coronary blood volume change with alterations of heart rate. Cardiovasc Res 27:1127–1134
40. Scheel KW, Williams SE, Parker JB (1990) Coronary sinus pressure has a direct effect on gradient for coronary perfusion. Am J Physiol 258:H1739–H1744
41. Bellamy RF, Lovensohn HS, Ehrlich W, Baer RW (1980) Effect of coronary sinus occlusion on coronary pressure-flow relations. Am J Physiol 239 (Heart Circ Physiol 8):H57–H64
42. Saito T, Mitsugi M, Saitoh S, Sato M, Maruyama Y (1993) Effects of coronary collateral circulation and outflow pressure elevation on coronary pressure-flow relationships in the beating and non-beating states. In: Maruyama Y, Kajiya F, Hoffman JIE, Spaan JAE (eds) Recent advances in coronary circulation. Springer-Verlag, Berlin, Heidelberg, New York, Tokyo, pp 114–122
43. Pantley GA, Bristow JD, Ladley HD, Anselone CG (1988) Effect of coronary sinus occlusion on coronary flow, resistance, and zero flow pressure during maximum vasodilation in swine. Cardiovasc Res 22:79–86
44. Izrailtyan I, Frasch F, Kresh JY (1994) Effects of venous pressure on coronary circulation and intramyocardial fluid mechanics. Am J Physiol 267:1002–1009
45. Maruyama Y, Ashikawa K, Isoyama S, Kanatsuka H, Ino-oka E, Takishima T (1982) Mechanical interactions between four heart chambers with and without the pericardium in canine hearts. Circ Res 50:86–100
46. Mitsugi M, Saito T, Saitoh S, Sato M, Maruyama Y (1996) Effects of synchronized retroperfusion on coronary arterial pressure-flow relationship. Cardiovasc Res (in press)
47. Goto M, Tsujioka K, Ogasawara Y, Wada Y, Tadaoka S, Hiramatsu O, Yanaka M, Kajiya F (1990) Effects of blood filling in intramyocardial vessels on coronary arterial inflow. Am J Physiol 258:H1042–H1048
48. Mohl W, Punzengruber C, Moser M, Kenner T, Heimisch W, Haendchen R, Meerbaum S, Maurer G, Corday E (1985) Effect of pressure controlled intermittent coronary sinus occlusion on regional ischemic myocardial function. J Am Coll Cardiol 5:939–947
49. Toggart EJ, Nellis SH, Liedtke AJ (1987) The efficacy of intermittent coronary sinus occlusion in the absence of coronary artery collaterals. Circulation 76(3):667–677
50. Sato M, Saito T, Mitsugi M, Saito S, Niitsuma T, Maehara K, Maruyama Y (1996) Effects of cardiac contraction and coronary sinus pressure elevation on collateral circulation. Am J Physiol (in press)
51. Manor D, Williams S, Ator R, Bryant K, Scheel KW (1994) Reduced collateral perfusion is a direct consequence of elevated right atrial pressure. Am J Physiol 267:H1151–H1156

Intraventricular Interaction

Yoichi Goto

Summary. It has been recognized that, in an acutely ischemic heart, systolic wall motion of a nonischemic region increases. Previous studies attributed this hyperkinesis to increased sympathetic activities or a compensatory operation of the Frank–Starling mechanism. However, recent studies have indicated that this hyperkinesis can occur without the operation of the Frank–Starling mechanism or an enhancement of regional myocardial contractility, and that the essential mechanism of this phenomenon is regional afterload reduction resulting from intraventricular mechanical interaction between ischemic and nonischemic regions. In contrast, systolic segment shortening of a region is depressed by increased regional afterload when myocardial contractility of the opposite wall of the left ventricle is enhanced by intracoronary administration of dobutamine. Thus, the extent of regional systolic shortening is significantly affected by intraventricular regional mechanical interaction through alterations of regional afterload generated by myocardial contractile force of the opposite wall.

Key words. Hyperkinesis—Myocardial contraction—Contractility—Regional myocardial function—Pressure–volume relation—Nonischemic region

Introduction

It has been fully recognized that mechanical interaction between the left and right ventricles or between an atrium and a ventricle substantially influences left-ventricular performance under normal or pathophysiological conditions [1–3]. On the other hand, the pathophysiology of mechanical interaction between regions within a ventricle has not been fully elucidated. One of the manifestations of intraventricular regional interaction is an increase in systolic wall motion (hyperkinesis) in a nonischemic region, which has been observed in an acutely ischemic heart both in experimental animals and in patients [4]. In this chapter, the mechanism and pathophysiological significance of intraventricular regional interaction will be reviewed with special attention to hyperkinesis in a nonischemic region of an acutely ischemic heart. In addition, another piece of experimental evidence for intraventricular inter-

Division of Cardiology, Department of Medicine, National Cardiovascular Center, Osaka, 565 Japan.

action, i.e., "hypokinesis" of the opposite wall during an enhancement of regional myocardial contractility, is also presented.

Hyperkinesis in a Nonischemic Region

The mechanisms of hyperkinesis in a nonischemic region have been ascribed to either increased sympathetic activity [5], the Frank–Starling mechanism [6], or a combination of the Frank–Starling mechanism and mechanical unloading caused by intraventricular regional interaction [7]. Nakano [5] first described an increase in contractile force in a nonischemic region and related it to increased sympathetic activity. Theroux et al. [6] demonstrated increased systolic segment shortening in a nonischemic region that correlated well with an increase in end-diastolic segment length and interpreted the increased shortening as a manifestation of a compensatory operation of the Frank–Starling mechanism. Thereafter, an increase in systolic segment shortening or wall thickening in a nonischemic region has been shown by many investigators and has been ascribed to either a compensatory operation of the Frank–Starling mechanism or increased sympathetic activity. More recent studies by Lew et al. [7] and Smalling et al. [8] have attributed the increased segment shortening to a combination of the Frank–Starling mechanism and mechanical unloading resulting from an intraventricular interaction between the ischemic and nonischemic regions.

However, all these studies were performed in hearts in situ in which acute ischemia is usually accompanied by compensatory operation of the Frank–Starling mechanism and myocardial function is regulated by various neurohumoral mechanisms. This has made it difficult to separate the effects of the Frank–Starling mechanism and neurohumoral factors from other mechanisms that might be responsible for the hyperkinesis. Therefore, we assessed regional myocardial function of a nonischemic region in an isolated heart of which the left-ventricular volume was accurately controlled by a servopump [9].

Analysis of Regional Function and Regional Work

To assess global left-ventricular contractile state, external work, and total mechanical energy, the end-systolic pressure–volume relation (ESPVR), a pressure–volume loop, and pressure–volume area (PVA) have been widely used [10–12]. By the analogy of the pressure–volume relation in the global ventricle, we have recently proposed and validated a new approach assessing regional mechanics and energetics using the wall tension–regional area (T-A) relation [13–15]. According to this method, regional myocardial contractility, regional work, and regional total mechanical energy expenditure of the left ventricle can be reliably quantified by the end-systolic T-A relation, the area within a T-A loop, and a specific area named "tension-area area" (TAA) surrounded by the end-systolic and end-diastolic T-A relations and the systolic segment of T-A trajectory, respectively.

We performed experiments in excised, cross-circulated dog left ventricles connected to a volume servopump [9]. Left-ventricular short axis diameter and two pairs of orthogonal segment lengths in the anterior and posterior wall regions were measured with pairs of ultrasonic crystals. Wall tension (T, in mmHg·cm) was calculated according to the force equilibrium relation of a thick-walled sphere, assuming that the left-ventricular contractile force is supported solely by the endocardial layer. Namely, T = 1/4 D P, where D is left-ventricular short-axis diameter (cm) and P is left-ventricular pressure (mmHg). Regional area (A, in cm^2) of the diamond-shaped region surrounded by the four crystals was calculated as $A = 1/2\ L_1\ L_2$, where L_1 and L_2 are circumferential and longitudinal segment lengths (cm), respectively. Regional work (mmHg·ml or dyne·cm) was then assessed from the integral of wall tension with respect to regional area, i.e., area within a T-A loop. Measurements were made before and after regional ischemia produced by an occlusion of either the left anterior descending or circumflex coronary artery with a snare.

Hyperkinesis Without the Frank–Starling Mechanism

Figure 1 shows left-ventricular pressure–volume loops and T-A loops in the ischemic and nonischemic regions obtained before and after coronary artery occlusion at a constant left-ventricular volume in a representative heart. After coronary artery

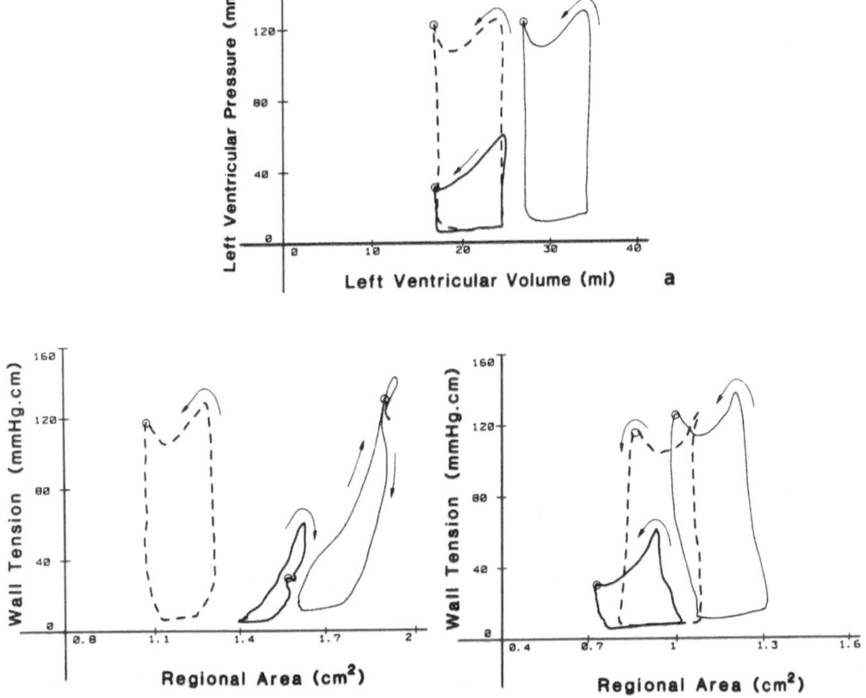

FIG. 1a–c. Representative examples of left-ventricular pressure–volume loops (a) and wall tension–regional area (T-A) loops in ischemic (b) and nonischemic regions (c) during the control period (*dashed line*), regional ischemia at constant left-ventricular end-diastolic and stroke volumes (*thick solid line*), and regional ischemia with increased end-diastolic volume (*thin solid line*). Increased systolic area shrinkage in the nonischemic region, not associated with an increased regional preload, corresponds to the systolic bulging in the ischemic region, indicating intraventricular regional interaction. (From [8] with permission)

occlusion (see thick solid lines), left-ventricular systolic pressure and stroke work moderately decreased at the same end-diastolic and stroke volumes (Fig. 1a). In the ischemic region (Fig. 1b), end-diastolic regional area increased despite the same end-diastolic wall tension, suggesting creep or passive stretch of the ischemic myocardium. The direction of rotation of the T-A loop was reversed, resulting in a negative regional work. On the average, systolic area shrinkage markedly decreased by 109% ± 30%.

In contrast, end-diastolic regional area in the nonischemic region decreased during ischemia (Fig. 1c). With this decreased regional preload, the T-A loop in the nonischemic region showed a noticeable area shrinkage during the isovolumic contraction period, corresponding reciprocally to the area expansion in the ischemic region. Because the whole left-ventricular volume was kept constant during the isovolumic contraction period, this reciprocal behavior of the ischemic and nonischemic regions are evidence of intraventricular regional interaction between weak and strong myocardium. As a result, systolic area shrinkage of the nonischemic region increased by 33% ± 41% despite the apparent lack of utilization of the Frank–Starling mechanism. Although there is no necessary reason that intraventricular regional interaction should occur only during the isovolumic period, we did not observe an increase in area shrinkage during the ejection period.

When left-ventricular end-diastolic volume was increased (Fig. 1, thin solid line) to simulate a clinical condition in which left-ventricular stroke work is compensated for by increased ventricular preload, systolic area shrinkage in the nonischemic region further increased at the expense of a further decrease in area shrinkage in the ischemic region despite the constant stroke volume.

Reduced Regional Work Despite Hyperkinesis

Another interesting finding in this study was that regional work of the nonischemic region significantly decreased (11 ± 3 to 5 ± 2 mmHg·ml) during regional ischemia at constant end-diastolic and stroke volumes, despite the presence of hyperkinesis. This mimics an unloaded whole ventricle in which a decrease in afterload results in an increase in ventricular stroke volume (or shortening) and a concomitant reduction in stroke work. Thus, hyperkinesis is not a "compensatory" mechanism for the decreased contractile performance of the ischemic region, but a natural consequence of mechanical unloading of the nonischemic region.

Unchanged Regional Contractility Despite Hyperkinesis

To examine whether enhanced regional myocardial contractility in the nonischemic region is responsible for hyperkinesis, the end-systolic T-A relation was assessed in the ischemic and nonischemic regions during the control period and during regional ischemia (Fig. 2). Compared to the marked rightward shift of the end-systolic T-A relation in the ischemic region, the end-sytolic T-A relation in the nonischemic region was almost unchanged during regional ischemia. This finding indicates that myocardial contractility in the nonischemic region is unchanged, and thus it is unlikely that hyperkinesis of the nonischemic region is caused by an enhanced regional myocardial contractility.

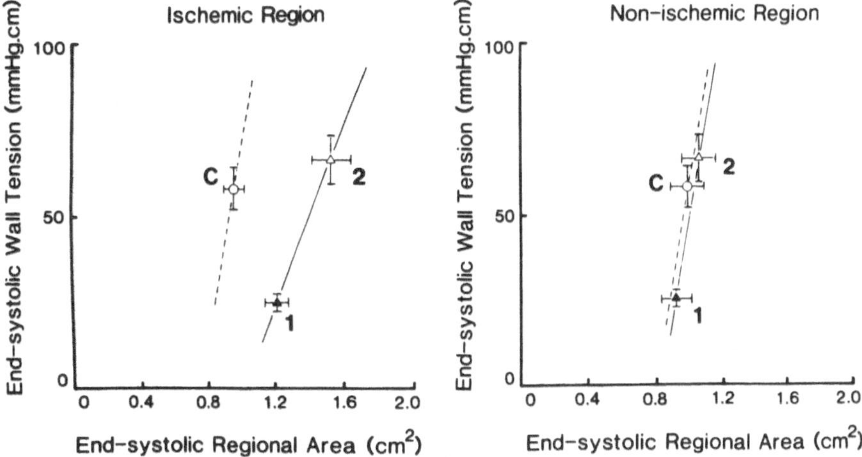

FIG. 2. The end-systolic wall tension–regional area relation in the ischemic (*left*) and nonischemic (*right*) regions during the control period (*open circle* and *dashed line*) and regional ischemia (*triangles* and *solid line*). The end-systolic tension–area relation in the ischemic region shifted markedly to the right after coronary occlusion while the relation in the nonischemic region was almost unchanged. Means ± SEM are indicated. (From [8] with permission)

In contrast to this result, however, a decreased myocardial contractility indicated by a right- and downward shift of the end-systolic pressure–length relation in a nonischemic region has been reported in an open-chest dog study [16]. This apparent discrepancy may be explained by the difference in coronary perfusion pressure (i.e., constant perfusion pressure in the isolated heart study versus decreased perfusion pressure in the open-chest animal study), because the downward shift of the end-systolic relation has been reported to be abolished by maintaining coronary perfusion pressure constant [17]. Thus, these lines of evidence indicate that regional afterload reduction caused by an intraventricular mechanical interaction between ischemic and nonischemic regions, rather than the Frank–Starling mechanism or increased regional myocardial contractitity, may be the primary mechanism of hyperkinesis in the nonischemic region.

Another point to be discussed is the determinant of the magnitude of hyperkinesis. In open-chest dog studies, Lew and colleagues [18,19] have demonstrated that a lower preload and a larger ischemic zone size cause hyperkinesis of a greater magnitude. However, in our isolated heart study [9], increased ventricular volume resulted in a greater hyperkinesis, indicating that not only preload but also whole ventricular loading conditions including afterload may be important. Also, Marino et al. [20] reported a greater hyperkinesis with posterior wall ischemia than with anterior wall ischemia at a similar ischemic zone size, indicating that ventricular geometry, fiber orientation, and tethering may be additional influencing factors [2,21–24].

Hypokinesis Caused by Intraventricular Interaction

As a next step, we hypothesized that, if hyperkinesis occurs in a nonischemic region as a result of intraventricular interaction between two regions with different contractility, then increased myocardial contractility of a ventricular region should result in an acute reduction of segmental shortening, or hypokinesis, of the opposite wall of the ventricle even in the absence of regional ischemia. To test this hypothesis, left-ventricular (LV) global and regional function of an anterior wall region were measured during enhancement of regional myocardial contractility of the posterior wall [25].

Figure 3 shows LV pressure-volume loops and T-A loops of the anterior wall before and during dobutamine infusion into the left circumflex coronary artery in a representative heart. When regional myocardial contractility of the posterior wall region was increased with dobutamine (DOB-LCX), the anterior wall region showed an area expansion during the isovolumic contraction period, resulting in a decrease in systolic area shrinkage. This is a phenomenon mimicking hyperkinesis in a nonischemic region, but opposite in direction.

Because regional preload was unchanged, this decreased area shrinkage cannot be explained by insufficient utilization of the Frank–Starling mechanism. Furthermore,

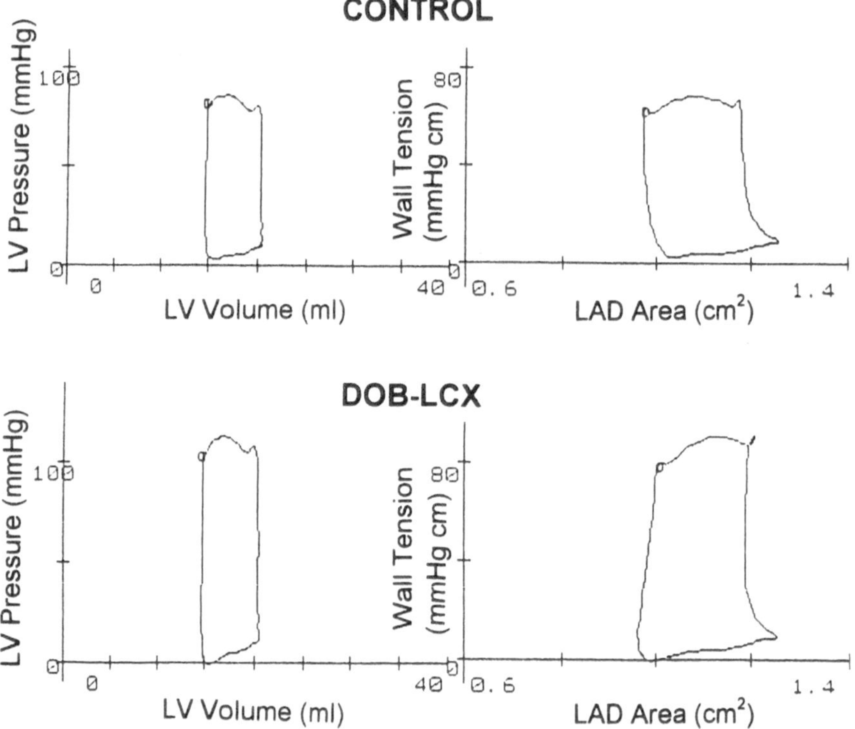

FIG. 3. Representative example of left-ventricular pressure–volume loops (*left*) and wall tension–regional area (T-A) loops (*right*) in the left anterior coronary artery (*LAD*) region during the *control* period (*upper panels*) and dobutamine (*DOB*) infusion into the left circumflex coronary (*LCX*) artery (*DOB-LCX*; *lower panels*). During dobutamine infusion into the LCX artery, left-ventricular (*LV*) contractility was enhanced, as indicated by a higher end-systolic pressure for the constant end-systolic volume. In contrast, the T-A loop in the *LAD* region was deformed, and the end-systolic regional area increased against the elevated end-systolic wall tension, resulting in a decrease in systolic area shrinkage from 21% to 17%

the end-systolic T-A relation of the anterior wall region was unchanged during dobutamine infusion into the circumflex coronary artery, indicating that regional myocardial contractility of the anterior wall region was constant despite the decrease in area shrinkage. Therefore, this hypokinesis is attributable to an increased regional afterload generated by the enhanced contraction of the opposite region with higher contractility, through the mechanism of intraventricular regional interaction. This phenomenon is opposite in direction to hyperkinesis in a nonischemic region, and analogous to the behavior of weak and strong muscles connected in series [26].

Chronic Intraventricular Interaction

Hyperkinesis in a nonischemic region is usually observed only in acutely ischemic hearts. In most patients with acute myocardial infarction, hyperkinesis disappears in the chronic phase. This can be explained by the thought that in the chronic phase, progressively increased myocardial stiffness caused by scar formation in the ischemic region [27] becomes able to counteract wall tension (i.e., regional afterload) generated by the nonischemic region. However, when ventricular aneurysm persists in the chronic phase after myocardial infarction, an augmented shortening in the noninfarcted region and a corresponding dyskinetic outward bulging in the aneurysmal region are observed in the systolic phase, resulting in a reduced forward stroke volume [27,28].

Another interesting issue is hypertrophy in a noninfarcted region in the chronic phase after myocardial infarction. In an experimental study, cell volume of the surviving myocardium was reported to increase by as much as 78% after a large infarction [29]. This hypertrophy of the noninfarcted region may represent a beneficial chronic adaptation to the loss of myocardium and contribute to maintaining cardiac ejection performance after myocardial infarction [30]. However, at a later stage or even with a greater infarction, expansion of the infarct region as well as stretch and elongation of the noninfarct region occurs, leading to ventricular dilatation and dysfunction, called ventricular remodeling [31]. The precise mechanisms of ventricular remodeling remain to be elucidated but the magnitude of remodeling has been thought to be determined by infarct size, infarct healing, and ventricular wall stresses [31], suggesting that intraventricular interaction between infarct and noninfarct regions may play a role in the development of ventricular remodeling.

Conclusion

In conclusion, an acute reduction or enhancement in regional myocardial contractility of a local ventricular wall results in reciprocal changes in regional myocardial shortening in the opposite wall of the ventricle because of regional afterload alterations through the mechanism of intraventricular mechanical interaction. Utilization of the Frank–Starling mechanism or alteration of regional myocardial contractility is not necessarily involved in this phenomenon. The extent of the affected zone size, ventricular loading conditions, and ventricular geometry and architecture may influence the magnitude of this phenomenon. The impact of intraventricular interaction on long-term outcome remains to be elucidated.

Acknowledgment. This work was supported in part by a Research Grant for Cardiovascular Diseases from the Ministry of Health and Welfare of Japan (7A-3). I thank all coauthors in may previous publications for their technical assistance and helpful discussion, which enabled completion of this article.

References

1. Maruyama Y, Ashikawa K, Isoyama S, Kanatsuka H, Ino-oka E, Takishima T (1982) Mechanical interactions between four heart chambers with and without the pericardium in canine hearts. Circ Res 50:86–100
2. Goto Y, Slinker BK, LeWinter MM (1989) Nonhomogeneous left ventricular regional shortening during acute right ventricular pressure overload. Circ Res 65:43–54
3. Slinker BK, Goto Y, LeWinter MM (1989) Systolic direct ventricular interaction affects left ventricular contraction and relaxation in the intact dog circulation. Circ Res 65:307–315
4. Sheehan FR, Mathey DG, Schofer J, Krebber H, Dodge HT (1983) Effect of interventions in salvaging left ventricular function in acute myocardial infarction: a study of intracoronary streptokinase. Am J Cardiol 52:431–438
5. Nakano J (1966) Effect of changes in coronary arterial blood flow on the myocardial contractile force. Jpn Heart J 7:78–86
6. Theroux P, Franklin D, Ross J Jr, Kemper WS (1974) Regional myocardial function during acute coronary artery occlusion and its modification by pharmacologic agents in the dog. Circ Res 35:896–908
7. Lew WYW, Chen Z, Guth B, Covell JW (1985) Mechanisms of augmented segment shortening in nonischemic areas during acute ischemia of the canine left ventricle. Circ Res 56:351–358
8. Smalling RW, Ekas RD, Felli PR, Binion L, Desmond J (1986) Reciprocal functional interaction of adjacent myocardial segments during regional ischemia. J Am Coll Cardiol 7:1335–1346
9. Goto Y, Igarashi Y, Yamada O, Hiramori K, Suga H (1988) Hyperkinesis without the Frank-Starling mechanism in a nonischemic region of acutely ischemic excised canine heart. Circulation 77:468–477
10. Suga H, Sagawa K (1974) Instantaneous pressure-volume relationships and their ratio in the excised, supported canine left ventricle. Circ Res 35:117–126
11. Sagawa K, Maughan L, Suga H, Sunagawa K (eds) (1988) Cardiac contraction and the pressure-volume relationship. Oxford University Press, New York
12. Suga H (1990) Ventricular energetics. Physiol Rev 70:247–277
13. Goto Y, Suga H, Yamada O, Igarashi Y, Saito M, Hiramori K (1986) Left ventricular regional work from wall tension-area loop in canine heart. Am J Physiol 250:H151–H158
14. Goto Y, Igarashi Y, Yasumura Y, Nozawa T, Futaki S, Hiramori K, Suga H (1988) Integrated regional work equals total left ventricular work in regionally ischemic canine heart. Am J Physiol 254:H894–H904
15. Goto Y, Futaki S, Kawaguchi O, Hata K, Takasago T, Saeki A, Nishioka T, Taylor TW, Suga H (1993) Coupling between regional myocardial oxygen consumption and contraction under altered preload and afterload. J Am Coll Cardiol 21:1522–1531
16. Meyer TE, Perlini S, Foex P (1994) Regional nonischemic performance as assessed by end-systolic measures of shortening and thickening. J Am Coll Cardiol 24:1797–1805
17. Aversano T, Marino PN (1990) Effect of ischemic zone size on nonischemic zone function. Am J Physiol 258:H1786–H1795
18. Lew WYW, Ban-Hayashi E (1985) Mechanisms of improving regional and global ventricular function by preload alterations during acute ischemia in the canine left ventricle. Circulation 72:1125–1134

19. Lew WYW (1987) Influence of ischemic zone size on nonischemic area function in the canine left ventricle. Am J Physiol 252:H990–H997
20. Marino PN, Kass DA, Becker LC, Lima JAC, Weiss JL (1989) Influence of site of regional ischemia on nonischemic thickening in anesthetized dog. Am J Physiol 256:H1417–H1425
21. Thomas CE (1957) The muscular architecture of the ventricles of hog and dog hearts. Am J Anat 101:17–58
22. Gallagher KP, Osakada G, Hess OM, Koziol JA, Kemper WS, Ross J Jr (1982) Subepicardial segmental function during coronary stenosis and the role of myocardial fiber orientation. Circ Res 50:352–359
23. Freeman GL, LeWinter MM, Engler RL, Covell JW (1985) Relationship between myocardial fiber direction and segment shortening in the midwall of the canine left ventricle. Circ Res 56:31–39
24. Lew WYW, LeWinter MM (1986) Regional comparison of midwall segment and area shortening in the canine left ventricle. Circ Res 58:678–691
25. Goto Y, Yaku H, Ohgoshi Y, Kawaguchi O, Hata K, Saeki A, Nishioka T, Takasago T, Suga H (1993) Intraventricular regional interaction: depressed myocardial shortening resulting from enhanced contractility of the opposite wall of the ventricle (abstract). Circulation 88:I-277
26. Wiegner AW, Allen GJ, Bing OHL (1978) Weak and strong myocardium in series: implications for segmental dysfunction. Am J Physiol 235:H776–H783
27. Bogen DK, Rabinowitz SA, Needleman A, McMahon TA, Abelmann WH (1980) An analysis of the mechanical disadvantage of myocardial infarction in the canine left ventricle. Circ Res 47:728–741
28. Parmley WW, Chuck L, Kivowitz C, Matloff JM, Swan HJC (1973) In vitro length-tension relations of human ventricular aneurysms. Am J Cardiol 32:889–894
29. Anversa P, Beghi C, Kikkawa Y, Olivetti G (1986) Myocardial infarction in rats. Circ Res 58:26–37
30. Ginzton LE, Conant R, Rodrigues DM, Laks MM (1989) Functional significance of hypertrophy of the noninfarcted myocardium after myocardial infarction in humans. Circulation 80:816–822
31. Pfeffer MA, Braunwald E (1990) Ventricular remodeling after myocardial infarction. Circulation 81:1161–1172

Adaptation of the Cardiac Cell to Loading: Function in Heart Failure

HENK E.D.J. TER KEURS

Summary. The inability of the heart in failure to eject a sufficient amount of blood to meet the needs of the body is thought to result from molecular changes in cardiac cells causing decreased active (systolic) force and impaired (diastolic) relaxation together with a greater stiffness of the remodeled ventricular wall. It has been suggested that the failure to generate a forceful contraction is a consequence of derailment of the processes in the failing cardiac cells to manipulate Ca^{2+} ions despite the increased stimulus from nervous and hormone systems to enhance cardiac performance. Because of the lack of adequate release and uptake of Ca^{2+} ions, the amount of mechanical work that can be put out by the heart muscle is diminished and the heart may fail. Uptake of Ca^{2+} ions by the intracellular store (the sarcoplasmic reticulum) is impaired in congestive heart failure (CHF), probably as result of inadequate gene expression. In consequence the amount of Ca^{2+} that is released during each heartbeat is less than normal, thus force is reduced. In normal heart muscle, the response of the contractile filaments to Ca^{2+} ions depends strongly on sarcomere length, thus explaining Starling's law of the heart. Recent evidence suggests that this sensitivity is largely lost in CHF, thereby reducing the effectiveness of stretch of cardiac cells on the mechanical output. The effect of longstanding sympathetic drive to the heart during the development of heart failure induces a loss of sensitivity of the myocardium to catecholamines by loss of β_1-receptors and partial uncoupling of the β-receptors from production of cyclic adenosine monophosphate; hence the effect of sympathetic activation is diminished and the heart has to rely more on Starling's law. Increase of the filling pressure of the left ventricle may in part accommodate the ongoing demands of the body. However, in the case of a stenosed coronary arterial system, the increased end-diastolic pressure carries the substantial risk of aggravating pre-existent myocardial ischemia.

Key words. Cardiac function—Heart failure—Muscle mechanics—Excitation contraction coupling

University of Calgary, Calgary, Alberta T2N 4N1, Canada.

Introduction

Coronary artery disease and myocardial damage caused by ischemia form the most common etiology of congestive heart failure. It is striking, however, that the final stage of congestive heart failure is remarkably similar irrespective of the initial cause of the syndrome. At normal filling pressures, the failing heart is unable to eject a sufficient amount of blood to meet the needs of the body, apparently because the cardiac muscle cannot generate a forceful contraction or relax rapidly after contraction. Lack of force of the heartbeat is predominantly caused by derailment of the processes in the failing cardiac cells to manipulate Ca^{2+} ions, despite the drive from regulatory systms to enhance cardiac performance.

In this chapter, we discuss some of the effects of development of heart failure on excitation-contraction coupling. Cardiac contraction requires that, at the cellular level, electrical excitation of the myocyte surface membrane is translated into force development and sliding of the contractile filaments. The resultant force can be modulated by inotropic interventions and by control of the length at which the cardiac fibers contract. The rate of relaxation, in turn, is modulated by interventions that determine the rate of Ca^{2+} extrusion from the cell. Regulatory mechanisms operate in the intact heart both through the cellular effects of catecholamines and via regulation of the interval between heartbeats; variation of fiber length is brought about by control of the filling pressure of the ventricles via control of venous system compliance.

Excitation–Contraction Coupling: Structural Aspects

Figure 1 shows a typical longitudinal electron microscopical section through a rapidly fixed trabecula of the right ventricle of a rat heart. The cell border is delineated by a glycoprotein layer overlying the sarcolemma, which invaginates the cell near the Z-

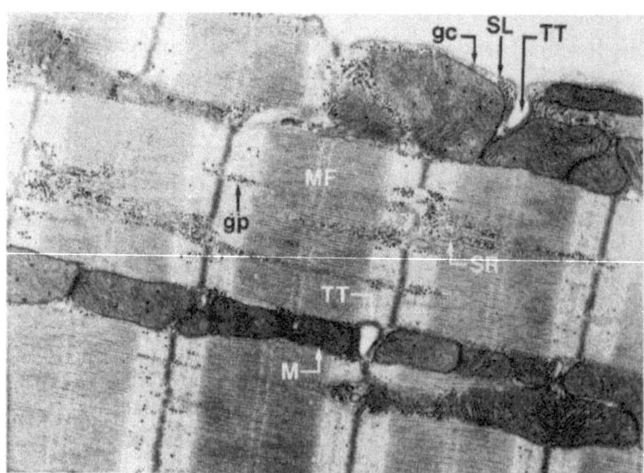

FIG. 1. Longitudinal electron microscopic section of a cardiac cell of the rat. The close spatial interrelationships between the surface membrane, the T-tubuli (*TT*), and the sarcoplasmic reticulum (*SR*) are prominent. The SR is seen to envelop the myofibrils, which constitute the major (60%) organelle in the cell; the mitochondria occupy another 40% of the intracellular space. The granules are glycogenolytic particles (*gp*). *SL*, Sarcolemma; *gc*, glycoprotein; *M*, mitochondria; *MF*, myofibril

lines of the myofibrils. These resultant transverse tubuli (T-tubuli) are rich in dihydropyridine-sensitive Ca^{2+} channels and Na^{+}/ Ca^{2+} exchange molecules. T-tubuli make contact with a longitudinal compartment contained in a lipid membrane, the sarcoplasmic reticulum (SR), which is a prominent Ca^{2+}-accumulating organelle in mammalian myocardium. The longitudinal component of the SR envelops the myofibrils and is densely covered by Ca^{2+} ATPase molecules, which drive Ca^{2+} into the SR where it is buffered by calsequestrin. The pump rate of the SR-Ca^{2+} pump molecules depends on [Ca^{2+}] and is modulated by the degree of phosphorylation of phospholamban [1]. The terminal cisternae of the SR, which abut the T-tubuli, contain Ca^{2+} channels (recognized by their high affinity for ryanodine) that mediate Ca^{2+} release from the SR. Sixty percent of the intracellular space is occupied by the contractile proteins arranged in the array of sarcomeres in the myofibrils. The remainder of the cell is virtually completely occupied by mitochondria adjacent to the sarcomeres.

Excitation–Contraction Coupling: Functional Aspects

It is well accepted that, during the action potential, Ca^{2+} ions enter the cell through the dihydropyridine-sensitive channels. Most of this Ca^{2+} is immediately bound to the Ca^{2+} pump in the SR. The amount of Ca^{2+} that enters the cell per second will therefore depend on the duration of the action potential and on the heart rate. The Ca^{2+} ions that enter through the T-tubuli start the process of excitation–contraction coupling by triggering release of Ca^{2+} from the SR. The amount of Ca^{2+} that is released is proportional to the Ca^{2+} content of the SR and dictates the force of the cardiac contraction [2].

The released Ca^{2+} activates the contractile machinery and, with the Ca^{2+} that entered the cell through the Ca^{2+} channels during the action potential, is eliminated from the cytosol via two pathways: (1) sequestration by the SR [3], and (2) the removal of the remaining Ca^{2+} from the cell, probably mostly extruded through the cell membrane by the low-affinity, high-capacity Na^{+}/Ca^{2+} exchanger immediately after activation of contraction. The low-capacity, high-affinity Ca^{2+} pump in the cell membrane further lowers the cytosolic Ca^{2+} level during the diastolic interval [1]. Evidently, Ca^{2+} efflux through the membrane balances the influx during the action potential in the steady state.

It follows that the amount of Ca^{2+} which is accumulated in the SR depends on the heart rate and on the duration of the action potentials and the rate of sequestration by the SR. It also follows from the foregoing that a fraction of the Ca^{2+} ions which are involved in the activation of contraction during the heartbeat recirculate into the SR and become available for activation of the next beat, so that the force of the heartbeat depends on the force of the previous beat. Also, it takes some time before the Ca^{2+} release process recovers completely from the last release and sequestered Ca^{2+} can again be released from the SR. The force of the heartbeat will, therefore, depend strongly on heart rate and on the interval preceding an individual beat as well as on the duration of the action potential [4]. These properties can be summarized in a model of the cardiac cell as in Fig. 2 [5]. Notably, the force of contraction of the quietly beating heart is probably only approximately one-third of the maximal force that can be generated by the contractile filaments, implying that the normal heart has a wide margin before its contractile reserves are exhausted.

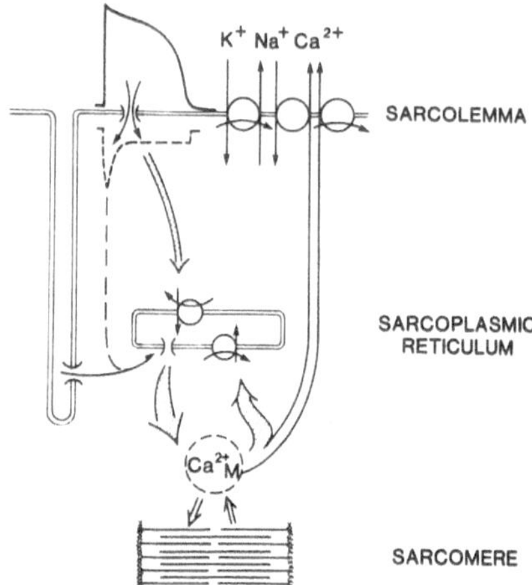

FIG. 2. Diagram of the excitation–contraction coupling system in the cardiac cell. During the action potential, Ca^{2+} enters the cells as a rapid influx followed by a maintained component of the slow inward current (*dashed line*). Ca^{2+} entry does not lead directly to force development because the Ca^{2+} ions that enter are rapidly bound to binding sites on the SR, which envelops the myofibils. The rapid influx of Ca^{2+} via the T-tubuli is thought to induce release of Ca^{2+} from a release compartment (RC) in the SR, by triggering opening of Ca^{2+} channels in the terminal cisternae, thus activating the contractile filaments to contract. Relaxation follows because the cytosolic Ca^{2+} is sequestered again in an uptake compartment (UC) of the SR and partly extruded through the cell membrane by the Na^+/Ca^{2+} exchanger and by the low-capacity, high-affinity Ca^{2+} pump. The force of contraction is thus determined by the circulation of Ca^{2+} from the SR to the myofilaments and back to the SR and by the amount of Ca^{2+} that has entered during the preceding action potential. The relaxation rate of the twitch depends on the rate of Ca^{2+} dissociation from the myofilaments and on the rates of Ca^{2+} sequestration and extrusion. It is important to note that the process of Na^+/Ca^{2+} exchange is electrogenic so that Ca^{2+} extrusion through the exchanger leads to a depolarizing current

It is possible to load the SR excessively with Ca^{2+} ions. This may occur following damage of cardiac cells [6,7] or following exposure to interventions that increase the intracellular Ca^{2+} levels (digitalis, high $[Ca^{2+}]_0$, high stimulus rate). Ca^{2+} overload of the SR leads to spontaneous Ca^{2+} release, which may be linked to the normal heartbeat in the form of after-contractions. Spontaneous uncoordinated Ca^{2+} release between heartbeats can be observed as spontaneous contractions of small groups of sarcomeres in cells of the myocardium, and gives rise to fluctuations of the light-scattering properties of the muscle [8,9]. Spontaneous Ca^{2+} release not only increases the diastolic force generated by the contractile filaments but also reduces systolic twitch force development [10,11]. It is clear from this model of excitation–contraction coupling that spontaneous release of Ca^{2+} ions is likely to lead to depolarization of the sarcolemma. This may result from activation of Ca^{2+}-dependent channels such as transient inward currents [12,13] or by the electrogenic action of the Na^+/Ca^{2+} exchanger itself.

Cytosolic Ca^{2+} Transients and Force Development

Figure 3 shows force and the estimated cytosolic $[Ca^{2+}]$ as a function of time during a twitch at an external $[Ca^{2+}]_0$ of 1 m*M* in a preparation loaded by microinjection of Fura-2 salt [14]. The results are representative of contractions at long and short end-systolic sarcomere length, i.e., at the extremes of the function curve of mammalian cardiac muscle. Figure 3 shows typical behavior of cardiac muscle; i.e., peak and time-course of the Ca^{2+} transients are remarkably independent of length, albeit the relaxation phase differs between the short and long muscle.

The interpretation of the $[Ca^{2+}]_i$ transients and their relationship to force development requires some caution, because it is known that full activation of the contractile system requires saturation of the Ca^{2+} sites on all troponin-C molecules for which approximately 70 µ*M* of Ca^{2+} is needed, with simultaneous binding of another 50 µ*M* of Ca^{2+} to calmodulin [15]. Thus, even activation of the muscle at only 25% of maximum, such as in this example, is accompanied by a Ca^{2+} turnover of nearly 30 µ*M*. It is clear that only a small fraction of this Ca^{2+} is "visible" in the cytosol. It is also clear that the kinetic aspects of Ca^{2+} release and binding to ligands in the cytosol require further study to understand the relationship between Ca^{2+} transport and force devel-

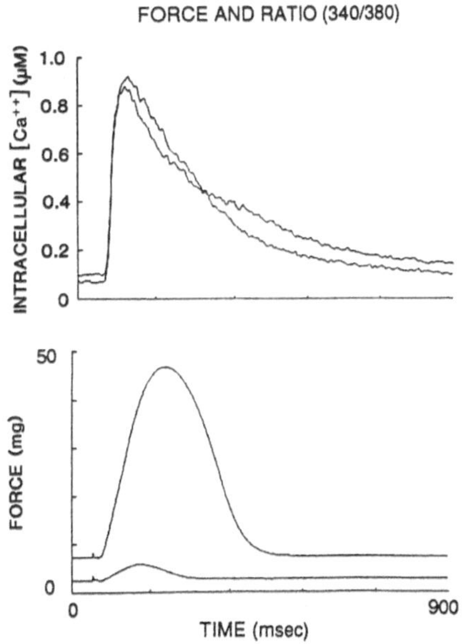

Fig. 3. *Top*: Ratio of the fluorescence at 340-nm excitation to that at 380-nm excitation during a twitch at two sarcomere lengths. The corresponding forcest at the two sarcomere lengths are shown the *bottom panel*: the larger force corresponds to that recorded at an end-systolic sarcomere length of 2.15 µm and the smaller force corresponds to a sarcomere length of 1.6 µm. The Ca^{2+} transient at the longer sarcomere length relaxes more rapidly in the early phase of the relaxation than that at the shorter sarcomere length. Later in the relaxation process, however, the Ca^{2+} transient at the longer sarcomere length relaxes more slowly than that at the shorter sarcomere length. As a result, the Ca^{2+} transients at the two sarcomere lengths cross over during the relaxation period

opment in the cardiac cell [16]. With these considerations in mind, we believe that the cytosolic Ca^{2+} transients can teach us about a number of important properties of the excitation–contraction coupling process in the heart.

It is unlikely that stretch of the muscle results only in a major change in the rate or amount of Ca^{2+} release; in that case, the rate of rise and the peak of the Ca^{2+} transient should increase with stretch. Rather, the changes in the kinetics of the transient with stretch are consistent with the hypothesis that the force–length relationship (as shown in Fig. 3) is determined by the length-dependent sensitivity of the contractile system (Fig. 4), which resides in Ca^{2+} affinity of troponin for Ca^{2+} as a function of stretch [17,18]. The mechanism underlying the effect of stretch on the sensitivity of the contractile system for Ca^{2+} ions has not been resolved yet; Babu et al. [19] have proposed that troponin-C plays an essential role, but this has been disputed by the group of Moss [20]. An alternative explanation is that the apparent affinity of the sarcomere for Ca^{2+} ions depends on the maximal number of (force-generating) cross-bridges [18,21]. This hypothesis, however, also implies that in the stretched myocardium much more Ca^{2+} is bound, as has been experimentally confirmed [22].

One can explain the observation that the peak amplitude of the $[Ca^{2+}]_i$ transient is unchanged if one assumes also that more Ca^{2+} has been released by the SR while more Ca^{2+} is bound to the contractile filaments following stretch. The relaxation phase of the Ca^{2+} transient depends also on the rate of binding of Ca^{2+} ions to the sarcolemmal Na^+/ Ca^{2+} exchanger and to the Ca^{2+} pump of the SR and their respective transport rates. Thus, studies at fixed muscle length allow evaluation of the influence of interventions on the rate of transport by the two latter mechanisms. Such evaluations have indicated that the rate of decline of the Ca^{2+} transient depends about equally on the activity of the Na^+/Ca^{2+} exchanger and the SR-Ca^{2+} pump [23]; the latter is modulated by phosphorylation of phospholamban caused by second-messenger activation and probably by Ca^{2+}-activated calmodulin action [24].

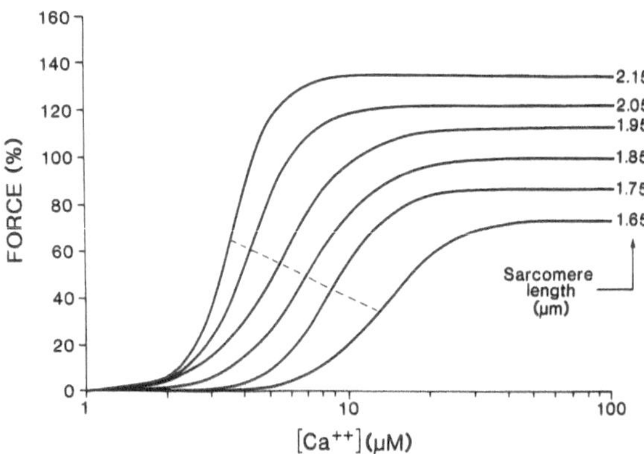

FIG. 4. Relationship between active force and $[Ca^{2+}]$ in skinned muscles at various sarcomere lengths (*SL*) (given in μm next to the appropriate curves) from data obtained by replotting the curves of Fig. 3 (bottom panel). The solid lines are best fits to the modified Hill equation: Force = Maximum force × $[Ca^{2+}]^n/(K^n + [Ca^{2+}]^n)$. The *dashed line* joins the points corresponding to half-maximal activation for each curve. It can be seen that increasing the SL raises both maximum force production and the Ca^{2+} sensitivity of the myofibrils. (After [18])

Relationship Between Force Development, $[Ca^{2+}]_i$, and Sarcomere length: Implications for Starling's Law of the Heart

On arrival of Ca^{2+} ions in the vicinity of the regulatory proteins on actin, contraction starts as a result of Ca^{2+} binding to troponin-C. Figure 4 shows that the response of the contractile apparatus to Ca^{2+} ions consists of a classical dose–response curve of a sigmoidal shape. It is now known that these dose–response curves shift toward lower $[Ca^{2+}]_i$ with stretch of the sarcomeres [17,18]; this results in the length dependence of force development (Figs. 4 and 5).

Studies that have shown the length-dependent sensitivity of the contractile system to Ca^{2+} ions have reported an EC_{50} (i.e., the $[Ca^{2+}]$ at which force is half-maximal) ranging from 3 to 13 μ*M* [17,18]. More recently studies on intact muscle have revealed that in the intact cell the sensitivity of the contractile apparatus to Ca^{2+} is an order of magnitude higher (0.65 μ*M*) [16]. We have observed recently that length-dependent Ca^{2+} sensitivity in intact muscle is also pronounced (EC_{50}, 0.2 and 0.5 μ*M* at long and short sarcomere lengths, respectively) [24a].

We and others have observed that the force–sarcomere length relationships that have been obtained at the peak of the twitch are rather insensitive to the way in which the muscle contracted before the peak of force; i.e., the same relationships were obtained following isometric contractions compared with contractions in which substantial shortening occurred. This unique character of the "end-systolic" force–length

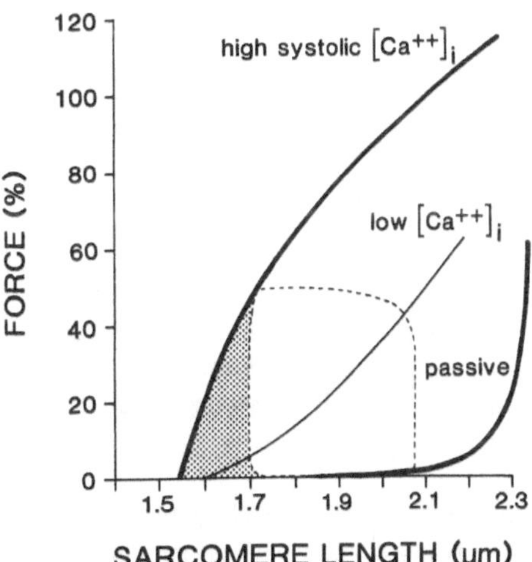

Fig. 5. The force–sarcomere length relationship observed at a low (*light line*) and at a high (*heavy line*) $[Ca^{2+}]_0$ in intact cardiac trabeculae. The relationships coincide with the relationships predicted by the curves in Fig. 4. The *dotted line* indicates the force–sarcomere length loop (FSLL) during cardiac systole; the *dotted area* together with the FSLL form the basis for the pressure–volume area in the heart; the *heavy dashed line* is the passive force–sarcomere length relationship indicating that stretch of the cardiac sarcomere is limited to a length of ~2.3–2.4 μm. Although these relationships were derived from the rat heart, they were also observed in other mammalian hearts

relationship has been found in many mammalian species [25]. The same unique relationship has been observed in the intact heart in the form of the end-systolic pressure–volume relationship (ESPVR) [26]. The existence of a fixed ESPVR predicts that the stroke volume of a ventricle which ejects blood against a fixed aortic pressure will increase in linear proportion to end-diastolic volume. The resultant stroke volume end-diastolic volume relationship has an intercept with the abscissa at a volume at which the contracting ventricle can just maintain aortic pressure during isovolumic contraction. This behavior of the heart has been known as Starling's law of the heart since the beginning of this century [27].

It is important to realize, with respect to the failing heart, that conventional clinical knowledge states that the Starling curve shows a biphasic behavior with a descending limb at high end-diastolic pressures. To understand the relationship between this behavior and the properties of the contractile elements at the level of the cell, let us return to the force–sarcomere length relationship in Fig. 5, which shows that the unstimulated passive muscle displays elastic behavior; i.e., with stretch, an elastic force is set up. It is clear from Fig. 5 that this passive force rises steeply when the sarcomeres in mammalian myocardium are stretched to a length of approximately 2.3 μm, thus preventing any further stretch of the myocardium. At this length the overlap between the contractile filaments is close to optimal.

The consequence of the steep rise of the passive force–length relationship is that the sarcomeres in the heart cannot be stretched beyond a length that corresponds with the length at which optimal overlap of the contractile filaments is present. Hence, the systolic force–sarcomere length relationship increases with stretch of the sarcomeres to ~2.3–2.4 μm and simply stops when further stretch becomes impossible. Similarly, the ESPVR shows a monotonic rise of systolic pressure with end-systolic volume without a declining phase.

The predicted stroke volume–end-diastolic volume relationship also stops at high end-diastolic volumes because of the limit to further stretch. However, before this happens the ventricle must operate at such high end-diastolic pressures that subendocardial ischemia is likely to occur, which causes a decreased performance of the heart and thus a reduction of stroke volume. Moreover, at high end-diastolic pressure in the left ventricle the pressure in the pulmonary artery rises, which decreases the output of the right ventricle and thus total cardiac output. It is evident that the fall of cardiac output at high end-diastolic pressure is secondary to factors which are extrinsic to the contractile elements in the cardiac muscle. Evidently, the development of mitral regurgitation at high end-diastolic left-ventricular pressure further aggravates the reduction of cardiac output in the severely compromised heart.

Another mechanism limiting left-ventricular function in coronary artery disease at high end-diastolic pressures consists of subendocardial ischemia. Ischemia of the subendocardium may compromise the mechanical function of the failing ventricle by ultimately limiting the ability of the myocytes to generate high-energy phosphates at an adequate rate [28]. It is important to recognize this mechanism because adequate early treatment of ischemia by coronary vasodilator drugs or revascularization and by lowering end-diastolic left-ventricular pressure (P_{LVED}) may restore the function of otherwise stunned or hibernating myocardium.

The Force–Interval and Force–Frequency Relationships

The force of contraction of normal mammalian myocardium is strongly dependent on the frequency of stimulation. Force is minimal at intervals below the physiological range (~12/min) and increases to a maximum at a heart rate of 200/min. An increase in heart rate is accompanied by a gradual increase of force of contraction to a new steady state; this phenomenon has been recognized as the staircase phenomenon of the heart since Bowditch's work [29]. The mechanism underlying this staircase follows from the foregoing model of excitation–contraction coupling; i.e., with regular stimulation at an increased heart rate, the SR accumulates an increased amount of Ca^{2+} while Na^+/Ca^{2+} exchange will gradually slow down as a result of accumulation of Na^+ ions during each cardiac cycle because of Na^+ entry in exchange for Ca^{2+} ions [30].

When the heart beats at a steady rate and an extra-systole occurs at a short interval, the Ca^{2+} that enters the cell through the T-tubuli will meet Ca^{2+} release channels which have not yet fully recovered from the last release process. This explains why the extra-systole has a small amplitude. In contrast, the action potential during the extra-systole may have a normal duration or may be even longer than the control action potential and is accompanied by a larger than normal Ca^{2+} influx into the cell causing post-extra-systolic potentiation. If the heart rate returns to a regular rate after the extra-systole, one observes that first post-extra-systole is potentiated, which is followed by an exponential decline of the degree of post-extra-systolic potentiation. The decline of post-extra-systole potentiation is assumed to reflect recirculation of Ca^{2+} ions via the SR during elimination of Ca^{2+} from the cytosolic space (see foregoing). The rate at which post-extra-systolic potentiation declines to the steady state is determined by the fraction of Ca^{2+} that recycles via the SR [31].

Autonomic Regulation of Contractile Force

Autonomic control over cardiac function employs interaction between parasympathetic and sympathetic nerves and modulates second-messenger systems in the cardiac muscle cells. Important segments of this process fail during heart failure; therefore, we provide a brief summary here with emphasis on the inotropic effects. In the normal individual at rest, activity of the sympathetic nerves is relatively low compared to the parasympathetic activity. The parasympathetic nerves influence the heart directly and influence the sympathetic effects on the heart at two levels; at a prejunctional level the parasympathetic nerve activity inhibits the release of sympathetic transmitter, which is especially noticeable when the sympathetic activity is prominent; at the postjunctional level, the two systems interactively control the second-messenger status of the cardiac myocyte.

The human heart contains both β-receptors and α_1-receptors, although the density of the latter is rather low and they cause only a modest level of phosphatidylinositol hydrolysis [32]. The density of the β_1-receptors is threefold higher than that of the β_2-receptors in the human heart [33]; their activation initiates production of cyclic adenosine monophosphate (cAMP) by adenylcyclase via the stimulatory G protein G_s. This effect is counteracted by activation of the inhibitory G protein G_s, which is activated by binding of acetylcholine α_1-agonists and opioids to their respective receptors. Binding of 25% of the β-receptors to their agonist leads to 50% maximal cAMP production in the human heart, indicating that the reserve of the β-receptor system in the human heart is modest compared to that of other species [34,35].

Protein phosphorylation resulting from cAMP-dependent protein kinase-A (PKA) leads to activation of sarcolemmal Ca^{2+} channels, enhanced Ca^{2+} uptake by the SR from phosphorylation of phospholamban, and a reduced sensitivity of the contractile filaments to Ca^{2+} ions as a result of phosphorylation of troponin-I [36]. The consequence of these events in the cardiac cell is enhanced Ca^{2+} loading of the SR and increased Ca^{2+} release so that the force of contraction is increased despite the reduced filament sensitivity. The rate of rise of contraction is increased because of the faster force generation by the actomyosin interactions. Relaxation is also accelerated by the faster dissociation of Ca^{2+} ions from the filaments as well as the increased rate of uptake of Ca^{2+} by the SR, so that the duration of systole is shortened and the diastolic interval is longer.

Activation of α_1-receptors leads to G-protein-mediated activation of phospholipase C and production of inositol trisphosphate (IP_3) and diacylglycerol (DAG). The role of IP_3 in myocardial excitation–contraction coupling has not been fully elucidated; it may facilitate Ca^{2+}-induced Ca^{2+} release [37]. However, this effect cannot be large because the $[Ca^{2+}]_i$ transient is unaffected by exposure of myocytes to α_1-agonists [38], while the Ca^{2+} current is unchanged by α_1-agonists [38]. Activation of α_1-receptor-mediated IP_3 and DAG formation is positive inotropic despite the aforementioned inhibitory effect on adenylcyclase activity.

The increase in force is mediated by several mechanisms: α_1-receptor activation causes protein kinase C- (PKC-) independent stimulation of the Na^+/H^+ antiporter, inducing alkalinization of the cytosol because of the enhanced exchange of H^+ for Na^+. The precise molecular mechanism of the effect on the antiporter is still unknown, but the alkalinization causes higher sensitivity of the filaments to Ca^{2+} ions [39]. The latter effect is enhanced by DAG-dependent phosphorylation of the contractile proteins, causing a further increase of the myofilament sensitivity for Ca^{2+} so that a larger force is developed for the same Ca^{2+} release from the SR. The mechanism of DAG-dependent positive inotropism has not been resolved yet, but seems to be shared by receptor activation by α_1-agonists and by muscarinic receptor activation [40].

Anomalies of Excitation–Contraction Coupling in Heart Failure

Increased sympathetic activity is generally considered to be of prime importance in the failling heart [41]. The increase is accompanied by a decrease of the cardiac norepinephrine (NE) content with an increase of the content of the precursor dopamine [42–46] as well as an increase of the spillover of NE in the coronary effluent [47]. These features are restricted to the ventricle that is involved in failure [48–50] and lead to a decrease of the β_1-receptor density [51–54]. The response to sympathetic stimulation is also reduced as a result of the downregulation of the β_1-receptors [55]. This phenomenon is quite well known from the inability of patients with congestive heart failure to increase their heart rate on exertion to the same rate as a normal individual [56].

The density of the high-affinity muscarinic M-receptors [57] also decreases, but no significant change occurs in the receptor density for β_2-agonists [33], adenosine (A_1) [58], or α_1-agonists [32]. Coupling of the β-receptors with adenylcyclase is decreased by an increase of the inhibitory G protein [54,59,60] without a significant change in G_s [61,62] or phosphodiesterase III or IV activity [58,63,64]. These changes lead to a decreased activity of adenylcyclase and decreased cAMP levels [65,66], while the inhibitory effect of activation of the muscarinic receptors is reduced [57]. Reduced cAMP levels and DAG levels are accompanied by reduced force development, as one would expect on the basis of the effects of cellular protein phosphorylation just described.

In addition to the changes in sympathetic regulation of force development, there is also ample evidence that the histological and cellular composition of the heart changes considerably during the development of cardiac failure. Systemic and local angiotensin II generation have been shown to play an important role in this process, and the development of failure can be substantially delayed by treatment with angiotensin-converting enzyme inhibitors [67]. An important change at the histological level is remodeling of the heart, accompanied by increased production of collagen, which underlie changes in the passive compliance of the myocardium [37,68].

The stimuli that are involved in the induction of growth and gene expression during overload of the heart cause significant changes in the protein isoforms participating in excitation–contraction coupling. In the foregoing we have indicated the changes in the sarcolemmal receptor system, which controls second-messenger production. Further sarcolemmal changes consist of a reduction of the number of Ca^{2+} channels with a decrease of the channel mRNA concentration [59,69,70]. There is a modest reduction of the density of the Na^+/K^+ pump [71,72], which is accompanied by an increase of action potential duration [73].

The changes in the SR in the failing heart are probably of outstanding importance. It has convincingly been shown that the diastolic Ca^{2+} levels in the cell are elevated and that the relaxation of the Ca^{2+} transient is delayed in heart failure [73]. Simultaneously, the threshold for Ca^{2+} release is increased [29] and the amount of Ca^{2+} released by the SR is reduced [73]. These observations indicate changes in the Ca^{2+} pump of the SR and in the ryanodine-sensitive Ca^{2+} channels. Several studies have shown that indeed the rate of Ca^{2+} uptake by the SR is reduced [73–75]; this may be based on a reduction of the density of the Ca^{2+} pumps in the SR, as would be expected on the basis of the reduced mRNA level for the SR Ca^{2+} protein [69]. On the other hand, the reduction of the Ca^{2+} pump activity is disputed [76,77].

Modified gene expression in heart failure also leads to several changes in the contractile protein system. The overall concentrations of actin and myosin have been reported to be substantially reduced [78], although this observation needs further confirmation [79]. The physicochemical structure of actin has been reported to be modified in heart failure, leading to impaired formation of filamentous actin [80,81]. Changes of the regulatory proteins of the contractile apparatus have also been found in the form of expression of the fetal form of troponin T, i.e., TnT_2 [82], while TnJ and TnC are not significantly changed [83]. The ATPase activity of actin-activated myosin is reduced [79,84], indicating that the interaction between actin and myosin is modified despite the fact that no new myosin is expressed in the human heart. In the rodent heart, the development of cardiac overload and failure is always accompanied by the change of myosin from the fast V_1 form to the slower V_3 form that is exclusively present in the human heart.

Reduced Sarcomere Response to Ca^{2+} Ions in Heart Failure

Cross-Bridge Kinetics

Alpert and collaborators have concluded from myothermal measurements that the cross-bridge in failing myocardium spends a substantially longer time in the attached/force-supporting state than in normal myocardium. The larger force–time integral of the cross-bridges from failing heart compensates in part for the decreased Ca^{2+} cycle [85,86] and is accompanied by a lower ATP cost [79,84] but leads to a lower maximal velocity of shortening of failing muscle [87]. The slower kinetics of the cross-bridge [88] have been found in human myocardium, which contains only V_3 isomyosin; hence, it is unlikely that isoform shifts of myosin are responsible for the observed changes to cross-bridge kinetics. Other changes in the contractile or regulatory proteins may explain both the low force output and the long force–time integral; e.g., modified concentrations of actin and myosin [78,79], a modified physicochemical structure of actin [80,81], or changes in regulatory proteins of the contractile apparatus such as expressions of the fetal form of troponin T, i.e., TnT_2 [82] or myosin light chains [89].

Effects of Stretch

Since the work of Jewell's group [17] and our work [18], it has been known that these dose-response curves shift toward lower $[Ca^{2+}]_i$ with stretch of the sarcomeres [17]. This phenomenon results in length dependence of force development [90], and underlies the well-known "end-systolic" pressure–volume relationship (ESVPR) in the mammalian heart [26]. It has recently been shown by Schwinger and colleagues that skinned muscle from the failing human heart is less sensitive to stretch than muscle from a normal heart [91]. The mechanism underlying this anomaly in the failing heart is unknown; however, it is not likely to reside in the properties of troponin-C because there is little or no change in this protein in failure [83]. The model proposed by Landesberg and Sideman [21], which couples the kinetic properties of the cross-bridges with the kinetics of Ca^{2+} binding, might provide an attractive link between the changes in the kinetics of the cross-bridges observed by Alpert and colleagues and the reduced effect of stretch on Ca^{2+} sensitivity [91].

Shortening Velocity

As has been indicated, the velocity of shortening of the sarcomeres is lower in heart failure than in the normal heart [87]. The velocity of shortening clearly depends on the load to the muscle relative to the muscle isometric force. One expects, therefore, that the velocity in the failing muscle is reduced because a lesser force is developed in failure. In the absence of an external load, it is expected that the velocity depends on internal loads to the contractile apparatus [92]. We have previously observed that internal elastic restoring forces decrease the maximal velocity of sarcomere shortening [93], and we have shown that, even when the elastic forces are negligible, the velocity of sarcomere shortening is limited by the viscous properties of the muscle [94]. It is known that the elastic properties of failing heart are changed, but little or nothing is known about the viscosity of failing myocardium.

Effects on Contraction of the Failing Heart

The combined effect of the changes that take place at the cellular and at the tissue level in the failing heart result in the modified passive and active properties of the failing myocardium. The passive compliance of the myocardium is decreased. Active force development is reduced even at a low heart rate but even more strikingly at a higher heart rate [95,96]. The lack of response to the sympathetic drive in vivo is important in this respect, but the negative force–frequency relationship is also observed in isolated tissue, implying that it probably is a result of functional impairment of the SR. Inability of the SR to rapidly eliminate Ca^{2+} ions from the cytosol results in slow ventricular relaxation, or lusitropic impairment, causing further elevation of diastolic pressure. An important consequence of the susceptibility of the SR to Ca^{2+} overload and spontaneous Ca^{2+} release is that spontaneous elevation of the cytosolic Ca^{2+} concentration may lead to depolarization of the surface membrane [6,7]. Induction of extra-systoles may, then, set up reentry tachycardias, which are notorious in the clinical presentation of heart failure.

References

1. Carafoli E (1987) Intracellular calcium homeostasis. Annu Rev Biochem 56:395–433
2. Stern MD (1992) Theory of excitation-contraction coupling in cardiac muscle. Biophys J 63:497–517
3. Bers DM (1991) Excitation-contraction coupling and cardiac contractile force. Kluwer, Boston
4. Wohlfart B (1979) Relationship between peak force, action potential duration, and stimulus interval in rabbit myocardium. Acta Physiol Scand 106:395–409
5. Schouten VJA, Deen JK, de Tombe PP, Verveen AA (1987) Force-interval relationship in heart muscle of mammals. A calcium compartment model. Biophys J 51:13–26
6. Mulder BJM, de Tombe PP, ter Keurs HEDJ (1989) Spontaneous and propagated contractions in rat cardiac trabeculae. J Gen Physiol 93:943–961
7. Daniels MCG, Fedida D, Lamont C, ter Keurs HEDJ (1991) Role of the sarcolemma in triggered propagated contractions in rat cardiac trabeculae. Circ Res 68:1408–1421
8. Kort AA, Lakatta EG (1984) Calcium-dependent mechanical oscillations occur spontaneously in unstimulated mammalian cardiac tissues. Circ Res 54:396–404
9. Lakatta EG, Lappe DL (1981) Diastolic scattered light fluctuation, resting force and twitch force in mammalian cardiac muscle. J Physiol 315:369–394
10. Lakatta EG, Jewell BR (1977) Length-dependent activation. Its effect on the length-tension relation in cat ventricular muscle. Circ Res 40:251–257
11. Stern MD, Capogrossi MC, Lakatta EG (1988) Spontaneous calcium release from the sarcoplasmic reticulum in myocardial cells: mechanisms and consequences. Cell Calcium 9:247–256
12. Colquhoun D, Neher E, Reuter H, Stevens CF (1981) Inward current channels activated by intracellular calcium in cultured cardiac cells. Nature 294:752–754
13. Han X, Ferrier GR (1992) Ionic mechanisms of transient inward current in the absence of Na^+/Ca^{2+} exchange in rabbit cardiac Purkinje fibres. J Physiol 456:19–38

14. Backx PH, ter Keurs HEDJ (1993) Fluorescent properties of rat cardiac trabeculae microinjected with fura-2 salt. Am J Physiol 264:H1098–H1110
15. Wier WG, Yue DT (1986) Intracellular calcium transients underlying the short-term force-interval relationship in ferret ventricular myocardium. J Physiol 376:507–530
16. Backx PH, Gao WD, Azan-Backx MD, Marbon E (1995) The relationship between contractile force and intracellular [Ca^{2+}] in intact rat cardiac trabeculae. J Gen Physiol 105:1–19
17. Hibberd MG, Jewell BR (1982) Calcium- and length-dependent force production in rat ventricular muscle. J Physiol 329:527–540
18. Kentish JC, ter Keurs HEDJ, Ricciardi L, Bucx JJJ, Noble MIM (1986) Comparison between the sarcomere length-force relations of intact and skinned trabeculae from rat right ventricle. Circ Res 58:755–768
19. Babu A, Sonnenblick EH, Gulati J (1988) Molecular basis for the influence of muscle length on myocardial performance. Science 240:74–76
20. Hofmann PA, Hartzell HC, Moss RL (1991) Alterations in Ca-sensitive tension due to partial extraction of C-protein from rat skinned cardiac myocytes and rabbit skeletal muscle fibres. J Gen Physiol 97:1141–1163
21. Landesberg A, Sideman S (1994) Mechanical regulation in the cardiac muscle by couping calcium kinetics with crossbridge cycling; a dynamic model. Am J Physiol 267:H779–H795
22. Hofmann PA, Fuchs F (1988) Bound calcium and force development in skinned cardiac muscle bundles: effect of sarcomere length. J Mol Cell Cardiol 20:667–677
23. Bridge JHB, Spitzer KW, Ershler PR (1988) Relaxation of isolated ventricular cardiomyocytes by a voltage-dependent process. Science 241:823–825
24. Morris GL (1993) The regulatory interaction between phospholamban and SR (Ca^{2+}-Mg^{2+}) ATPase. Thesis, The University of Calgary, Alberta, Canada
24a. Hollander EH, ter Keurs HEDJ (1992) Sarcomere length and the force–Ca^{2+} relationship in rat myocardium (abstract). Circ 92–8:1254
25. ter Keurs HEDJ, Kentish JC, Bucx JJJ (1987) On the force-length relation in myocardium. In: ter Keurs HEDJ, Tyberg JV (eds) Mechanics of the circulation. Martinus Nijhoff, Dordrecht, The Netherlands, pp 91–105
26. Sagawa K, Suga H, Shoukas AA, Bakalar KM (1977) End-systolic pressure-volume ratio: a new index of ventricular contractility. Am J Cardiol 40:748–753
27. Patternson SW, Piper H, Starling EH (1914) The regulation of the heart beat. J Physiol 48:465–513
28. Regitz V, Fleck E (1992) Myocardial adenine nucleotide concentrations and myocardial norepinephrine content in patients with heart failure secondary to idiopathic dilated or ischemic cardiomyopathy. Am J Cardiol 69:1574–1580
29. Bowditch HP (1871) Uber die Eigenhumlichkeiten der reizbarkeit, welche die muskelfasern des herzens Zeigen. Ber Sachs Ges Wiss 23: 652–689
30. Boyett MR, Franpton JE, Harrison SM, Kirby MS, Levi AJ, McCall EM, Milner DR, Orchard CH (1992) The role of intracellular calcium, sodium and pH in rate-dependent changes of cardiac contractile force. In: Noble MIM, Seed WA (eds) The Interval-force relationship of the heart; Bowditch revisited. Cambridge University Press, Cambridge, pp 111–172
31. ter Keurs HEDJ (1992) Post-extrasystolic potentiation and its decay. In: Noble MIM, Seed WA (eds) The interval-force relationship of the Heart; Bowditch revisited. Cambridge University Press, Cambridge, pp 259–276
32. Bristow MR, Minobe W, Rasmussen R, Hershberger RE, Hoffman BB (1988) Alpha-1 adrenergic receptors in the nonfailing and failing human heart. J Pharmacol Exp Ther 247:1039–1045
33. Bristow MR, Ginsburg R, Umans V, Fowler M, Minobe W, Rasmussen R, Zera P, Menlove R, Shah P, Jamieson S (1986) Beta 1- and beta 2-adrenergic-receptor subpopulations in nonfailing and failing human ventricular myocardium: coupling of both receptor subtypes to muscle contraction and selective beta 1-receptor down-regulation in heart failure. Circ Res 59:297–309

34. Brodde OE, Zerkowski HR, Borst HG, Maier W, Michel MC (1989) Drug- and disease-induced changes of human cardiac beta 1- and beta 2-adrenoceptors. Eur Heart J 10(suppl B):38–44
35. Brodde OE, Hillemann S, Kunde K, Vogelsand M, Zerkowski HR (1992) Receptor systems affecting force of contraction in the human heart and their alterations in chronic heart failure. J Heart Lung Transplant 11:S164–S174
36. Solaro R, Moir AJG, Perry SV (1976) Phosphorylation of troponin 1 and the inotropic effect of adrenaline in the perfused rabbit heart. Nature 262:615–617
37. Fabiato A (1986) Inositol P_3-induced release of calcium from the sarcoplasmic reticulum of skinned cardiac cells. Biophys J 49:190a
38. Terzic A, Puceat M, Clement O, Scamps F, Vassort G (1992) $Alpha_1$- adrenergic effects on intracellular pH and calcium and on myofilaments in single rat cardiac cells. J Physiol 447:275–292
39. Puceat M, Clement-Chomienne O, Terzic A, Vassort G (1993) $Alpha_1$-adrenoceptor and purinoceptor agonists modulate Na-H antiport in single cardiac cells. Am J Physiol 33:H310–H319
40. Puceat M, Terzic A, Clement O, Scamps F, Vogel SM, Vassort G (1992) Cardiac $alpha_1$-adrenoceptors mediate positive inotropy via myofibrillar sensitization. Trends Pharmacol Sci 13:263–265
41. Rose CP, Burgess JH, Cousineau D (1985) Tracer norepinephrine kinetics in coronary circulation of patients with heart failure secondary to chronic pressure and volume overload. J Clin Invest 76:1740–1747
42. Regitz V, Leuchs B, Bossaller C, Sehested J, Rappolder M, Fleck E (1991) Myocardial catecholamine concentrations in dilated cardiomyopathy and heart failure of different origins. Eur Heart J 12(suppl D):171–174
43. Yoshikawa T, Handa S, Yamada T, Wainai Y, Suzuki M, Tani M, Nakamura Y (1990) [Role of adrenergic-neural regulation in failing heart due to aortic regurgitation in rabbits] (in Japanese). Kokyu To Junkan (Respir Circ) 38:153–158
44. Regitz V, Sasse S, Bossaller C, Strasser R, Schuler S, Hetzer R, Fleck E (1989) [Myocardial catecholamine content in heart failure—I: Regional distribution in explanted hearts. Comparison between dilated cardiomyopathy and coronary heart disease] (in German). Z Kardiol 78:751–758
45. Regitz V, Sasse S, Fleck E (1989) [Myocardial catecholamine content in heart failure—II: Measurement in endomyocardial biopsies, reference systems, normal values] (in German). Z Kardiol 78:759–763
46. Pierpont GL, Francis GS, DeMaster EG, Olivari MT, Ring WS, Goldenberg IF, Reynolds S, Cohn JN (1987) Heterogeneous myocardial catecholamine concentrations in patients with congestive heart failure. Am J Cardiol 60:316–321
47. Eichorn EJ, Bedotto JB, Malloy CR (1990) Effects of β-adrenergic blockade on myocardial function and energetics in congestive heart failure. Circulation 82:473–483
48. Vescovo G, Jones SM, Harding SE, Poole-Wilson PA (1989) Isoproterenol sensitivity of isolated cardiac myocytes from rats with monocrotaline-induced right-sided hypertrophy and heart failure. J Mol Cell Cardiol 21:1047–1061
49. White CW, Mirro MJ, Lund DD, Skorton DJ, Pandian NG, Kerber RE (1986) Alterations in ventricular excitability in conscious dogs during development of chronic heart failure. Am J Physiol 250:H1022–H1029
50. Rouleau JL, Juneau C, Stephens H, Shenasa H, Parmley WW, Brutsaert DL (1989) Mechanical properties of papillary muscle in cardiac failure: importance of pathogenesis and of ventricle of origin. J Mol Cell Cardiol 21:817–828
51. Fowler MB, Bristow MR (1985) Rationale for beta-adrenergic blocking drugs in cardiomyopathy. Am J Cardiol 55:120D–124D
52. Liang CS, Frantz RP, Suematsu M, Sakamoto S, Sullebarger JT, Fan TM, Guthinger L (1991) Chronic beta-adrenoceptor blockade prevents the development of beta-adrenergic subsensitivity in experimental right-sided congestive heart failure in dogs. Circulation 84:254–266

53. Vatner DE, Vatner SF, Fujii AM, Homcy CJ (1985) Loss of high affinity cardiac beta-adrenergic receptors in dogs with heart failure. J Clin Invest 76:2259–2264
54. Perreault CL, Shannon RP, Komamura K, Vatner SF, Morgan JP (1992) Abnormalities in intracellular calcium regulation and contractile function in myocardium from dogs with pacing-induced heart failure. J Clin Invest 89:932–938
55. Fowler MB, Laser JA, Hopkins GL, Minobe W, Bristow MR (1986) Assessment of the beta-adrenergic receptor pathway in the intact failing human heart: progressive receptor down-regulation and subsensitivity to agonist response. Circulation 74:1290–1302
56. Yamabe H, Kobayashi K, Takata T, Fukuzaki H (1987) Reduced chronotropic reserve to the metabolic requirement during exercise in advanced heart failure with old myocardial infarction. Jpn Circ J 51:259–264
57. Vatner DE, Lee DL, Schwarz KR, Longabaugh JP, Fujii AM, Vatner SF, Homcy CJ (1988) Impaired cardiac muscarinic receptor function in dogs with heart failure. J Clin Invest 81:1836–1842
58. Hershberger RE, Feldman AM, Bristow MR (1991) A_1-Adenosine receptor inhibition of adenylate cyclase in failing and nonfailing human ventricular myocardium. Circulation 83:1343–1351
59. Denniss AR, Colucci WS, Allen PD, Marsh JD (1989) Distribution and function of human ventricular beta adrenergic receptors in congestive heart failure. J Mol Cell Cardiol 21:651–660
60. Eschenhagen T, Mende U, Nose M, Schmitz W, Scholz H, Haverich A, Hirt S, Döring V, Kalmar P, Höppner W (1992) Increased messenger RNA level of the inhibitory G protein alpha subunit G_i alpha-2 in human end-stage heart failure. Circ Res 70:688–696
61. Longabaugh JP, Vatner DE, Vatner SF, Homcy CJ (1988) Decreased stimulatory guanosine triphosphate binding protein in dogs with pressure-overload left ventricular failure. J Clin Invest 81:420–424
62. Schnabel P, Böhm M, Gierschik P, Jakobs KH, Erdmann E (1990) Improvement of cholera toxin-catalyzed ADP-ribosylation by endogenous ADP-ribosylation factor from bovine brain provides evidence for an unchanged amount of G_s alpha in failing human myocardium. J Mol Cell Cardiol 22:73–82
63. Bethke T, Klimkiewicz A, Kohl C, von der Leyen H, Mehl H, Mende U, Meyer W, Neumann J, Schmitz W, Scholz H (1991) Effects of isomazole on force of contraction and phosphodiesterase isoenzymes I-IV in nonfailing and failing human hearts. J Cardiovasc Pharmacol 18:386–397
64. Gristwood RW, English TA, Wallwork J, Sampford KA, Owen DA (1987) Analysis of responses to a selective phosphodiesterase III inhibitor, SK&F 94120, on isolated myocardium, including human ventricular myocardium from "end-stage" failure patients. J Cardiovasc Pharmacol 9:719–727
65. Urasawa K, Sato K, Igarashi Y, Kawaguchi H, Yasuda H (1992) A mechanism of catecholamine tolerance in congestive heart failure—alterations in the hormone-sensitive adenylyl cyclase system of the heart. Jpn Circ J 56:456–461
66. Feldman MD, Copelas L, Gwathmey JK, Phillips P, Warren SE, Schoen FJ, Grossman W, Morgan JP (1987) Deficient production of cyclic AMP: pharmacologic evidence of an important cause of contractile dysfunction in patients with end-stage heart failure. Circulation 75:331–339
67. Pfeffer JM, Pfeffer MA (1988) Angiotensin converting enzyme inhibition and ventricular remodeling in heart failure. Am J Med 84:37–44
68. Weber KT, Anversa P, Armstrong PW, Brilla CG, Burnett JC Jr, Cruickshank JM, Devereux RB, Giles TD, Korsgaard N, Leier CV (1992) Remodeling and reparation of the cardiovascular system. J Am Coll Cardiol 20:3–16
69. Takahashi T, Allen PD, Lacro RV, Marks AR, Dennis AR, Schoen FJ, Grossman W, Marsh JD, Izumo S (1992) Expression of dihydropyridine receptor (Ca^{2+} channel) and calsequestrin genes in the myocardium of patients with end-stage heart failure. J Clin Invest 90:927–935
70. Dixon IM, Lee SL, Dhalla NS (1990) Nitrendipine binding in congestive heart failure due to myocardial infarction. Circ Res 66:782–788

71. Norgaard A, Bjerregaard P, Baandrup U, Kjeldsen K, Reske-Nielsen E, Thomsen PE (1990) The concentration of the Na,K-pump in skeletal and heart muscle in congestive heart failure. Int J Cardiol 26:185–190
72. Allen PD, Schmidt TA, Marsh JD, Kjeldsen K (1992) Na,K-ATPase expression in normal and failing human left ventricle. Basic Res Cardiol 87(suppl 1):87–94
73. Beuckelmann DJ, Nabauer M, Erdmann E (1992) Intracellular calcium handling in isolated ventricular myocytes from patients with terminal heart failure [see comments]. Circulation 85:1046–1055
74. Limas CJ, Olivari MT, Goldenberg IF, Levine TB, Benditt DG, Simon A (1987) Calcium uptake by cardiac sarcoplasmic reticulum in human dilated cardiomyopathy. Cardiovasc Res 21:601–605
75. Maeba T (1985) Calcium-binding of cardiac sarcoplasmic reticulum and diastolic hemodynamics in volume overloaded canine hearts. Jpn Circ J 49:163–170
76. Movsesian MA, Bristow MR, Krall J (1989) Ca^{2+} uptake by cardiac sarcoplasmic reticulum from patients with idiopathic dilated cardiomyopathy. Circ Res 65:1141–1144
77. Movsesian MA (1987) Calcium uptake by sarcoplasmic reticulum and its modulation by cAMP-dependent phosphorylation in normal and failing human myocardium. Basic Res Cardiol 87(suppl 1):277–284
78. Wiegand V, Schuler S, Figulla H, Warnecke H, Kreuzer H (1986) Contractile proteins in human dilatative cardiomyopathy. J Mol Cell Cardiol 18(suppl):74a
79. Pagani ED, Alousi AA, Grant AM, Older TM, Dziuban SW Jr, Allen PD (1988) Changes in myofibrillar content and Mg-ATPase activity in ventricular tissues from patients with heart failure caused by coronary artery disease, cardiomyopathy, or mitral valve insufficiency. Circ Res 63:380–385
80. Karsanov NV, Nizharadze GI, Pirtskhalaishvili MP, Eristavi JJ, Khundadze OS, Kuchava LE, Pavlenishvili IV, Shengelia LV (1985) Superprecipitation of hybrid actomyosin containing pathologic actin from failing hearts of adults and infants. Gen Physiol Biophys 4:417–423
81. Karsanov NV, Pirtskhalaishvili MP, Semerikova VJ, Losaberidze NS (1986) Thin myofilament proteins in norm and heart failure. I. Polymerizability of myocardial Straub actin in acute and chronic heart failure [published erratum appears in Basic Res Cardiol (1986) 81(6):646]. Basic Res Cardiol 81:199–212
82. Anderson PAW, Malouf NN, Oakeley AE, Pagani ED, Allen PD (1991) Troponin T isoform expression in humans; a comparison between normal and failing adult heart, fetal heart, fetal heart, and adult and fetal skeletal muscle. Circ Res 69:1226–1233
83. Hunkeler NM, Kullman J, Murphy A (1991) Troponin-I isoform expression in human heart. Circ Res 69:1409–1414
84. Alousi AA, Grant AM, Etzler JR, Cofer BR, Van der Bel-Kahn J (1990) Reduced cardiac myofibrillar Mg-ATPase activity without changes in myosin isozyme patterns in patients with end-stage heart failure. Mol Cell Biochem 96:79–88
85. Schwinger RHG, Bohm M, Erdmann E (1992) Inotropic and lusitropic dysfunction in myocardium from patients with dilated cardomyopathy. Am Heart J 123:116–128
86. Mulieri LA, Hasenfuss G, Leavitt B, Allen PD, Alpert NR (1992) Altered myocardial force frequency relation in human heart failure. Circulation 85:1743–1750
87. Hajjar RJ, Gwathmey JK (1992) Cross-bridge dynamics in human ventricular myocardium; regulation of contractility in the failing heart. Circulation 86:1819–1826
88. Hasenfuss G, Mulieri LA, Leavitt BJ, Allen PD, Haeberle JR, Alpert NR (1992) Alteration of contractile function and excitation-contraction coupling in dilated cardiomyopathy. Circ Res 70:1225–1232
89. Margossian SS, White HD, Caulfield JB, Norton P, Taylor S, Slayter HS (1992) Light $chain_2$ profile and activity of human ventricular myosin during dilated cardiomyopathy; identification of a causal agent for impaired myocardial function. Circulation 85:1720–1733
90. ter Keurs HEDJ, Bucx JJJ, de Tombe PP, Backx P, Iwazumi T (1987) The effects of sarcomere length on force and velocity of shortening in cardiac muscle. In: Sugi H, Pollack GH (eds) Molecular mechanisms of muscle contraction. Plenum, New York

91. Schwinger RHG, Bohm M, Koch A, Schmidt U, Morano I, Eissner H-J, Ueberfuhr P, Reichart B, Erdman E (1994) The failing human heart is unable to use Frank Starling's mechanism. Circ Res 74(5):959–969
92. Chiu Y-L, Ballow EW, Ford LE (1982) Internal viscoelastic loading in cat papillary muscle. Biophys J 40:109–120
93. Daniels MCG, Noble MIM, ter Keurs HEDJ, Wohlfart B (1984) Velocity of sarcomere shortening in rat cardiac muscle: relationship to force, sarcomere length, calcium, and time. J Physiol 355:367–381
94. de Tombe PP, ter Keurs HEDJ (1992) An internal viscous element limits unloaded velocity of sarcomere shortening in rat myocardium. J Physiol 454:619–642
95. ter Keurs HEDJ, Backx PH, Banijamali H, Maclntosh B, Gao WD (1992) Calcium release and force development in rat myocardium. In: Frank GB, Bianchi CP, ter Keurs HEDJ (eds) Excitation-contraction coupling in skeletal, cardiac, and smooth muscle. Plenum, New York, pp 199–212
96. Nozawa T, Yasumura Y, Futaki S, Tanaka N, Igarashi Y, Goto Y, Suga H (1987) Relation between oxygen consumption and pressure-volume area of in situ dog heart. Am J Physiol 253:31–40

Transient Hemodynamic Perturbations: An Approach to the Study of Ventricular Interaction and Restoring Forces

Martin M. LeWinter

Summary. In the intact circulation, it is often difficult to delineate the significance and magnitude of complex, interactive phenomena. To overcome this problem, we have developed approaches using transient (single-beat) hemodynamic perturbations that eliminate the confounding effects of series interactions and filling variations. In experimental studies in open-chest dogs, we have employed rapid changes in right heart volume to quantitate direct right-to-left diastolic ventricular interaction gain. Using transient increases in right-ventricular afterload, we have delineated the magnitude of left-to-right systolic interaction and documented an effect of increased right-ventricular afterload that resulted in slowing of left-ventricular relaxation. Using transient perturbations in left heart loading we have correspondingly demonstrated left-to-right systolic interaction and an effect of left-ventricular afterload on venous return to the right heart. Using a servomotor connected to the left atrium that allows abrupt clamping of the left-atrial pressure at a predetermined value, we have produced single-beat, nonfilling, left-ventricular diastolic intervals that allow measurement of the fully relaxed left-ventricular pressure with the ventricle in the end-systolic configuration, thereby providing an estimate of the presence and magnitude of restoring forces. Under physiological conditions in open-chest dogs, we have documented the presence of significant restoring forces at the lower levels of physiological filling. Finally, parallel clinical studies suggest that the ability to generate restoring forces may play a role in determining the level of exercise tolerance in patients with systolic ventricular dysfunction.

Key words. Ventricular interaction—Ventricular loading—Restoring forces—Ventricular function—Diastolic function

Introduction

In the heart, contractile performance and relaxation/filling phenomena are often considered as separate processes. However, numerous prior observations indicate that these are closely linked. Thus, for example, it has been appreciated for many years that the level of passive filling of one ventricle influences the passive filling of the contralateral ventricle (direct interaction), in turn influencing contractile perfor-

Cardiology Unit, University of Vermont College of Medicine, Burlington, VT 05401, USA.

mance via the Frank–Starling mechanism [1–4]. Similarly, the magnitude and timing of systolic loading are important determinants of relaxation rate [5–7]. For a number of years, we have employed the approach of using transient (single-beat) hemodynamic perturbations to shed light on both ventricular interaction and systolic–diastolic relationships in the intact circulation. This method in effect temporarily renders the circulation an open-loop system, eliminating confounding effects of series interactions and filling variations. In this chapter, we review selected aspects of this work.

Methods

Experimental Studies

All the studies to be described have been performed in anesthetized, open-chest dogs. In many studies, the specific sinus node inhibitor zatebradine, or ULFS49 (Boehringer-Ingelheim, Ridgefield, CT, USA), has been used to slow the sinus rate [8], allowing atrial pacing at constant simulation frequency during the experimental protocols. Intravascular pressures have been measured with MILLAR (Houston, TX, USA) micromanometers. Ventricular volumes have been measured with a conductance catheter (Leycom, Oegstgeest, The Netherlands) and regional segment lengths with sonomicrometers. For experiments assessing the generation of ventricular restoring forces, a servomotor system [9,10] connected to the left atrium was employed to rapidly modify and then clamp the left-atrial pressure at a specified value below the left-venricular diastolic pressure during a single contraction, thus causing a reversed ventriculoatrial pressure gradient during the subsequent diastole and preventing ventricular inflow, i.e., a nonfilling diastole during which the ventricle is allowed to fully relax. A negative pressure in the fully relaxed ventricle is considered to indicate both the presence and magnitude of restoring forces generated during contraction; i.e., the ventricle is below its equilibrium volume (V_{eq}), the volume at which the pressures inside and outside the chamber are equal [11].

All data were recorded on line, digitized, averaged, and processed using custom-designed software programs.

Clinical Studies

The clinical studies described employed commercially available echocardiographic Doppler technology and standard views and formulas for ventricular volumes and ejection fraction. Ventricular shape is expressed as eccentricity, the ratio of the long to short left-ventricular axis. Exercise echocardiography was accomplished using upright bicycle ergometry, employing sequential 25-W increments in workload.

Results and Discussion

Ventricular Interaction

We have analyzed the magnitude of direct diastolic ventricular interaction by producing changes in ventricular volume during a single beat [12]. The methodology allows studies to be accomplished with both an intact and absent pericardium. Figure 1 shows the effects on left-ventricular diastolic pressure of a single-beat decrease in

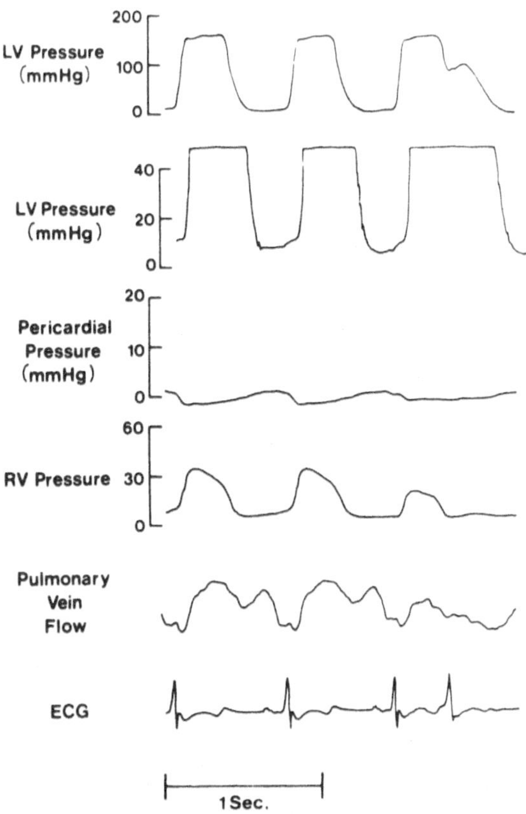

FIG. 1. Example of right-to-left direct diastolic interation. During the second diastolic interval, a rapid decrease in right heart volume is produced. This is associated with a realtively large decrease in right-ventricular (*RV*) diastolic pressure and a smaller decrease in left-ventricular (*LV*) diastolic pressure. Pulmonary venous flow is unchanged until the next systolic interval. Note also a rapid, small decrease in pericardial pressure in association with the change in right heart volume. A premature beat occurs following the abrupt change in volume. *ECG*, electrocardiogram. (From [15] with permission)

right-ventricular diastolic pressure caused by the combination of sudden bicaval occlusion and rapid suction of blood through a large cannula inserted into the right ventricle via the right atrium. Note that there is an immediate decrease in right-ventricular diastolic pressure and a simultaneous small decrease in left-ventricular diastolic pressure in the absence of a simultaneous change in pulmonary venous flow. This decrease in left-ventricular diastolic pressure thus occurred before the effects of decreased right heart volume influenced the filling of the left heart via series interaction. However, changes in pulmonary venous flow occur exceedingly rapidly and are evident on the next beat. Thus, to quantitate direct diastolic interaction in the intact heart without the confounding effects of a systolic, series interaction requires such a single-beat, transient perturbation.

On average, we have found that right-to-left interaction gain, expressed as the ratio of the change in left-ventricular end-diastolic pressure caused by a given change in right-ventricular end-diastolic pressure, averages about 0.33 over a wide range of filling pressures. Also shown in Fig. 1 is the pericardial pressure over the right ventricle measured with a flat balloon. Note that the pericardial pressure also changes rapidly during the transient perturbation and is important for coupling the changes in diastolic pressure related to ventricular interaction. In our preparation, direct diastolic ventricular interaction is markedly attenuated after removal of the pericardium.

In addition to changes in right-ventricular diastolic loading, we have also found that altered right-ventricular systolic loading influences left-ventricular diastolic function [13]. In Fig. 2, an abrupt increase in right-ventricular afterload produced by a diastolic pulmonary artery constriction is seen to result in immediate augmentation of left-ventricular systolic pressure and stroke work (direct right-to-left systolic interaction) and in addition slows left-ventricular isovolumic pressure decline in association with considerable regional inhomogeneity during left-ventricular isovolumic relaxation [14]. We suspect that the latter is related to alterations in septal contraction because of the higher right ventricular afterload. In any case, this phenomenon demonstrates that increased right-ventricular systolic load directly augments left-ventricular systolic performance but slows left-ventricular relaxation.

Abrupt perturbations in left-ventricular loading produced somewhat more complex effects than similar perturbations produced on the right side of the heart. A rapid decrease in left heart volume produced by suction of blood from the left ventricle during a single diastole, analogous to the previously described decrease in right heart volume, resulted in a decrease in right-ventricular diastolic pressure on the same beat and an increase in right-ventricular stroke work on the following contraction [15]. However, in contrast to the lack of an immediate effect of decreased right heart volume on pulmonary venous flow, a similar decrease in left heart volume is associated with an immediate increase in caval flow. We suspect that the latter is caused by a rapid shift in the interventricular septum to the left caused by the abrupt reduction in left heart volume, rendering the right ventricle more distensible and thereby increasing right heart inflow via a suction effect.

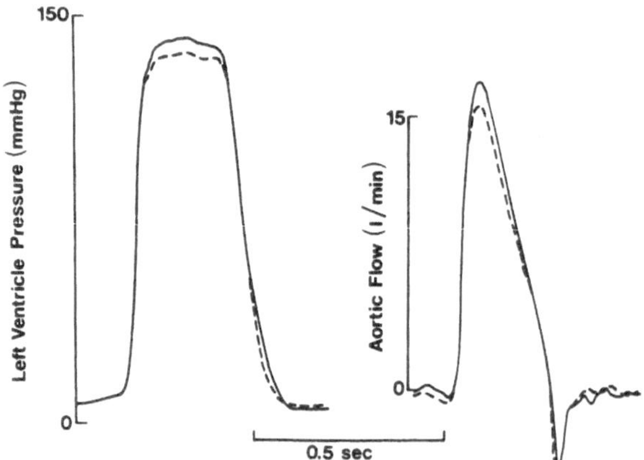

FIG. 2. Left-ventricular pressure (*left*) and aortic flow (*right*) immediately before (*dashed line*) and then after (*solid line*) an abrupt pulmonary artery constriction (*PAC*). Note the increase in left-ventricular systolic pressure and aortic flow (direct systolic interaction) and the slowing of left-ventricular pressure fall in association with inreased right-ventricular afterload. (From [13] with permission)

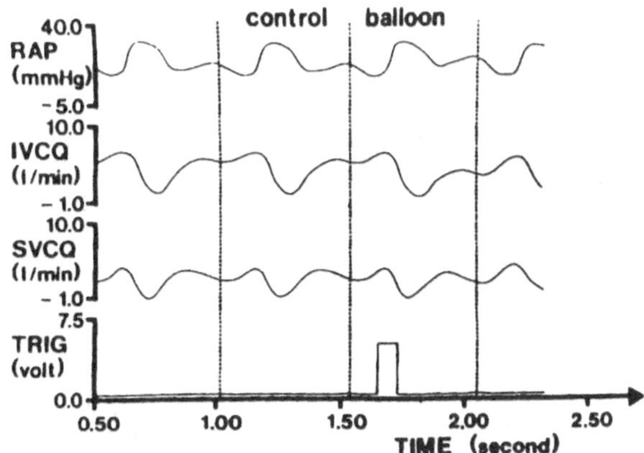

FIG. 3. Right atrial pressure (*RAP*) and inferior and superior vena caval flow (*IVCQ*, *SVCQ*) before and immediately after an abrupt increase in left-ventricular afterload (marked on *TRIG* channel at bottom). A small decrease in caval flow is associated with increased left-ventricular afterload

A perturbation physiologically more realistic and relevant can be produced by abruptly increasing left-ventricular afterload by rapid inflation of a balloon in the aorta during diastole. The subsequent right-ventricular contraction has a slightly higher systolic pressure, a manifestation of direct left-to-right systolic interaction related to increased left-ventricular pressure generation. Of note, during this contraction a small but consistent decrease in caval flow is observed, simultaneous with the left-ventricular afterload increase (Fig. 3). Thus, variations in left-ventricular systolic loading can actually influence venous return to the right heart without invoking a change in right-ventricular diastolic distensibility. The mechanism of this phenomenon is as yet unresolved, but a logical hypothesis is that in some way the higher left-ventricular afterload and resultant change in contraction mechanics physically alter right atrial or caval distensibility.

It is obvious from the foregoing that direct interaction effects, which in turn represent links between systolic and diastolic function, are intrinsically small in magnitude when studied during the course of transient hemodynamic perturbations, which once again allow separation of direct from series effects in the intact circulation. This does not mean that these effects, which buffer changes in loading conditions, are unimportant, because small unbalanced changes in left or right heart output obviously would have major and disastrous consequences if they were cumulative over a number of heartbeats. Thus, it is our view that these small effects are important in beat-to-beat modulation of left and right heart output in response to constantly varying loading conditions.

Restoring Forces

By definition, when the ventricle contracts to a volume less then its V_{eq}, the pressure in the fully relaxed chamber is negative with reference to its surroundings and restoring forces are present. The ability of the ventricle to generate restoring forces has interested physiologists for many years because dissipation of these forces during relaxation/filling causes diastolic suction, in effect allowing filling to occur at lower levels of ventricular pressure than otherwise would have been present [16,17]. However, quantitative assessment of these forces is exceedingly difficult in the intact circulation. To overcome this problem, Nikolic, Yellin, and colleagues studied restoring forces in anesthetized open-chest dogs using a technique in which the native mitral valve was replaced with an electrically controlled prosthesis. By maintaining the prosthesis in a closed position following ventricular systole, filling was prevented and the ventricle allowed to fully relax. The fully relaxed ventricular pressure during these nonfilling diastoles was considered to indicate the presence of restoring forces when its value was less than 0, i.e., the volume was less than V_{eq}. Using this procedure, it was concluded that under relatively physiological loading conditions, substantial restoring forces are ordinarily present with fully relaxed ventricular pressures of the order of −5 to −10 mmHg. It was also noted that the process of filling retards left-ventricular pressure fall when compared to nonfilling diastoles.

Although these important studies represent a novel approach to the study of restoring forces, replacement of the native mitral valve with a prothesis has effects on ventricular function that could themselves alter the magnitude of restoring forces [18]. The servomotor system we have described allows us to generate nonfilling diastoles without the need to remove the mitral valve. Our initial studies using this system [10] were designed to determine the fully relaxed left-ventricular pressure over a normal range of filling conditions in the open-chest, anesthetized dog. An example of a left-atrial pressure clamp beat resulting in a nonfilling diastole is shown in Fig. 4. Note that the signal obtained from the left-ventricular conductance catheter confirms the absence of filling. In this example, the left-ventricular end-diastolic pressure was approximately 5 mmHg and the fully relaxed pressure was slightly negative.

In these studies, we varied ventricular filling by manipulating venous return (caval occlusions, volume infusions). We found that at left-ventricular end-diastolic pressures in the low physiological range [average, 4.0 ± 1.5 (SD) mmHg], the average fully relaxed pressure was −2.1 ± 1.9 mmHg. At a midphysiological range of left-ventricular

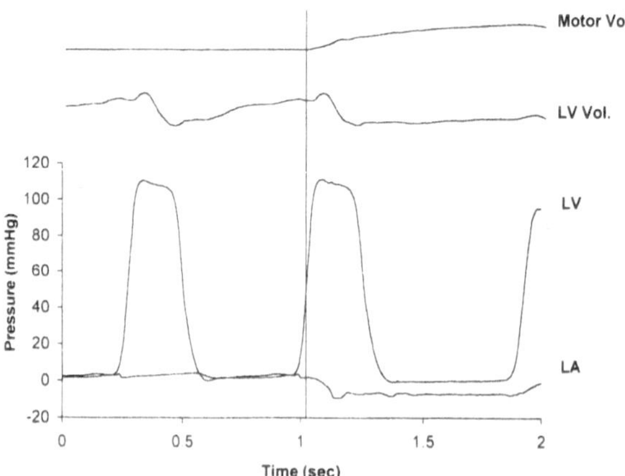

FIG. 4. Left-ventricular (*LV*) and left-atrial (*LA*) pressure and LV volume (*vol.*) (conductance catheter) immediately before and during a nonfilling diastole produced by abruptly clamping the LA pressure below the LV diastolic pressure

end-diastolic pressure (−8.1 ± 0.9 mmHg), the fully relaxed pressure averaged 0.2 ± 2.0 mmHg, very close to 0 mmHg, defining the V_{eq}. At left-ventricular end-diastolic pressure averaging 12.8 ± 2.1 mmHg, the fully relaxed pressure was significantly positive, averaging 1.1 ± 3.2 mmHg. Thus, using this approach we found that restoring forces were consistently present at the low range of physiological left-ventricular filling, but their magnitude appears to be smaller than that estimated in prior studies using a prosthetic mitral valve. Despite these smaller values for negative fully relaxed pressures, restoring forces of this magnitude are likely significant with respect to ventricular filling because they are not small in relation to the usual magnitude of gradient across the mitral valve early in diastole. Similar to Nikolic et al. [11], we also found that filling slows left-ventricular pressure fall.

In addition to delineating the level of restoring forces present over a physiological range of filling, we also assessed regional deformation during nonfilling diastoles. Using mutually orthogonal sonomicrometers implanted in the anterior free wall of the left ventricle, we found that when restoring forces are present, i.e., when the fully relaxed pressure is negative, there is considerable early diastolic regional deformation despite the absence of filling. As filling pressure was increased, the amount of diastolic regional deformation decreased but remained significant even with fully relaxed pressures greater than zero. This suggests the possibility that there are regional variations in equilibrium muscle length which result in regional deformation even though the overall ventricular volume is greater than the equilibrium value. Additional studies are required to test this hypothesis.

While producing comparable transient perturbations is not feasible in patients, recent clinical echocardiographic-Doppler studies from our laboratory suggest the possibility that the ability to generate restoring forces may be an important factor in patients with systolic left-ventricular dysfunction [19]. Experimental studies have demonstrated that as the left ventricle becomes more eccentric (a greater ratio of long to short axis), greater restoring forces are generated during contraction [20]. In addition, restoring forces may be linked to ventricular geometry by virtue of left-ventricular torsional deformation ("twist") [21]. The occurrence of torsional deformation may represent one mechanism by which the energy generated during contraction is actually stored (and then dissipated as restoring forces), once again facilitating filling at a lower pressure during the subsequent diastole.

In patients with predominantly or exclusively systolic dysfunction resulting from both ischemic and nonischemic disease processes, we have related the shape (eccentricity) of the left ventricle to exercise tolerance. We found that patients with more eccentric ventricles have better exercise tolerance. This finding could not be related to differences in end-diastolic volume, end-systolic volume, or ejection fraction, none of which was significantly correlated with exercise tolerance in this group of patients. We speculated that the mechanism of this observation was related to a shape-associated, increased ability to generate restoring forces, in turn allowing filling to occur at lower diastolic pressures during the stress of exercise. However, this study provided no direct proof of this hypothesis.

Accordingly, we recently investigated a separate group of subjects with systolic dysfunction and related ventricular shape to mitral valve Doppler flow velocity profiles [22]. We hypothesized that if a more eccentric ventricle is better able to generate restoring forces this would favor filling early during diastole. We found that there was a highly significant correlation between ventricular eccentricity and the occurrence of a greater proportion of filling early as opposed to late during diastole. As in the previous study, we could not demonstrate a significant correlation between end-diastolic or end-systolic volume or ejection fraction and Doppler velocity patterns. Thus, while not constituting a definitive proof, these observations support the general hypothesis that the ability to generate restoring forces during contraction is a determinant of both diastolic filling pattern and exercise tolerance in patients with chronic contractile dysfunction. We believe that additional studies to test this hypothesis are warranted.

Conclusions

Our results demonstrate the importance of distinguishing direct from indirect effects of alterations in loading on cardiac performance and the power of the use of transient hemodynamic perturbations to separate the two and provide insights into mechanics in intact preparations. This approach has allowed us to begin to delineate the complex effects of altered loading on the contractile function of both ventricles mediated by interaction. While it is well known that diastolic interaction can influence filling and contraction of the contralateral ventricle in a straightforward fashion via the Frank-Starling relation, our work also demonstrates that systolic loading has complex effects on contralateral systolic and diastolic function. Using the left-atrial servomotor system to produce transient perturbations in left-ventricular filling, we have been able to estimate the ability of the normal left ventricle to generate restoring forces over a physiological range of filling pressure in the open-chest, anesthetized dog. In future studies using this method, we hope to provide further clarification of the significance of these forces, especially in relation to pathophysiological states. Finally, our clinical studies suggest that the ability to generate restoring forces may be an important determinant of exercise tolerance in patients with left-ventricular contractile dysfunction who are otherwise indistinguishable with respect to ventricular volumes and ejection fraction.

References

1. Taylor RR, Covell J, Sonnenblick EH, Ross J Jr (1967) Dependence of ventricular distensibility on filling of the opposite ventricle. Am J Physiol 213:711–718
2. Elzinga G, van Grondelle R, Westerhof N, van den Bos GC (1974) Ventricular interference. Am J Physiol 226:941–947
3. Santamore WP, Lynch PR, Meier G, Hechman J, Bove AA (1976) Myocardial interaction between the ventricles. J Appl Physiol 41:362–368
4. Maughan WL, Kallman CH, Shoukas A (1991) The effect of right ventricular filling on the pressure-volume relationship of the ejecting canine left ventricle. Circ Res 49:382–388
5. Karliner JS, LeWinter MM, Mahler F, Engler R, O'Rourke RA (1977) Pharmacologic and hemodynamic influences on the rate of isovolumic left ventricular relaxation in the normal conscious dog. J Clin Invest 60:511–521
6. Gaasch WH, Carroll JD, Blaustein AS, Bing OHL (1986) Myocardial relaxation: effects of preload on the time course of isovolumetric relaxation. Circulation 73:1037–1041
7. Brutsaert DL, Sys SU (1989) Relaxation and diastole of the heart. Physiol Rev 69:1228–1315
8. Breall JA, Watanabe J, Grossman W (1993) Effect of Zatebradine on contractility, relaxation and coronary blood flow. J Am Coll Cardiol 21:471–477

9. Ingels NB Jr, Daughters GT II, Nikolic SD, DeAnda A, Moon MR, Bolger AF, Komeda M, Derby GC, Yellin EL, Miller DC (1994) Left atrial pressure-clamp servomechanism demonstrates LV suction in canine hearts with normal mitral valves. Am J Physiol 267:H354–H362
10. Bell SP, Fabian J, Higashiyama A, Chen Z, Tischler MD, Watkins MW, LeWinter MM (1996) Restoring forces assessed with left atrial pressure clamps. Am J Physiol 270:H1015–H1020
11. Nikolic S, Yellin EL, Tamura K, Vetter H, Tamura T, Meisner JS, Frater RWM (1988) Passive properties of canine left ventricle: diastolic stiffness and restoring forces. Circ Res 62:1210–1222
12. Slinker BK, Goto Y, LeWinter MM (1989) Direct diastolic interaction gain measured with sudden hemodynamic transients. Am J Physiol 256:H567–H573
13. Slinker BK, Goto Y, LeWinter MM (1989) Systolic direct ventricular interaction affects left ventricular contraction and relaxation in the intact dog circulation. Circ Res 65:307–315
14. Goto Y, Slinker BK, LeWinter MM (1989) Non-homogeneous left ventricular regional shortening during acute right ventricular pressure overload. Circ Res 65:43–54
15. LeWinter MM, Tischler MD (1994) Pericardium and ventricular interaction: effects on diastolic filling. In: Lorell BH, Grossman W (eds) Diastolic relaxation of the heart, 2nd edn. Kluwer, Boston, pp 233–241
16. Ebstein E (1904) Die diastole des herzens. Ergeb Physiol 3:123–194
17. Brecher GA (1958) Gritical review of recent work on ventricular diastolic suction. Circ Res 6:554–566
18. Hansen DE, Cahill PD, DeCampli WM, Harrison DC, Derby GC, Mitchell RS, Miller DC (1986) Valvular-ventricular interaction: importance of the mitral apparatus in canine left ventricular systolic performance. Circulation 73:1310–1320
19. Tischler MD, Niggel J, Borowski DT, LeWinter MM (1993) Relation between left venricular shape and exercise capacity in patients with left ventricular dysfunction. J Am Coll Cardiol 22:751–757
20. Nikolic SD, Yellin EL, Dahm M, Pajaro O, Frater RWM (1990) Relation between diastolic shape (eccentricity) and passive elastic properties in canine left ventricle. Am J Physiol 259:H457–H463
21. Beyar R, Ying FCP, Hausknecht M, Weisfeldt ML, Kass DA (1989) Dependence of left ventricular twist-radial shortening relations on cardiac cycle phase. Am J Physiol 257:H1119–H1126
22. Tischler MD, Ashikaga T, LeWinter M (1995) Relation between left ventricular shape and Doppler filling parameters in patients with left ventricular dysfunction secondary to coronary artery disease. Am J Cardiol 76: 553–556

Extrinsic Versus Intrinsic Determinants of the Diastolic Pressure–Volume Relation

David A. Kass

Summary. The left-ventricular diastolic pressure–volume relation (LV-DPVR) is determined by intrinsic properties of the chamber as well as by forces that are external to the heart, principally resulting from pericardial and right heart loads. Intrinsic factors can be divided into passive and active components. Passive elements include myocardial viscoelasticity, caused by structural proteins within the sarcomere, extracellular matrix proteins, and coronary vascular turgor, and chamber geometric factors such as wall thickness, cavity shape, and filling deformation. Active components of chamber stiffness are principally related to calcium handling and neurohumoral activation. This can manifest as delayed isovolumic relaxation that influences the early portion of the LV-DPVR, or altered diastolic "tone" which influences overall chamber distensibility. Among these factors, the dominant determinants of the LV-DPVR appear to be structural, because the relation is generally unaltered by acute pharmacological or physiological manipulations. External influences on the LV-DPVR are primarily direct ventricular interaction and constraining forces from the pericardium. These external loads are more easily altered by acute interventions and likely underlie many reported upward or downward shifts in the LV-DPVR from various interventions (e.g., vasodilators, channel blockers, exercise, ischemia). External constraining forces contribute a substantial percent (≈35%) of the resting diastolic pressures in humans, and thereby play an important role in determining the net dependence of cardiac output on filling pressure. In this chapter, we review the determinants of the LV-DPVR and focus on the relative influence of intrinsic and extrinsic factors.

Key words. Distole—Compliance—Left ventricle—Pericardium—Stiffness—Chamber geometry—Ventricular interaction

Introduction

The left-ventricular diastolic pressure–volume relation (LV-DPVR) characterizes the pressures required to fill a heart so as to generate cardiac output. The relation results from an interaction of both passive and active components intrinsic to the LV wall, as

Division of Cardiology, Department of Internal Medicine, Johns Hopkins Medical Institutions, Baltimore, MD 21287, USA

well as to external loads and constraining forces. Passive chamber properties are related to structural elements within the chamber wall, including myocytes, extracellular matrix proteins, coronary arterial and venous filling, and chamber geometry. Active processes are related to calcium handling and neurohumoral mediators such as intracardiac renin–angiotensin that can contribute to basal diastolic "tone." In general, passive structural factors appear to dominate, because acute changes in the LV-DPVR are relatively difficult to achieve while chronic disease conditions and therapies that result in chamber remodeling can substantially alter the relation. The LV-DPVR is also influenced by external forces, primarily from right heart and septal loading and to the pericardial constraint. These extrinsic factors are more easily manipulated by acute interventions and likely play a major role in mediating many reported acute changes in chamber distensibility.

One of the difficulties of evaluating the relative role of intrinsic and extrinsic contributors to the LV-DPVR has been the common practice of measuring data from single or steady-state cardiac cycles. This can obscure pressure–volume properties of the chamber within those caused by external constraints [1,2]. The common practice of measuring right-atrial pressure to index external load [3,4] is helpful but not necessarily adequate. The right heart is compliant, and substantial changes in filling and thus wall stretch (and cardiac output) are often associated with very small changes in pressure, differences that can easily fall within the noise range of many recording systems.

One approach to this problem is to determine LV-DPVRs from many cardiac cycles at varying preloads, measured before and during transient obstruction of the inferior vena cava [5,6]. An example of this technique is displayed in Fig. 1. Figure 1a shows resting (steady-state) pressure–volume loops. The lower boundary is the resting LV-DPVR, and this curvilinear relation is typical of the data reported in most studies. However, this dependence of chamber pressure on volume is not unique. When right heart load is abruptly diminished by inferior vena cava (IVC-)inflow obstruction, this

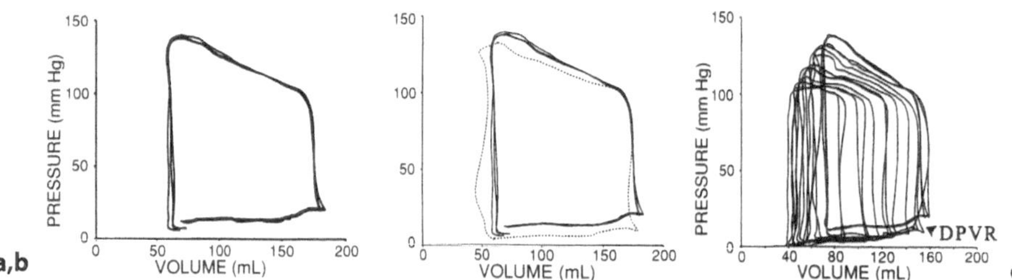

Fig. 1a–c. Paradigms for analysis of the diastolic pressure-volume relation (*DPVR*). **a** Pressure–volume loops at resting steady-state conditions. The lower boundary is the DPVR from this rest condition (see **c**). This curve measured throughout cardiac filling from a single condition is the most commonly obtained data point used for left-ventricular DPVR (LV-DPVR) analysis. **b** When right heart loading is abruptly diminished, in this instance by sudden obstruction of inferior vena cava inflow, there is a downward displacement of the LV-DPVR with minimal change in chamber volume (*dashed loop*). This reflects removal of external loading constraints and points to their often substantial influence on the resting curve. **c** If preload reduction is allowed to precede in the LV, a series of beats are obtained at gradually declining volumes. The LV-DPVR from these beats has a lower offset pressure and remains displaced downward. This latter relation is closer to characterizing intrinsic LV chamber properties. (Modified from [6])

relation suddenly shifts downward (Fig. 1b) with only minimal change in volume. This rapid pressure decline is not caused by changes in LV filling but rather by the sudden release of external loading forces. As venous inflow continues to decline, LV preload gradually falls along with chamber pressures, defining a new relationship displaced downward from the original resting LV-DPVR (Fig. 1c). This latter LV-DPVR is a closer estimate of intrinsic LV chamber properties.

This methodology has been employed to study influences on the LV-DPVR from regional ischemia [7], pharmacological agents used to treat hypertrophic [8] and dilated heart failure [9–12], adenosine-induced changes in coronary turgor [13], tachycardia pacing [14], and supine leg exercise [15]. The results have built a strong case for the notion that the relation is primarily dependent on passive structural factors which are not easily acutely altered. There are exceptions, such as severe calcium loading or global ischemia, and perhaps, as recently suggested, inhibition of cardiac angiotensin-converting enzyme (ACE) in hypertrophied hearts [16]. However, the preponderance of data indicates that active processes play a less prominent role. We also recently quantified the contribution of external forces to resting diastolic pressures [7], finding this to be rather high in patients with elevated end-diastolic pressures. This chapter reviews these studies and examines current understanding regarding the determinants of the LV-DPVR.

Structural Elements: Myocytes Versus Extracellular Matrix

While much attention has been focused on the role of the extracellular matrix proteins, particularly collagen, in determining chamber stiffness and the LV-DPVR [17–20], recent studies have shown that a substantial proportion of this stiffness stems from structures within the myocyte and sarcomeres themselves [21–23]. Force-length data in the operating range of sarcomere lengths (1.6–2.3 μm) from skinned myocytes and isolated myofibrils are very similar to those measured in isolated papillary muscle [21] (Fig. 2). This is intriguing, because only the latter includes the contributions of the extracellular matrix. Myocyte hypertrophy and the addition of sarcomeres in series can thereby increase stiffness. Recent studies have shown that regression of hypertrophy after treatment of hypertension is associated with significant increases in chamber compliance and reduction in LV mass, without consistent reduction in collagen content [24]. Collagen content and the type of collagen (i.e., increased type I relative to type III) [25,26] observed in hypertrophied and failure states likely contributes to chamber diastolic stiffness, particularly at higher sarcomere lengths, and some studies have directly correlated myocardial collagen content with diastolic DPVR steepening [27,28].

Structural proteins within myocytes that have been implicated in contributing to stiffness and viscosity include the sarcomeric protein titin and cytosolic microtubules. Titin is a large protein that contributes to the structural integrity of the sarcomere, linking the Z-bands. Recent data suggest that titin gene expression is reduced in heart failure, and this could contribute to chamber dilation [29]. It remains very difficult to biochemically manipulate titin, so its exact contribution to myocyte stiffness remains unknown. A second factor that is more easily altered is microtubules. Cooper and colleagues [30,31] have shown that microtubules are markedly increased in hypertrophied myocytes and that this rise is directly related to a fall in maximal unloaded cell

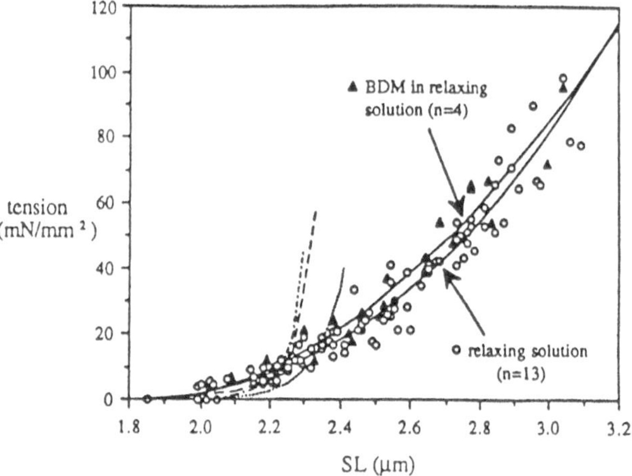

FIG. 2. Sarcomere length (*SL*) versus muscle tension plots from isolated myofibrils (*circles, triangles*) compared to curves obtained from intact trabeculae (*dashed lines*). The myofibril data are obtained in a relaxing solution, (i.e., passive data), with or without further inhibition of cross-bridge interaction with 2,3-butanedione monoxime (*BDM*) (*triangles*). At sarcomere lengths between 1.8 and 2.3 µm, which spans the normal operating range, passive stiffness of the sarcomere itself is very similar to that observed in more intact muscle. This indicates that much of the stiffness resides in intracellular proteins (such as titin). At higher sarcomere lengths, the intact muscle stiffness rises much faster, likely reflecting the influence of extracellular matrix proteins. (From [21], with permission)

shortening. This work has shown that the microtubules primarily influence the viscous properties of cells, and this could play an important role in elevation of the LV-DPVR during rapid filling in hypertrophied hearts.

Coronary Vascular Filling

Vogel et al. [32] reported that isolated rabbit hearts exposed to intracoronary adenosine display a reduction in chamber compliance and an upward and leftward DPVR shift. The dominant mechanism was thought to be an increase in coronary vascular engorgement (erectile effect), which increased the stiffness of the myocardial wall. In isolated hearts and tissues, basal coronary vasomotor tone is typically reduced from the intact state, so that changes in coronary perfusion can more easily lead to increased stiffness. However, in intact human left ventricles, vascular integrity and basal coronary tone are greater, and as we recently reported adenosine-induced coronary dilation does not similarly alter chamber stiffness [15].

In the recent study of Nussbacher et al. [13], adenosine was administered i.v. at 140 µg kg^{-1} min^{-1} for just over 6 min in nine patients with normal LV and coronary arteries. This dose had been shown to elevate pulmonary capillary wedge pressure by 3–12 mmHg, which was previously attributed to an "erectile effect" [33,34]. However, in our study [13], the rise in wedge pressure from 13.6 to 16.2 mmHg (and LV end-diastolic pressure from 16.6 to 21.1 mmHg) was linked to volume increases rather than changes in chamber compliance. Figure 3 displays pressure–volume relations

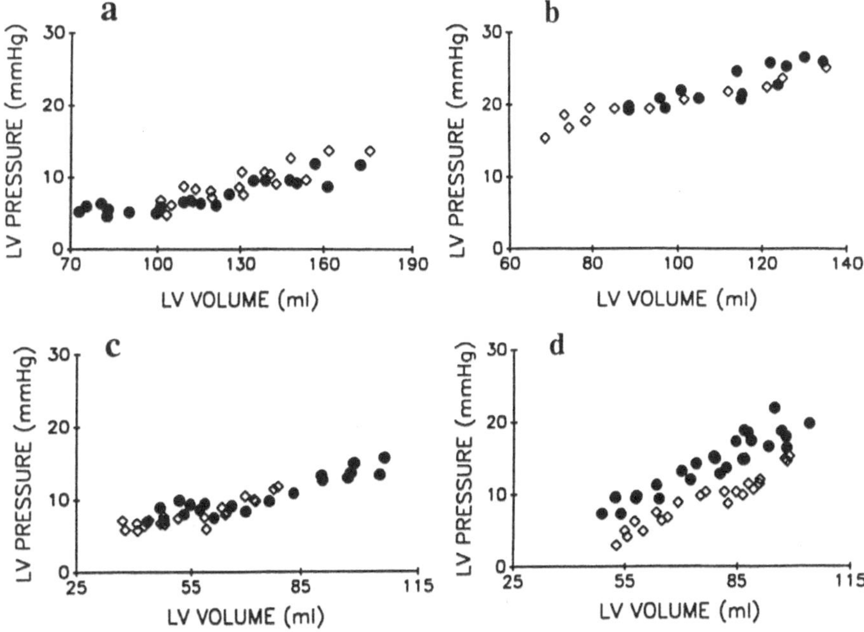

FIG. 3a–d. LV-DPVRs from four example patients before and after 6 min of continuous i.v. adenosine infusion (*circles*). There was generally no change from baseline (*diamonds*) in the LV-DPVR despite marked coronary vasodilation induced by adenosine. This contrasts to some experimental studies in which coronary dilation has been reported to increase chamber stiffness. (From [13], with permission)

derived by the IVC-inflow obstruction method in four patients from this study. The methodology used to derive these relations is as shown in Fig. 1, with the data being derived as in Fig. 1c. The LV-DPVRs were generally unchanged by adenosine. In a small minority (example *d*), there was a residual parallel upward shift in the data, which could be attributed to the nearly 40% rise in right heart pressures and thus residual external constraining forces. The lack of diastolic stiffening was not caused by an inadequate vasodilator response, because coronary sinus oxygen saturation more than doubled despite similar workload. More likely, these data highlight the importance of a well-regulated and intact coronary vascular bed toward limiting vascular engorgement with vasodilation, reducing the influence of coronary turgor on the LV-DPVR.

Chamber Geometry

In addition to cellular and subcellular structural components within the myocardial wall, three-dimensional geometry of the heart and changes in geometry during filling markedly influence the LV-DPVR. The steepness of the LV-DPVR is inversely dependent on chamber size, which itself often results in chamber stiffness being lower for enlarged chambers such as dilated cardiomyopathies as compared to control normal-sized ventricles. Alternatively, reverse remodeling of dilated hearts in response to therapy (such as ACE inhibition) can lead to steepening of the LV-DPVR. For example, Pouleur et al. [35] reported on chamber remodeling in patients with dilated cardiomyelopathy (DCM) treated with or without enalapril for 1 year. The placebo group displayed further chamber dilation with increases in end-diastolic and systolic volumes and a downward and rightward shift of the LV-DPVR. The monoexponential chamber stiffness coefficient declined from 0.044 to 0.032 in this group, which was interpreted as an increase in chamber distensibility. However, in the enalapril-treated group, both systolic and diastolic volumes declined, and the LV-DPVRs shifted upward and to the right (the chamber monoexponential stiffness coefficient was unchanged, but the shift suggested reduced distensibility). Although it is impossible to be certain, this chamber-level behavior most likely reflected geometric remodeling, rather than implying specific changes in the viscoelastic properties of the myocardial wall. These data supported earlier animal model experiments [36].

The influence of active changes in chamber geometry during filling on the LV-DPVR has been highlighted by Gibson et al. [37,38]. During filling, the ventricular shape normally changes from an elongated elliptical end-systolic contour to a more spherical (end-diastolic) one. This shape transition enables the endocardial surface of the heart to expand relatively less for any given increase in cavity volume than if shape had remained constant. This follows because a sphere provides the largest volume relative to a given surface area. Thus, simply by changing shape, the heart can increase volume without stretching the myocardium. To the extent that the passive LV-DPVR depends on viscoelastic behavior of the myocardial wall, a lack of stretch caused by normal cavity shape change would enhance net chamber compliance.

In DCM, the heart retains a spherical geometry throughout the cardiac cycle, so that volume increases are more directly linked to myocardial stretch and chamber stiffness increases. The opposite may apply to hypertrophied hearts that often display extreme geometric changes during filling [39]. Early filling could occur by an unfolding of the walls, making the LV pressure primarily dictated by left-atrial filling pressure rather than by stress–strain properties of the wall. As the cardiac shape becomes more spherical to accommodate volume, the pressure would remain fairly constant. Only at the end of diastole when the wall was stretched would intrinsic stiffness contribute to the LV-DPVR. The result would be a flat LV-DPVR, despite marked chamber hypertrophy, and an exacerbated dependence of pressures on mean left-atrial pressures (i.e., venous filling pressures).

This prediction was recently confirmed in a study of patients with marked hypertrophic cardiomyopathy [40]. The hearts in these patients had ejection fractions exceeding 80%, with near-total distal cavity obliteration at the start of diastole leading to more rounded contours at end-diastole.

Figure 4 (upper left) shows an example of these data. There were two striking findings. First, for each beat, the LV-DPVR was remarkably flat throughout filling, much flatter than anticipated on the basis of wall thickness and notions of increased stiffness caused by hypertrophy. Second, when the IVC-inflow obstruction maneuver was performed and chamber filling (atrial pressures) declined, there was a parallel downward shift of the loops. This "staircase-like" behavior was not observed in patients with acquired hypertensive hypertrophy (upper right) or in normals or subjects with DCM (lower panels). Quantitative analysis of relaxation delay, viscous properties, and the pericardial constraint suggested that these factors did not explain the disparity. However, this behavior was consistent with geometric unfolding of the chamber during initial filling, enhancing a change in volume with less change in

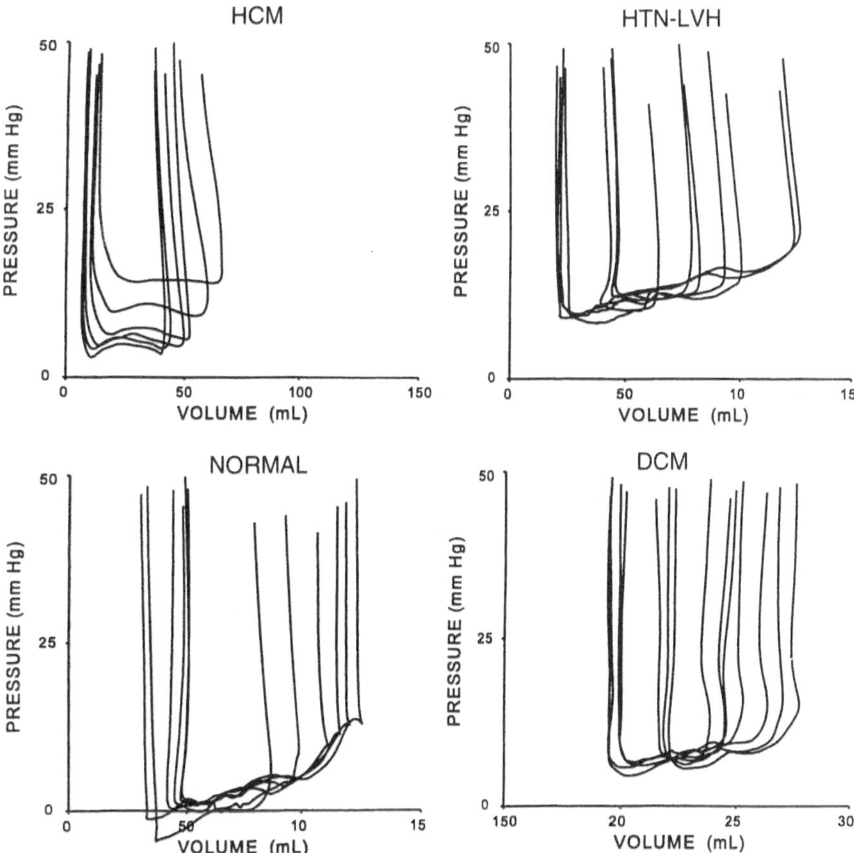

FIG. 4. LV-DPVRs from a patient with primary hypertrophic cardiomyopathy (*HCM*), compared with those with acquired hypertrophy caused by hypertension (*HTN-LVH*), *normal* controls, and patients with dilated cardiomyopathy (*DCM*). Each set of data was obtained using the inferior vena caval inflow obstruction method to derive diastolic curves from multiple beats (i.e., Fig. 1c). HCM patients displayed a unique LV-DPVR behavior with very flat relations that shifted downward in parallel as load was reduced. All other patient groups showed data that lie more or less on a single curve. These examples were typical of group data. (From [40], with permission)

pressure. Another mechanism that likely contributed was enhanced right–left heart interaction from septal configurational changes.

These results provide a clear example of how steady-state LV-DPVR data do not necessarily define a fixed dependence of chamber diastolic pressure on volume. In primary hypertrophic cardiomyopathy (HCM) patients, the same chamber volume was associated with markedly different pressures depending on the state of loading of the venous filling circuit (e.g., right heart → left atrial pressure). This behavior could well contribute to observations in HCM patients in which pharmacological agents that diminish vascular loading have a disproportionately large influence on shifting the LV-DPVR downward.

Pharmacological Influences on the LV-DPVR

The extent to which the LV-DPVR is determined by active cellular processes, particularly those involving calcium handling, remains somewhat controversial. Active control of the LV-DPVR is generally considered to be mediated by processes of relaxation and resting diastolic tone. However, most pharmacological manipulations that would be anticipated to alter such tone or relaxation, such as adrenergic stimulation or blockade, inhibition of L-type calcium channels, alteration of phosphodiesterase activity, or inhibition of angiotensin-II production, do not significantly alter chamber stiffness or the slope of the LV-DPVR. Many of these interventions have been found to displace the relation downward; however, this could involve extrinsic loading changes because most of these agents simultaneously lower peripheral vascular and notably right heart chamber loading.

There are many examples of a *lack* of direct pharmacological influence on the LV-DPVR. Both verapamil and beta-adrenergic blocking agents have been proposed to improve lusitropic behavior in patients with HCM. In particular, verapamil and nifedipine have both been reported to enhance chamber distensibility [41,42]; this is primarily manifest by an acute downward displacement of the LV-DPVR relation. However, such displacements may not reflect changes in the intrinsic LV-DPVR. Figure 5a displays LV-DPVR from a recent study of patients with ventricular hypertrophy and exertional or rest dyspnea [8]. The patient example shows data measured before and after i.v. verapamil both at resting steady state (solid lines) and derived from multiple cardiac cycles after IVC-inflow obstruction (i.e., as in Fig. 1c; symbols). Verapamil induced a downward shift of the resting LV-DPVR data, very similar to results previously reported [14,42]. However, when data were measured *after* most of the external constraint was removed (i.e., after IVC-inflow obstruction), these data both declined to a lower pressure and were essentially superimposable. This latter result was observed in virtually all patients studied [8].

Similarly, many pharmacological agents used to treat dilated heart failure have relatively little direct influence on the intrinsic chamber LV-DPVR. The most commonly reported observation is that of a parallel downward (and/or leftward) shift. However, shifts are nearly always associated with simultaneous reduction in volumes and external constraining forces. One recent example in which more specific LV-DPVR analysis was performed was in the recent study of Feldman et al. [9] of a novel quinolinone derivative OPC-18790. This agent has both phosphodiesterase-III inhibitory activity and influences repolarizing postassium currents. At $5\,\mu g\,kg^{-1}\,min^{-1}$, the drug is a potent venodilator, and it becomes a mixed arterial/venous dilator at twice this dosage. It had modest inotropic effects at both dosages and shortened the isovolumic relaxation time by approximately 20%. Interestingly, however, OPC-18790 did not alter LV-DPVRs (Fig. 5b).

Recently, two investigative groups studied the influence of ACE inhibition on relaxation and the LV-DPVR in patients with ventricular hypertrophy. Haber et al. [43] administered i.c. enalaprilat to 25 patients with and without left-ventricular hypertrophy (LVH) caused by essential hypertension. Enalaprilat shortened isovolumic relaxation time constant only in the LVH group (from 56 to 44 ms), and the extent of reduction was directly proportional to the degree of LVH ($r = .92$; $P < .001$). LV end-diastolic pressure declined from 23 ± 2 to 15 ± 1 mmHg; $P < .01$), but this was related to simultaneous declines in LV volume, with a leftward shift along the same pressure–volume relation and no change in chamber stiffness. Importantly,

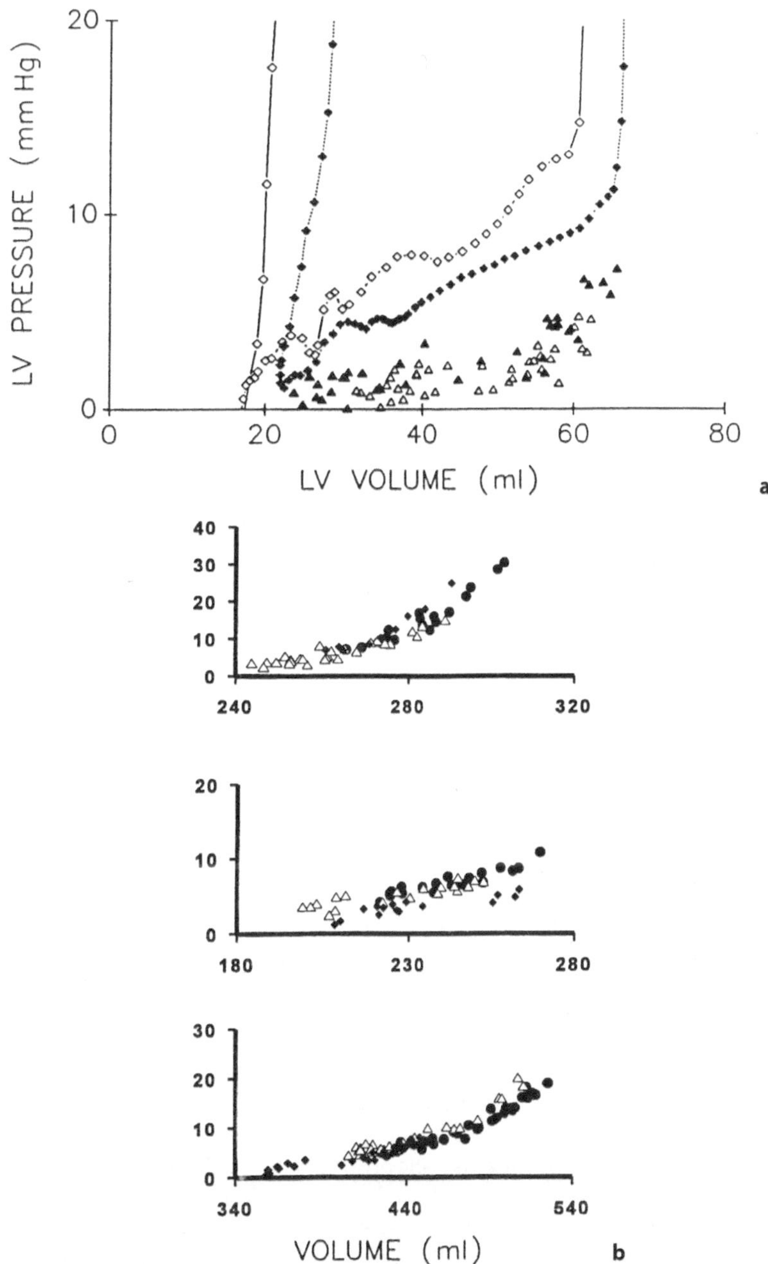

Fig. 5. **a** LV-DPVRs before and after i.v. verapamil in a patient with acquired HCM and dyspnea. Verapamil induced (*solid diamonds*) a downward shift in the resting (steady-state, *open diamonds*) LV-DPVRs. However, when relations were measured by the inferior vena cava (IVC) obstruction method (e.g., Fig. 1c), the data shifted downward and became essentially superimposible. *Open diamonds*, Pre-verapamil; *solid triangles*, post-verapamil. Thus, the downshift from verapamil in the steady-state loops was more likely related to removal of an external constraint. **b** Three examples of pressure–volume data in patients with DCM at baseline (*circles*), and after 30 (*diamonds*) and 60 (*triangles*) min of i.v. OPC-18790. The drug is a positive inotrope and vasodilator; accelerated chamber relaxation but had no effect on the LV-DPVR. (From [9], with permission)

these investigators evaluated LV-DPVRs by the multiple-beat method with data measured during IVC-inflow obstruction. In another study, Friedrich et al. [16] infused enalaprilat (0.05 mg/min) for 15 min into the left coronary arteries of 20 patients with LVH caused by aortic stenosis and compared the response to that observed in 10 patients with DCM. LVEDP also declined from 25 ± 2 to 20 ± 2 mmHg ($P < .05$); however, the LV-DPVR was shifted downward without a significant decline in right-atrial pressure. This was interpreted as indicating improved chamber distensibility.

Patients with DCM displayed but a similar decline in LV end-diastolic pressure and downward shift in the LV-DPVR, as right-atrial pressure also fell by 3 mmHg, this was taken as a reflection of a fall in the pericardial constraint. These data suggested that a component of resting LV-DPVR in patients with hypertrophy was related to active "tone" mediated by the cardiac renin–angiotensin system. This result supported animal data from hypertrophied rat hearts exposed to ACE inhibitors [44,45]. A limitation of these LV-DPVR data, however, was that the analysis was based on a single steady-state condition, making accurate dissection of extrinsic and intrinsic contributors difficult if not impossible.

The lack of LV-DPVR change despite substantial changes in isovolumic relaxation rates, as observed with beta receptor and calcium channel blockers [9] and PDE-III inhibitors [11], is not that surprising. More often than not, these relaxation changes are quantitatively small relative to the diastolic filling period, and while they may alter early filling pressures, they cannot explain sustained parallel shifts of the entire LV-DPVR. Such data call into question an often-assumed linkage between these two components of diastolic behavior. Interventions that yield downward parallel shifts of LV-DPVRs, most notably those inducing some degree of arterial vasodilation, are also frequently associated with shortening of relaxation time. Stimuli that elevate vascular and extrinsic loads, such as increased arterial resistance and ischemia, often prolong relaxation and shift the LV-DPVR upward. However, the cause-and-effect relation between relaxation and the LV-DPVR in the majority of instances is far from clear.

Heart rate and the LV-DPVR

The LV-DPVR is influenced by reduction in cardiac cycle length through a number of mechanisms. As the diastolic filling period is shortened, relaxation extends further into early filling, raising diastolic pressures. If baseline relaxation is prolonged and heart rate increases sufficiently, relaxation may remain incomplete at end-diastole, shifting the entire LV-DPVR upward. This behavior was first revealed in isolated hearts [46] and has been observed in animal studies of LVH in which heart rates are markedly increased [47]. It is less common in patients, because even with prolonged relaxation times in dilated and hypertrophied myopathic hearts diastolic time is adequate for relaxation to precede to completion at most physiological heartrates.

Rapid pacing also increases myocardial oxygen consumption and reduces the diastolic perfusion period. If baseline myocardial flow is compromised because of coronary artery disease or an imbalance of myocardial flow supply and demand (e.g., with LVH), rapid pacing can induce myocardial ischemia resulting in increased chamber stiffness and elevation of the LV-DPVR [48,49]. Slowing of the kinetics of calcium cycling caused by abnormalities in sarcoplasmic reticular function also contribute to LV-DPVR abnormalities by delaying relaxation and potentially increasing diastolic calcium [50,51].

Elevation of the LV-DPVR because of rapid pacing-induced ischemia in hearts with coronary artery disease has been ascribed to abnormal calcium handling, regional wall motion heterogeneity and disparate relaxation rates [52], and rapid stretch-activated channels increasing chamber stiffness [53,54]; this last mechanism was recently proposed by Glantz and colleagues [53,54]. These investigators placed a servocontrolled mitral valve in the inflow orifice to enable measurement of LV-DPVRs with or without chamber filling. With normal filling beats, pacing-induced ischemia resulted in an upward shift of the LV-DPVR. This could not be attributed to extrinsic factors, because the pericardium was removed, the chest opened, and data measured during IVC-inflow obstruction to reduce ventricular interaction. However, when similar LV-DPVRs were constructed from beats in which the heart underwent diastolic relaxation but was not permitted to fill (the mitral valve was held shut), LV-DPVR elevation during pacing ischemia was reduced. The LV-DPVR elevation was also shown to be diminished by administration of gadolinium [53], an agent that blocks stretch-activated calcium channels. This suggested that early rapid stretch at fast pacing rates could increase the calcium loading of myocytes and stiffen the ventricle.

Rapid pacing also results in upward LV-DPVR shifts and a rise in LVEDP in hypertrophied ventricles. This has been extensively studied in animal models of pressure-load hypertrophy [48], and recently in humans with LVH [16,55]. An example of the latter was reported by Cannon et al. [55], who found that HCM patients undergoing rapid atrial pacing had increased coronary sinus lactate formation and elevation of the ventricular end-diastolic pressure. These patients had primarily presented with angina-like chest pain. In contrast, Liu et al. [14] reported data from ten patients with concentric LVH and exertional or rest dyspnea. In these patients, rapid pacing neither significantly increased the ventricular end-diastolic pressure nor altered chamber compliance. Interestingly, although isovolumic relaxation was significantly longer at baseline in these LVH patients, it shortened considerably with rapid heart rate, achieving values similar to the control group at heart rates of 150 min^{-1}. The difference in the diastolic response in these two studies may relate to the specific patient population, i.e., HCM patients with angina versus those with symptoms of congestive failure, and to the way the LV-DPVR data were constructed. Liu et al. used the IVC-inflow obstruction method to generate the relations at each heart rate; Cannon et al., and all prior studies of rapid rate responses in LVH, reported data at one steady state. This may be important, because rapid pacing elevates mean atrial pressure, and this could contribute an external load to the LV-DPVR.

Lastly, rapid heart rate also alters the timing of atrial contraction relative to diastolic filling, and this may influence the LV-DPVR. In a study of canine LVH induced by perinephritis/renal artery hypertension, Berger et al. [56] showed that rapid atrial pacing at rates of 200–220 min^{-1} increased left-ventricular end-diastolic pressure (LVEDP) from 5.6 to 22.6 mmHg in the LVH group versus 3 to 4.6 mmHg in controls. Corresponding LV-DPVRs shifted upward with rapid pacing throughout diastolic filling (Fig. 6a). Similar responses have been reported by others in models of LVH induced by aortic banding and ascribed to abnormalities in myocardial relaxation and subendocardial ischemia [48,49]. However, neither mechanism could explain the

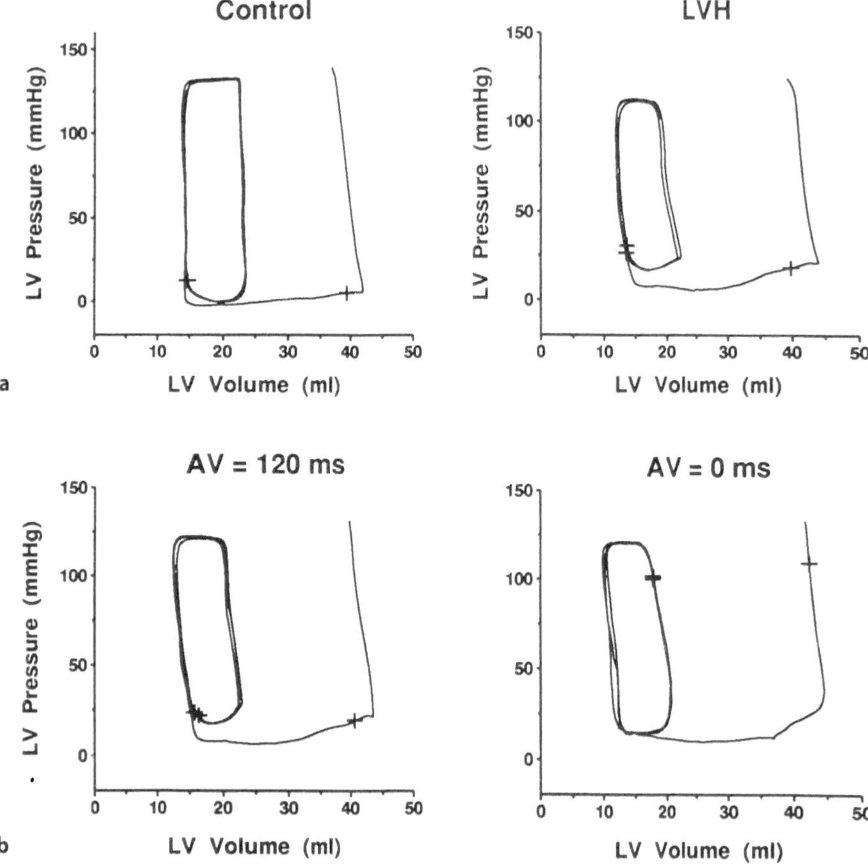

FIG. 6a,b. Influence of atrial contraction on heart rate-mediated changes in LV-DPVR. **a** Pressure–volume (PV) loops are shown during rapid atrial pacing (small loops at *left*) and immediately after terminating pacing (*lower boundary* with larger filling volume). The time of atrial peak systole is denoted by plus (+) symbols for each beat. LVH hearts displayed an upward shift of the DPVR data during rapid pacing. However, when pacing was abruptly stopped, this elevation disappeared within a few milliseconds. Atrial contraction occurred near the onset of filling with rapid pacing, as compared to late diastole for normal beats (i.e., postpaced beat shown in figure). This likely contributed to the elevated LV-DPVR. **b** Atrioventricular (*AV*) sequential pacing data in LVH heart. When AV delay was normal (120 ms), the data reproduced the elevated LV-DPVR previously shown in Fig. 6a (*right*). However, when the AV delay was set to 0 (i.e., synchronous contraction of atrium and ventricle), the LV-DPVR was no longer elevated, despite the fact that the pacing rate of the LV was identical. These data support a role for mechanical interaction of atrial contraction and the LV-DPVR to explain the elevation during rapid pacing, rather than ischemic or calcium-mediated mechanisms. (From [56], with permission)

results in Berger's study. This is demonstrated in Fig. 6a, which shows the results of abrupt cessation of pacing. There was an immediate return to the baseline LV-DPVR relation, which took place before ventricular activation would have occurred had pacing been continued. Thus, the LV-DPVR elevation was rapidly reversible and not directly related to the LV activation rate.

Terminating pacing, however, did alter the timing at which atrial contraction occurred. The time of atrial systole is noted on each set of PV loops by a plus (+) in Fig. 6. During rapid pacing, atrial systole occurred synchronous with the *onset* of ventricular filling rather than at end-diastole. When pacing was terminated, atrial contraction was delayed, which also delayed the onset of ventricular filling. It was during this very brief period that diastolic PV data fell to baseline. Left-atrial contraction at the onset of filling could alter ventricular chamber stiffness by triggering stretch-activated mechanisms, as proposed by Takano and Glantz [54], or by direct interaction between atrial systolic stiffening and the ventricle. The importance of atrial contraction timing to the LV-DPVR elevation was further tested by changing the atrioventricular delay (Fig. 6b). When atrial and ventricular excitation were synchronous, atrial systole now occurred during isovolumic contraction, and no longer contributed to acute volume increases and pressure rise during early diastole. The result was a downward shift in the LV-DPVR, despite the identical ventricular pacing rate. These data further suggest that the elevated LV-DPVRs in this study were not related to altered properties of the LV itself, but rather to the timing interaction of atrial and ventricular contraction.

Influence of Ischemia

In addition to rapid pacing-induced ischemia in hearts with underlying coronary artery stenoses, ischemia generated by total occlusion of a coronary artery has also been shown to alter the LV-DPVR. Studies first performed in patients undergoing coronary artery angioplasty revealed an upward elevation of the LV-DPVR after 60–90s of total artery occlusion [57,58]. This shift was ascribed to a decline in chamber distensibility caused by ischemia-induced regional wall motion and relaxation heterogeneity and to prolonged net relaxation. However, when similar conditions were studied using the methods of IVC-inflow obstruction to derive multiple-beat LV-DPVRs, the results were quite different [7]. At steady state, the LV-DPVR shifted upward during occlusion, just as observed in earlier studies. However, similar to the response to verapamil shown in Fig. 5a, the relations derived during IVC-inflow obstruction did not reveal any elevation. Thus, the LV-DPVR change during coronary occlusion appeared to result primarily from changes in extrinsic chamber loading rather than an intrinsic diastolic LV chamber abnormality. The chamber load effect could result from a dyskinetic wall motion pushing against the pericardial surface, sympathetic stimulation leading to increased right heart loading, and elevated heart rate and mean atrial pressures [52].

The role of the pericardium in mediating extrinsic loading effects during coronary occlusion was subsequently tested by Applegate et al. [59]. These authors found that the upward shift of the LV-DPVR during coronary artery occlusion was minimized after pericardiectomy. An example from similar experiments performed in our laboratory is shown in Fig. 7. With the pericardium intact, acute occlusion of the LAD resulted in an upward shift of the LV-DPVR, while there was purely a rightward displacement and much greater volume increase when the pericardium was removed. Data derived with the IVC-inflow obstruction method (right hand panels) revealed no significant shifts under either condition, but the curve was much flatter after pericardial removal, indicating its contribution to intact "chamber" stiffness.

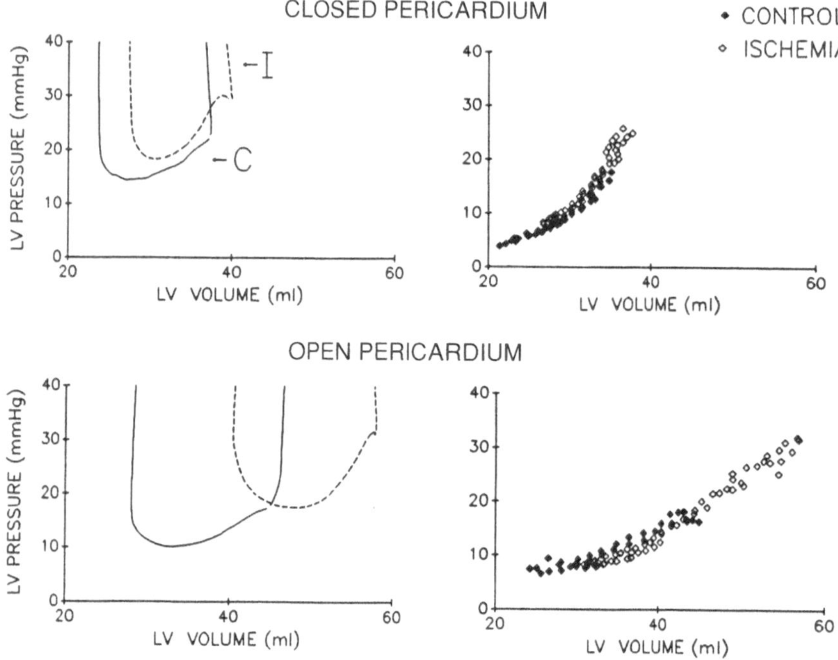

FIG. 7. Influence of pericardium on acute coronary occlusion-mediated changes in the LV-DPVR. Data are from a canine model before (*solid line-C, solid diamonds*) and after (*dashed line-I, open diamonds*) angioplasty of the left anterior descending artery. With a closed pericardium (*top*), there was a slight increase in volume but primarily an upward shift of the LV-DPVR data at rest. However, multiple beat-derived LV-DPVRs (*right panel*) showed no significant difference in the relations. When the pericardium was opened (*bottom*), the same coronary occlusion resulted in a marked volume increase and little upward shift. Again, the multiple-beat LV-DPVRs fell along a single curve. The difference in the steepness of the curves shown in the two corresponding right panels reflects the influence of the pericardial constraint

Influence of Exercise

In patients with dilated hypertrophied or myopathic hearts, exercise results in a rapid rise in the LV end-diastolic pressure, generally with minimal increase in volume [60–62]. Several mechanisms have been suggested for this response, including abnormal relaxation, ischemic stiffening, and wall motion heterogeneity. Hori et al. [63] recently reported that acute β-blockade in patients with dilated cardiomyopathy limited the upward LV-DPVR shift observed during bicycle exercise. When LV pressures were corrected for right-atrial pressure to account for pericardial constraint effects, an effect of β-blockade was still observed. These data suggested that exercise might exert detrimental effects on the failing heart via increases in sympathetic tone that could be inhibited by β-blockers, potentially explaining some of the benefit of these agents in heart failure patients.

However, sympathetic stimulation during exercise is also associated with a redistribution of venous volume into the heart, increasing right heart and pericardial constraining forces. This could also influence the LV-DPVR. This mechanism is supported by data obtained by the IVC-inflow obstruction methodology employed in patients during supine exercise [15]. In a study of nine patients, we found that while exercise increased LVEDP with no apparent change in LVEDV (Fig. 8a), the primary mechanism appeared to be a substantial rise in right heart filling pressures that led to increased external constraints. Dilated hearts operate near the pericardial constraint [61,64], so that even small increases in chamber volume resulted primarily in elevations in LV pressure. Figure 8b shows PV data from an example patient at peak bicycle exercise. Exercise induced an upward LV-DPVR shift. However, upon IVC-inflow obstruction, this elevation fell rapidly to baseline, and the resulting LV-DPVR was very similar to control. Similar results were obtained in most of the patients.

Quantitation of External Force Contribution to Resting LV-DPVR

The consistent finding that upward or downward shifts in the LV-DPVR induced by pharmacological or physiological interventions were most often caused by changes in external loading forces rather than alterations in intrinsic chamber stiffness led us to question the extent to which the extrinsic load contributed to basal diastolic pressures [6]. As shown in Fig. 1b, when IVC inflow was abruptly impeded, right heart pressures and volumes declined rapidly. In many patients, this occurred fast enough to define

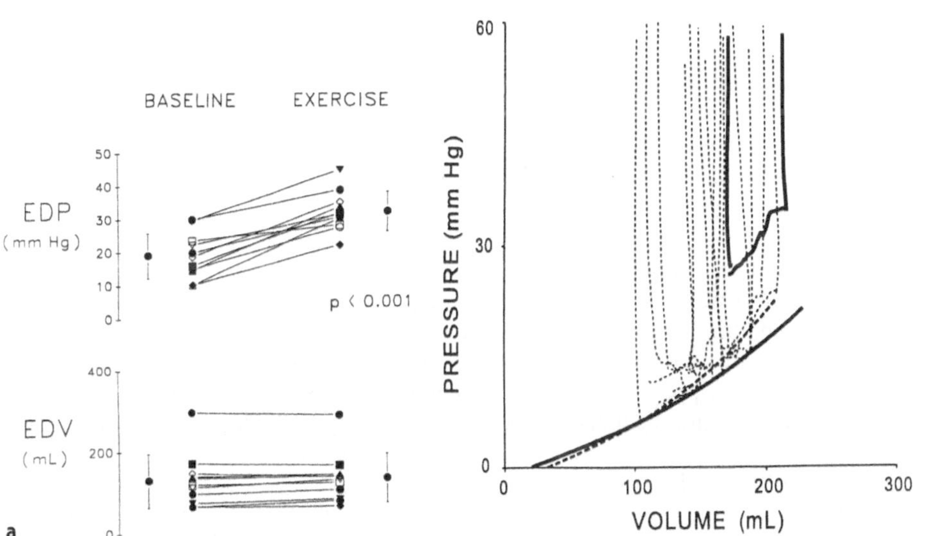

FIG. 8. **a** Effect of supine exercise in patients with dilated or hypertrophied hearts. Exercise induced a consistent marked elevation in the end-diastolic pressure (*EDP*), while end-diastolic volume (*EDV*) was virtually unchanged. **b** The primary mechanism for this change is shown in this example patient. The baseline LV-DPVR is shown by the *lower solid curve*. With bicycle exercise, the steady-state PV loop shifted upward (*dark solid loop*) markedly. However, when IVC-inflow obstruction was initiated, this loop returned downward toward baseline, and subsequent PV loop diastolic data fell much closer to the baseline relation (*dashed curve*). Thus, the elevated pressures were primarily the result of increased external forces, rather than an increase in intrinsic chamber stiffness

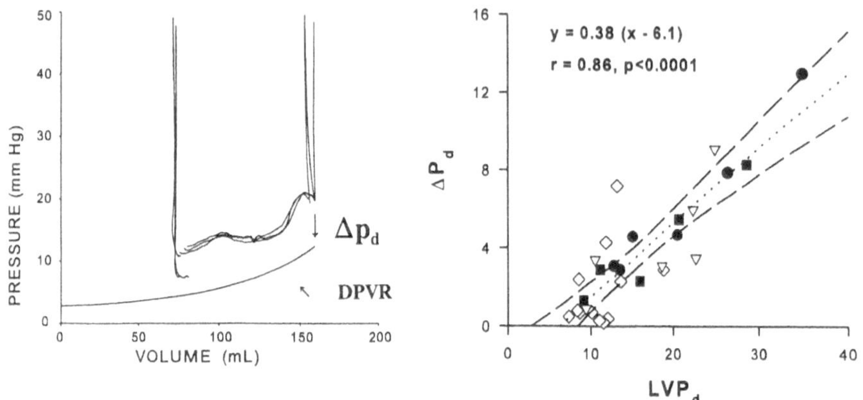

FIG. 9a,b. Quantification of influence of external constraint on resting LV-DPVR. **a** With the acute onset of right heart unloading caused by IVC-inflow obstruction, the LV-DPVR shifted downward (ΔP_d). This drop, which could be determined at the identical diastolic volume, reflected the influence of external forces on the resting LV-DPVR. **b** Plot of ΔP_d versus resting diastolic pressure (LVP_d) from 29 patients. Patient symbols are normals (*open diamonds*), HCM (*solid squares*), DCM (*open triangles*), and ischemic heart disease (*solid circles*). The slope (0.38) indicates that above an initial pressure of ~6 mmHg, 38% of the resting LV diastolic pressure resulted from forces external to the LV. (From [6], with permission)

the external constraining force before any significant change in LV preload volume. This analysis was performed in 29 patients with normal hearts or common cardiac diseases (LVH, DCM, and ischemic heart disease). At rest, these patients exhibited a broad range of end-diastolic pressures, from 8 to 35 mmHg. For each patient, we measured the initial rapid drop in the LV-DPVR data within the first few seconds after IVC-obstruction was initiated (ΔP_d), and compared this shift to the resting value of the diastolic pressure (Fig. 9a).

Figure 9b displays the plot of ΔP_d, which indexed external load effects on LV chamber pressure, versus the resting diastolic pressure (LVP_d). There was a significant correlation ($r = .86$), with a slope of 0.38 and intercept of 6.1 mmHg. Once LVP_d exceeded a nominal level of 6 mmHg, the proportion of the resting diastolic pressure that was related to the chamber itself was only 62%. The remaining fraction of the resting LV diastolic pressure was ascribed to pressure loads extrinsic to the LV. Given the substantial magnitude of the external effect, it becomes less surprising that drugs and other interventions that alter this influence can substantially shift the LV-DPVR upward or downward.

The primary value of quantifying this relation was that it helped define the magnitude of diastolic pressure reduction that could be achieved without compromising LV filling volume and thus cardiac output. Prior studies had shown that the LV-DPVR could be shifted downward in parallel with drugs such as nitroprusside [65], and that in patients with heart failure, careful titration of this drug could achieve a lower LVEDP without compromising CO [66]. Theoretically, careful titration of an agent that more selectively dilated the venous capacitance bed might facilitate this effect and induce a shift in the PV data between the solid and dashed beats shown in Fig. 1b. This would achieve the same stroke volume (cardiac output) but at a lower filling pressure.

Conclusion

Although the LV-DPVR stems from a complex interaction of intrinsic myocardial, vascular, interstitial, active biochemical, and external mechanical forces, it is the latter that consistently appears to dominate many of the acutely observed changes in intact patients. Intrinsic properties and geometric factors as well as external forces all substantially contribute to chronic changes associated with chronic disease remodeling. The development of the IVC-inflow obstruction methodology for examining LV-DPVRs has helped to clarify pathophysiology in a wide variety of disease conditions by better defining the relative influence of extrinsic and intrinsic factors.

References

1. Tyberg JV, Misbach GA, Glantz SA, Moores WY, Parmley WW (1978) A mechanism for shifts in the diastolic, left ventricular, pressure-volume curve: the role of the pericardium. Eur J Cardiol Suppl 7:163–175
2. Ross J Jr (1979) Acute displacement of the diastolic pressure-volume curve of the left ventricle: role of the pericardium and the right ventricle. Circulation 59:32–37
3. Smiseth OA, Refsum H, Tyberg JV (1984) Pericardial pressure assessed by right atrial pressure: a basis for calculation of left ventricular transmural pressure. Am Heart J 108:603–608
4. Tyberg JV, Taichman GC, Smith ER, Douglas NWS, Smiseth O, Keon WJ (1986) The relation between pericardial pressure and right atrial pressure: an intraoperative study. Circulation 73:428–432
5. Kass DA, Midei M, Graves W, Brinker JA, Maughan WL (1988) Use of a conductance (volume) catheter and transient inferior vena caval occlusion for rapid determination of pressure-volume relationships in man. Catheterization Cardiovasc Diagn 15:192–202
6. Dauterman K, Pak PH, Nussbacher A, Arie S, Liu CP, Kass DA (1995) Contribution of external forces to left ventricle diastolic pressure: implications for the clinical use of the Frank–Starling law. Ann Intern Med 122:737–742
7. Kass DA, Midei M, Brinker J, Maughan WL (1990) Influence of coronary occlusion during PTCA on end-systolic and end-diastolic pressure-volume relations in humans. Circulation 81:447–460
8. Kass DA, Wolff MR, Ting CT, Liu CP, Lawrence W, Chang MS, Maughan WL (1993) Diastolic compliance of hypertrophied ventricle is not acutely altered by pharmacologic agents influencing active processes. Ann Intern Med 119:466–473
9. Feldman MD, Pak PH, Wu CC, Haber HL, Heesch CM, Bergin JD, Powers ER, Coward TD, Johnson W, Feldman WM, Kass DA (1996) Acute cardiovascular effects of OPC-18790 in patients with congestive heart failure. Time- and dose-dependence analysis based on pressure-volume relations. Circulation 93:474–483
10. Asanoi H, Ishizaka S, Kameyama T, Ishise H, Sasayama S (1994) Disparate inotropic and lusitropic responses to pimobendan in conscious dogs with tachycardia-induced heart failure. J Cardiovasc Pharmacol 23:268–274
11. Ishihara H, Yokota M, Sobue T, Saito H (1994) Relation between ventriculoarterial coupling and myocardial energetics in patients with idiopathic dilated cardiomyopathy. J Am Coll Cardiol 23:406–416
12. Ishizaka S, Asanoi H, Kameyama T, Sasayama S (1991) Ventricular-load optimization by inotropic stimulation in patients with heart failure. Int J Cardiol 31:51–58
13. Nussbacher A, Arie S, Kalil R, Horta P, Feldman MD, Bellotti G, Pileggi F, Ellis M, Johnson W, Camarano GB, Kass DA (1995) Mechanism of adenosine-induced elevation of pulmonary capillary wedge pressure in humans. Circulation 92:371–379

14. Liu CP, Ting CT, Lawrence W, Maughan WL, Chang MS, Kass DA (1993) Diminished contractile response to increased heart rate in intact human left ventricular hypertrophy: systolic versus diastolic determinants. Circulation 88(pt 1):1893–1906
15. Kass DA, Wolff M, Maughan WL (1992) Mechanism of exercise-induced elevations of left ventricular end-diastolic pressure in human cardiomyopathy (abstract). Circulation 86:I-2044
16. Friedrich SP, Lorell BH, Rousseau MF, Hayashida W, Hess OM, Douglas PS, Gordon S, Keighley CS, Benedict C, Krayenbuehl HP, Gossmann W, Pouleur H (1994) Intracardiac angiotensin-converting enzyme inhibition improves diastolic function in patients with left ventricular hypertrophy due to aortic stenosis. Circulation 90:2761–2771
17. Weber KT, Jalil JE, Janicki JS, Pick R (1989) Myocardial collagen remodeling in pressure overload hypertrophy. Am J Hypertens 2:931–940
18. Weber KT, Janicki JS, Shroff SG, Pick R, Chen RM, Bashey RI (1988) Collagen remodeling of the pressure-overloaded, hypertrophied nonhuman primate myocardium. Circ Res 62:757–765
19. Weber KT, Sun Y, Tyagi SC, Cleutjens JP (1994) Collagen network of the myocardium: function, structural remodeling and regulatory mechanisms. J Mol Cell Cardiol 26:279–292
20. Klappacher G, Franzen P, Haab D, Mehrabi M, Binder M, Plesch K, Pacher R, Grimm M, Pribill I, Eichler HG, Glogar D (1995) Measuring extracellular matrix turnover in the serum of patients with idiopathic or ischemic dilated cardiomyopathy and impact on diagnosis and prognosis. Am J Cardiol 75:913–918
21. Linke WA, Popov VI, Pollack GH (1994) Passive and active tension in single cardiac myofibrils. Biophys J 67:782–792
22. Brady AJ (1991) Length dependence of passive stiffness in single cardiac myocytes. Am J Physiol 260:H1062–H1071
23. Le Guennec JY, Peineau N, Argibay JA, Mongo KG, Garnier D (1990) A new method of attachment of isolated mammalian ventricular myocytes for tension recording: length dependence of passive and active tension. J Mol Cell Cardiol 22:1083–1093
24. Schraeger JA, Canby CA, Rongish BJ, Kawai M, Tomanek RJ (1994) Normal left ventricular compliance after regression of hypertrophy. J Cardiovasc Pharmacol 23:349–357
25. Marjianowski MM, Teeling P, Mann J, Becker AE (1995) Dilated cardiomyopathy is associated with an increase in the type I/type III collagen ratio: a quantitative assessment. J Am Coll Cardiol 25:1263–1272
26. Mukherjee D, Sen S (1993) Alteration of cardiac collagen phenotypes in hypertensive hypertrophy: a role of blood pressure. J Mol Cell Cardiol 25:185–196
27. Doering CW, Jalil JE, Janicki JS, Pick R, Aghili S, Abrahams C, Weber KT (1988) Collagen network remodelling and diastolic stiffness of the rat left ventricle with pressure overload hypertrophy. Cardiovasc Res 22:686–695
28. Narayan S, Janicki JS, Shroff SG, Pick R, Weber KT (1989) Myocardial collagen and mechanics after preventing hypertrophy in hypertensive rats. Am J Hypertens 2:265–682
29. Hein S, Scholz D, Fujitani N, Rennollet H, Brand T, Friedl A, Schaper J (1994) Altered expression of titin and contractile proteins in failing human myocardium. J Mol Cell Cardiol 26:1291–1306
30. Tsutsui H, Tagawa H, Kent RL, McCollam PL, Ishihara K, Nagatsu M, Cooper G IV (1994) Role of microtubules in contracile dysfunction of hypertrophied cardiocytes. Circulation 90:533–555
31. Tsutsui H, Ishihara K, Cooper G IV (1993) Cytoskeletal role in the contractile dysfunction of hypertrophied myocardium. Science 260:682–687
32. Vogel P, Mark W, Apstein CS, Briggs LL, Gaasch WH, Ahn J (1982) Acute alterations in left ventricular diastolic chamber stiffness: role of the "erectile" effect of coronary arterial pressure and flow in normal and damaged hearts. Circ Res 51:465-478
33. Ogilby JD, Iskandrian AS, Untereker WJ, Heo J, Nguyen TN, Mercuro J (1992) Effect of intravenous adenosine infusion on myocardial perfusion and function: hemodynamic/angiographic and scintigraphic study. Circulation 86:887–895

34. Haywood GA, Sneddon JF, Bashir Y, Jennison SH, Gray HH, McKenna WJ (1992) Adenosine infusion for the reversal of pulmonary vasoconstriction in biventricular failure: a good test by a poor therapy. Circulation 86:896–902
35. Pouleur H, Rousseau MF, van Eyll C, Stoleru L, Hayashida W, Udelson JA, Dolan N, Kinan D, Gallagher P, Ahn S, Benedict CR, Yusuf S, Konstam M (1993) Effects of long-term enalapril therapy on left ventricular diastolic properties in patients with depressed ejection fraction. SOLVD Investigators. Circulation 88:481–491
36. Pfeffer JM, Pfeffer MA, Braunwald E (1985) Influence of chronic captopril therapy on the infarcted left venticle of the rat. Circ Res 57:84–95
37. Yettram AL, Grewal BS, Dawson JR, Gibson DG (1992) Factors influencing left-ventricular stiffness. J Biomed Eng 14:21–26
38. Gibson DG, Brown DG (1974) Relation between diastolic left ventricular wall stress and strain in man. Br Heart J 36:1066–1077
39. Gibson DG, Brown DJ (1975) Continuous assessment of left ventricular shape in man. Br Heart J 37:904–910
40. Pak PH, Maughan WL, Baughman KL, Kass DA (1996) Marked discordance between dynamic and passive pressure-volume relation in idiopathic hypertrophic cardiomyopathy. Circulation 94:52–60
41. Bonow RO, Ostrow HG, Rosing DR, Cannon RO III, Lipson LC, Maron BJ, Kent KM, Bacharach SL, Green MV (1983) Effects of verapamil on left ventricular systolic and diastolic function in patients with hypertrophic cardiomyopathy: pressure-volume analysis with a nonimaging scintillation probe. Circulation 68:1062–1073
42. Paulus WJ, Lorell BH, Craig WE, Wynne J, Murgo JP, Grossman W (1983) Comparison of the effects of nitroprusside and nifedipine on diastolic properties in patients with hypertrophic cardiomyopathy: Altered left ventricular loading or improved muscle inactivation? J Am Coll Cardiol 2:879–886
43. Haber HL, Powers ER, Gimple LW, Wu CC, Subbiah K, Johnson WH, Feldman MD (1994) Intracoronary angiotensin-converting enzyme inhibition improves diastolic function in patients with hypertensive left ventricular hypertrophy. Circulation 89:2616–2625
44. Kromer EP, Riegger GAJ (1988) Effects of long-term angiotensin-converting enzyme inhibition on myocardial hypertrophy in experimental aortic stenosis in the rat. Am J Cardiol 62:161–164
45. Eberli FR, Apstein CS, Ngoy S, Lorell BH (1992) Exacerbation of left ventricular ischemic diastolic dysfunction by pressure overload hypertrophy: prevention by specific inhibition of intrinsic cardiac renin-angiotensin system. Circ Res 70:931–943
46. Weisfeldt ML, Weiss JL, Frederiksen JT, Yin FCP (1980) Quantification of incomplete left ventricular relaxation: relationship to the time constant for isovolumic pressure fall. Eur Heart J Suppl A:119–129
47. Fujii AM, Gelpi RJ, Mirsky I, Vatner SF (1988) Systolic and diastolic dysfunction during atrial pacing in conscious dogs with left ventricular hypertrophy. Circ Res 62:462–470
48. Zhang J, Merkle H, Hendrich K, Garwood M, From AHL, Ugurbil K, Bache RJ (1993) Bioenergetic abnormalities associated with severe left ventricular hypertrophy. J Clin Invest 92:993–1003
49. Bache RJ, Dai X, Alyono D, Vrobel TR, Homans DC (1987) Myocardial blood flow durng exercise in dogs will left ventricular hypertrophy produced by aortic banding and perinephritic hypertension. Circulation 76:835–842
50. Morgan JP, Erny RE, Allen PD, Grossman W, Gwathmey JK (1990) Abnormal intracellular calcium handling, a major cause of systolic and diastolic dysfunction in ventricular myocardium from patients with heart failure. Circulation 81(suppl III): III-21–III-32
51. Gwathmey JK, Warren SE, Briggs GM, Copelas L, Feldman MD, Phillips PJ, Callahan M Jr, Schoen FJ, Grossman W, Morgan JP (1991) Diastolic dysfunction in hypertrophic cardiomyopathy. J Clin Invest 87:1023–1031
52. Paulus WJ (1990) Upward shift and outward bulge. Divergent myocardial effects of pacing angina and brief coronary occlusion. Circulation 81:1436–1439

53. Shitani H, Glantz SA (1994) Influence of filling on left ventricular diastolic pressure-volume curve during pacing ischemia in dogs. Am J Physiol 266:H1373–H1385
54. Takano H, Glantz SA (1995) Gadolinium attenuates the upward shift of the left ventricular diastolic pressure-volume relation during pacing-induced ischemia in dogs. Circulation 91:1575–1587
55. Cannon RO III, Rosing DR, Maron BJ, Leon MB, Bonow RO, Watson RM, Epstein SE (1985) Myocardial ischemia in patients with hypertrophic cardiomyopathy: contribution of inadequate vasodilator reserve and elevated left ventricular filling pressures. Circulation 71:234–243
56. Berger RD, Wolff MR, Anderson JH, Kass DA (1995) Role of atrial contraction in diastolic pressure elevation induced by rapid pacing of hypertrophic canine ventricle. Circ Res 77:163–173
57. Bertrand ME, LeBlanche JM, Fourrier JL, Traisnel G, Mirsky I (1988) Left ventricular systolic and diastolic function during acute coronary artery balloon occlusion in humans. J Am Coll Cardiol 12:341–347
58. Wijns W, Serruys PW, Slager CJ, Grimm J, Krayenbuehl HP, Hugenholtz PG, Hess OM (1986) Effect of coronary occlusion during percutaneous transluminal angioplasty in humans on left ventricular chamber stiffness and regional diastolic pressure-radius relations. J Am Coll Cardiol 7:455–463
59. Applegate RJ, Walsh RA, O'Rourke RA (1990) Comparative effects of pacing-induced and flow-limited ischemia on left ventricular function. Circulation 81:1380–1392
60. Kitzman DW, Higginbotham MB, Cobb FR, Sheikh KH, Sullivan MJ (1991) Exercise intolerance in patients with heart failure and preserved left ventricular systolic function: failure of the Frank-Starling mechanism. J Am Coll Cardiol 17:1065–1072
61. Janicki JS (1990) Influence of the pericardium and ventricular interdependence on left ventricular diastolic and systolic function in patients with heart failure. Circulation 81(suppl):III15–III1120
62. Roubin GS, Anderson SD, Shen WF, Choong CY, Alwyn M, Hillery S, Harris PJ, Kelly DT (1990) Hemodynamic and metabolic basis of impaired exercise tolerance in patients with severe left ventricular dysfunction. J Am Coll Cardiol 15:986–994
63. Sato H, Hori M, Ozaki H, Yokoyama H, Imai K, Morikawa M, Takeda H, Inoue M, Kamada T (1993) Exercise-induced upward shift of diastolic left ventricular pressure-volume relation in patients with dilated cardiomyopathy. Effects of beta-adrenergic blockade. Circulation 88:2215–2223
64. Yamamoto K, Masuyama T, Tanouchi J, Uematsu M, Doi Y, Naito J, Hori M, Tada M, Kamada T (1992) Decreased and abnormal left ventricular filling in acute heart failure: role of pericardial constraint and its mechanism. J Am Soc Echocardiogr 5:504–514
65. Ludbrook PA, Byrne JD, Kurnik PB, McKnight RC (1977) Influence of reduction of preload and afterload by nitroglycerin on left ventricular diastolic pressure-volume relations and relaxation in man. Circulation 56:937
66. Carroll JD, Lang RM, Neumann AL, Borow KM, Rafer SI (1986) The differential effects of positive inotropic and vasodilator therapy on diastolic properties in patients with congestive cardiomyopathy. Circulation 74:815–825

Endocardial–Myocardial Interaction

Gilles W. De Keulenaer[1], Luc J. Andries[1], Paul F. Fransen[1], Puneet Mohan[1], Gregory Kaluza[1], Jean L. Rouleau[2], Dirk L. Brutsaert[1], and Stanislas U. Sys[1]

Summary. In 1986, Brutsaert et al. observed that the endocardial endothelium (EE) directly controls performance of the myocardium. From this observation, the existence of an endocardium-mediated intracavitary autoregulation of cardiac performance has been postulated. Recent discoveries on morphology and function of the endocardium have substantiated this hypothesis. *Morphologically*, phenotypic expression of several receptors, tight and gap junctions, adhesion molecules, intercellular clefts, and a specific organization of the cytoskeleton distinguishes the EE from other endothelial subtypes and emphasizes a structural and functional specialization. *Functionally*, the EE imparts a twitch-prolonging and positive inotropic effect on cardiac muscle, apparently by enhancing the responsiveness of myocardial contractile proteins to Ca^{2+}. EE-mediated control of myocardial performance resembles length/volume-mediated control and interacts with neurohumoral-mediated control. EE cells respond to physical and humoral stimuli with release of several substances that have important inotropic actions on the myocardium. Specific electrophysiological properties determine the intracellular Ca^{2+} increase in response to external stimuli and modulate the release of these substances. An asymmetrical distribution of ion channels between the luminal and abluminal membrane of the EE suggests a specific transendothelial transport of ions over the endocardium. Accordingly, the endocardium is highly specialized for communication with and regulation of the myocardium. It will be interesting to evaluate endocardium-mediated control of cardiac performance in pathophysiological conditions.

Key words. Endocardium—Endothelium—Myocardium—Contractility—Relaxation

Introduction

Following the observations by Furchgott and Zawadski [1] about the obligatory role of vascular endothelium in vasomotor tone, Brutsaert et al. [2] observed that the endocardial endothelium (EE), which lines the innermost surface of both ventricles, appeared to have important effects on myocardial contraction and relaxation. Following EE damage in papillary muscles, the isometric twitch was shorter in duration by an

[1] Department of Physiology and Medicine Antwerp University, Antwerp, Belgium.
[2] Department of Medicine, University of Montreal, Montreal, Canada.

earlier onset of isometric force decline. This twitch abbreviation was accompanied by a slight decrease in peak twitch force development but with no significant changes in the early rate of rise of the twitch [2]. This response pattern differs from virtually all other negative inotropic interventions, such as decreasing extracellular calcium or reduction of cAMP-mediated effects. Furthermore, EE damage altered contractile responsiveness of the myocardium to several hormones and transmitters.

These fundamental observations on the function of EE have subsequently been confirmed in cardiac muscle from different animal species and also in the intact animal in vivo [3,4]. From these observations, the existence of an endocardium-mediated intracavitary autoregulation of cardiac performance was postulated [5,6]. In this chapter, we summarize recent discoveries on morphological and functional features underlying the interaction between endocardium and ventricular myocardium.

Functional Morphology of the Endocardium

In vertebrates and humans, the luminal surface of the heart has a complex structure consisting of furrows, cylinder- and sheetlike trabeculae, and papillary muscles. The complex cavitary surface of the ventricular wall is completely covered by the endocardium, delineated by a delicate innermost layer of cells, the endocardial endothelium (EE). We emphasize here a number of differences between the EE and other cardiac endothelia (Fig. 1).

Intercellular Communication in Endocardial Endothelium

The labyrinth of trabeculae and furrows, and on a microscopical scale the presence of numerous microvilli on the luminal surface of the EE cells, augment the available contact surface area that exposes the EE to the superfusing blood. A large contact surface area suggests an important sensory function of the EE. Indeed, EE possesses receptors for various substances circulating in the blood. These receptors have been shown to initiate the release of endocardial inotropic factors via receptor-mediated mechanisms. Morphological techniques have demonstrated EE receptors for atrial natriuretic peptide [7], mineralocorticoids [8], endothelin A and B receptors [9,10], and the presence of angiotensin-converting enzyme [11]. The distribution of these receptors was found to be highly nonuniform within the ventricles. Moreover, the degree of nonuniformity differed between ventricular EE and EE from cardiac valves and between EE and microvascular endothelium.

In cardiac endothelium, adhesion molecules such as the intracellular adhesion molecule (ICAM-1) and the vascular cell adhesion molecule (VCAM), as well as major histocompatibility complex (MHC) molecules, also appear to have a distinct heterogeneous distribution [12–15]. Endothelium of myocardial capillaries expresses components essential for interactions with lymphocytes, suggesting that these endothelial cells are active in immune responses. By contrast, EE and endothelium of larger vessels do not have these properties.

Transmission electron microscopy of EE demonstrated the presence of gap junctions in intercellular clefts [16–18]. Gap junctions are important cell membrane spe-

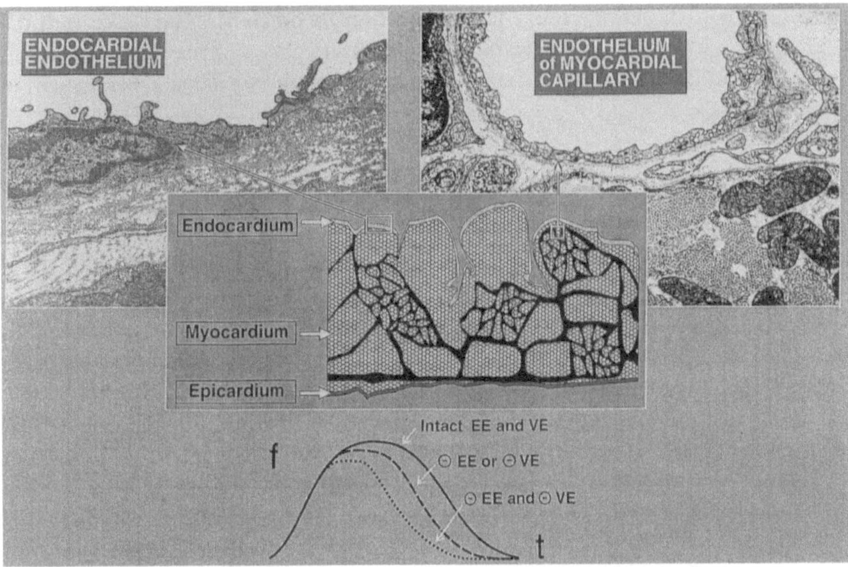

FIG. 1. Structural organization and localization of endocardial and microvascular endothelium in the heart. Central diagram shows part of the ventricular wall, which is delineated by the epicardium, a smooth outer layer of mesenchymal cells and interstitial tissue, and the endocardium, which includes a luminal layer of endothelial cells, the endocardial endothelium (*EE*), and the underlying basement membrane and fibroelastic layer. The myocardium contains a complex system of coronary blood vessels and sinusoids, all of which are lined by an intimal layer of vascular endothelial (*VE*) cells. Transmission electron micrographs show a detail of the EE (*upper left*) and VE of a myocardial capillary (*upper right*). Effects of the presence of intact EE and VE and the absence of one or both endothelial types are summarized in the graph at the *bottom*. The time (*t*) curves represent control (intact EE and VE) and experimental isometric force (*f*) twitches from isolated papillary muscles in Krebs-Ringer solution with 1.25 m*M* Ca^{2+}. In one group of experiments, the EE of isolated papillary muscles was selectively removed by a quick immersion in Triton X-100 (–EE). In a second group of experiments, the VE in Langendorff hearts was first damaged or made dysfunctional by a bolus injection of Triton X-100 in the coronary arteries; the EE in these hearts was still intact. From the Triton-treated Langendorff hearts, papillary muscles (–VE) were isolated and their contractile characteristics were compared with control muscles. Finally, the EE of the –VE muscles was damaged by quick immersion in Triton X-100 (–EE and –VE). The summarizing graphs illustrate that removal of VE has similar effects as removal of EE. Removal of EE from –VE muscles resulted in further shortening of twitch duration and a further decrease of total tension. Both endothelial types have an additive or complementary contraction-prolonging effect on myocardium

cializations involved in cell communication. *En face* confocal scanning laser microscopy showed that EE, like aortic endothelial cells, were delineated by numerous spots containing the gap junctional protein connexin 43 [16]; this is in accordance with the suggested sensor function of the EE. Following activation of a single EE cell by a receptor-mediated mechanism, second messengers can traverse gap junctions, which can then activate neighboring EE cells and amplify the output capacity, i.e., the release of endocardial factors. Immunolabeling of connexin 43 was not observed in endothelium of myocardial capillaries, suggesting a more local regulatory function of these cells.

Accordingly, the EE possesses various phenotypic differences compared with the endothelium from other compartments in the heart. These differences could result from adaptations to the specific anatomical position and functional environment of the EE.

Cell Shape, Cytoskeleton, and Permeability of the Endocardial Endothelium

The EE is subjected to considerable physicomechanical stress resulting from the large differences in hydrostatic pressure and from the large variations in shape of the heart walls during the cardiac cycle. An important regulator of cell shape is the cytoskeleton. Actin filaments form a peripheral actin band or are organized as axially aligned stress fibers. The peripheral actin band is associated with the zonula adherens and participates in the regulation of paracellular permeability [19–21]. Stress fibers play a role in maintaining the integrity of the endothelium, presumably by enhancing cell-substratum adhesion (for a review, see [22]) and by strengthening the endothelial cell surface [23]. In most EE cells in the left and right ventricle of the rat, F-actin staining was restricted to the peripheral actin band, thereby outlining a polygonal shape without centrally located stress fibers. Stress fibers were found in the outflow tract, but their occurrence was highly variable. Because shear stress is expected to be high in the outflow tract, it could explain the presence of stress fibers in EE of this region. However, the presence of numerous stress fibers in small EE cells along the tendon end of papillary muscles and along the proximal border of the atrioventricular valve in the right ventricle [16,24] cannot be explained by shear stress alone. Because these substrata are highly elastic, cyclical stretch and cell shape changes may explain the numerous stress fibers in these EE cells [25–27].

Paracellular transport occurs through the intercellular spaces or clefts and is limited by the presence of tight junctions and by the structural organization of the clefts. In EE, tight junctions have a simple structure with one or two junctional contact points. Intercellular clefts in ventricular EE are usually more complex and deeper than in endothelium of myocardial capillaries [18,28,29]. On the other hand, EE cells have a larger surface area and hence a lower value for *en face* overall circumferential length of the intercellular clefts than vascular endothelium. Because permeability is inversely related to the depth of intercellular clefts and is proportional to circumferential length, the overall paracellular transport through the EE might be expected to be lower than through endothelium of myocardial capillaries. The ultrastructure of EE intercellular clefts and the large surface area of EE cells may constitute and adaptation to limit diffusion driven by high hydrostatic pressure in the heart [28].

Endocardial Control of Myocardial Performance

Ventricular performance of the heart relies largely on autoregulatory mechanisms (Fig. 2). Autoregulation is accomplished through intrinsic feedback by the loading conditions and the length of the muscle fibers (heterometric autoregulation), and through extrinsic feedback by the activity of the neurohumoral system (homeometric autoregulation). In the following paragraphs we explore the currently available experimental evidence for an additional intracavitary autoregulation by the EE.

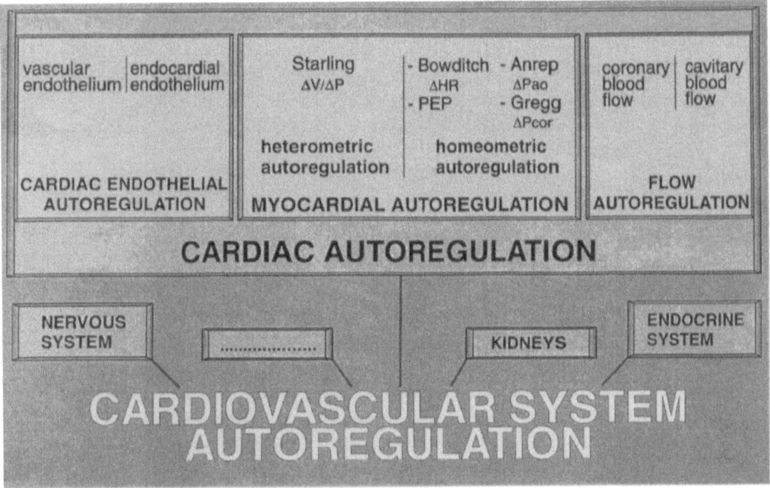

FIG. 2. Integrated configuration of myocardial, cardiac, and cardiovascular system autoregulation. Cardiovascular performance relies largely on autoregulatory mechanisms, including modulation by various other systems in the body. At the myocardial level, heterometric autoregulation is accomplished through Starling mechanisms (ΔV, end-diastolic volume, or ΔP, pressure changes) and homeometric autoregulation through changes in the pattern of stimulation (*Bowditch* ΔHR, changes in heart rate; *PEP*, post-extrasystolic potentiation) and through changes in aortic pressure (*Anrep* ΔPao) or coronary perfusion pressure (*Gregg* $\Delta Pcor$). The endocardial and (micro)vascular endothelium contribute to the intermediate level of cardiac autoregulation through a direct and sustained inotropic action on the myocardium (cardiac endothelial autoregulation), through modulation or exclusive mediation of the myocardial effect of substances in the (su)perfusing blood, and through direct sensitivity to flow of cavitary or coronary blood (flow autoregulation)

Effects of Endocardial Damage on Isolated Cardiac Muscle Performance

The observations on the effects of endocardial damage on myocardial contraction (Figs. 1 and 3) were initially made on isolated cat papillary muscle at 2.5 mM $[Ca^{2+}]_0$ and 29°C [2]. The EE was damaged by a 1-s exposure of the muscle (stretched at lmax, optimal muscle length) to 1% Triton X-100, a mild detergent dissolved in a Krebs–Ringer solution, and followed by an immediate and abundant wash. Morphological observations confirmed that this procedure sufficed (in subsequent experiments only 0.5% Triton was used) to selectively damage the EE. Subsequently, the effects of EE damage on the mechanical performance of isolated cardiac muscle were confirmed in different animal species [30–32] at different stimulation frequencies, muscle lengths, temperatures, and extracellular calcium concentrations [2] and in Langendorff-perfused hearts [33]. In particular, the effect of EE damage was more pronounced at physiological temperature (35°C) and physiological $[Ca^{2+}]_0$ (1.25 m*M*).

The validity of these results critically depends on the selectivity of endothelial damage in the muscle preparation. *Morphological integrity* of the subjacent myocardium after Triton immersion was demonstrated by transmission electron microscopy and confocal scanning laser microscopy showing intact cardiomyocytes with normal

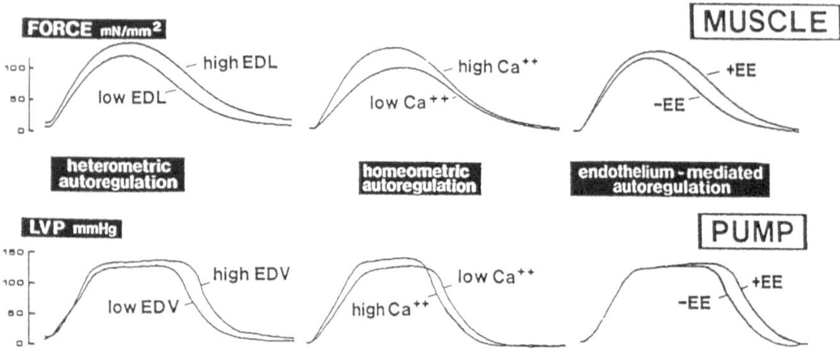

Fig. 3. Pattern and relative importance of cardiac endothelium-mediated autoregulation compared to heterometric (length or volume) and homeometric (changes in $[Ca^{2+}]_0$) autoregulation in vitro in isolated cat papillary muscle, *upper*, and in vivo in intact ventricle of anesthetized open-chest dogs, *lower*. Notice the striking resemblance in pattern and amplitude between the changes induced by increasing initial muscle length from 96% to lmax in vitro or by increasing ventricular volume in vivo [end-diastolic pressure (EDP) from 8.5 mmHg to 13 mmHg, (*left*)] and the changes induced by intact endothelium (*right*). Prolongation of contraction with small changes on the initial phase of contraction is typical for both endothelium-mediated and length- or volume-mediated positive inotropic effect. In contrast, positive inotropic response induced by $[Ca^{2+}]_0$ (*middle*) is characterized by large changes in the initial phase of contraction with abbreviation of twitch duration when $[Ca^{2+}]_0$ is increased. A similar response is obtained with beta-agonists, frequency potentiation, etc. (Modified from [5]). *LVP*, Left-ventricular pressure; *EDL*, end-diastolic length; *EDV*, end-diastolic volume

striation pattern and exclusion of fluorochromes such as lucifer yellow, propidium iodide, and ethidium homodimer. *Functional integrity* of the myocardium after Triton was confirmed because (i) peak isometric force at high calcium was unaltered [2,30,34]; (ii) there was no correlation between muscle thickness and the decrease in twitch duration induced by EE impairment [2]; and (iii) resting $[Ca^{2+}]_i$ in the myocardium was unaltered [34]. Furthermore, the effect on cardiac muscle twitch induced by EE damage was similar for EE damage induced by Triton, ultrasound, air drying, or other pharmacological damaging procedures.

Mechanisms of Endocardium-Dependent Myocardial Positive Inotropism

Wang and Morgan [34] reported the effects of Triton-induced EE damage on $[Ca^{2+}]_i$ transient and peak isometric twitch force in ferret papillary muscles loaded with the Ca^{2+}-regulated bioluminescent indicator aequorin. After EE damage, peak isometric twitch force was reduced despite a slight increase in the amplitude of the $[Ca^{2+}]_i$ transient. These observations confirmed the hypothesis, formulated by Brutsaert and colleagues [2], that decreased contractile performance after EE damage resulted from a decreased responsiveness of the myocardial contractile proteins to $[Ca^{2+}]_i$. Accordingly, the presence of intact endothelial cells in the endocardium enhances the responsiveness of the myocardial contractile proteins to $[Ca^{2+}]_i$ and thus increases cardiac muscle and ventricular developed force or pressure and the duration of the contraction–relaxation cycle.

Endocardium-Mediated Control Versus Length- and Neurohumoral-Mediated Control of Cardiac Performance

Intracavitary autoregulation by the endocardium (see Figs. 1 and 3) may add a unique type of controlling system to the heart in which endothelial cells from the endocardium may sense homeostatic variations in the superfusing blood, transmit them to the underlying myocardium, and participate in regional and global adjustments of the subjacent myocardium to these variations. How does endocardium-mediated autoregulation relate to other autoregulatory control mechanisms of cardiac performance, e.g., through volume loading (Starling; heterometric autoregulation) or through neurohumoral activation (homeometric autoregulation)?

When left-ventricular end-diastolic pressure was elevated from low (4.1 ± 0.9 mmHg) to baseline (10.6 ± 1.5 mmHg) or from baseline to high (17.9 ± 1.8 mmHg) in anesthetized open-chest dogs, time to peak –dP/dt increased by a mean value of 12.2% and 4.3%, respectively [3]. Comparison of these variations in systolic duration over a reasonably wide range of heterometric autoregulation to the variations by endocardium-mediated autoregulation in vivo suggests that both types of autoregulation may be nearly as powerful in day-to-day adaptations of cardiac performance. Interestingly, it thus seems that the two phylogenetically and embryologically most primitive autoregulatory control systems would act over a similar range and both through variations in the responsiveness of the contractile proteins to $[Ca^{2+}]_i$.

Elevations in responsiveness of the contractile proteins are manifested mechanically as changes in the onset of relaxation and the duration of systole, with or without changes in the rate of relaxation. Increased responsiveness does not necessarily imply slower relaxation, as the direction of the change in rate of relaxation is determined by the interaction of the loading conditions and inactivation processes (off-rate constant of Ca^{2+}-Troponin C and actin–myosin detachment) [35]. The hemodynamic consequences of altering the onset of relaxation, and thus of systolic duration, are only beginning to be explored. Modulation of the onset of relaxation will influence both ventricular ejection and diastolic filling and may have important consequences for pump function, especially in conditions of impaired relaxation. Although slowed or incomplete systolic relaxation is part of a pathophysiological process that is always deleterious in the long run, prolongation, or conversely shortening, of the systole through modulation of the onset of relaxation may within limits and at physiological heart rates be regarded as physiological, compensatory, and not deleterious per se.

A quantitative comparison between neurohumoral (homeometric) control and endothelium-mediated control is more difficult, as the net effect on cardiac function by neurohumoral control is the result of the impact on cardiac and peripheral system by a variety of hormones and transmitters, the individual concentration of which may fluctuate continuously. In vitro studies have shown that the EE may participate in the inotropic response to several of these substances. For example, the negative (and contraction-shortening) inotropic response to atrial natriuretic peptide (ANP) was dependent on a morphologically intact EE [36]; the positive inotropic (and contraction-prolonging) response to low concentrations of the $alpha_1$-adrenoceptor agonist phenylephrine [37] or to a short burst of reactive oxygen species [38] also required the presence of an intact EE. For other substances, e.g., serotonin [39], vasopressin [40], acetylcholine [41], angiotensin (I and II) [42], and endothelin [43], the EE was shown to modify the inotropic response. At high in vitro levels of epinephrine, vasopressin, serotonin, or atrial natriuretic peptide, the EE was selectively damaged. Thus, there seems to be in vitro evidence that the EE mediates the effect of extrinsic (homeometric) cardiac compensatory mechanisms. Because plasma levels of most of these substances are increased in chronic heart failure, one wonders to what extent elevated plasma levels could create a point of no return in the pathogenesis of the disease, possibly by converging to some state of irreversible cardiac endothelial dysfunction both in the endocardium and in the microvasculature [44].

Mechanisms of the Interaction Between Endocardium and Myocardium

How signals from the endocardial endothelium (EE) are transduced to the underlying myocytes is still under investigation. Two mechanisms of signal transduction have been proposed [45] (Fig. 4). On the one hand, EE cells may be stimulated (or activated) to secrete endothelium-derived factors that alter the contractile state of the cardiomyocytes (stimulus–secretion–contraction coupling). On the other hand, the EE may act as a physicochemical barrier, controlling the specific ionic composition of the interstitial milieu of the heart muscle cells and thereby modulating cardiac performance (blood–heart barrier).

Stimulus–Secretion Coupling

Increase in EE $[Ca^{2+}]_i$ after physical or humoral stimulation, a crucial step in the stimulus–secretion coupling, occurs in two phases [46,47]. After an initial transient

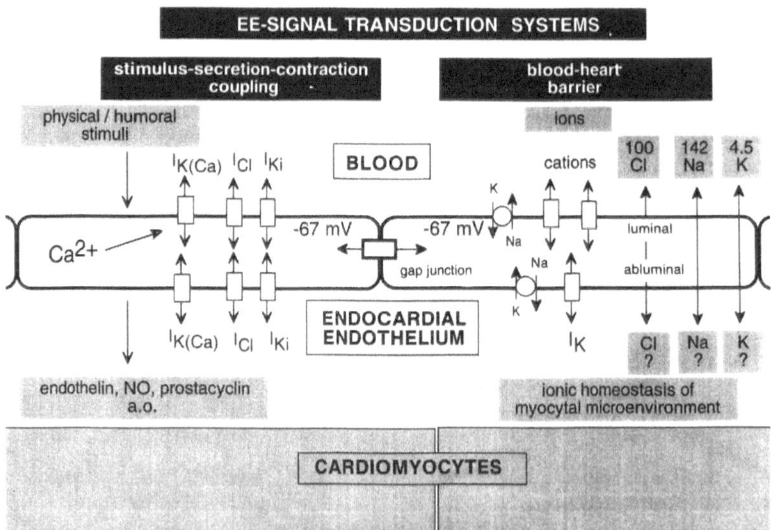

FIG. 4. Diagrammatic representation of the electrophysiological properties of endocardial endothelial cells in signal transduction to cardiomyocytes, including the role of mediators such as endothelin, nitric oxide, and prostacyclin

increase caused by the release of Ca^{2+} ions from intracellular stores, $[Ca^{2+}]_i$ remains elevated because of Ca^{2+} influx from the extracellular space. The Ca^{2+} influx depends on the driving force for Ca^{2+} ions and thus on the resting membrane potential (V_m) of the cells. V_m is, therefore, an important parameter in EE Ca^{2+} homeostasis and the release of EE-derived factors [47–50].

V_m values in EE cells are comparable to values reported in intact arterial endothelium and isolated cultured macrovascular endothelial cells [51]. As expected, in nonstimulated cultured EE cells of the porcine right ventricle, the main membrane current under whole-cell, voltage-clamp conditions is the inwardly rectifying K^+ current (I_{Ki}) and the negative value of V_m (−67 mV) in EE cells is dependent on $[K^+]_0$. V_m of EE cells is, however, always more depolarized than the equilibrium potential for K^+ ions (−85 mV), suggesting that other ionic conductances are involved in setting the V_m of EE cells. Interestingly, Cl^- ions, which play no role in setting the V_m of vascular endothelial cells [52,53], clearly influence V_m in EE cells. These observations indicate that nonstimulated EE cells, besides I_{Ki}, also display a Cl^--sensitive current and that both currents determine the V_m of EE cells. As the amplitude of both currents is small between V_m and 0 mV, the V_m of EE cells is dependent on a fine balance between K^+ and Cl^- currents. As a consequence, the intraendothelial Ca^{2+} homeostasis and the release of EE-derived factors are expected to depend on both the K^+ and Cl^- concentration.

Physical stimulation of cultured EE cells by membrane stretch, simulated by cell swelling or application of pressure to the patch pipette, activated a Cl^- current and concomitant depolarization of V_m to potentials around the equilibrium potential for Cl^- ions [54]. The role of the swelling-activated Cl^- current in EE cells is still speculative. It might be involved in the complex process of cell volume regulation and regulation of transendothelial permeability. In addition, activation of this current by shear stress or membrane stretch is expected to influence V_m and to play an important role in the cellular Ca^{2+} homeostasis and release of EE-derived factors. Through this current and the cyclical stretch of the ventricular wall, EE cells might perhaps display some kind of a cyclical transmembraneous potential change (action potential?) in which EE cells would switch during each cardiac cycle from the equilibrium potential for potassium to the equilibrium potential for chloride.

Secretion–Contraction Coupling: Cardiac Endothelial Mediators

There is increasing experimental evidence that EE cells release bioactive mediators that may affect the inotropic state of the myocyte. Whether these cardiac endothelial mediators can fully explain the positive inotropic activity imparted by an intact endothelium is still controversial, however. Earlier experiments demonstrated that cardiac endothelium mediates or modulates the in vitro inotropic response to various substances and cells (see foregoing). Subsequent experiments with the effluent from superfused cultured porcine EE cells [55] showed that EE cells tonically released (i) an unstable endothelium-derived relaxing factor-like (EDRF-like) substance causing endothelium-independent vasodilation of the coronary artery and (ii) a stable unidentified substance with contraction-prolonging and positive inotropic activity on isolated cardiac muscle. Furthermore, coronary venous effluent from isolated perfused hearts was observed to increase contraction of isolated trabeculae. This increase depended on the degree of oxygen saturation of the effluent and on the rate of coronary flow [56]. Accordingly, on the basis of these experiments, it was hypothesized that cardiac endothelial cells may release agents that influence the contraction of the underlying cardiac muscle, e.g., NO, prostaglandins, and endothelin.

Role of Nitric Oxide-Cyclic Guanosine Monophosphate (gmp)

In the past 5 years several laboratories, including our own, have provided evidence that NO, resulting from the activity of constitutive and inducible NO synthase in endocardial [57] and coronary endothelial cells or within the cardiomyocytes themselves [58], directly modulates myocardial contractility. The main subcellular action of NO is activation of guanylate cyclase, which elevates intracellular cGMP [59].

The inotropic effects of NO and cGMP are complex. From recent experiments in our and other laboratories, it seems that the inotropic activity of NO and cGMP depends on the baseline myocardial cGMP content. Myocardial cGMP content may be determined by the integrity of the EE and (micro)vascular cardiac endothelium (similarly as in the vessel wall), cardiac vagal tone, and other cGMP-modulating factors such as constitutive and inducible NO synthase activity in the cardiomyocytes. From these experiments, we postulate that an increase of myocardial cGMP from low to normal (such as probably induced by basal release of endogenous NO by the EE in physiological conditions) will increase intracellular Ca^{2+} and induce a positive inotropic effect [60]. A further increase of cGMP in the myocardium, however, will induce premature myocardial relaxation and a decrease in twitch force [61–63], which results from a cGMP-induced reduction in the responsiveness of the myofilaments to intracellular Ca^{2+} [64].

Accordingly, basal myocardial cGMP is likely to support cardiac contractility and may thus be important in preserving cardiac function in physiological states. Excess myocardial cGMP stimulated by endothelium-derived NO in conditions of increased endothelial shear stress, or by cytokines in conditions of septic shock, or in dilated cardiomyopathy [65], would lead to an earlier onset of left-ventricular relaxation and an increase in diastolic distensibility.

Role of Prostaglandins

Tissue cyclooxygenase has been found to be twice as high in the endocardium as compared to the myocardium and located specifically in the endothelial fraction [66]. Recently, production of both prostacyclin and prostaglandin E_2 from cultured bovine right and left ventricular EE cells was reported, with prostacyclin production tenfold higher than prostaglandin E_2 [67]. It is also of interest that EE cells have been reported to produce far larger amounts of prostacyclin than vascular endothelial cells [68].

The inotropic properties of prostaglandins are poorly understood, however. Previous studies utilizing exogenous prostaglandins to examine their direct myocardial actions were unclear. Recent observations in our laboratory suggested that the inotropic activity of endogenous prostaglandins may depend on the simultaneous capability of cardiac endothelial cells to release NO [69], which may explain some of the controversies around inotropic effects of prostaglandins.

Thus, in addition to their regulatory role of vascular tone in blood vessels, basal release of NO and prostaglandins and their interaction may modulate myocardial relaxation in the heart. The inotropic activity as well as the subcellular mechanism(s) involved in myocardial effects of prostaglandins, however, still need to be explored further.

Role of Endothelin

Endothelin is a recently discovered 21-amino-acid peptide released from endothelial cells. It displays a high density of receptors (endothelin A and endothelin B receptors) on myocytes, conduction tissue, and vascular and EE cells [70]. The demonstration of mRNA for endothelin-1 in the EE [71] and the release of endothelin from EE in culture [71] and from EE in isolated cardiac muscle [38,72] suggest that paracrine or autocrine activity of endothelin in the heart participates in endocardium-mediated regulation of cardiac function.

Endothelin is one of the most potent positive inotropic agents known so far in isolated cardiac tissue and in myocytes from animals and human. The positive inotropic effect of exogenous endothelin in isolated papillary muscle from various animal species resembles the characteristic positive inotropic effect induced by the cardiac endothelium [34]. In addition, similarly as for the cardiac endothelium, the positive inotropic activity of endothelin has been, at least partly, explained by an enhanced responsiveness of the contractile proteins to calcium [73] through activation of the sarcolemmal Na^+/H^+ exchanger [74]. Besides the inotropic effects, endothelin has potent chronotropic effects [75], is a potent vasoconstrictor, and can induce proliferation of vascular smooth muscle and cardiac muscle [76].

Until recently, a role for endothelin in normal cardiac physiology appeared puzzling because endothelin released from cardiac endothelium would enhance oxygen consumption of myocytes by its potent inotropic properties and decrease oxygen supply by its vasoconstrictive actions. There is, however, growing evidence that in vivo endothelin at basal rates of secretion acts in an autocrine way, by binding at the endothelin-B receptor on the endothelial surface, thereby stimulating the release of the vasodilating mediators NO and prostaglandin [77], rather than directly promoting vasoconstriction. Furthermore, endothelin has been shown to beneficially reverse acidosis-induced negative inotropic and lusitropic effects [78] and to directly counter the arrhythmogenic effects of catecholamines [79,80]. Accordingly, endothelin may be involved in the control of cardiac contraction and rhythm under physiological conditions and in disease states.

Blood–Heart Barrier

Similar to brain tissue, the myocardium is also a highly excitable tissue, the ionic homeostasis of which is vital not merely to the function of the organ but also to the entire body. In the heart it is, therefore, also essential that the ionic environment of the cardiomyocytes be well controlled. Alterations in the extracellular concentrations of Ca^{2+}, K^+, Cl^-, HCO_3^-, and Na^+, for example, have been shown to profoundly influence normal rhythmicity and contractility of the heart muscle cells. The unique position of the EE at the interface between blood and cardiomyocytes makes it a suitable candidate to regulate the specific ionic environment of the cardiomyocytes. The following paragraph speculates on observations in favor of a blood–heart barrier function of EE cells.

Morphologically, EE cells form a thin monolayer of closely apposed cells that are larger and have a more polygonal shape than the endothelial cells of heart blood vessels. Between the EE cells, there are complex interdigitations, leading to extensive overlap at the junctional edges. Tight junctions are present and are always located at the luminal side of the EE. At this luminal side, the glycocalyx is better developed than at the basolateral side below the tight junctions. Hence, the cells display morphological asymmetry. Gap junctional coupling between EE cells has been demonstrated by connexin 43 labeling and, recently, by the spreading of intracellularly applied lucifer yellow between neighboring cells [81].

Perhaps the strongest evidence for a blood–heart barrier function of the EE can be derived from the observation that the luminal and abluminal sides of the EE cell membrane display a different population of ion channels [82]. This asymmetrical distribution of ion channels and transporters makes a transendothelial transport of ions possible. It could somehow participate in the regulation of the ionic composition of the interstial fluid in the heart for optimal electrical activity of the cardiomyocytes. Further investigation is required, especially on the presence of ionic channels, active pumps, and transporters in the abluminal membrane, to unequivocally demonstrate the presence of a blood–heart barrier constituted by the endocardium.

Acknowledgments. We wish to acknowledge the Belgian Program on Interuniversity Poles of Attraction, initiated by the Belgian State, Prime Minister's Office, Science Policy Programming.

References

1. Furchgott RF, Zawadski JV (1980) The obligatory role of endothelial cells in the relaxation of arterial smooth muscle by acetylcholine. Nature 288:373–376
2. Brutsaert DL, Meulemans AL, Sipido KR, Sys SU (1988) Effects of damaging the endocardial surface on the mechanical performance of isolated cardiac muscle. Circ Res 62:357–366
3. De Hert SG, Gillebert TC, Brutsaert DL (1993) Alteration of left ventricular endocardial function by intracavitary high-power ultrasound interacts with volume, inotropic state, and alpha$_1$-adrenergic stimulation. Circulation 87:1275–1285
4. Gillebert TC, De Hert SG, Andries LJ, Jageneau AH, Brutsaert DL (1992) Intracavitary ultrasound impairs left ventricular performance: presumed role of endocardial endothelium. Am J Physiol 263:H857–H865
5. Brutsaert DL (1993) Endocardial and coronary endothelial control of cardiac performance. News Physiol Sci 8:82–86
6. Brutsaert DL, Andries LJ (1992) The endocardial endothelium. Am J Physiol 263: H985–H1002
7. Wilcox JN, Augustine A, Goeddel DV, Lowe DG (1991) Differential regional expression of three natriuretic peptide receptor genes within primate tissues. Mol Cell Biol 11:3454–3462
8. Lombès M, Oblin ME, Gasc JM, Baulieu EE, Farman N, Bonvalet JP (1992) Immunohistochemical and biochemical evidence for a cardiovascular mineralocorticoid receptor. Circ Res 71:503–510
9. Molenaar P, O'Reilly G, Sharkey A, Kuc RE, Harding DP, Plumpton C, Gresham GA, Davenport AP (1993) Characterization and localization of endothelin receptor subtypes in the human atrioventricular conducting system and myocardium. Circ Res 72:526–538
10. Wharton J, Rutherford RA, Gordon L, Moscoso G, Schiemberg I, Gaer JA, Taylor KM, Polak JM (1991) Localization of endothelin binding sites and endothelin-like immunoreactivity in human fetal heart. J Cardiovasc Pharmacol 7(suppl 17):S378–S384

11. Yamada H, Fabris B, Allen AM, Jackson B, Johnston CI, Mendelsohn AO (1991) Localization of angiotensin-converting enzyme in rat heart. Circ Res 68:141–149
12. Page C, Rose M, Yacoub M, Pigott R (1992) Antigenic heterogeneity of vascular endothelium. Am J Pathol 141:673–683
13. Tanaka H, Sukhova GK, Swanson SJ, Cybulsky MI, Schoen FJ, Libby P (1994) Endothelial and smooth muscle cells express leukocyte adhesion molecules heterogeneously during acute rejection of rabbit cardiac allografts. Am J Pathol 144:938–951
14. Taylor PM, Rose ML, Yacoub MH, Pigott R (1992) Induction of vascular adhesion molecules during rejection of human cardiac allografts. Transplantation 54(3):451–457
15. Yamazaki T, Seko Y, Tamtani T, Myasaka M, Yagita H, Okumura K, Nagai R, Yazaki Y (1993) Expression of intercellular adhesion molecule-1 in rat heart with ischemia/reperfusion and limitation of infarct size by treatment with antibodies against cell adhesion molecules. Am J Pathol 143:410–418
16. Andries LJ (1994) Endocardial endothelium: functional morphology. Landes, Austin, p 143
17. Andries LJ, Brutsaert DL (1991) Differences in functional structure between endocardial endothelium and vascular endothelium. J Cardiovasc Pharmacol 17(suppl 3): S243–S246
18. Anversa P, Giacomelli F, Wiener J (1975) Intercellular junctions of rat endocardium. Anat Rec 183:477–484
19. Rotrosen D, Gallin JI (1986) Histamine type I receptor occupancy increases endothelial cytosolic calcium, reduces F-actin, and promotes albumin diffusion across cultured endothelial monolayers. J Cell Biol 103:2379–2387
20. Schnittler HJ, Wilke A, Gress T, Suttorp N, Drenckhahn D (1990) Role of actin and myosin in the control of paracellular permeability in pig, rat and human vascular endothelium. J Physiol (Camb) 431:379–401
21. Wysolmerski RB, Lagunoff D (1988) Inhibition of endothelial cell retraction by ATP depletion. Am J Pathol 132:28–37
22. Gotlieb AI, Langille L, Wong MKK, Kim DW (1991) Biology of disease. Structure and function of the endothelial cytoskeleton. Lab Invest 65:123–137
23. White GE, Gimbrone MA, Fujiwara K (1983) Factors influencing the expression of stress fibers in vascular endothelial cells in situ. J Cell Biol 97:417–424
24. Andries LJ, Brutsaert DL (1993) Endocardial endothelium in rat heart: cell shape and organization of the cytoskeleton. Cell Tissue Res 273:107–117
25. Dartsch PC, Betz E (1989) Response of cultured endothelial cells to mechanical stimulation. Basic Res Cardiol 84:268–281
26. Iba T, Sumpio BE (1991) Morphological response of human endothelial cells subjected to cyclic strain in vitro. Microvasc Res 42:245–254
27. Shirinsky VP, Antonov AS, Birukov KG, Sobolevsky AV, Romanov YA, Kabaeva NV, Antanova GN, Smirnov VN (1989) Mechano-chemical control of human endothelium orientation and size. J Cell Biol 109:331–339
28. Andries LJ, Brutsaert DL (1994) Endocardial endothelium: junctional organization and permeability. Cell Tissue Res 277:391–400
29. Melax H, Leeson TS (1967) Fine structure of the endocardium in adult rats. Cardiovasc Res 1:349–355
30. Li K, Rouleau JL, Andries LJ, Brutsaert DL (1993) Effect of dysfunctional vascular endothelium on myocardial performance in isolated papillary muscles. Circ Res 72:768–777
31. Li K, Rouleau JL, Calderone A, Andries LJ, Brutsaert DL (1993) Endocardial function in pacing-induced heart failure in the dog. J Mol Cell Cardiol 25:529–538
32. Shah AM, Smith JA, Lewis MJ (1991) The role of endocardium in the modulation of contraction of isolated papillary muscles of the ferret. J Cardiovasc Pharmacol 17:S251–S257
33. Fort S, Lewis MJ, Shah AM (1993) The role of endocardial endothelium in the modulation of myocardial contraction in the isolated whole heart. Cardioscience 4:217–222

34. Wang J, Morgan JP (1992) Endocardial endothelium modulates myofilament Ca^{2+} responsiveness in aequorin-loaded ferret myocardium. Circ Res 70:754–760
35. Brutsaert DL, Sys SU (1989) Relaxation and diastole of the heart. Physiol Rev 69:1228–1315
36. Meulemans AL, Sipido KR, Sys SU, Brutsaert DL (1988) Atriopeptin III induces early relaxation of isolated mammalian papillary muscle. Circ Res 62:1171–1174
37. Meulemans AL, Andries LJ, Brutsaert DL (1990) Endocardial endothelium mediates positive inotropic response to $alpha_1$-adrenoreceptor agonist in mammalian heart. J Mol Cell Cardiol 22:667–685
38. De Keulenaer GW, Andries LJ, Sys SU, Brutsaert DL (1995) Endothelin-mediated positive inotropic effect by free radicals in isolated cardiac muscle. Circ Res 76:878–884
39. Shah AM, Andries LJ, Meulemans AL, Brutsaert DL (1989) Endocardium modulates inotropic response to 5-hydroxytryptamine. Am J Physiol 257:H1790–H1797
40. Schoemaker IE, Meulemans AL, Andries LJ, Brutsaert DL (1990) Role of the endocardial endothelium in the positive inotropic action of vasopressin. Am J Physiol 259:H1148–H1151
41. Mohan P, Brutsaert DL, Sys SU (1994) Positive inotropic effect of acetylcholine; role of endocardial endothelium, cGMP, nitric oxide, prostaglandins. Circulation 90:I–649
42. Meulemans AL, Andries LJ, Brutsaert DL (1990) Does endocardial endothelium mediate positive inotropic response to angiotensin I and angiotensin II? Circ Res 66:1591–1601
43. Li K, Stewart DJ, Rouleau JL (1991) Myocardial contractile actions of endothelin-1 in rat and rabbit papillary muscles. Role of endocardial endothelium. Circ Res 69:301–312
44. Brutsaert DL (1991) Role of endocardium in cardiac overloading and failure. Eur Heart J 11:8–16
45. Brutsaert DL (1989) The endocardium. Annu Rev Physiol 51:263–273
46. Adams DJ, Rusko J, Van Slooten G (1993) Calcium signalling in vascular endothelial cells: Ca^{2+} entry and release. In: Weir EK, Hume JR, Reeves JI (eds) Ion flux in pulmonary vascular control. NATO ASI series, vol 251. Plenum, New York, pp 259–275
47. Laskey RE, Adams DJ, Johns A, Rubanyi GM, Van Breemen C (1990) Regulation of $[Ca^{2+}]_i$ in endocardial cells by membrane potential. In: Rubanyi GM, Vanhoutte PM (eds) Endothelium-derived relaxing factors. Basel, Karger, pp 28–35
48. Busse R, Hecker M, Fleming I (1994) Control of nitric oxide and prostacyclin synthesis in endothelial cells. Arzneim-Forsch 44:392–396
49. Laskey RE, Adams DJ, Cannell M, Van Breemen C (1992) Calcium entry-dependent oscillations of cytoplasmic calcium concentration in cultured endothelial cell monolayers. Proc Natl Acad Sci USA 89:1690–1694
50. Laskey RE, Adams DJ, Johns A, Rubanyi GM, Van Breemen C (1990) Membrane potential and Na^+-K^+ pump activity modulate resting and bradykinin-stimulated changes in cytosolic free calcium in cultured endothelial cells from bovine atria. J Biol Chem 265:2613–1629
51. Adams DJ (1994) Ionic channels in vascular endothelial cells. Trends Cardiovasc Med 4:18–26
52. Adams DJ, Barakeh J, Laskey R, Van Breemen C (1989) Ion channels and regulation of intracellular calcium in vascular endothelial cells. FASEB J 3:2389–2400
53. Northover BJ (1980) The membrane potential of vascular endothelial cells. Adv Microcirc 9:135–160
54. Fransen PF, Demolder MJM, Brutsaert DL (1995) Whole-cell membrane currents in cultured pig endocardial endothelial cells from the porcine right ventricle. Am J Physiol 258:H2036–H2047
55. Smith JA, Shah AM, Lewis MJ (1991) Factors released from endocardium of the ferret and pig modulate myocardial contraction. J Physiol (Lond) 439:1–14

56. Ramaciotti C, McClellan G, Sharkey A, Rose D, Weisberg A, Winegrad S (1993) Cardiac endothelial cells modulate contractility of rat heart in response to oxygen tension and coronary flow. Circ Res 72:1044–1064
57. Schulz R, Smith JA, Lewis MJ, Moncada S (1991) Nitric oxide synthase in cultured endocardial cells of the pig. Br J Pharmacol 104:21–24
58. Schulz R, Nava E, Moncada S (1992) Induction and potential biological relevance of Ca^{2+}-independent nitric oxide synthase in the myocardium. Br J Pharmacol 105:575–580
59. Moncada S, Palmer RMJ, Higgs EA (1991) Nitric oxide: physiology, pathophysiology and pharmacology. Pharmacol Rev 43:109–142
60. Mohan P, Brutsaert DL, Paulus WJ, Sys SU (1996) Myocardial contractile response to nitric oxide and cGMP. Circulation 93:1223–1229
61. Grocott-Mason R, Anning P, Evans H, Lewis MJ, Shah AM (1994) Modulation of left ventricular relaxation in isolated heart by endogenous nitric oxide. Am J Physiol 267:H1804–H1813
62. Paulus WJ, Vantrimpont PJ, Shah AM (1994) Acute effects of nitric oxide on left ventricular relaxation and diastolic distensibility in humans. Assessment by bicoronary sodium nitroprusside infusion. Circulation 89:2070–2078
63. Shah AM, Lewis MJ, Henderson AH (1991) Effects of 8-bromo-cyclic GMP on contraction and on inotropic response of ferret cardiac muscle. J Mol Cell Cardiol 23:55–64
64. Shah AM, Spurgeon HA, Sollot SJ, Talo A, Lakatta EG (1994) 8-Bromo-cGMP reduces the myofilament response to Ca^{2+} in intact cardiac myocytes. Circ Res 74:970–978
65. De Belder AJ, Radomski MW, Why HJF, Richardson PJ, Bucknall CA, Salas E, Martin JF, Moncada S (1993) Nitric oxide synthase activities in human myocardium. Lancet 341:84–85
66. Brandt R, Nowak J, Sonnenfeld T (1984) Prostaglandin formation from exogenous precursor in homogenates of human cardiac tissue. Basic Res Cardiol 79:135–141
67. Mebazaa A, Martin LD, Robotham JL, Maeda K, Gabrielson EW, Wetzel RC (1993) Right and left ventricular cultured endocardial endothelium produces prostacyclin and PGE_2. J Mol Cell Cardiol 25:245–248
68. Mebazaa A, Wetzel R, Cherian M, Abraham M (1995) Comparison between endocardial and great vessel endothelial cells: morphology, growth, and prostaglandin release. Am J Physiol 268:H250–H259
69. Mohan P, Brutsaert DL, Sys SU (1995) Myocardial performance is modulated by interaction of cardiac endothelium-derived nitric oxide and prostaglandins. Cardiovasc Res 29:637–640
70. Davenport AP, Nunez DJ, Hall A, Kaumann AK, Brown MJ (1989) Autoradiographical localization of binding sites for porcine (^{125}I)endothelin-1 in humans, pigs, and rats: functional relevance in humans. J Cardiovasc Pharmacol 161:1252–1259
71. Mebazaa A, Mayoux E, Maeda K, Martin LD, Lakatta EG, Robotham JL, Shah AM (1993) Paracrine effects of endocardial endothelial cells on myocyte contraction mediated via endothelin. Am J Physiol 265:H1841–H1846
72. Evans HG, Lewis MJ, Shah AM (1994) Modulation of myocardial relaxation by basal release of endothelin from endocardial endothelium. Cardiovasc Res 28:1694–1699
73. Wang J, Paik G, Morgan J (1991) Endothelin enhances myofilament Ca^{2+} responsiveness in aequorin-loaded ferret myocardium. Circ Res 69:582–589
74. Kramer BK, Smith TW, Kelly RA (1991) Endothelin and increased contractility in adult rat ventricular myocytes: role of intracellular alkalosis induced by activation of the protein kinase C-dependent Na^+-H^+-exchanger. Circ Res 68:269–279
75. Ono K, Eto K, Sakamoto A, Masaki T, Shihata K, Sada T, Hashimoto K, Tsujimoto G (1995) Negative chronotropic effect of endothelin-1 mediated through ET-A receptors in guinea pig atria. Circ Res 76:284–292
76. Ito H, Hirata Y, Hiroe M, Tsujino M, Adachi S, Takamoto T, Nitta M, Taniguchi K, Marumo F (1991) Endothelin-1 induces hypertrophy with enhanced expression of muscle specific genes in cultured neonatal rat cardiomyocytes. Circ Res 69:209–215

77. Lüscher TF, Boulanger CM, Yang Z, Noll G, Dohi Y (1993) Interactions between endothelium-derived relaxing and contracting factors in health and cardiovascular disease. Circulation 87(suppl V):V36–V44
78. Wang J, Morgan JP (1992) Endothelin reverses the effects of acidosis on the intracellular Ca^{2+} transient and contractility in ferret myocardium. Circ Res 71:631–639
79. James AF, Xie L, Fujitani Y, Hayashi S, Horie M (1994) Inhibition of cardiac protein kinase A-dependent chloride conductance by endothelin-1. Nature 370:297–300
80. Ono K, Tsujimoto G, Sakamoto A, Eto K, Masaki T, Ozaki Y, Satake M (1994) Endothelin-A receptor mediates cardiac inhibition by regulating calcium and potassium currents. Nature 370:301–304
81. Fransen P, Andries LJ, Van Bedaf D, Demolder M, Sys SU (1996) Cell coupling in cultured endocardial endothelium. Eur J Physiol 431:R323
82. Ito H, Matsuda H, Noma A (1993) Ion channels in the luminal membrane of endothelial cells of the bull-frog heart. Jpn J Physiol 43:191–206

Part 3
Ventricular–Vascular Interaction

Arterial Compliance as Load on the Heart

NICO WESTERHOF[1], NIKOS STERGIOPULOS[2], O.S. RANDALL[3], and GERARD C. VAN DEN BOS[1]

Summary. Total arterial compliance and peripheral resistance form the main part of the load on the heart. Together with stroke volume and heart rate (or heart period), these are the major determinants of systolic and diastolic pressure in the ascending aorta. A decrease in total arterial compliance will, in the nonregulated cardiovascular system, decrease cardiac output and thus mean pressure while increasing pulse pressure. The result is a large decrease in diastolic pressure and a small increase in systolic pressure. In the intact animal with cardiovascular regulation, a (moderate) decrease in total arterial compliance causes little change in stroke volume, so that mean pressure also remains the same. This results in a decrease in diastolic pressure and an increase in systolic pressure. If total arterial compliance decreases, as in atherosclerosis or aging, peripheral resistance may increase so that both diastolic pressure and systolic pressure increase. However, a change in compliance alone can only result in a decrease of diastolic blood pressure. Several methods are available to derive total arterial compliance from aortic pressure and flow. The pulse pressure method, based on the two-element windkessel model, was shown to be superior to all other methods. This method fits the pulse pressure derived from the two-element windkessel model to the measured pulse pressure. However, waveforms of pressure and flow are not realistic in this model. The methods based on the three-element windkessel all perform poorly in the estimation of total arterial compliance, but the wave shapes produced by this model are very close to the measured wave shapes. Thus, both windkessels have shortcomings and it is concluded that they need to be revised.

Key words. Peripheral resistance—Systolic and diastolic pressure—Heart period/heartrate—RC time—Windkessel—Compliance derivation

[1] Laboratory for Physiology, ICaR-VLL Free University of Amsterdam, 1081 BT Amsterdam, The Netherlands.
[2] Biomedical Engineering Laboratory, Swiss Federated Institute of Technology, 1015 Lausanne, Switzerland.
[3] Department of Medicine, Division of Cardiovascular Diseases, Howard University Hospital, Washington, DC, USA.

Introduction

Total arterial compliance and peripheral resistance form the main load on the heart [1]. Peripheral resistance is the ratio of mean (aortic) pressure minus venous pressure and mean ascending aortic flow and can easily be determined. However, total arterial compliance is not so easily obtained in the in vivo situation. It turns out that models are needed to derive arterial compliance. Local compliance, compliance per length of artery, or area compliance is the ratio of the change in vascular cross-sectional area caused by a change in transmural pressure. When arterial length is constant, the area compliance times length equals the volume compliance of that segment. Addition of all local volume compliances of the arterial system gives the total arterial compliance. Area compliance depends on the mechanical properties of the vessel wall (Young modulus) and on vessel size, i.e., wall thickness and diameter. The mechanical properties differ little between arteries and species, but the sizes of the arteries do differ considerably, and local compliance increases strongly with vessel size. Most of the arterial compliance is therefore located in the large arteries, with about 50% of total arterial compliance attributed to the ascending and proximal descending aorta. Several studies have been performed to obtain information about the effects of decreases in total arterial compliance on cardiovascular function [1–4]. The compliance changes are thought to be a model for events in atherosclerosis and aging [5].

In this chapter, we discuss two related subjects. First, we discuss the effects of changes in total arterial compliance on blood pressure, and then we review methods to obtain total arterial compliance.

Total Arterial Compliance and Blood Pressure

Total arterial compliance and peripheral resistance form the main load on the heart and are therefore important determinants of aortic pressure. Aortic pressure is also determined by cardiac pump function, the main variables being stroke volume and heart rate. Thus, in good approximation, aortic pressure is determined by two cardiac variables, stroke volume and heart rate, or the R–R interval (their product or ratio, giving cardiac output), and two arterial variables, total arterial compliance ($C = \Delta V/\Delta P$) and peripheral resistance ($R = P_m/F_m$, mean pressure divided by mean flow). The product of the latter two variables gives the RC time of the arterial system, which determines the decay time of diastolic aortic pressure.

Total Arterial Compliance and Mean Systemic Pressure

Mean pressure is the product of cardiac output and peripheral resistance. However, compliance also affects mean pressure. Decreased compliance in the isolated heart, i.e., in a condition without humoral and nervous control mechanisms present, will result in decreased stroke volume [1]. With heart rate and arterial resistance kept constant, this decrease in stroke volume results in a decrease in mean pressure. In the intact animal, however, cardiac output changes very little with a moderate reduction in total arterial compliance [3] so that mean pressure remains virtually constant. The long-range effects of a decrease in total arterial compliance on cardiac output and systolic and diastolic pressure in the ascending aorta have not been studied in detail.

Total Arterial Compliance and Pulse Pressure

The rate of decrease of aortic pressure in diastole is determined by the product of peripheral resistance and total arterial compliance, the RC time of the arterial system. The diastolic aortic pressure depends on this rate of the decrease in diastole and on the duration of diastole. Because the duration of diastole is linked to the duration of the heart period [6], diastolic pressure depends on the RC time of the arterial system and on heart rate or heart period. When the RC time of the arterial system increases, the rate of decay of pressure decreases and diastolic pressure will be higher. When heart rate decreases, i.e., the heart period increases, aortic pressure will decay for a longer period of time and diastolic pressure will be lower. Thus, pulse pressure, the difference between systolic and diastolic pressure, is determined by the ratio of RC time and heart period. In the ascending aorta, it holds that mean pressure equals the average of systolic and diastolic pressure [7]. Thus, in good approximation and at constant mean pressure a decrease in diastolic pressure relates to a similar increase in systolic pressure.

The foregoing analysis implies that for a decrease in compliance, when heart period, stroke volume, and peripheral resistance remain the same, mean pressure will remain the same as well but diastolic pressure will decrease and systolic pressure will increase with about the same amount. This reasoning is supported by the experimental findings presented in Fig. 1, where, in the anesthetized closed-chest dog during cardiac pacing, the aorta is replaced by a stiff tube to decrease total arterial compliance. The preparation and conduct of this type of experiment is given in detail elsewhere [3]. Stroke volume and peripheral resistance did not change significantly; we indeed saw that systolic pressure increases and diastolic pressure decreases. Thus, with a decrease in total arterial compliance alone, systolic pressure increases and diastolic pressure decreases [3].

When compensatory mechanisms are not present, as is the case in the isolated paced heart with constant filling pressure, stroke volume will decrease with reduced compliance so that a decrease in diastolic pressure is accompanied with a very small increase in systolic pressure [1]. Therefore, the connection between progression of atherosclerosis and decreased diastolic blood pressure, as recently suggested by Witteman et al. [8], may be in agreement with the foregoing. However, in atherosclerosis other cardiac and arterial parameters change as well, thereby complicating the picture. If in atherosclerosis systolic and diastolic pressure both are found to increase, this implies that more cardiac parameters and arterial system parameters have changed than total arterial compliance alone.

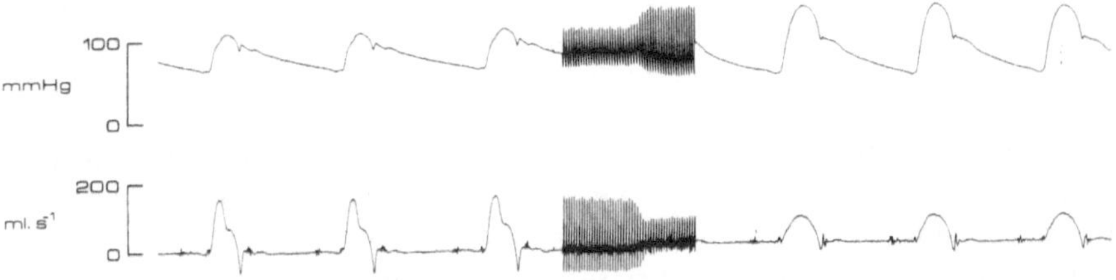

Fig. 1. Pressure and flow in the ascending aorta during control conditions and after decreased total arterial compliance in the dog. The heart is paced, stroke volume is decreased slightly (just under 10%), and peripheral resistance is not changed. Systolic pressure increases and diastolic pressure decreases

Increased Total Arterial Compliance

The relation between total arterial compliance and systolic and diastolic pressure is strongly nonlinear. For low compliance values, small changes have a large effect on both systolic and diastolic pressure, but for normal and large compliance values, changes in compliance have a moderate to small effect on pressure. Increases in total arterial compliance, have, as far as we know, not been studied in vivo. Model studies have shown that increases in compliance from the control value decrease pulse pressure only slightly [9]. We suggest that this means that if pulse pressure were reduced, total arterial compliance would have to increase disproportionately. With such a very large total arterial compliance, and thus long RC time, pressure regulation would become impossibly slow.

Peripheral Resistance and Blood Pressure

When peripheral resistance increases, mean pressure in the arterial system will rise. However, the increase in pressure and peripheral resistance will, in general, not be proportional because with increased peripheral resistance cardiac output will decrease [1]. With increased peripheral resistance the RC time of the arterial system will be larger and the rate of decay of diastolic pressure will be slower, making pulse pressure smaller. The studies done so far show that mean pressure and systolic and diastolic aortic blood pressure are more sensitive to changes in peripheral resistance than to changes in total arterial compliance. This can best be seen from the experiments of Elzinga and Westerhof [1].

The qualitative reasoning just given needs to be worked out further. It is necessary that studies be undertaken to derive the sensitivity of systolic and diastolic pressure to arterial (peripheral resistance, total arterial compliance) and cardiac (heart rate, contractility, filling) parameters. If sensitivity of systolic and diastolic pressure to individual parameters and combinations of parameters were known, predictions based on cardiac and arterial changes would be possible.

Implications of Decreased Total Arterial Compliance

One wonders if decreased compliance with increased systolic but decreased diastolic pressure could have a deleterious effect on cardiac function. Kelly et al. [4] studied this in the reflex-blocked anesthetized dog. Like Randall et al. [3], they replaced the aorta with a stiff conduit, and found an increase in systolic pressure from 110 to 162 mmHg and a decrease in diastolic pressure from 76 to 62 mmHg without statistically significant changes in dP_{lv}/dt and the end-systolic pressure–volume relation (E_{max}), suggesting little change in cardiac contractility. Cardiac oxygen consumption increased, probably because of the increase in systolic pressure and wall stress. These authors concluded that vascular stiffening results in little decrease in cardiac function but limits the reserve capacity under conditions of increased demand [4]. In a later study, the supply–demand relation of the heart was studied in the same preparation [10]. The supply is the product of local coronary flow and arterial oxygen content; the demand was derived from oxygen consumption or the pressure–volume area [11]. Although diastolic coronary perfusion decreased as a result of the reduced aortic pressure in diastole, systolic coronary flow was augmented. Systolic flow augmentation could be explained on basis of the varying coronary elastance theory [12]. These results show that a stiff arterial system did not compromise supply-demand, contrary to general notion.

Cardiac hypertrophy may result from systolic hypertension brought about by decreased total arterial compliance alone. If, however, heart rate or peripheral resistance change as well and mean pressure increases, the systolic pressure may increase more than with arterial stiffening alone, so that the stimulus for hypertrophy is augmented.

The Arterial System and Body Size

Total arterial compliance and peripheral resistance are determined by body size. The peripheral resistance decreases and the total arterial compliance increases with body size. The RC time in mammals from rat to man was shown to be proportional to body mass to the power +0.29 [6]. Body length also relates to body mass with approximately the same exponent. The RC time thus is proportional to body length [13]. Heart rate exhibits the same relation to body mass as the RC time, the exponent being 0.27, so that the ratio of RC time and heart rate is the same in mammals. This leads to similar pulse pressures in mammals (see earlier). Mean pressure is the same in mammals, probably to guarantee brain perfusion.

Cardiac output is determined by the metabolic needs of the body and is approximately proportional to body mass. Peripheral resistance, being the ratio of mean pressure and cardiac output, is thus inversely proportional to body mass. Total arterial compliance is determined by the elastic properties of the blood vessels and the anatomy of the vasculature. Elastic properties (Young modulus) are similar for all mammalian arteries. Thus, the differences in total arterial compliance between mammals arises from the differences in size of the arteries only. In other words, body size prescribes total arterial compliance. To make systolic and diastolic pressure the same in mammals, heart rate has to adjust to the RC time of the arterial system. We therefore suggested that heart rate is linked to body length so that the combination of heart and arterial system gives pulse pressures that are similar among mammals [13].

Determination of Total Arterial Compliance

Several methods are available to derive total arterial compliance, but a single generally accepted method does not yet exist. The problem in evaluating the different methods to estimate total arterial compliance is that its in vivo determination cannot be carried out directly. Simple addition of all local compliances, each with their pressure dependence [14], is not feasible because of the local differences and the large number of arterial segments. Plugging the periphery and then determining a pressure-volume relation of the entire systemic bed is also not feasible in the living animal.

Stergiopulos et al. [15] recently reviewed the methods available to derive total arterial compliance. They relied on the information contained in an extensive model of the systemic arterial tree and tested the suggested methods against this extensive model. The models used in the estimation of total arterial compliance are all based on the two-element [16] or on the three-element windkessel model [17]. The simplest method is the fit of diastolic aortic pressure with a single exponential. This method is based on both the two-element and three-element windkessel models and can lead to errors as great as 40% [15]. Others used the area under the diastolic pressure curve to derive total arterial compliance [18]. In all methods, cardiac output and the wave shape of the ascending aortic pressure must be known. In the more sophisticated methods, the wave shape of flow is also required in the analysis.

One of the important findings of Stergiopulos et al. [15] was that the use of the three-element windkessel in estimating total arterial compliance leads to a consistent overestimation of this parameter by more than 15% [15]. Comparison of the methods also showed that those based on the two-element windkessel lead to more accurate compliance estimates than those based on the three-element windkessel. Measured flow when fed to the three-element windkessel can result in an aortic pressure wave form that is very close to measured aortic pressure, but total arterial compliance used to obtain the best fit deviates strongly from the actual value. On the other hand, the wave shape of aortic pressure predicted by the two-element windkessel model when measured aortic flow is used as input is very different from the measured aortic pressure. However, with the correct compliance value, the predicted and measured pulse pressures were very similar.

This finding led Stergiopulos et al. [19] to suggest a very simple method to estimate total arterial compliance from the two-element windkessel. They first calculated peripheral resistance from mean pressure and cardiac output. The measured flow wave in the aorta subsequently was used as input to the two-element windkessel model to predict pulse pressure. This predicted pulse pressure was fitted to the measured pulse pressure by varying total arterial compliance. The best fit of pulse pressure gave a value of total arterial compliance that was very accurate (errors less than 5%) under very different conditions such as decreased compliance, increased and decreased heart rate, and exercise. The method was called the pulse pressure method, PPM [19]. The explanation is that for low frequencies the input impedance of the two-element windkessel is very close to the actual input impedance, while the three-element windkessel better agrees at the higher frequencies.

The poor high-frequency behavior of the two-element windkessel results in a delayed and smoothed aortic pressure but with correct pulse pressure. The poor low-frequency behavior of the three-element windkessel gives an overestimation of the pulse pressure but a good wave shape. When a total arterial compliance larger than the actual value is used, the wave shape and the magnitude of aortic pressure agree with the measured pressure. The poor low-frequency behavior of the three-element windkessel results from the introduction of the characteristic impedance as a resistance term. The characteristic impedance of the proximal aorta accounts for local compliance and inertia and pertains to oscillartory pressure and flow only. By introducing it as a resistance, it also has an effect on the relation between pressure and flow at 0 Hz (i.e., mean values) and low frequencies. The unsatisfactory behavior of the three-element windkessel at low frequencies suggests that this model needs to be improved.

After comparison of all methods to derive total arterial compliance it was concluded that the pulse pressure method is superior to all other methods previously published [15].

Conclusion

We have shown that a decrease in total arterial compliance alone increases systolic pressure and decreases diastolic pressure when stroke volume is not affected. When stroke volume decreases with decreased compliance, diastolic pressure decreases strongly and systolic pressure changes little. However, in the intact animal and in the human, the quantitative effects of compliance changes and the quantitative effects of changes in cardiac pump function on systolic and diastolic pressure have not been worked out. It seems important to perform studies to quantify these effects. A first attempt to study the effects of cardiac and arterial parameters on systolic and diastolic pressure was recently carried out by Stergiopulos et al. [20].

The methods to obtain arterial compliance have been compared [15]. On one hand, the two-element windkessel can be used to estimate total arterial compliance but does not produce acceptable wave shapes of pressure and flow. On the other hand, the three-element windkessel can mimic pressure and flow waveforms well but cannot be used in the estimation of total arterial compliance. This situation is very unsatisfactory. Therefore, the windkessels need to be improved in the sense that a lumped model should emerge that can not only mimic the arterial load correctly but can do so with the correct parameter values.

References

1. Elzinga G, Westerhof N (1973) Pressure and flow generated by the left ventricle against different impedances. Circ Res 32:178–186
2. O'Rourke MF (1967) Steady and pulsatile energy losses in the systemic circulation under normal conditions and in simulated arterial disease. Cardiovasc Res 1:313–326
3. Randall OS, van den Bos GC, Westerhof N (1984) Systemic compliance: does it play a role in the genesis of essential hypertension? Cardiovasc Res 18:455–462
4. Kelly RP, Tunin RS, Kass DA (1992) Effect of reduced aortic compliance on cardiac efficiency and contractile function of in situ canine left ventricle. Circ Res 71:490–502
5. Avolio AP, Chen S-G, Wang R-P, Zhang C-L, LI M-F, O'Rourke MF (1983) Effects of aging on changing arterial compliance and left ventricular load in northern Chinese urban community. Circulation 68:50–58
6. Westerhof N, Elzinga G (1991) Normalized input impedance and arterial decay time over heart period are independent of body size. Am J Physiol 261:R126–R133
7. Gevers M, Hack WWM, Ree RF, Lafeber HN, Westerhof N (1993) Calculated mean arterial pressure in critically ill neonates. Basic Res Cardiol 88:80–85
8. Witteman JCM, Grobbee DE, Valkenburg HA, van Hemert AM, Stijnen T, Burger H, Hofman A (1994) J-Shaped relation between change in diastolic blood pressure and progression of aortic atherosclerosis. Lancet 343:504–507
9. Beger DS, Li JK-J, Noordergraaf A (1994) Differential effects of wave reflections and peripheral resistance on aortic blood pressure: a model based study. Am J Physiol 266:H1626–H1646
10. Saeki A, Recchia F, Kass DA (1995) Systolic flow augmentation in hearts ejecting into a model of stiff aging vasculature. Circ Res 76:132–141
11. Suga H (1990) Ventricular energetics. Physiol Rev 70:247–277
12. Krams R, Sipkema P, Zegers J, Westerhof N (1989) Contractility is the main determinant of coronary systolic flow impediment. Am J Physiol 257:H1936–H1944
13. Westerhof N (1994) Heart period is proportional to body length. Cardioscience 5:283–285
14. Langewouters J, Wesseling KH, Goedhard WJA (1984) The static elastic properties of 45 human thoracic and 20 abdominal aortas in vitro and the parameters of a new model. J Biomech 17:425–435
15. Stergiopulos N, Meister J-J, Westerhof N (1995) Evaluation of methods for estimation of total arterial compliance. Am J Physiol 268:H1540–H1548
16. Frank O (1899) Die grundform des Arteriellen Pulses. Z Biol 37:483–526

17. Westerhof N, Elzinga G, Sipkema P (1971) An artificial arterial system for pumping hearts. J Appl Physiol 31:776–781
18. Yin FCP, Lin Z (1989) Estimating arterial resistance and compliance during transient conditions in humans. Am J Physiol 257:H190–H197
19. Stergiopulos N, Meister J-J, Westerhof N (1994) Simple and accurate way for estimating total and segmental arterial compliance: the pulse pressure method. Ann Biomed Eng 22:392–397
20. Stergiopulos N, Meister J-J, Westerhof N. Determination of stroke volume and systolic and diastolic aortic pressure. Am J Physiol (in press)

Optimal Afterload that Maximizes External Work and Optimal Heart that Minimizes O_2 Consumption

MASARU SUGIMACHI and KENJI SUNAGAWA

Summary. We examined the optimality of ventriculoarterial coupling from the point of view of energy efficiency. We defined the optimal afterload as that which maximizes external work. A theoretical analysis indicated, for a given end-diastolic volume, that external work becomes maximal when effective arterial elastance (E_a) equals end-systolic elastance (E_{es}). The ratio of external work to its maximum value was used to indicate the optimality of the afterload, Q_{load}. We defined the optimal heart as that which minimizes myocardial oxygen consumption for a given mean arterial pressure and cardiac output under a constant end-diastolic volume. This is accomplished when E_a/E_{es} is about 0.4. The ratio of minimum oxygen consumption, which is derived theoretically, to actual oxygen consumption is used to indicate the optimality of the heart, Q_{heart}. In chronically instrumented dogs, E_a/E_{es} ranged from 0.39 to 1.00 (0.69 ± 0.26 [SD]). Q_{load} was 0.93 ± 0.08, suggesting that the external work of the left ventricle was nearly maximum at this condition. Similarly, Q_{heart} was 0.98 ± 0.01, indicating that myocardial oxygen consumption was almost minimum. When dogs ran on a treadmill, their heart rate and cardiac output doubled. E_{es} and E_a increased moderately; E_a/E_{es}, however, remained unchanged. Thus, despite dramatic changes in hemodynamic conditions, external work remained nearly maximum and oxygen consumption was minimum. In contrast, experimental left heart dysfunction increased E_a/E_{es}, which, in turn, decreased Q_{heart}. We concluded that the ventriculoarterial coupling is nearly optimal under physiological stress. It is, however, no longer optimal for pathological conditions.

Key words. Ventriculoarterial coupling—External work—Myocardial oxygen consumption—Time-varying elastance—Pressure-volume area—Optimality index

Introduction

When the cardiovascular system generates cardiac output to meet peripheral demand, the combination of preload, contractility, heart rate, and afterload is not unique; the regulatory system chooses a specific combination of these. How the regulatory system determines the unique combination, however, remains unknown. As the heart contin-

---

Department of Cardiovascular Dynamics, National Cardiovascular Center Research Institute, Osaka, 565 Japan.

ues to pump blood throughout the lifetime, it is conceivable that energetic efficiency of ventricular ejection is one of the major concerns of the regulatory system. Thus, we hypothesized that the regulatory system is preprogrammed to optimize energetic efficiency. To examine this hypothesis, we defined two optimality indexes in terms of the afterload and heart. We defined the optimal afterload as the afterload that extracts maximal external work from a given heart under a constant preload. We defined the optimal heart as the heart that minimizes oxygen consumption while sufficing peripheral demand of pressure and flow. We examined how these optimality indexes are preserved when significant hemodynamic changes occur. We also examined how pathological conditions affected them. If these indexes are optimal for physiological conditions but are suboptimal for pathological conditions, these indexes may, at least in part, represent one of the rules that govern cardiovascular regulation.

In the following sections, we first formulate the indexes of optimality on the basis of recent knowledge of ventriculoarterial coupling and cardiac energetics. We then evaluate these indexes under physiological conditions and pathological conditions.

Optimal Heart and Optimal Afterload

Ventriculoarterial Coupling Using the Elastance Concept

Suga et al. [1–3] characterized the left ventricle by a time-varying elastance. They found that end-systolic elastance (E_{es}) is relatively insensitive to changes in loading conditions but sensitive to changes in contractility. Using end-systolic pressure (P_{es}) and stroke volume (SV), the P_{es}–SV relation is expressed as

$$P_{es} = E_{es}\left(V_{ed} - V_0 - SV\right) \tag{1}$$

Where V_0 is the volume intercept of the end-systolic pressure–volume relationship. Equation 1 states that end-systolic pressure is negatively and linearly related to stroke volume.

The mechanical properties of the afterload are accurately expressed by aortic input impedance. Sunagawa et al. [4] demonstrated, however, that peripheral resistance is the major determinant of stroke volume; capacitance and characteristic impedance play a minimal role. If mean arterial pressure approximates end-systolic pressure, P_{es} is a function of resistance (R) and cardiac output (F) as

$$P_{es} \approx R \cdot F \tag{2}$$

Because cardiac output is ejected volume per unit time, it can be obtained by dividing SV by one cardiac cycle length (T), i.e., $F = SV/T$. Substituting F in Eq. 2 with SV and T yields

$$P_{es} = \frac{R}{T} SV \tag{3}$$

Equation 3 suggests that end-systolic pressure is a linear function of SV. The slope of the relationship (R/T) has been called effective arterial elastance (E_a), i.e., $E_a = R/T$. E_a represents functional elastance of the afterload. Substituting R/T in Eq. 3 with E_a yields

$$P_{es} = E_a SV \tag{4}$$

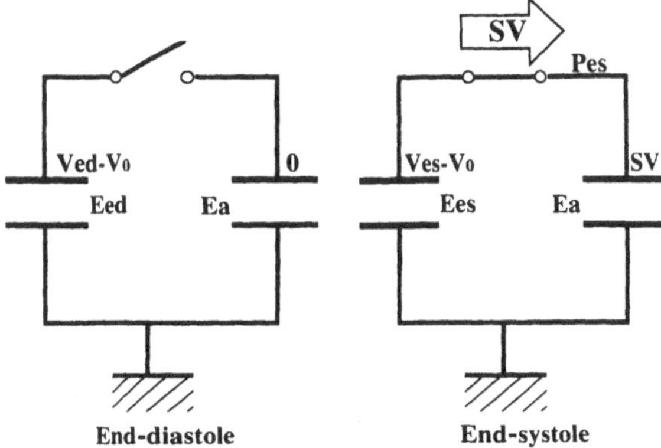

FIG. 1. Schema of volume shift when ventricular elastance and effective arterial elastance are coupled. The blood volume $V_{ed} - V_0$ at end-diastole would be distributed between two elastic chambers so that the pressure would equilibrate. From this, stroke volume can be calculated. E_{ed}, End-diastolic left-ventricular elastance; E_{es}, end-systolic left-ventricular elastance; E_a, effective arterial elastance; V_{ed}, end-diastolic left-ventricular volume; V_{es}, end-systolic left-ventricular volume; V_0, volume intercept of left-ventricular end-systolic pressure-volume relationship; P_{es}, end-systolic left-ventricular pressure; SV, stroke volume

From Eqs. 1 and 4, we can derive SV algebraically as

$$SV = \frac{E_{es}}{E_{es} + E_a}\left(V_{ed} - V_0\right) \tag{5}$$

This is equivalent to saying that once the arterial system is characterized by effective elastance, stroke volume can be calculated as the volume shift when the two elastic chambers are connected. In reference to Fig. 1, the left ventricle is filled with the blood of $V_{ed} - V_0$ in diastole. This blood would then be redistributed in end-systole between the ventricle and the effective arterial elastance so that the pressure levels at the two chambers are the same. Sunagawa et al. [1,5,6] demonstrated that one could predict stroke volume reasonably accurately with this framework.

Optimal Afterload

The fundamental function of the cardiovascular system is to provide sufficient flow to achieve adequate perfusion of the peripheral tissues. In addition, there would be an optimal or a normal level of blood pressure to ensure metabolite transport. Thus, the optimal afterload would provide maximal cardiac output as well as sufficiently high pressure. We defined the optimal afterload as the arterial resistance (R) that maximizes external work for a given heart (characterized by E_{es}, $V_{ed} - V_0$, and heart rate, HR) [7,8]. Using the ventriculoarterial coupling framework, for the quasi-isobaric ejecting contraction, external work (EW) can be approximated as

$$EW = \frac{E_a E_{es}^2}{\left(E_{es} + E_a\right)^2}\left(V_{ed} - V_0\right)^2 \tag{6}$$

Differentiating EW with respect to E_a gives

$$\frac{\partial EW}{\partial E_a} = \frac{E_{es}^2(E_{es} - E_a)}{(E_{es} + E_a)^3}(V_{ed} - V_0)^2 \tag{7}$$

Equation 7 indicates that EW becomes maximal when E_a equals E_{es}. Thus, resistance to give maximal EW would be

$$R = T \cdot E_{es} \tag{8}$$

Maximal external work (EW_{max}) under the optimal afterload is

$$EW_{\max} = \frac{E_{es}}{4}(V_{ed} - V_0)^2 \tag{9}$$

Shown in Fig. 2a are pressure–volume loops under optimal and suboptimal afterload conditions. The value of E_a is given by the slope of the lines connecting the end-systolic point to corresponding V_{ed} at zero pressure. Loop *A* gives large SV at the expense of low ejecting pressure. In contrast, loop *C* gives high ejecting pressure at the expense of small SV. None of these loops, however, gives maximal EW. Loop *B* is obtained when E_a equals E_{es}. It is clear that EW is maximum at this condition. Plotting EW as a function of E_a gives Fig. 2b. External work is maximal when E_a equals E_{es} and becomes smaller when E_a becomes smaller or larger than this optimal value.

If the optimal point provides either suboptimal cardiac output or suboptimal pressure, it might not suffice for peripheral demand (Fig. 3). Figure 3a describes a case in which the optimal afterload satisfies the necessary flow and pressure as well as maximizing EW. In Fig. 3b, the optimal afterload suffices necessary flow but has compromised pressure. In this case, the cardiovascular system might operate at the afterload

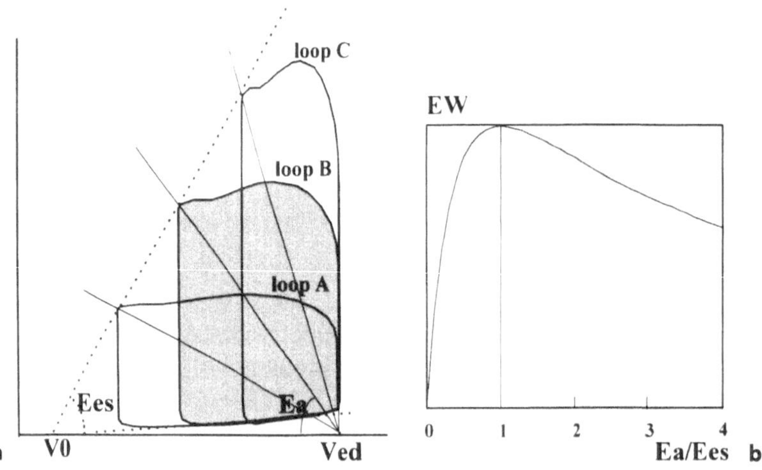

Fig. 2. **a** Various afterloads coupled with a given heart. *Loop A* originates from low afterload, *loop B* from the optimal, and *loop C* from high afterload. Effective arterial elastance (E_a) is the slope connecting the end-systolic point and the point of end-diastolic volume (V_{ed}) and zero pressure. E_{es} and V_0, slope and volume intercept of the end-systolic pressure–volume relationship, respectively. **b** Change of external work (EW) as a function of the ratio E_a/E_{es}. EW becomes maximal for the unity E_a/E_{es}

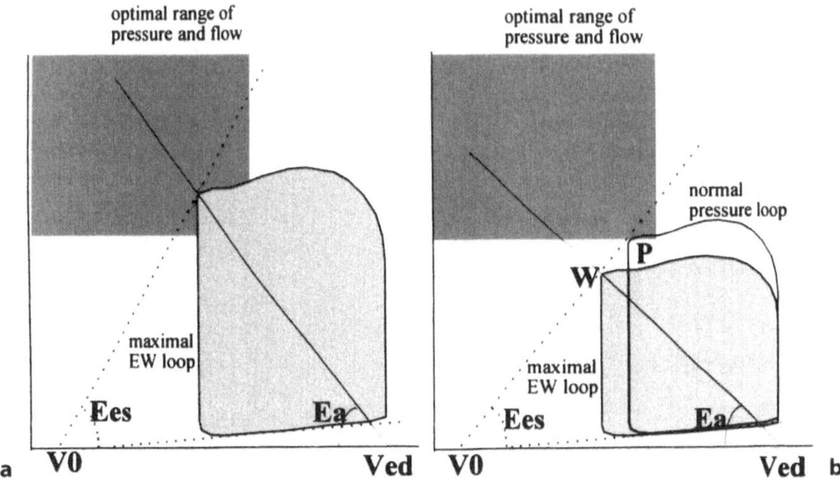

FIG. 3a,b. Relationship between the loop with maximal external work (EW) and optimal range of pressure and flow. In (**a**), maximal EW loop gives appropriate pressure and flow, but in (**b**), pressure is suboptimal. V_{ed}, End-diastolic volume; E_{es} and V_0, slope and volume intercept of the end-systolic pressure–volume relationship, respectively; E_a, effective arterial elastance; P, loop for appropriate pressure but suboptimal EW; W, loop for maximal EW but suboptimal pressure

that gives the appropriate pressure at the expense of EW (at loop P rather than W). Therefore, we quantified the degree of optimality of the afterload (Q_{load}) by the following formula:

$$Q_{load} = \frac{EW}{EW_{max}} = \frac{4E_aE_{es}}{\left(E_{es}+E_a\right)^2} \tag{10}$$

Q_{load} is unity when EW becomes maximal, i.e., when E_a equals E_{es}.

Optimal Heart

When the periphery is perfused with sufficient pressure and flow, the second requirement of the cardiovascular system would be to minimize its cardiac oxygen consumption. We defined the optimal heart, for a given afterload, as the heart that uses minimum oxygen [8]. As the pressure–volume area (PVA) represents myocardial oxygen consumption accurately [9–12], we can estimate oxygen consumption for each of the possible combinations of preload, contractility, and heart rate. Because stressed end-systolic volume is the ratio of P_{es} to E_{es} (Fig. 4), stressed diastolic volume is

$$V_{ed} - V_0 = P_{es}/E_{es} + F/HR \tag{11}$$

Within this constraint, PVA can be calculated as

$$PVA = P_{es}\left(V_{ed} - V_0 - P_{es}/2E_{es}\right) \tag{12}$$

Using the linear relationship between PVA and myocardial oxygen consumption per beat, one can estimate myocardial oxygen consumption per minute (mVO_2) as

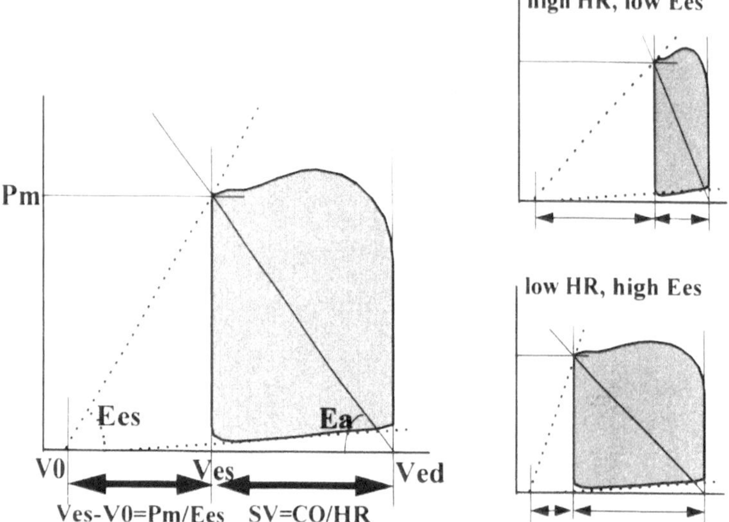

FIG. 4. Various hearts coupled with a given afterload. The heart can maintain cardiac output (*CO*) by either increasing heart rate (*HR*) (*right upper panel*) or increasing contractility (E_{es}) (*right lower panel*) under a constant pressure (*Pm*) and preload ($V_{ed} - V_0$). V_0, Volume intercept of the end-systolic pressure–volume relationship; E_a, effective arterial elastance; V_{es}, end-systolic volume; *SV*, stroke volume

$$
\begin{aligned}
mVO_2 &= \left(A \cdot PVA + B \cdot E_{es} + C\right)HR \\
&= A \cdot P_{es}\left(F + P_{es}HR/2E_{es}\right) + B \cdot HR \cdot E_{es} + C \cdot HR
\end{aligned}
\tag{13}
$$

where *A*, *B*, and *C* are coefficients that are determined experimentally. Figure 5a illustrates the predicted mVO_2 as function of *HR* and E_{es} under a constant cardiac output and mean arterial pressure. The preload value ($V_{ed} - V_0$) is unique for a given combination of *HR* and E_{es}. When heart rate decreases, mVO_2 monotonically decreases, while $V_{ed} - V_0$ becomes larger and larger. Thus, the bradycardiac left ventricle spends less oxygen at the expense of ventricular enlargement. This is because every term relating to *PVA*, E_{es}, and VO_2 intercept depends on *HR*. On the other hand, for a given heart rate, as shown in Fig. 5a by parallel multiple curvilinear lines, there seems to be an E_{es} value that minimizes mVO_2. If E_{es} becomes smaller, mVO_2 increases. This is because oxygen consumption that depends on *PVA* overrides that on E_{es}. If E_{es} becomes larger, mVO_2 increases as oxygen consumption which depends on E_{es} overrides that on *PVA*. Myocardial oxygen consumption becomes minimal when the effects of these components balance.

In Fig. 5a, the relationship between mVO_2 and *HR* (and in this case, E_{es} would also be unique) is also shown for the constant preload. This relationship is redrawn in another format in Fig. 5b. The upper panel describes E_{es} as a function of *HR* to fulfill the condition of peripheral demand and preload (see Fig. 4, right panels). The lower panel of Fig. 5b describes the relationship between *HR* and mVO_2. As clearly shown, mVO_2 becomes minimal for a set of particular values of E_{es} and *HR*. We thus defined the optimal heart as that which uses minimal oxygen. The optimal heart rate and E_{es}

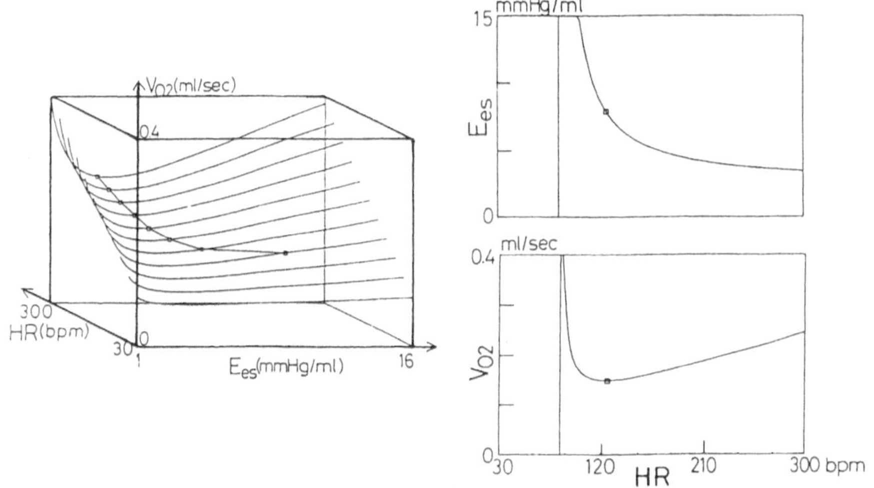

FIG. 5. **a** Estimated oxygen consumption per second (VO_2) for various hearts coupled with a given afterload. *Parallel multiple curvilinear lines* show that minimal oxygen is required at a certain contractility (E_{es}) for each heart rate (*HR*). VO_2, however, decreases monotonically with lowering HR. The *line connecting open squares* was obtained under a constant preload. **b** Relationship between VO_2 and HR is shown for a given afterload and a constant preload. To lower HR, E_{es} must increase (*upper panel*; see Fig. 4). VO_2 becomes minimal for a certain HR and E_{es} (*lower panel, open square*). (Panel **a** from [15], panel **b** from [8], with permission)

can be expressed in terms of E_a/E_{es}. Using reported values of A, B, and C and physiological $V_{ed} - V_0$ values, the optimal E_a/E_{es} is found to be about 0.4.

We quantified the degree of optimality of the heart (Q_{heart}) by the following formula:

$$Q_{heart} = \frac{mVO_{2\min}}{mVO_2} \tag{14}$$

where mVO_{2min} is the theoretically obtained minimal oxygen consumption for a given afterload. Q_{heart} is unity when mVO_2 becomes minimal.

Optimality of the Afterload and Heart During Exercise

We examined the optimality of the afterload and that of the heart during exercise as well as at rest [13–15]. To this end, we used six adult mongrel dogs pretrained to run on a treadmill. Under sterile conditions and anesthesia with sodium pentobarbital (30 mg/kg, i.v.), we instrumented these dogs with a high-fidelity pressure sensor (Model P-7, Konigsberg, Pasadena, CA, USA) in the left ventricle, an electromagnetic flowmeter (Model MFV-1200, Nihon-Koden, Tokyo, Japan) around the aortic root, and a water-filled catheter in the aortic arch. In addition we inserted a water-filled catheter for the calibration of left-ventricular pressure measurement. After full recovery from surgery (about 10 days after surgery), we measured these variables at rest while the animals were quietly lying on the floor. Then, we imposed various levels of exercise stress using a treadmill, and recorded hemodynamics for about 10 s after it reached a steady state. We increased the level of exercise up to the speed of 7 km/hr and the slope of 20%.

We estimated E_{es} by a single-beat estimation method [16–18]. Briefly, we fitted the isovolumic portions of the left-ventricular pressure wave to a sinusoidal wave with offset and predicted the maximal pressure that might be reached if ejection were not allowed. We calculated ejected volume at each moment by integrating the flow signal. Connecting the isovolumic peak pressure at zero ejecting volume and the right shoulder of the pressure-ejected volume loop yielded the estimation of E_{es}. Asoh et al. [17] reported that the E_{es} value obtained by this method was fairly close to that obtained by aortic clamping. We calculated E_a by the ratio of end-systolic pressure to stroke volume.

With increasing level of exercise, heart rate increased from 108 ± 17 to 199 ± 31 beats/min (+84%). Cardiac output also increased from 1.8 ± 0.5 to 3.5 ± 1.1 l/min (+94%) to a similar extent. Arterial pressure increased less (from 77 ± 4 to 110 ± 22 mmHg; +43%) because of the decrease in arterial resistance associated with exercise (−25%).

Figure 6 illustrates the summary of the change of E_{es}, E_a, and E_a/E_{es} associated with exercise. As shown in the left panel, E_{es} increased from 7.6 ± 1.7 to 10.9 ± 2.6 mmHg/ml (+43%). The middle panel shows that E_a also increased, from 4.9 ± 1.4 to 6.7 ± 1.8 mmHg/ml (+36%). As a result, E_a/E_{es} was 0.69 ± 0.26 at rest and remained fairly constant during exercise (0.63 ± 0.21). Thus, in spite of the drastic hemodynamic change with exercise, E_a/E_{es} changed little.

As discussed in the previous sections, E_a/E_{es} is the sole determinant of the optimality of the afterload and is also the major determinant of the optimality of the heart. Thus, we could calculate these optimalities from the E_a/E_{es} values obtained at the each level of exercise. To be more accurate, we also took the change in the preload into account in calculating Q_{heart}. Figure 7 demonstrates the optimality indexes at rest and during exercise. The upper panels show the optimality index at various levels of exercise, while the lower panels show the averaged index at rest and at a maximal level of exercise. As Q_{load} (left panels) was 0.93 ± 0.08 at rest and was kept close to unity also at maximal exercise (0.92 ± 0.09), at each level of exercise afterload is selected by the regulatory system so as to maximize external work. In addition, as shown in the right panels, Q_{heart} was also close to unity during exercise (0.99 ± 0.01) as well as at rest (0.98 ± 0.01). Therefore, the heart is regulated to minimize oxygen consumption both at rest and during exercise.

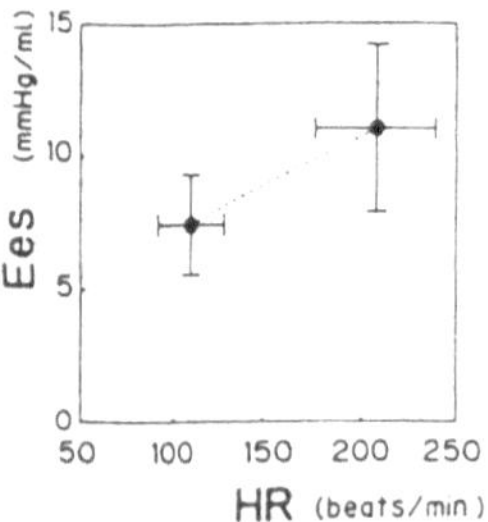

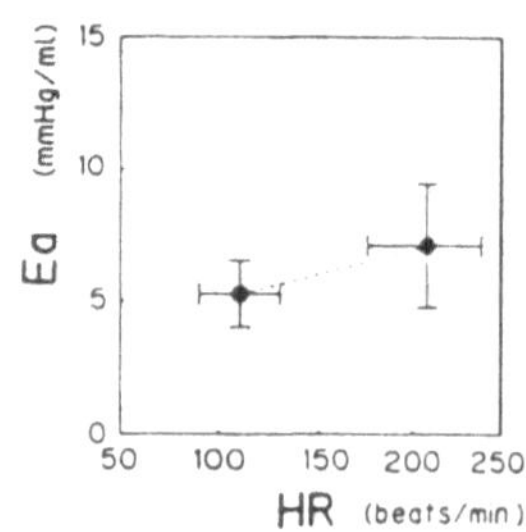

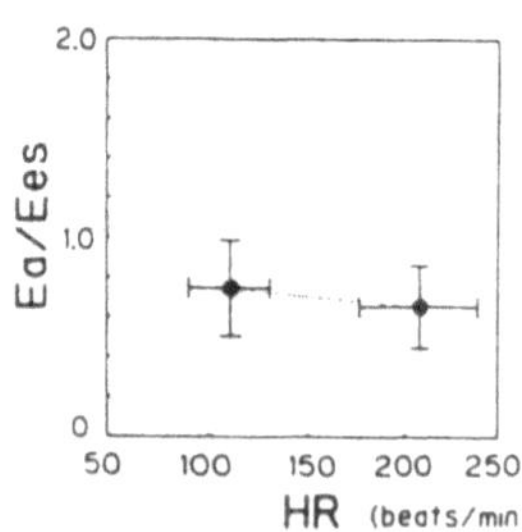

FIG. 6. Change of ventricular elastance (E_{es}; *left panel*), effective arterial elastance (E_a; *middle panel*), and their ratio (E_a/E_{es}; *right panel*) in response to exercise. E_a/E_{es} was kept relatively constant during a dramatic hemodynamic change. (From [16], with permission)

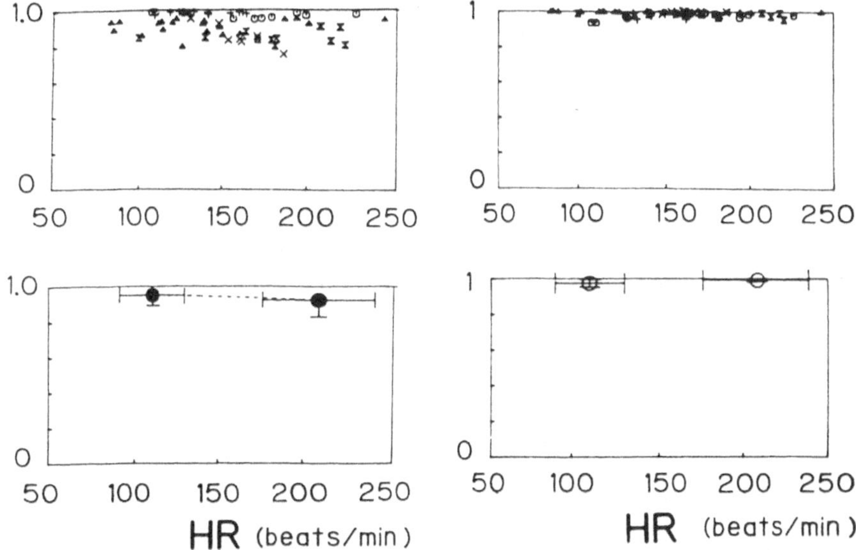

FIG. 7. Optimality index of afterload (*left*) and heart (*right*) at various levels of exercise. In the *upper panels*, these indexes are plotted against exercise level (expressed by heart rate, *HR*). Different symbols indicate different animals. In *lower panels*, pooled results for rest and maximal level of exercise are shown. (From [15], with permission)

Optimality of the Afterload and the Heart During Hemorrhage and Infusion

We examined the optimality of the afterload and the heart during blood volume change, namely during hemorrhage and volume infusion in chronically instrumented dogs [14,19]. On separate days, we either infused dextran (in 4–6 steps, up to 35 ml/kg) or drew blood (in 4–6 steps, up to 35 ml/kg) via a catheter placed in the aorta. Blood drawn was carefully stored in a sterile bottle with anticoagulation with heparin and reinfused after the recordings were finished. We intentionally avoided taking data for both infusion and hemorrhage on the same day to exclude the long-lasting effect of one procedure on the other.

Hemorrhage (30 ml/kg) decreased arterial pressure by 18 mmHg and decreased cardiac output by 40%. Heart rate rose and reached 120 beat/min. Arterial resistance was increased by 43%. Effective arterial elastance was increased by 78%, reflecting the increase in arterial resistance and heart rate. On the other hand, E_{es} changed little in response to volume depletion. Thus, E_a/E_{es} increased from 0.80 to 1.55 (+93%). As shown in Fig. 8, Q_{load} was kept close to unity (0.97 ± 0.03 to 0.96 ± 0.02; left panel). In contrast, Q_{heart} was close to unity but gradually decreased by hemorrhage (from 0.92 ± 0.04 to 0.80 ± 0.04, $P < .01$; right panel). From these data, we could say that the afterload extracted nearly maximal external work even during hemorrhage, but the heart consumed more oxygen than the most efficient heart by excessive hemorrhage.

Dextran infusion (30 ml/kg) increased cardiac output dramatically (+86%). Arterial pressure increased about 20 mmHg. Arterial resistance decreased, while heart rate

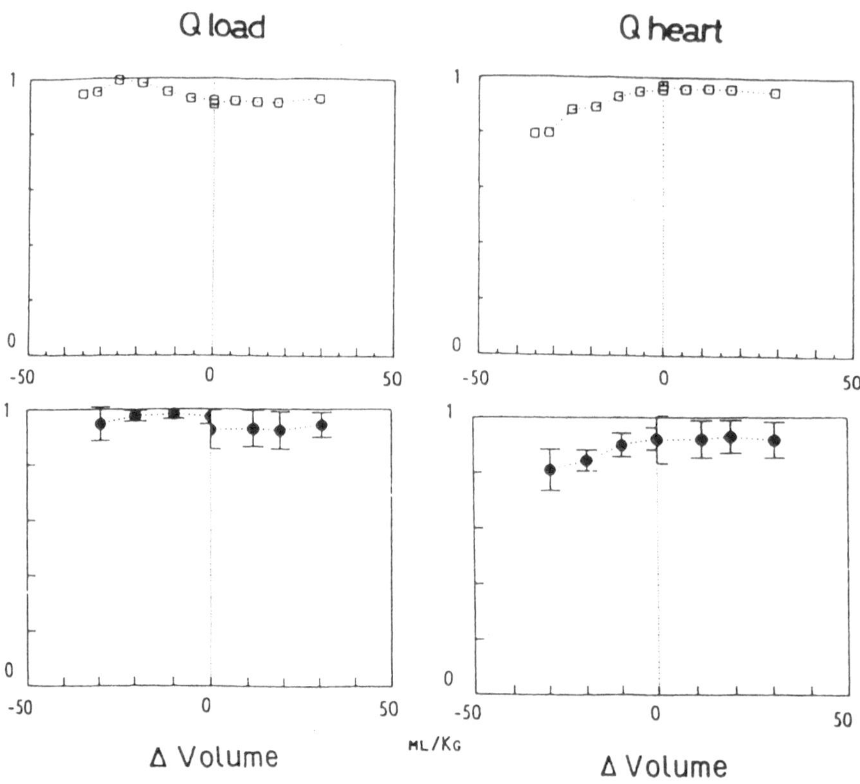

FIG. 8. Optimality index of afterload (Q_{load}; *left*) and heart (Q_{heart}; *right*) at various levels of hemorrhage and dextran infusion. *Upper panels* describe a representative case and *lower panels* show pooled results. (From [15], with permission)

increased up to 120 beat/min. Decreases in arterial resistance and increases in heart rate slightly decreased E_a. As E_{es} also changed little, E_a/E_{es} was slightly decreased. This made Q_{load} as well as Q_{heart} close to unity (Q_{load} from 0.93 ± 0.07 to 0.94 ± 0.04, Q_{heart} from 0.92 ± 0.09 to 0.92 ± 0.06; Fig. 8). From these data, we concluded that the afterload extracted nearly maximal external work and the heart consumed minimum oxygen during volume infusion.

Optimality of the Afterload and the Heart of Animals with Left-Ventricular Dysfunction

We examined the optimality of the afterload and heart also in animals with left-ventricular dysfunction. We studied this condition as a pathological condition and as compared with physiological stimuli such as exercise and volume change. We also used similar chronically instrumented dogs [14,20]. After recording control hemodynamics, we created left-ventricular dysfunction by infusing microspheres into the left main coronary artery via the right carotid artery. We repeatedly injected 15-μm microspheres until left-atrial pressure reached 30 mmHg. On the next day, after the animals became fully conscious, we recorded hemodynamic variables.

On the average, microembolization resulted in the decrease of cardiac output by 20%, while arterial pressure was unchanged and arterial resistance increased slightly. Heart rate increased by 40%. E_{es} did not decrease significantly, and E_a increased as a result of increases in arterial resistance and heart rate. E_a/E_{es} increased from 0.7 to 1.5 ($P < .01$). Ventricular dysfunction was relatively mild and thus could be counterbalanced by the compensatory mechanism. The activation of such a compensatory mechanism was evidenced by the high heart rate and arterial resistance.

Mild ventricular dysfunction did not alter Q_{load} (0.95 ± 0.08 vs. 0.96 ± 0.02; NS). If Eq. 13 holds for ventricular dysfunction, Q_{heart} decreased from 0.98 ± 0.02 to 0.88 ± 0.05 ($P < .001$). Thus, in animals with mild left-ventricular dysfunction the afterload could also extract nearly maximal external work. It seems, however, the heart failed to minimize its oxygen consumption.

Discussion

We have shown that, in various conditions, the afterload was regulated to extract nearly maximal external work while the heart was regulated to minimize oxygen consumption. These conditions include a resting condition, exercise, and a mild degree of volume perturbation. During moderate hemorrhage and mild left-ventricular dysfunction, the afterload still extracted nearly maximal external work. The heart, however, could not minimize its oxygen consumption.

Simultaneous Optimization of the Afterload and the Heart

We defined two optimality indexes in terms of the afterload and the heart. If these conditions coincide, the regulatory system could optimize the afterload and the heart simultaneously. As discussed in the previous section, the optimal afterload occurs at unity E_a/E_{es} while the optimal heart was obtained at E_a/E_{es} of about 0.4. Therefore, strictly speaking, the optimal afterload and the optimal heart do not coincide. However, we demonstrated that the cardiovascular system nearly optimized the afterload and the heart at the same time. This is because both optimalities are relatively insensitive to the E_a/E_{es} values, especially around the physiological E_a/E_{es} values. Figure 9 depicts the relationship between Q_{load} and E_a/E_{es}, and Q_{heart} and E_a/E_{es}. As shown in the

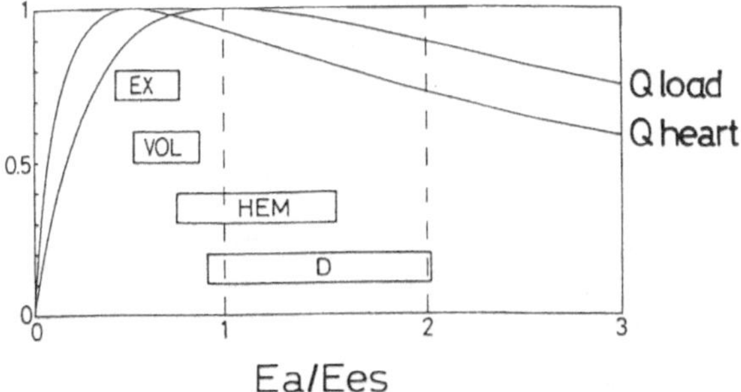

Fig. 9. Optimality of afterload (Q_{load}) and the heart (Q_{heart}) plotted against the ratio E_a/E_{es}. As shown, Q_{load} is unity for unity E_a/E_{es}, and Q_{heart} is unity for E_a/E_{es} of about 0.4. Four ranges of E_a/E_{es} are shown with experimentally obtained values (mean ± SD) for exercise, including rest (*EX*), hemorrhage (*HEM*), volume infusion (*VOL*), and ventricular dysfunction (*D*)

figure, Q_{load} was larger than 0.9 if E_a/E_{es} was between 0.52 and 1.92. Q_{heart} was larger than 0.9 for E_a/E_{es} between 0.17 and 0.89. Thus, both these indexes exceeded 0.9 when E_a/E_{es} was within the range of 0.52 to 0.89. This happened to correspond to the physiological range of E_a/E_{es}.

E_a/E_{es} as the Determinant of Optimality Indexes

We demonstrated that E_a/E_{es} is the sole determinant of Q_{load} and is the major determinant of Q_{heart}. The preload level and coefficients of the PVA-E_{es}-mVO_2 relationships affect Q_{heart} but to a lesser extent. The fact that E_a/E_{es} is the major determinant of optimality indexes is important because E_a/E_{es} can be related to effective ejection fraction. If we define effective ejection fraction (EF_e) by the ratio of stroke volume to stressed end-diastolic volume, EF_e can be expressed as a function of E_a/E_{es} as

$$EF_e = \frac{E_{es}}{E_{es} + E_a} = \frac{1}{1 + E_a/E_{es}} \quad (15)$$

In reference to Fig. 10a, unity E_a/E_{es} corresponds to EF_e of 50% and E_a/E_{es} of 0.4 corresponds to EF_e of 71%. Conversely, EF_e of 30% reflects E_a/E_{es} of 2.3, and EF_e of 90% reflects E_a/E_{es} of 0.1. The difference between conventional and effective ejection fraction is not large when V_0 is a small positive value. Using this formula, one can estimate E_a/E_{es} from the EF_e value and thus the optimality indexes. In Fig. 10b, the relationship between Q_{load} and EF_e, and Q_{heart} and EF_e is illustrated. Although mild ventricular dysfunction affected Q_{heart} slightly (as shown by range D in Fig. 10b), theoretical analysis indicated that in patients with severe heart failure, Q_{load} and Q_{heart} might be 0.36 and 0.24, respectively, if EF_e is 10%.

Other Optimality Indexes

It is apparent that Q_{load} and Q_{heart} are not the only optimality indexes. In fact, these optimality indexes are not preserved during large volume loss. In this case, Q_{heart} is

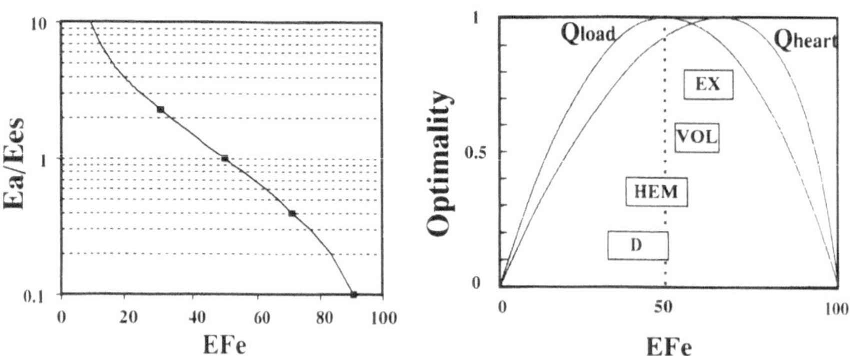

FIG. 10. a Theoretical relationship between effective ejection fraction (EF_e) and the ratio of E_a/E_{es}. The vertical axis is logarithmically scaled. *Closed squares* indicate sets of EF_e and E_a/E_{es} values; E_a/E_{es} is 1 for 50% EF_e, 0.4 for 71% EF_e, 2.3 for 30% EF_e, and 0.1 for 90% EF_e. **b** Optimality of afterload (Q_{load}) and the heart (Q_{heart}) plotted against effective ejection fraction (EF_e). As shown, Q_{load} is unity for 50% EF_e, and Q_{heart} is unity for 71% EF_e. Four ranges of EF_e are shown with experimentally obtained values (mean ± SD) for exercise, including rest (*EX*), hemorrhage (*HEM*), volume infusion (*VOL*), and ventricular dysfunction (*D*). (From [14], with permission)

more compromised than Q_{load}. We conjectured, however, that the cardiovascular system operated to restore normal blood pressure rather than to achieve the optimal states that we defined in this situation. Asanoi et al. [22] also pointed out in patients with heart failure that energetically optimal end-systolic pressure (corresponding to our optimal afterload and Burkhoff's [21] optimal afterload) is lower than the normal blood pressure as well as the actual measured pressure. They reported that this manifested more as the end-systolic elastance became smaller. They also conjectured that maintenance of normal pressure by the autonomic nervous system dominated the energetic optimization.

Burkhoff and Sagawa [21] used a similar but different optimality criterion. They also maximized EW/mVO_2 per beat as we did. The difference arises because they optimized by changing afterload while fixing the heart, but we changed the heart and fixed the afterload. As a result, the E_a/E_{es} value that maximized Q_{heart} was slightly smaller (0.4) than their optimal value (0.5).

Despite the energetic optimality observed in conscious animals and in conscious humans, we still do not exclude the possibility that this might be only a coincidence. Further investigations are required to fully disclose the mechanism that governs the operation point of the ventriculoaterial coupling.

Conclusion

We conclude that the afterload is regulated to extract nearly maximal external work, while the heart is regulated to minimize oxygen consumption in a resting condition, during exercise, and during a mild degree of volume perturbation. During excessive hemorrhage and mild left-ventricular dysfunction, the afterload still extracted nearly maximal external work. The heart, however, could not minimize oxygen consumption.

References

1. Sagawa K, Maughan WL, Suga H, Sunagawa K (1988) Cardiac contraction and the pressure-volume relationship. Oxford, New York
2. Suga H, Sagawa K (1974) Instantaneous pressure-volume relationships and their ratio in the excised, supported canine left ventricle. Circ Res 35:117–126
3. Suga H, Sagawa K, Shoukas AA (1973) Load independence of the instantaneous pressure-volume ratio of the canine left ventricle and effects of epinephrine and heart rate on the ratio. Circ Res 32:314–332
4. Sunagawa K, Maughan WL, Sagawa K (1985) Stroke volume effect of changing arterial impedance over selected frequency range. Am J Physiol 248:H477–H484
5. Sunagawa K, Maughan WL, Burkhoff D, Sagawa K (1983) Left ventricular interaction with arterial load studied in isolated canine left ventricle. Am J Physiol 245:H773–H780
6. Sunagawa K, Sagawa K, Maughan WL (1984) Ventricular interaction with the loading system. Ann Biomed Eng 12:163–189
7. Sunagawa K, Maughan WL, Sagawa K (1985) Optimal arterial resistance for the maximal stroke work studied in isolated canine left ventricle. Circ Res 56:586–595
8. Sugimachi M, Todaka K, Sunagawa K, Nakamura M (1990) Optimal afterload for the heart vs. optimal heart for the afterload. Front Med Biol Eng 2:217–221
9. Suga H (1979) Total mechanical energy of a ventricle model and cardiac oxygen consumption. Am J Physiol 236:H498–H505
10. Khalafbeigui F, Suga H, Sagawa K (1979) Left ventricular pressure-volume area correlates with oxygen consumption. Am J Physiol 237:H566–H569

11. Suga H, Hayashi T, Shirahata M (1981) Ventricular pressure-volume area as predictor of cardiac oxygen consumption. Am J Physiol 240:H39–H44
12. Suga H, Yamamura Y, Nozawa T (1987) Prospective prediction of O_2 consumption from pressure-volume area (PVA) in dog heart. Am J Physiol 252:H1258–H1268
13. Hayashida K, Sunagawa K, Noma M, Sugimachi M, Ando H, Nakamura M (1992) Mechanical matching of the left ventricle with the arterial system in exercising dogs. Circ Res 71:481–489
14. Sunagawa K, Hayashida K, Sugimachi M, Todaka K, Kubota T, Itaya R, Chishaki A, Takeshita A (1991) Optimal left ventricle versus optimal afterload. In: Sasayama S, Suga H (eds) Recent progress in failing heart syndrome. Springer-Verlag, Tokyo, pp 161–186
15. Sunagawa K, Hayashida K, Sugimachi M, Noma M, Ando H, Tajimi T, Tomoike H, Nose Y, Nakamura M (1989) Ventriculoarterial matching in exercising dogs. In: Sideman S, Beyar R (eds) Analysis and simulation of the cardiac system—ischemia. CRC, Boca Raton, pp 89–98
16. Sunagawa K, Yamada A, Senda Y, Kikuchi Y, Nakamura M (1980) Estimation of the hydromotive source pressure from ejecting beats of the left ventricle. IEEE Trans Biomed Eng 27:299–305
17. Asoh T, Oe M, Morita S, Tanaka J, Tokunaga K, Sunagawa K (1987) Single beat estimation of end-systolic pressure-volume relation of in situ dog heart (abstract). Jpn Circ J 51:767
18. Takeuchi M, Igarashi Y, Tomimoto S, Odake M, Hayashi T, Tsukamoto T, Hata K, Takaoka H, Fukuzaki H (1991) Single beat estimation of the slope of the end-systolic pressure-volume relation in the human left ventricle. Circulation 83:202–212
19. Todaka K, Sugimachi M, Sunagawa K, Ando H (1989) Do the heart and arterial system of conscious dogs operate at the optimal point in hemorrhage and volume load? (abstract). Circulation 80(suppl II):II–248
20. Sugimachi M, Sunagawa K, Todaka K, Hayashida K, Noma M, Egashira S, Nose Y (1988) Ventriculo-arterial matching in conscious dogs with left ventricular dysfunction (abstract). Circulation 78(suppl II):II–523
21. Burkhoff D, Sagawa K (1986) Ventricular efficiency predicted by an analytical model. Am J Physiol 250:R1021–R1027
22. Asanoi H, Kameyama T, Ishizaka S, Nozawa T, Inoue H (1996) Energetically optimal left ventricular pressure for the failing human heart. Circulation 93:67–73

Influence of Aortic Impedance on the Development of Pressure-Overloaded Left-Ventricular Hypertrophy in Rats

Masunori Matsuzaki, Shigeki Kobayashi, Michihiro Kohno, Masafumi Yano, Tsutomu Ryoke, Tomoko Ohkusa, Yuji Hisamatsu, and Masakazu Obayashi

Summary. Aortic impedance, which represents left-ventricular (LV) afterload, is considered as a major determinant for the development of pressure-overloaded LV hypertrophy. To test whether the sustained change in aortic impedance might affect the mode of the development of LV hypertrophy, coarctation of either the ascending (group 1) or abdominal aorta (group 2) was performed in 6-week-old rats. Four weeks after the operation, both groups developed LV hypertrophy. Although peak LV pressure and total systemic resistance (TSR) were comparable, time to peak LV pressure was shorter in group 1 (early systolic loading) than in group 2 (late systolic loading). The aortic input impedance spectrum was characterized by an increase in characteristic impedance (Zc) in group 1 and an increase in arterial wave reflection in group 2. The ex vivo passive pressure–volume relation was shifted toward the volume axis in group 1 compared with group 2, suggesting eccentric LV hypertrophy in group 1. Histological analysis revealed that either myocyte diameter or connective tissue volume fraction (CVF) was higher in group 2 than in group 1, suggesting reduced hypertrophy in early systolic loading versus late systolic loading. However, there was no sign of contractile or relaxation dysfunction in either group. The sustained early systolic loading was accompanied by eccentric, reduced hypertrophy while the sustained late systolic loading was accompanied by concentric, adequate hypertrophy. Such different modes of cardiac hypertrophy were associated with the differences in aortic impedance spectra between pressure-overloaded groups.

Key words. Aortic impedance—LV hypertrophy—Pressure overload—LV remodeling

Introduction

The left ventricle (LV) is generally thought to adapt to sustained arterial hypertension by developing concentric hypertrophy [1]. As a compensatory ventricular response to a chronic pressure overload, ventricular wall thickness increases to normalize the wall stress, and LV dilatation represents a late transition toward myocardial failure. Although evidence has accumulated as to the role of hemodynamic load or the myocar-

The Second Department of Internal Medicine, Yamaguchi University School of Medicine, Ube, Yamaguchi, 755 Japan.

dial contractile state [2,3], little information is available about the influence of the properties of the arterial system on the development of LV hypertrophy and their possible pathophysiological mechanisms.

Aortic input impedance, which depends directly on the geometry and mechanical properties of the arterial network, represents LV afterload and thus is considered to be a major determinant for the development of LV hypertrophy [4]. In this connection, some reports [5,6] have emphasized the important role of arterial wall characteristics for the development of LV hypertrophy. However, the exact inter-relationship between aortic impedance and the development of pressure-overloaded LV hypertrophy remains to be elucidated. In this study, by varying the coarctation site of the aorta, we produced a sustained increase in the aortic impedance in two different ways. Our objective was to investigate the influence of the different patterns of the aortic impedance and thus of the different systolic loading sequences on the mode of the development of LV hypertrophy.

Material and Methods

The experiments were done on 6-week-old Wistar rats of 130–150 g body weight at the beginning of the study. Early systolic loading was produced by coarctation of the aortic arch (group 1) and late systolic loading by coarctation of the abdominal aorta (group 2). In age-matched control rats, sham operations were performed without inducing the aortic constriction for either group 1 or group 2. Because there was no significant difference in all variables between sham control rats for group 1 and those for group 2, both sham control rats were combined and expressed as sham controls (Sham).

Surgical Procedure

In group 1, each rat was anesthetized with ether and endotracheally intubated under direct visualization. The left thorax was opened at the level of the third intercostal space to expose the aortic arch under articifical ventilation with oxygen and ether (Model 680, Harvard Apparatus, South Natick, MA, USA). To produce a constriction, a 1.2-mm-diameter wire was placed alongside the aortic arch and tightly fixed with surgical silk between the first and second branch of the aortic arch. The wire was removed leaving the aortic arch constricted to an outer diameter equivalent to the diameter of the wire (Fig. 1). The lung was inflated with a positive end-expiratory pressure of 10 cm H_2O, and the chest was closed with a silk thread. The tracheal tube was then removed.

In group 2, each rat was anesthetized with an intraperitoneal injection of 3.6% chloral hydrate. After the midline abdominal incision, a 0.5-cm segment of aorta above the renal arteries was dissected free. A 0.8-mm-diameter wire was then placed around the abdominal aorta, and a constriction similar to that done in group 1 was performed. Mortality of the rats was 10% with coarctation of the aortic arch but 0% with coarctation of the abdominal aorta; 90% of these rats died within a day after the surgical procedure.

Hemodynamic Studies

Four weeks after the operation, hemodynamic measurements were performed. After induction of anesthesia with intraperitoneal pentobarbital (50 mg/kg), the trachea was

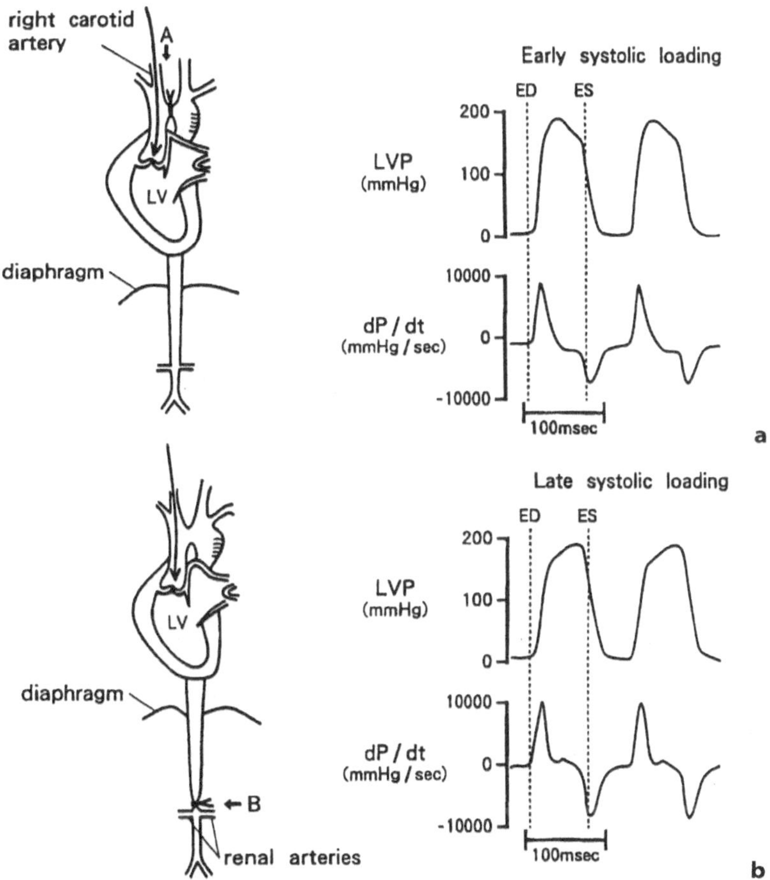

FIG. 1a,b. *Left panel*: Ultraminiature catheter pressure transducer was inserted via the right carotid artery. Constriction of the arch aorta was performed for early systolic loading (*A*) and constriction of the abdominal aorta for late systolic loading (*B*). *Right panel*: Typical traces of left-ventricular pressure (*LVP*) and first derivative of LVP (*dP/dt*) with early systolic loading (**a**) and late systolic loading (**b**). *ED*, End-diastole; *ES*, end-systole

cannulated and ventilated with a rodent respirator. Two French ultraminiature catheter pressure transducer (Model PR 249, Millar Instruments, Houston, TX, USA) was positioned into the LV via the right carotid artery to measure LV pressure. The time constant (tau, τ) of isovolumetric LV pressure decay from peak (−)dP/dt to 10 mmHg above the minimal LV pressure was calculated by the Weiss semilogarithmic method [7].

After left-sternal thoracotomy, LV anterior wall thickness (Wth) was measured by a single ultrasonic transducer (Wall tracker module, WT-20, Crystal Biotech, Hopkinton) attached to the midportion of the epicardium of the LV anterior wall and triggered by the peak (+)dP/dt of LV pressure [8] (Fig. 2).

The Millar ultraminiature catheter pressure transducer was pulled back and positioned in the ascending aorta to measure aortic pressure. The ascending aorta was isolated, and the ultrasonic transit time flow probe (T-106, Transonic Systems, Ithaca, NY, USA) was then placed to measure the phasic instantaneous aortic blood flow. The

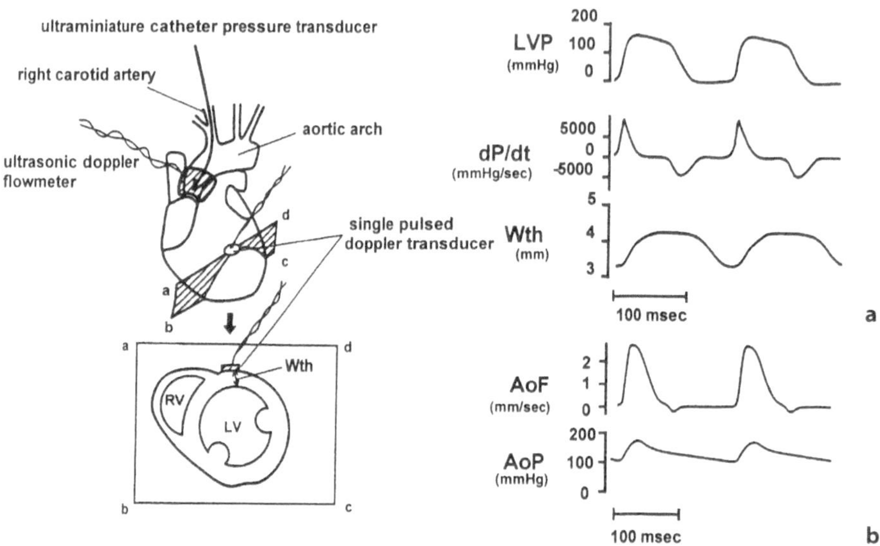

FIG. 2a,b. *Left panel*: Anatomical scheme of whole heart and the position of a single epicardial probe that measures wall-thickening dynamics at any selected depth in the myocardium by a pulsed doppler technique. The ultrasonic doppler probe was placed around the ascending aorta to measure mean and phasic instantaneous aortic blood flow. *Right panel*: **a** Original recordings of LV pressure (*LVP*), first derivative of LV pressure (*dP/dt*), and transmural wall thickening (*Wth*) measured by a pulsed doppler technique with range gate tracking. **b** Simultaneous recordings of ascending aortic blood flow (*AoF*) and aortic pressure (*AoP*)

frequency response of the pressure recording channel was flat from 0 to 100 Hz. Pressure was low-pass filtered with a corner frequency at 100 Hz. Flow-velocity signal was recorded at a 100-Hz filter setting (frequency response, −6 dB at 100 Hz). After the steady state was confirmed, hemodynamic parameters were recorded.

Ten percent of all the rats that underwent the catheterization, crystal attachment, and flowmeter implantation developed serious, persistent hypotension (<90 mmHg) or bradycardia (<200 beats/min); these rats were eliminated from the data analysis. All data were stored on analog magnetic tape (Sony Instrumentation Data recorder UN-61430, Japan) at a tape speed of 9.5 mm/s through the D/C amplifier (V4201, Electronics for Medicine VR-12). The analog signals were digitized by a 12-bit A/D converter connected to a microcomputer (PC 9801 RA, NEC, Japan) at 1-ms intervals and stored on disk. Ten consecutive cardiac cycles were sampled at each stage and averaged to provide the hemodynamic values.

End-diastole was defined by the beginning of (+)dP/dt of LV pressure, and end-systole was defined by the time of 10 ms preceding the peak (−)dP/dt of LV pressure. Aortic input impedance was computed using Fourier analysis of the phasic aortic pressure and flow waves. Aortic characteristic impedance (Zc) was estimated by averaging impedance moduli between 4 and 10 harmonics [9]. The first harmonic of the impedance modulus (Z1) was used as an index of arterial wave reflection [10]. Total systemic resistance (TSR) was determined as input resistance (the modulus of the impedance at zero harmonic).

Zc was also estimated from the aortic pressure–flow relationship during the early ejection period (4–14 ms after the onset of ejection), where arterial wave reflection can be neglected [11]. The pressure–flow (P–F) relation during this period was fitted to a linear equation, and the slope (P/F slope) was used as Zc in the time domain.

Pressure–Volume Relation

Pressure–Volume data were determined by methods described previously [12,13]. At the conclusion of hemodynamic measurements, potassium chloride was injected via the external jugular vein to arrest the heart. The heart was rapidly removed, and the right ventricle was incised and the atrioventricular groove isolated by a ligature. A double-lumen catheter, attached to a pressure transducer (Statham 23 Id) and an infusion pump, was passed into the left ventricle. After gentle aspiration of the LV cavity to remove any residual blood, normal saline was infused at 0.70 ml/min into the suspended left ventricle while pressure was recorded. Saline was infused until the pressure increased to 40 mmHg. This procedure was performed a minimum of two or a maximum of three times within 10 min of cardiac arrest before the onset of rigor mortis.

Morphology

After the physiological studies, the hearts were removed, and their atria, great vessels, and valvular structures were trimmed away. Right and left ventricles (plus septum) were separated. Coronal sections of the LV were obtained for light microscopy. These specimens were fixed in 10% formalin and embedded in paraffin; 4-μm-thick sections were cut and stained with hematoxylin and eosin for the morphological studies.

The diameter of myocardial cells was measured on a longitudinal section across the nucleus of the cells by a micrometer in the microscope. The mean value of cell diameter was calculated from 100 cells for each heart [14].

Connective tissue volume fraction was also assessed using a computer-assisted procedure [15]. Paraffin section 4 μm thick, fixed in 10% buffered formaldehyde and stained with Azan [16,17], were examined under a projection microscope. Each section was divided into four quadrants with the center of the section as the origin from each quadrant; 4 fields were randomly selected from the subendocardial and the subepicardial regions (2 fields each). Each field was then transferred to a digitizing pad connected to a cursor-computer assembly (LA-555, PIAS, Osaka, Japan). Segments representing connective tissue and muscle were identified and traced, and a computer program was used to calculate the area fraction of connective tissue and muscle. Connective tissue volume fraction was calculated as the sum of all connective tissue areas in the 16 fields divided by the sum of all connective tissue and muscle areas in all fields. Areas of connective tissue surrounding the intramyocardial coronary arteries were excluded from the calculation.

Statistical Analysis

An unpaired *t*-test was employed for statistical comparison of the data. Results are expressed as mean ± SD, and $P < .05$ was accepted as statistically significant.

Result

The ratio of LV weight to body weight increased significantly in both pressure-overloaded groups. The LV hypertrophy was particularly marked in group 2 (Table 1).

The parameters of hemodynamics and LV geometry are summarized in Table 2. Although peak LV pressure was similarly increased in both pressure-overloaded groups, time to peak LV pressure (TP_{max}) was shorter in group 1 than in group 2, indicating early systolic loading in group 1 and late systolic loading in group 2. The peak (+)dP/dt of LV pressure was increased in both pressure-overloaded groups. Time constant (tau, τ) of LV pressure decay remained unchanged in group 1 and even decreased in group 2. There was no significant difference among all groups in heart rate and LV end-diastolic pressure.

TABLE 1. Morphological characteristics 4 weeks after aortic banding.

Group	BW (g)	LVW (mg)	LVW/BW (mg/g)
Sham (n = 10)	255 ± 11	506 ± 27	2.0 ± 0.1
Group 1 (n = 8)	249 ± 8	586 ± 31*	2.4 ± 0.1*
Group 2 (n = 7)	255 ± 16	690 ± 73*,**	2.7 ± 0.2*,**

Values are mean ± SD.
BW, Body weight; LVW, left-ventricular weight; LVW/BW, LV weight/body weight ratio.

TABLE 2. Hemodynamics and left-ventricular (LV) geometry 4 weeks after aortic banding.

Values	Sham (n = 9–10)	Group 1 (n = 7–8)	Goup 2 (n = 6–7)
HR (bpm)	392 ± 38	418 ± 28	432 ± 39
PAoP (mmHg)	137 ± 7	188 ± 12*	201 ± 15*
DAoP (mmHg)	111 ± 6	130 ± 5*	153 ± 14*,**
PLVP (mmHg)	131 ± 13	191 ± 13*	208 ± 17*
LVEDP (mmHg)	5.0 ± 2.0	6.0 ± 3.0	5.7 ± 1.6
+dP/dt (mmHg/s)	5640 ± 638	8739 ± 977*	8964 ± 1301*
TPmax (ms)	53 ± 12	48 ± 5	64 ± 3*,**
τ (ms)	10.7 ± 1.8	9.7 ± 1.5	8.5 ± 1.7*
CO (ml/min)	24 ± 7	24 ± 6	23 ± 6
TSR (dyne·s·cm^{-5} × 10^3)	373 ± 48	502 ± 109*	511 ± 60*
Zc (dyne·s·cm^{-5} × 10^3)	14 ± 7	22 ± 5*	12 ± 3**
Z_1 (dyne·s·cm^{-5} × 10^3)	26 ± 8	31 ± 4	36 ± 5*
$\Delta P/\Delta F$ (mmHg·s/ml)	10.3 ± 5.9	17.7 ± 3.1*	8.7 ± 1.8**
Wed (mm)	3.1 ± 0.3	3.4 ± 0.3*	3.6 ± 0.3*

Values are mean 6 SD.
HR, Heart rate; PAoP, peak aortic pressure; DAoP, end-diastolic aortic pressure; PLVP, peak LV pressure; LVEDP, LV end-diastolic pressure; +dP/dt, maximal value of first derivative of LV pressure; TPmax, time to peak LV pressure from the beginning of +dP/dt of LV pressure; τ, time constant of isovolumetric LV pressure decay obtained from the semilogarithmic method; CO, cardiac output; TSR, total systemic resistance; Zc, aortic characteristic impedance; Z_1, modulus of aortic impedance at first harmonic; $\Delta P/\Delta F$, slope of the pressure–flow relation during early ejection period; Wed, end-diastolic LV wall thickness.
*, $P < .05$ vs. sham; **, $P < .05$ vs. group 1.

Although cardiac output was comparable among all groups, TSR increased to a similar extent in both group 1 and group 2. In group 1, Zc and P/F slope increased significantly, while in group 2, Z_1 increased (Table 2). In group 1, the LV end-diastolic wall thickness (Wed) increased by 9.7%, while it increased by 16% in group 2 (Table 2).

The average aortic input impedance spectra for each group are shown in Fig. 3. In the sham controls, the moduli of input impedance fell steeply from a high value at zero frequency (input resistance) to a minimum at 13–14 Hz, then rose to a peak at 27–29 Hz. In group 1, the minimum occurred at 10–11 Hz and in group 2, at 20–21 Hz. In group 1, the moduli of input impedance significantly increased from 10 to 26 Hz and tended to increase even above 25 Hz. In group 2, the moduli of input impedance significantly increased at low frequencies between 0 and 10 Hz; however, it did not change above 10 Hz. In sham controls, phase angle, which was initially negative, crossed zero at approximately the frequency of the modulus minimum to become positive (above 18 Hz), and crossed zero in the neighborhood of the maximum at about twice the frequency of the minimum to become negative (about 26 Hz). In

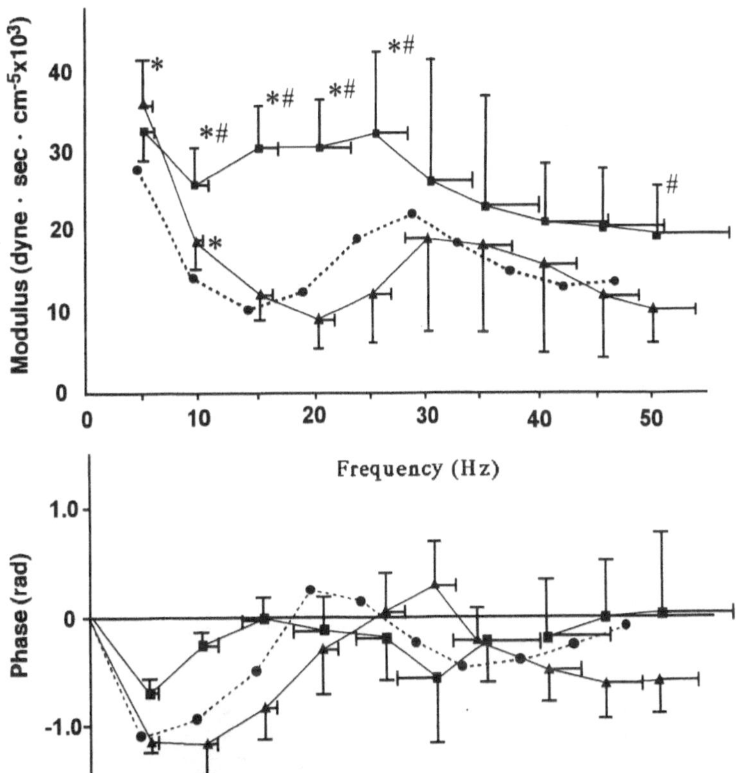

FIG. 3. Comparison of aortic input impedance among sham controls (*dotted line with closed circles*; $n = 10$), group 1 (*solid line with closed squares*; $n = 8$), and group 2 (*solid line with closed triangles*; $n = 7$). In group 1, the modulus of input impedance (*upper graph*) significantly increased from 10 to 26 Hz and tended to increase even above 26 Hz. In group 2, the modulus of input impedance significantly increased at low frequencies between 0 and 10 Hz but remained unchanged above 10 Hz. In group 1, phase angle (*lower graph*) was shifted to the left compared to sham controls, and in group 2 it was shifted to the right on the frequency scale. *, $P < .05$ vs. sham; #, $P < .05$ vs. group 2

group 1, phase angle was shifted to the left compared to the sham controls while in group 2 it was shifted to the right.

Figure 4 shows the ex vivo passive pressure–volume relationships. In group 1, the curve was shifted toward the volume axis compared with group 2 or sham controls, suggesting eccentric LV hypertrophy. In all groups, there was no sign of heart failure (e.g., ascites and pleural effusion). Also, aortic regurgitation could not be detected as evidenced by the lack of regurgitant flow on the aortic flow waveform.

The myocyte diameter of the LV in both group 1 (14.4 ± 0.8 μm; $P < .05$) and group 2 (16.1 ± 1.2 μm; $P < .05$) increased significantly compared to the sham controls (13.3 ± 0.4 μm). Myocyte diameter was larger in group 2 than in group 1 (Fig. 5).

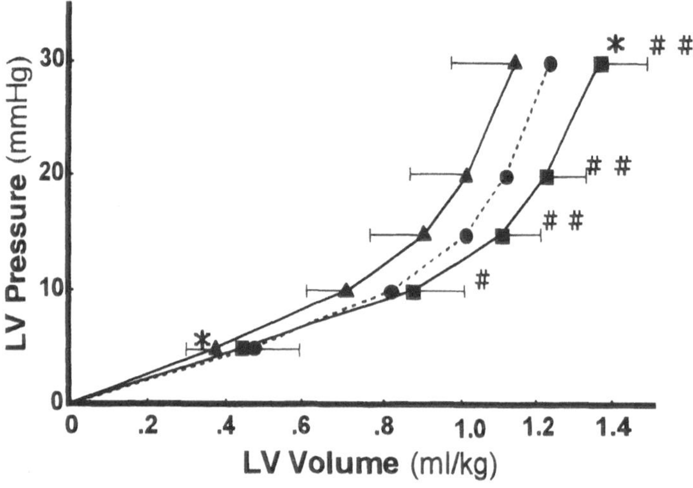

FIG. 4. Mean left-ventricular (*LV*) pressure–volume relations in sham controls (*closed circles with dotted line*; $n = 12$), group 1 (*closed squares with solid line*; $n = 9$), and group 2 (*closed triangles with solid line*; $n = 8$). In group 1 (early systolic loading), the relation curve was shifted to the right compared with group 1. Error bars, SD. *, $P < .05$ vs. sham; #, $P < .05$; ##, $P < .01$ vs. group 2

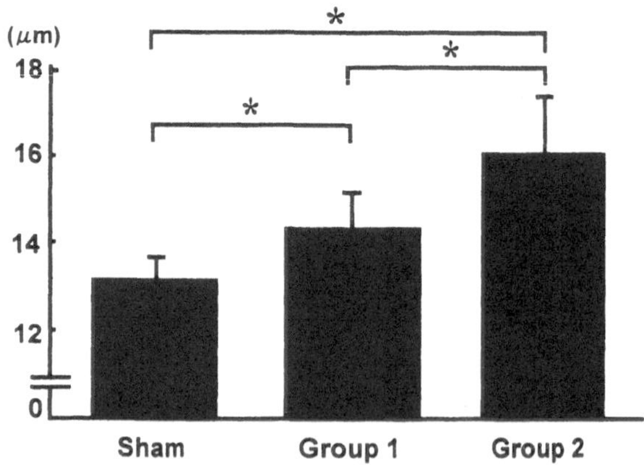

FIG. 5. Myocyte diameter 4 weeks after surgery. In both pressure-overloaded groups, myocyte diameter (μm) increased compared to sham controls; it was larger in group 2 than in group 1

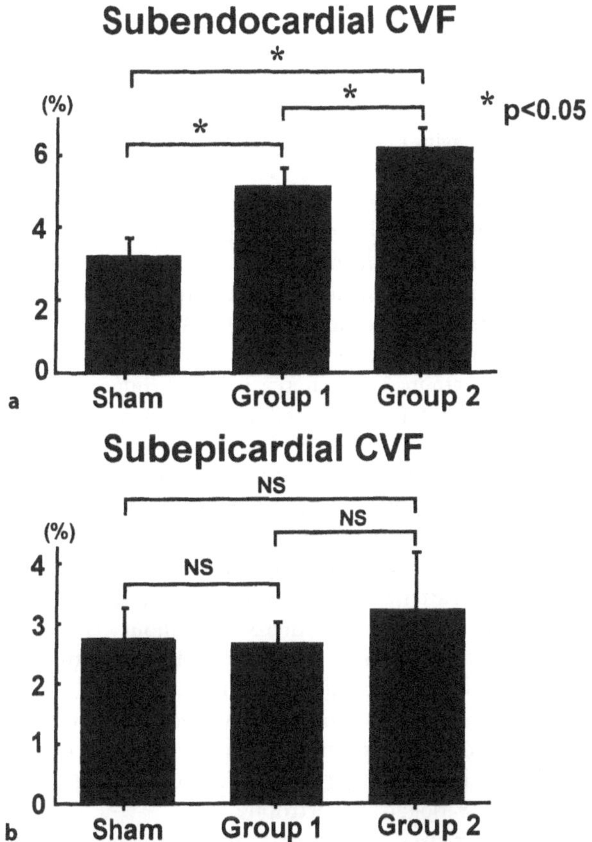

FIG. 6a,b. Connective tissue volume fractions (*CVF*) in subedocardial (**a**) and subepicaridal (**b**) region 4 weeks after surgery. Subendocardial CVF increased in both group 1 and group 2 rats as compared to sham controls. In group 2, it was more prominent than in group 1. There was no significant difference in subepicardial CVF among these three groups

Connective tissue volume fraction (CVF) in the subendocardial region was higher in group 2 (6.2% ± 6%) than in group 1 (5.2% ± 0.5%; $P < .05$), with no significant difference in the subepicaridal CVF (Fig. 6).

Discussion

The major finding of this study was that in a pressure-overloaded heart a sustained change in systolic loading sequence might influence the mode of the development of LV hypertrophy; in particular, early peaking pressure overload was associated with eccentric, reduced hypertrophy without a sign of LV systolic and diastolic dysfunction.

It is well known that the history of the LV pressure rise affects LV systolic and diastolic function when pressure overload is acutely applied to the LV [19,20].

However, little information is available as to the influence of the sustained change in systolic loading sequence on LV function and on the development of hypertrophy in a pressure-overloaded heart. Because the systolic loading sequence changes in a variety of clinical settings (e.g., the administration of vasoactive drugs, arteriosclerosis, or aging), such information might have important clinical implications.

Nichols and Pepine [21] demonstrated that coarctation of the aorta caused a significant change of the impedance spectra, resulting in increased resistance, elastic modulus, pulse wave velocity, and wave reflection. In this study, we also found a clear difference in the aortic input impedance spectra between group 1 (early systolic loading) and group 2 (late systolic loading) (see Fig. 3). In group 1, the impedance modulus was larger at high frequencies (>10 Hz), indicating a stiffening of the proximal aorta, while in group 2 the increase in the impedance modulus was prominent at low frequencies (<10 Hz), reflecting an increase in arterial wave reflection [22].

On the basis of the finding of aortic impedance spectra, we interpret the LV pressure profile with early or late systolic loading as follows. Because Zc plays a singificant role as a load against LV during early ejection phase [11,23], LV pressure produced a peak during early ejection phase with ascending aortic banding in which Zc increased with no singificant change in arterial wave reflection. However, in the case of abdominal aortic banding, arterial wave reflection increased without a change in Zc. Therefore, LV pressure produced a peak during late ejection phase reflecting the increase in arterial wave reflection probably from the constriction site.

In group 1, eccentric hypertrophy was observed as evidenced by the rightward shift of the ex vivo LV pressure-volume relations, suggesting that Zc as well as arterial wave reflection is one of the important factors for determining the mode of the development of pressure-overloaded hypertrophy.

Several investigators have stressed the importance of the relation between aortic stiffness and the LV pressure—volume relation [24], LV wall stress [25], and regional LV wall motion [26] in intact animals. From these previous findings, one can speculate that in group 1, Zc might act as an additional LV afterload and thus increase wall stress to a greater extent than in group 2. However, this effect does not simply explain the cause of dilatation of LV chamber without cardiac failure. If adequate hypertrophy adaptively had occurred in proportion to the increase in the wall stress in group 1, the size of the LV chamber would have been normalized as in group 2.

Histological analysis showed that the degree of hypertrophic response was less in group 1 than in group 2, evidenced by the finding that both myocyte diameter and connective tissue volume fraction were smaller in group 1 than in group 2. Taken together with these results, a reduced hypertrophic response might be a primary cause of the dilatation of the LV chamber in group 1, and the increase in Zc might promote the dilatation. However, the mechanism inducing reduced hypertrophy with early systolic loading is unclear. From the point of view of the difference in pressure profile or the impedance spectra between group 1 and group 2, we speculate that late systolic loading or the increase in arterial wave reflection may be necessary for the adequate development of LV hypertrophy.

Recent major concern has focused on the mechanism of the development of pressure-overloaded hypertrophy at the molecular level. Yamazaki et al. [27] proposed a stretch-induced pathway for the development of cardiac hypertrophy. Using cultured neonatal rat cardiocytes, they demonstrated that mechanical stretch can be a direct trigger for cell growth. In their study, mechanical stretch induced an increase in the rate of protein synthesis associated with the early expression of proto-oncogenes and fetal contractile proteins. Although they did not mention the time-course of the developed tension applied to the cell, it is interesting to consider whether the time-course of the tension may affect cell growth similar to our in vivo findings.

In pressure-overloaded LV hypertrophy, systolic function is usually preserved at a compensatory state, while diastolic function is often impaired even at the compensatory state [28]. In our study, systolic or diastolic function was even enhanced in both group 1 and group 2. These results are consistent with our previous report [29] in which both Ca^{2+} release and uptake from the sarcoplasmic reticulum were increased with compensatory, mild pressure-overloaded hypertrophy in rats.

Factors other than mechanical load should be taken into consideration for the development of LV hypertrophy. First, the renin–angiotensin–aldosterone system may possibly influence the development of LV hypertrophy because the end product, angiotensin II, acts as a regulator of myocyte growth independent of its vasoconstrictive effect [30]. In this connection, we measured the pressure below the constriction site in six rats 1 week after abdominal aortic banding. In all the rats, the poststenotic pressure was well maintained (peak pressure, 103 ± 7 mmHg, $n = 6$), associated with a significant pressure gradient (prestenotic peak pressure, 172 ± 16 mmHg, $n = 6$). Although we did not measure the poststenotic pressure in ascending aortic banding rats, Zierhut et al. [31] reported poststenotic pressure was maintained in a physiological range (prestenotic peak pressure, 231 ± 7 mmHg, $n = 18$; post-stenotic peak pressure, 149 ± 7 mmHg, $n = 18$).

In this regard, Baker et al. [32] reported that the activity of the circulating renin–angiotensin system was increased within a day after banding, and then returned to a control level by the third day after banding. These findings suggest that the circulating angiotensin II or aldosteone produced by the renin–angiotensin–aldosterone system does not play a major role in the development of pressure-overloaded LV hypertrophy. However, we cannot exclude the possible role of a local renin–angiotensin system, especially existing in myocytes [33]. Second, catecholamines are also known to induce cardiac hypertrophy [34]. Therefore, catecholamines may have triggered hypertrophy in myocardial cells and influenced the mode of development of LV hypertrophy in this study. However, because there was no significant difference in heart rate, peak (+)dP/dt of LV pressure, or TSR between group 1 and group 2, catecholamines may not play a critical role in determining the difference in the mode of hypertrophy between these two types of sustained pressure overload.

The role of baroreflex should also be considered in interpreting the results of our study. In our study, however, there was no significant difference in heart rate among three groups, although heart rate tended to be larger in aortic banding rats than in sham control rats. These results are compatible with the findings reported by Beznak [35]. Beznak investigated the time-course of a change in heart rate using the aortic banding rat model and demonstrated that heart rate gradually decreased until 1 week after aortic banding (control, 460 beats → 440 beats/min at 1 week), followed by the gradual increase to a higher level (500 beats/min) at 3 weeks after banding than control. Therefore, heart rate might be determined by a number of factors including baroreflex and sympathetic activity, which might change in a complicated fashion during the period of pressure overload.

We calculated Zc as an average of the impedance modulus from the 4th to 10th harmonics [9]. However, because fluctuations of the modulus were observed even at these high frequencies, we cannot eliminate the possibility that wave reflection may affect the calculated values of Zc. Therefore, we estimated Zc by measuring the P/F slope during the early ejection period (see *Methods*), where arterial wave reflection can be neglected [11]. The P/F slope so obtained was closely correlated with Zc in the frequency domain ($r = .92$, $P < .001$, $n = 19$), indicating the reasonable estimation of Zc in our study.

In conclusion, we demonstrated that sustained early systolic loading was accompanied by eccentric, reduced hypertrophy, while sustained late systolic loading was accompanied by concentric, adequate hypertrophy. Such different modes of cardiac hypertrophy were associated with the differences in aortic impedance spectra between pressure-overloaded groups.

Acknowledgments. The authors gratefully acknowledge Dr. Masakazu Obayashi and Dr. Yuji Hisamatsu for time and efforts related to this project.

References

1. Grossman W, Jones D, McLaurin LP (1975) Wall stress and patterns of hypertrophy in the human left ventricle. J Clin Invest 56:56–64
2. Campus S, Maravasi A, Gannau A (1987) Systolic function of the hypertrophied left ventricle. J Clin Hypertens 3:79–87
3. Ganau A, Devereux RB, Pickering TG, Roman MJ, Schnall PL, Santucci S, Spitzer MC, Laragh JH (1990) Relation of left ventricular hemodynamic load and contractile performance to left ventricular mass in hypertension. Circulation 81:25–36
4. Levy BI, Safar ME (1990) Ventricular afterload and aortic impedance. In: Research in cardiac hypertrophy and failure. Libbey, London, pp 521–529
5. Girerd X, Laurent S, Pannier B, Asmar R, Safar M (1991) Arterial distensibility and left ventricular hypertrophy in patients with sustained essential hypertension. Am Heart J 122:1210–1214
6. Levy BI, Babalis D, Lacolley P, Poitevin P, Safer ME (1988) Cardiac hypertrophy and characteristic impedance in spontaneously hypertensive rates. J Hypertens 6(suppl 4):110–111
7. Weiss JL, Frederiksen JW, Weisfeldt ML (1976) Hemodynamic determinants of the time-course of fall in canine left ventricular pressure. J Clin Invest 58:751–760
8. Zhu WX, Myers ML, Hartley CJ, Roberts R, Bolli R (1986) Validation of a single crystal for measurement of transmural and epicadial thickening. Am J Physiol 251:H1045–H1055
9. Levy BI, Michel JB, Salzmann JL, Azizi M, Poitevin P, Safar M, Camilleri JP (1988) Effects of chronic inhibition of converting enzyme on mechanical and structural properties of arteries in rat renovascular hypertension. Circ Res 63:227–239
10. Laskey WK, Kussmaul WG, Martin JL, Kleaveland JP, Hirshfeld JW, Shroff S (1985) Characteristics of vascular hydraulic load in patients with heart failure. Circulation 72:61–71
11. Dujardin JP, Stone DN (1981) Characteristic impedance of the proximal aorta determined in the time and frequency domain: a comparison. Med Biol Eng Comput 19:565–568
12. Raya TE, Gay RG, Lancaster L, Aguirre M, Moffett C, Goldman S (1988) Serial changes in left ventricular relaxation and chamber stiffness after large myocardial infarction in rats. Circulation 77:1424–1431
13. Litwin SE, Raya TE, Anderson PG, Litwin CM, Bressler R, Goldman S (1991) Induction of myocardial hypertrophy after cononary ligation in rats decreases ventricular dilatation and improves systolic function. Circulation 84:1819–1827

14. Sekiguchi M, Hiroe M, Morimoto S (1979) On the standardization of histopathological diagnosis and semiquantitative assessment of the endomyocardium obtained by endomyocardial biopsy. Bull Heart Inst Jpn 1979:55–85
15. Weber KT, Janicki JS, Shroff SG, Pick R, Chen RM, Bashey RI (1988) Collagen remodeling of the pressure-overloaded, nonhuman primate myocardium. Circ Res 62:757–765
16. Heidenhaim M (1915) Über die Mallory'sche Binde-gwebsfärbung mit Karmin und Azokarmin als Vorfarben. Z Wiss Mikrosk 32:361
17. Kojima M, Shiojima I, Yamazaki T, Komuro I, Yunzeng Z, Ying W, Mizuno T, Ueki K, Tobe K, Kadowaki T, Nagai R, Yazaki Y (1994) Angiotensin II receptor antagonist TCV-116 induces regression of hypertensive left ventricular hypertrophy in vivo and inhibits the intracellular signaling pathway of stretch-mediated cardiomyocyte hypertrophy in vitro. Circulation 89:2204–2211
18. Brilla CG, Pick R, Tan LB, Janicki JS, Weber KT (1990) Remodeling of the rat right and left ventricles in experimental hypertension. Circ Res 67:1355–1364
19. Hori M, Inoue M, Kitakaze M, Tsujioka K, Ishida Y, Fukunami M, Nakajima S, Kitabatake A, Abe H (1985) Loading sequence is a major determinant of afterload-dependent relaxation in intact canine heart. Am J Physiol 249:H747–H754
20. Yano M, Kohno M, Konishi M, Takahashi T, Seki K, Matsuzaki M (1994) Influence of left ventricular regional nonuniformity on afterload-dependent relaxation in intact dogs. Am J Physiol 267:H148–H154
21. Nichols WW, Pepine CJ (1982) Left ventricular afterload and aortic input impedance. Implications of pulsatile blood flow. Prog Cardiovasc Dis 24:293–306
22. Nichols WW, O'Rourke MF (1990) McDonald's blood flow in arteries, 3rd edn. Arnold, Baltimore, pp 251–269
23. O'Rourke MF, Kelly RP, Avolio AP (1992) The arterial pulse. Lea and Febiger, London, pp 146–176
24. Kelly RP, Tunin R, Kass DA (1992) Effect of reduced aortic compliance on cardiac efficency and contractile function of in situ canine left ventricle. Circ Res 71:490–502
25. Pouleur H, Covell JW, Ross J Jr (1979) Effects of alterations in aortic input impedance on the force-velocity-length relationship in the intact canine heart. Circ Res 45:126–135
26. Kohno M, Matsuzaki M, Ozaki M, Yano M, Katayama K, Fujii T, Kohtoku S, Ohtani N, Tateno S, Sakai H, Kusukawa R (1988) Influence of acute increase in aortic impedance on left ventricular regional wall motion during early ejection period. Proc IEEE Eng Med Biol Soc 10:249–251
27. Yamazaki T, Tobe K, Hoh E, Maemura K, Kaida T, Komuro I, Tanemoto H, Kadowaki T, Nagai R, Yazaki Y (1993) Mechanical loading activates mitogen-activated protein kinase and S6 peptide kinase in cultured rat cardiac myocytes. J Biol Chem 268:12069–12076
28. Lecarpentier Y, Waldenstrom A, Clergue M, Chemla D, Oliviero P, Martin JL, Swynghedauw B (1987) Major alterations in relaxation during cardiac hypertrophy induced by aortic stenosis in guinea pig. Circ Res 61:107–116
29. Ohkusa T, Yano M, Hisamatsu Y, Kobayashi S, Ryoke T, Kohno M (1994) Altered cardiac mechanism and sarcoplasmic reticulum function in hypertrophic rat hearts induced by aortic coarctation (abstract). Circulation 90(suppl I):I-483
30. Khairallah PA, Robertson AL, Davila D (1972) Effects of angiotensin II on DNA, RNA and protein synthesis. In: Genest J, Koiw R (eds) Hypertension. Springer, Berlin Heidelberg New York, pp 212–220
31. Zierhut W, Zimmer HG, Gerdes AM (1991) Effect of angiotensin converting enzyme inhibition on pressure-induced left ventricular hypertrophy in rats. Circ Res 69:609–617
32. Baker KM, Mitchell IC, Sally KW, Joseph FA (1990) Renin-angiotensin system involvement in pressure-overload cardiac hypertrophy in rats. Am J Physiol 259:H324–H332
33. Sadoshima J, Xu Y, Slayter HS, Izumo S (1993) Autocrine release of angiotensin II mediates stretch-induced hypertrophy of cardiac myocytes in vitro. Cell 75:977–984

34. Simpson P (1957) Stimulation of hypertrophy of cultured neonatal rat heart cells through an α-adrenergic receptor and induction of beating through an α_1- and β_1-adrenergic receptor interaction: evidence for independent regulation of growth and beating. Circ Res 56:884–894
35. Beznak M (1957) Cardiac output in rats during the development of cardiac hypertrophy. Circ Res 6:207–212

The Arterial Response to an Exercise Stress Test in Healthy Subjects, Hypertensives, and Patients with Left-Ventricular Dysfunction

Erez Nevo[1], Meir Marmor[2], and Alon Marmor[3]

Summary. Quantitative assessment of the arterial load response to an exercise stress test may add valuable insight to the assessment of cardiovascular performance in cardiac patients. Noninvasive pressure-volume measurements were taken during an exercise test in 42 subjects: 24 with left-ventricular dysfunction (LVD), 10 hypertensives, and 8 healthy subjects. Systemic vascular resistance (SVR), arterial elastance (Ea), and total and pulsatile power were calculated from the measured data. The LVD patients were subgrouped to patients with good exercise performance (3 stages; $n = 16$) and low exercise performance (2 stages; $n = 8$). Significantly different arterial response was found in patients with lower exercise capacity (Ea increased from 1.5 ± 0.3 to 2.6 ± 1.6 mmHg/ml at $P < .05$, contrasting with nonsignificant (NS) decrease in the other three groups). These changes were associated with inadequate increase of global cardiovascular performance in patients with low exercise performance (e.g., cardiac output increased from 6 ± 1 (rest) to 8 ± 4 (stress stage 2), $P = \text{NS}$, compared with 8 ± 4 increased to 15 ± 4 in healthy). The ratio between pulsatile and total power was similar in healthy and LVD patients, while the hypertensive patients had a higher pulsatile component. In conclusion, LVD patients with reduced exercise capacity demonstrated inadequate decrease of arterial load during exercise, suggesting a relationship between load reduction and impaired exercise capacity. Similar power ratios in healthy and in LVD patients suggested that energy transmission efficiency is not affected by heart failure. In hypertensives, however, a higher pulsatile component during rest and exercise implies higher energy waste for the same amount of work.

Key words. Power—Ventriculoarterial coupling—Arterial elastance—Systemic vascular resistance

Introduction

Arterial load, one of the major determinants of cardiovascular performance, is commonly quantified by systemic vascular resistance (SVR). While SVR can be easily

[1] Hadassah-Mt. Scopus Hospital, Jerusalem, Israel 91240.
[2] Faculty of Medicine, Technion, Haifa, Israel 32000.
[3] Rivka Ziv Hospital, Safed, Israel 73100.

measured, it does not account for the pulsatile component of pressure and flow. The pulsatile component has turned out to have greater effect in patients with hypertension or heart failure [1,2], and it was found to be a risk factor for cardiac morbidity and mortality [3,4]. A more comprehensive description of the arterial hydraulic load is provided by aortic input impedance [5,6], which is commonly determined by invasive measurements of aortic pressure and flow. Pressure measurement by applanation tonometry and flow measurement by doppler sonography have been utilized for the noninvasive estimation of input impedance during rest [7].

Arterial elastance was formulated by Sunagawa et al. [8,9] as a lumped description of the arterial load that also accounts for the pulsatile load. The arterial elastance can be estimated from end-systolic pressure and stroke volume (Pes/SV) and is thus more feasible for noninvasive evaluation. Using a modified Windkessel model of the arterial system [9], arterial elastance was shown to account for the principal elements of arterial load including peripheral resistance, lumped arterial compliance, and characteristic impedance. Kelly et al. [10] suggested and validated an approximation of the end-systolic pressure based on the systolic and diastolic arterial pressures (Pes = (2*Psys + Pdia)/3), which greatly facilitates the noninvasive measurement of the arterial elastance (Ea).

Directly related to the quantification of the load parameters are descriptors of energy transmission from the left ventricle to the arterial system. The total power represents the external left-ventricular hydraulic work per unit time. While total and mean hydraulic power depend predominantly on cardiac output, the pulsatile fraction of total hydraulic power is also a function of aortic distensibility and aortic characteristic impedance [11]. The ratio between pulsatile power and total hydraulic power provides an index of ventriculoarterial coupling and of the efficiency of energy transfer through the arterial system.

The objective of this study was to evaluate the response of arterial load indices and power components during exercise in patients with various degrees of left-ventricular dysfunction. A noninvasive technique for measurement of central arterial pressure and ventricular volume was used to evaluate the two indices of arterial load (Ea and SVR), the pulsatile power, the total power, and their ratio.

Methods

Study Population

The study population consisted of 24 patients (23 men, 1 woman; mean age, 63 ± 10 years) with angiographically documented coronary artery disease and with various levels of left-ventricular dysfunction (LVD) as assessed by rest radionuclide ejection fraction (EF = 47% ± 13%), 10 hypertensive patients (9 men, 1 woman; mean age, 54 ± 10 years) with systolic blood pressure >150 mmHg and diastolic blood pressure >90 mmHg (HTN), and 8 healthy subjects (7 men, 1 woman; age, 49 ± 9 years), who were considered as the control group (HEL). All LVD patients were asymptomatic for at least 1 year before the study and were capable of performing at least 2 stages of supine bicycle ergometry (NYHA class I–II). Of these patients, 8 received no cardiac medication, 9 received digoxin and/or diuretics, 7 were taking calcium antagonists, and 2 were taking beta blockers. In HTN patients, no evidence of ischemic heart disease was found by history, ECG, or exercise stress test. All antihypertensive medications were stopped 1 week before the study. Ethical considerations required that only patients with mild to moderate hypertension (systolic blood pressure not higher than 180 mmHg, diastolic blood pressure not higher than 110 mmHg) were included in the study.

Measurements

Equilibrium radionuclide angiography (ERNA) was used to assess baseline ventricular function as expressed by global left-ventricular ejection fraction (EF). The methodology of absolute ventricular volume measurements by ERNA during exercise was reported previously [12]. The methodology of arterial pressure waveform measurement was described in detail in a previous validation study [13]. Briefly, it consists of a sphygmomanometric cuff attached to an air pressure unit, a doppler transducer positioned above the brachial artery, and an ECG monitoring system. The cuff is automatically inflated until cuff pressure exceeds systolic pressure and the doppler signal disappears. As the cuff is deflated, the doppler signals reappear, and the time delay from the preceding R-wave of the ECG to the onset of brachial flow is measured. The cuff pressure is plotted as a function of the time delay, and the resulting curve is an estimation of the upstroke of the central arterial pressure wave.

Exercise Protocol

Supine rest ERNA, arterial pressure waveform, heart rate, and resting ECG were recorded as baseline data. Baseline hemodynamic variables were determined as average values of two data sets obtained immediately before exercise with legs elevated on ergometer pedals. All patients underwent up to 3 stages of supine bicycle exercise, each stage 3 min long, with a baseline level of 25 W and increments of 25 W/stage. Heart rate, blood pressure, ERNA, and ECG were continuously recorded during the test. Exercise was stopped after 3 stages or earlier if severe fatigue or dyspnea occurred. The pressure–volume measurements were recorded during the last 2 min of each stage of exercise to minimize the effect of transient response between stages.

Statistical Analysis

Data were expressed as mean ± standard deviation. One-way analysis of variance (ANOVA) was used to analyze differences between the groups. When ANOVA showed statistical significance, the Student–Newman–Keuls test was performed between the groups. A probability of $P < .05$ was considered significant for all tests.

Results

All patients completed at least 2 stages of supine exercise. In the LVD group, 8 patients were unable to perform more than 2 stages of exercise, and were considered as the group with low exercise capacity or LVD2. The remaining 16 patients completed the 3 stages of supine exercise and were defined as the group with preserved exercise capacity or LVD3. All healthy subjects and hypertensive patients completed at least 3 stages of exercise.

Baseline Hemodynamic Data

Cardiac output and ejection fraction were reduced in the LVD group (6 ± 2 l/min and 47% ± 13%, respectively) compared with the healthy group (8 ± 4 l/min, 66% ± 7%, both at $P < .05$) and with hypertensive patients (7 ± 3 l/min and 63% ± 10%). There was no significant difference in these parameters between the two LVD groups (Table 1).

Exercise Performance

The LVD3 patients reached a maximal double product of 21300 ± 7100 (mmHg*beats/min). The LVD2 patients reached a similar maximal value (21530 ± 5820, P = NS). None of the patients developed chest pain during exercise. There was no significant difference in double product between the patients receiving cardioactive medication and the rest of the patients (19700 ± 5950 vs. 22180 ± 7140). All hypertensive patients performed 3 stages of exercise and reached a double product of 27540 ± 3240. All healthy subjects completed at least 3 stages of exercise, 2 reached stage 4, and 2 reached stage 5. Although these controls had better exercise tolerance than the patients, they achieved a similar double product at stage 3 (24980 ± 4770, P = NS in all comparisons). As the purpose of the study was to assess behavior of peripheral vasculature at similar workloads in different populations, hemodynamic performance was compared in healthy subjects at stage 3 and in patients at the maximal stage (either 2 or 3).

Global Cardiovascular Performance During Exercise

Healthy subjects increased cardiac output substantially from 8 ± 4 to 17.5 ± 4 l/min at end of stage 3, while LVD patients increased it modestly (from 5.8 ± 2 l/min to 10 ± 3 l/min in LVD3 group, and from 6 ± 1 l/min to 8.1 ± 4 l/min in the LVD2 group; see Table 1). Stroke volume decreased from 84 ± 11 ml at rest to 66 ± 20 ml and 71 ± 27 ml at stage 1 and stage 2, respectively, in the LVD2 group, contrasting with the nonsignificant change observed in the LVD3 group at peak exercise (from 83 ± 24 ml to 89 ± 25 ml; Table 1). In healthy subjects and in hypertensives, the stroke volume increased substantially (114 ± 46 to 135 ± 33 ml in healthy; 89 ± 24 to 117 ± 39 ml in hypertensives). The small increase of mean blood pressure during the exercise was similar in all groups (Fig. 1).

Changes in Contractility

Changes in peak power, a descriptor of ventricular contractility, were reported in detail in a previous paper [12]. Briefly, in healthy subjects and in hypertensives, the peak power (normalized by end-diastolic volume) increased during exercise (healthy, 5.4 ± 0.8 to 11.4 ± 3.1 watts/ml, $P < .01$; hypertensives, 5.5 ± 0.9 to 12.3 ± 4 W/ml, $P < .001$). In the LVD3 group the peak power increase was smaller but still significant (4 ± 1.3 to 5.6 ± 2.1 W/ml, $P < .01$). In the LVD2 group, despite the fact that the patients performed only 2 stages, peak power increased to the same extent as in LVD3 group (3.9 ± 1.1 to 5.4 ± 2 W/ml, $P < .01$).

Arterial Load Response to Exercise

Arterial elastance and systemic vascular resistance at rest were elevated in the LVD and hypertensive group compared with the healthy subjects (see Table 1). The highest

TABLE 1. Comparison of hemodynamic variables during exercise in the four different groups. The presented values are average values and standard deviations; comparisons are between exercise stages for each group and between groups during the same stages.

Variable	HEL	LVD2	LVD3	HTN
CO (l/min)				
REST	8.0 ± 3.8	6.0 ± 1.0	5.8 ± 2.2	7.2 ± 2.9
STAGE1	12.7 ± 3.6*[b,c]	6.7 ± 2.4[a]	7.7 ± 3.1[a]	10.7 ± 4.7
STAGE2	15.2 ± 4.2*[b]	8.2 ± 3.9[a]	9.9 ± 3.8*	13.2 ± 7.8*
STAGE3	17.5 ± 4.3*[c]	—	10.0 ± 3.0*	13.9 ± 7.2*
SV (ml)				
REST	114 ± 46	84 ± 11	83 ± 24	90 ± 24
STAGE1	137 ± 41[b,c,d]	66 ± 20[a,d]	85 ± 22[a]	104 ± 32[a,b]
STAGE2	143 ± 44[b,c]	71 ± 27[a,d]	89 ± 29[a]	225 ± 45[b]
STAGE3	135 ± 33[c]	—	89 ± 25[a,d]	117 ± 39[c]
Ea (mmHg/ml)				
REST	1.23 ± 0.55	1.51 ± 0.29	1.74 ± 0.53	1.68 ± 0.55
STAGE1	0.96 ± 0.29[b,c,d]	2.34 ± 0.95*[a,c,d]	1.73 ± 0.43[a,b]	1.53 ± 0.49[a,b]
STAGE2	1.01 ± 0.28[b]	2.61 ± 1.60[a,d]	1.93 ± 1.03	1.40 ± 0.40[b]
STAGE3	1.13 ± 0.29	—	1.96 ± 1.0	1.51 ± 0.68
SVR (dyn*sec*cm^{-5})				
REST	1520 ± 740	1780 ± 510	2090 ± 770	1740 ± 550
STAGE1	820 ± 210[b,c]	1900 ± 750[a,d]	1650 ± 500*[a,d]	1210 ± 360*[b,c]
STAGE2	750 ± 220*[b,c]	1840 ± 980[a,c,d]	1350 ± 450*[a,b]	1050 ± 310*[b]
STAGE3	660 ± 120*[c,d]	—	1300 ± 330*[a,d]	1040 ± 340*[a,c]
Total power (mmHg*ml/s)				
REST	15370 ± 6900	12320 ± 2120	12950 ± 5750	17040 ± 6680
STAGE1	25890 ± 8660	15500 ± 6530	17850 ± 7850	27590 ± 14130
STAGE2	33440 ± 10540*	19870 ± 9120*	25060 ± 12080*	35370 ± 24640
STAGE3	42490 ± 11990*	—	25780 ± 9350*	38860 ± 25990
Pulsatile power (mmHg*ml/s)				
REST	2860 ± 1378	2162 ± 559[d]	2331 ± 1013[d]	3657 ± 1460[b,c]
STAGE1	5017 ± 1346*	3108 ± 2026[d]	3332 ± 1582*[d]	5969 ± 3144[b,c]
STAGE2	6665 ± 2285*	4496 ± 3113	4933 ± 2348*	7713 ± 5495
STAGE3	9861 ± 2967*[c]	—	5068 ± 1969*[a,d]	8515 ± 5802[c]
Power transmission efficiency (%)				
REST	19 ± 3	18 ± 3[d]	18 ± 2[d]	21 ± 0.5[b,c]
STAGE1	20 ± 3	19 ± 5	18 ± 2[d]	22 ± 0.5*[c]
STAGE2	20 ± 2	20 ± 5	20 ± 3*	22 ± 0.6*
STAGE3	23 ± 4*	—	20 ± 3*	22 ± 0.5*

HEL, Healthy subjects; LVD2, patients with left-ventricular dysfunction and low exercise capacity; LVD3, patients with left-ventricular dysfunction and preserved exercise capacity; HTN, hypertensives; CO, cardiac output; SV, stroke volume; Ea, arterial elastance; SVR, systemic vascular resistance.
Comparison between rest and exercise stages, using paired *t*-test: * $P < .05$.
Comparisons between groups, using ANOVA and Student-Newman-Keuls:
[a] $P < .05$ compared to HEL.
[b] $P < .05$ compared to LVD2.
[c] $P < .05$ compared to LVD3.
[d] $P < .05$ compared to HTN.

values were found in the LVD3 group: arterial elastance, 1.74 ± 0.5 mmHg/ml, and systemic vascular resistance 2090 ± 77 dynes*s/cm^5. During exercise, healthy subjects and hypertensives decreased both arterial resistance and arterial elastance (Table 1; Fig. 1). In contrast, the arterial elastance increased in LVD2 patients (from 1.5 ± 0.3 to

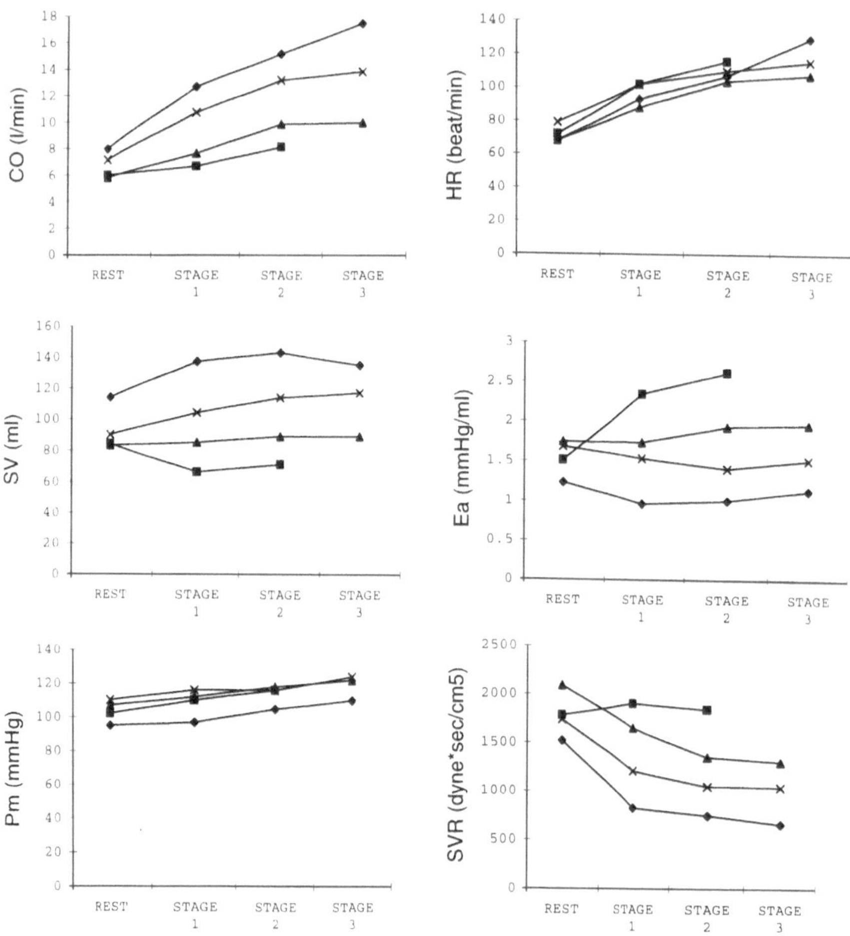

FIG. 1. Hemodynamic adaptation during exercise is demonstrated for the four groups: healthy subjects, HEL, *diamonds*; hypertensives, HTN, *crosses*; patients with left-ventricular dysfunction and preserved exercise capacity, LVD3, *triangles*; and patients with left-ventricular dysfunction and low exercise capacity, LVD2, *squares*. The global cardiovascular performance is quantified by (*left panel*, from top down) the cardiac output (*CO*), the stroke volume (*SV*), and the mean arterial pressure (*Pm*). Two of the four major determinants of the global cardiovascular performance are presented (*right panel*, from top down), the heart rate (*HR*) and the afterload indexes: the systemic vascular resistance (*SVR*) and the arterial elastance (*Ea*). Average values of the hemodynamic variables, standard deviations, and statistical comparisons between the different groups and the different stages are given in Table 1

2.3 ± 0.9, $P < .05$). The patients in group LVD3 demonstrated an intermediate response: the mean value of the arterial elastance increased from stage to stage without reaching statistical significance (Table 1; Fig. 1). In LVD2 patients, the arterial resistance did not change during exercise (1780 ± 510 to 1900 ± 750 dynes*sec/cm^5, P = NS), contrasting with significant resistance decrease in healthy, in hypertensives and in LVD3 patients (see Table 1).

Power

As expected, the total power increased during exercise in all groups. The pulsatile power increased as well, and the increase of the power ratio in healthy and LVD groups demonstrated that the relative increase of the pulsatile component was slightly larger than that of the total power. The power ratio was higher in HTN, compared with IVD and healthy. It was unchanged during the exercise but was still higher than the power ratio of the other groups during peak exercise (Table 1).

Discussion

The purpose of this study was to evaluate the response of the arterial load to exercise in different patient populations, and to relate the pattern of response to the level of LVD and to the demonstrated exercise performance capacity. Two widely accepted indexes of arterial load, the arterial elastance and the systemic vascular resistance, were assessed from noninvasively measured arterial pressure and left-ventricular stroke volume. The systemic vascular resistance is a major determinant of cardiovascular performance, and is often elevated in heart failure patients because of compensatory responses [14]. Arterial elastance accounts for the effect of pressure pulsations, which may increase the effective load imposed on the left ventricle. In many patients, most notably in heart failure patients, wave reflections affect the left-ventricular ejection and impose a higher effective load than described by the systemic vascular resistance [15]; however, the role of the pulsatile load versus the steady one is not yet defined.

Cardiac output and ejection fraction during rest clearly identify the group with left-ventricular dysfunction. Neither cardiac output nor ejection fraction predicted exercise capacity of the LVD subjects. Baseline load variables were similar in the two LVD groups having different exercise performance. The baseline contractility and the contractility response during exercise, as assessed by left-ventricular peak ejection power, were similar in the two LVD groups [12]. Consequently, the different exercise performance of the two groups could not be attributed to different hemodynamic state during rest or to different contractile response during exercise.

The pattern of change of arterial elastance and systemic vascular resistance was substantially different in the LVD2 group compared with the responses of the other groups. While the typical normal response was a decrease of resistance and minimal change of arterial elastance, the LVD2 group demonstrated minimal change in resistance and an increase in arterial elastance (see Table 1 and Fig. 1). Of particular interest is the different response of LVD2 patients compared with the LVD3 patients. The latter had a response pattern similar to of the healthy and hypertensive patients, even though the absolute values were different. These findings suggest a potential relationship between abnormal response of the arterial load and low exercise capacity.

The arterial elastance, as a more comprehensive descriptor of arterial load, lumps the steady arterial load (resistance) and the pulsatile load into a single index (in contrast with the input impedance, which presents the load for each frequency band). This complicates the physiological interpretation of the arterial elastance response during exercise, when simultaneous changes of resistance, heart rate, and blood pressure occur. Whether arterial elastance change results only from resistance and heart rate changes or from change in the mechanical properties of the arteries (i.e., arterial compliance) can be better appraised by monitoring the ratio between resistance–heart rate product and arterial elastance [10], which is equivalent to the pulsatile to total power ratio (see Appendix). In this study, the power ratio demonstrated a slight increase during exercise. This can be interpreted as a slight increase of pulsatile load that is not caused by heart rate or resistance changes. The much larger magnitude of exercise-induced change in the steady component of the arterial load (e.g., a 46% resistance decrease versus a 5% pulsatile load increase in the healthy group) places the decreased steady load component as the major hemodynamic adaptation mechanism of the arterial load during exercise. The demonstrated slight increase of the pulsatile load (indirectly presented by the increased power ratio) was not significantly different between the two LVD groups and the healthy subjects, and cannot be identified as the exercise-limiting factor in LVD2 patients.

The power components can be further utilized to evaluate the coupling between the left ventricle and the arterial load. The ratio between the pulsatile power (energy loss in the conductance process through the arteries) and the total power (the energy transferred from the ventricle to the arterial system) represents the relative loss of energy and is commonly used as a measure of energy transfer efficiency [11,16]. In the current study the pulsatile power and the total power increased significantly during exercise in healthy and, to a lesser extent, in LVD patients (see Table 1), but there were no significant differences in the pulsatile/total power ratio. The power ratio was higher in hypertensives, in whom the pulsatile component of arterial load is increased probably because of the increased stiffness of the arterial wall. In all groups the change of power ratio was small, meaning that change in conductance efficiency during exercise has a minimal role compared with the previously demonstrated role of ventricular efficiency [17].

Similar results were obtained invasively by Laskey et al. [16]. They reported that during rest the steady load component (resistance) was larger in idiopathic cardiomyopathy patients, while the pulsatile component (characteristic impedance) was similar to that of healthy subjects. During exercise, stroke volume increased in healthy subjects and not in patients, heart rate increased similarly in both groups, and resistance decreased to a lesser extent in the cardiomyopathy patients. In both groups the characteristic impedance did not change during exercise. Healthy subjects had a larger increase in total power, but the ratio between pulsatile power and total power remained similar in the two groups and did not change during exercise. The authors noted the similarity between the characteristic impedances and the oscillatory fraction of the total power. Similarly, Murgo et al. [18] found in normal subjects a decrease of arterial resistance and no change of characteristic impedance.

Our noninvasive results with hypertensive patients are similar to those of Kelly et al. [10] and Nichols et al. [19]. We found a larger arterial elastance in hypertensives compared with healthy subjects, although because of the large variance the difference was not significant (1.68 ± 0.55 vs. 1.23 ± 0.55). Kelly et al. reported similar values (1.65 ± 0.64 in hypertensives vs. 1.0 ± 0.22 in healthy), while Nichols et al. reported similar resistance and higher characteristic impedance in hypertensives. The power ratio, however, is only slightly higher in hypertensives, and the difference reached statistical significance only compared with LVD patients (see Table 1), while Nichols et al. [19] found a highly significant increased power ratio in hypertensives (18 ± 4 vs. 13 ± 3 in normotensives, $P < .001$).

The demonstrated association between inadequate systemic vascular resistance decrease and reduced exercise capacity is compatible with the local response of vascular resistance in exercising muscles. In heart failure patients, skeletal muscle perfusion was found to be decreased at rest and during exercise, and local vascular resistance was increased [20]. Either the decreased perfusion or altered metabolism in the muscle cells [21] result in early anaerobic metabolism, which is a major limiting factor of exercise capacity [22].

Conclusions

A typical response to exercise was defined in healthy subjects, hypertensives, and in patients with left-ventricular dysfunction. LVD patients with lower exercise performance experienced inadequate decrease of the arteial load, which resulted in blunted increase of global cardiovascular performance. The baseline parameters of these patients were similar to LVD patients with good exercise performance. Thus it appears that in LVD patients poor exercise capacity is associated with inadequate adaptation during exercise rather than baseline hemodynamic status. The ratio of pulsatile to total power (power transmission efficiency) was similar in the healthy persons and in the LVD patients during exercise, indicating a comparable energy transfer to the periphery in both groups. This ratio was somewhat increased in hypertensives, suggesting a larger energy loss in the arterial system.

Acknowledgments. We thank Dr. Barry Zaret, Dr. Frans Wackers, and Dr. Diwakar Jain, from the Section of Cardiovascular Medicine, Yale University School of Medicine, New Haven, Connecticut, for their help in data acquisition.

Appendix: Calculation of Hemodynamic Parameters

Systemic vascular resistance (R) is determined by:

$$R = Pm/CO = Pm/(SV * HR) \qquad [1]$$

Arterial elastance (Ea) is determined by:

$$Ea = Pes/SV \qquad [2]$$

End-systolic pressure (Pes) is determined by [10]:

$$Pes = (Psys * 2 + Pdia)/3 \qquad [3]$$

Total power is the ejected energy per time unit [15]:

$$\begin{aligned} \text{Total power} &= SW * HR \\ &= Pes * SV * HR = E_a * SV^2 * HR \end{aligned} \qquad [4]$$

Mean power is evaluated by [15]:

$$\begin{aligned} \text{Mean power} &= Pm * CO \\ &= Pm * SV * HR = R * SV^2 * HR^2 \end{aligned} \qquad [5]$$

Pulsatile power is the difference between the total and the mean power:

$$\text{Pulsatile power} = \text{total power} - \text{mean power} \qquad [6]$$

The power ratio represents power transmission efficiency:

$$\begin{aligned} \text{Efficiency} &= \text{mean power/total power} \\ &= \text{Pm/Pes} = \text{R} * \text{HR/Ea} \end{aligned} \qquad [7]$$

In these equations, Pm is mean arterial pressure, CO is cardiac output (= HR * SV), HR is heart rate, SV is stroke volume, Psys and Pdia are the systolic and diastolic pressure, respectively, and SW is stroke work.

References

1. Avolio AP, Chen SG, Wang RP, Zhang CL, Li MF, O'Rourke MF (1983) Effects of age on changing arterial compliance and left ventricular load in a northern Chinese urban community. Circulation 68:50–58
2. Nichols WW, O'Rourke MF, Avolio AP, Yaginuma T, Murgo JP, Pepine CJ, Conti CR (1985) Effects of age in ventricular/vascular coupling. Am J Cardiol 55:1179–1184
3. Rutan GH, Kuller LH, Neaton JD, Wentworth DN, McDonald RH, McFate-Smith W (1988) Mortality associated with diastolic hypertension and isolated systolic hypertension among men screened for Multiple Risk Factor Intervention Trial. Circulation 77:504–514
4. Kannel WB, Wolf PA, McGee DL, Dawber TR, McNamara P, Castelli WP (1981) Systolic blood pressure, arterial rigidity and risk of stroke. JAMA 245:125–129
5. Nichols WW, Conti CR, Walker WE, Milnor WR (1977) Input impedance of the systemic circulation in man. Circ Res 40:451–458
6. Murgo JP, Westerhof N, Giolma JP, Altobelli SA (1980) Aortic input impedance in normal man: relationships to pressure waveforms. Circulation 62:105–116
7. Kelly R, Fitchett D (1992) Noninvasive determination of aortic input impedance and external left ventricular power ouptut: a validation and repeatability study of a new technique. J Am Coll Cardiol 20:952–963
8. Sunagawa K, Maughan WL, Burkhoff D, Sagawa K (1983) Left ventricular interaction with arterial load studied in isolated canine ventricle. Am J Physiol 56:586–595
9. Sunagawa K, Maughan WL, Sagawa K (1985) Optimal arterial resistance for maximal stroke work studied in isolated canine ventricle. Circ Res 56:586–595
10. Kelly RP, Ting CT, Yang TM, Liu CP, Maughan WL, Chang MS, Kass DA (1992) Effective arterial elastance as index of arterial vascular load in humans. Circulation 86:513-521
11. O'Rourke MF, Yagimuma T, Avolio AP (1984) Physiological and pathophysiological implications of ventricular-vascular coupling. Ann Biomed Eng 12:119–134
12. Marmor A, Jain D, Cohen LS, Nevo E, Wackers FJT, Zaret BL (1993) Left ventricular peak power during exercise. A noninvasive approach for assessment of contractile reserve. J Nucl Med 34:1877–1885
13. Sharir T, Marmor A, Ting CT, Chen JW, Liu CP, Chang MS, Yin FCP, Kass DA (1993) Validation of a method for noninvasive measurement of central arterial pressure. Hypertension 21:74–82
14. Ferguson DW (1992) Autonomic response to heart failure. Prog Cardiol 5:45–58
15. Laskey WK, Kussmaul WG (1987) Arterial wave reflection in heart failure. Circulation 75:711–722
16. Laskey WK, Kussmaul WG, Martin JL, Kleaveland JP, Hirshfeld JW, Shroff S (1985) Characteristics of vascular hydraulic load in patients with heart failure. Circulation 72:61–71
17. Little WC, Cheng CP (1993) Effect of exercise on left ventricular-arterial coupling assessed in the pressure-volume plane. Am J Physiol 264:H1629–H1633
18. Murgo JP, Westerhof N, Giolma JP, Altobelli SA (1981) Effects of exercise on aortic input impedance and pressure wave forms in normal humans. Circ Res 48:334–343

19. Nichols WW, O'Rourke MF, Avolio AP, Yaginuma T, Pepine CJ, Conti R (1986) Ventricular/vascular interaction in patients with mild systemic hypertension and normal peripheral resistance. Circulation 74:455–462
20. Sullivan MJ, Knight JD, Higginbotham MB, Cobb FR (1989) Relation between central and peripheral hemodynamics during exercise in patients with chronic heart failure: muscle blood flow is reduced with maintenance of arterial perfusion pressure. Circulation 80:769–781
21. Wiener DH, Fink LI, Maris J, Jones RA, Chance B, Wilson JR (1986) Abnormal skeletal muscle bioenergetics during exercise in patients with heart failure: role of reduced muscle blood flow. Circulation 73:1127–1136
22. Weber KT, Janicki JS (1985) Lactate production during maximal and submaximal exercise in patients with chronic heart failure. J Am Coll Cardiol 6:717–724

Altered Left-Ventriculoarterial Coupling in Mitral Regurgitation

Mark R. Starling

Summary. Twenty-nine control patients and 41 patients with long-term mitral regurgitation were studied, using micromanometer left-ventricular catheters and radionuclide angiography, to test the hypothesis that the low-impedance left atrium obscures an impairment in left-ventriculoarterial coupling and forward left-ventricular pump efficiency in patients with long-term mitral regurgitation. Left-ventriculoarterial coupling relationships were calculated in patients with long-term mitral regurgitation both before mitral valve surgery at baseline and during steady-state vasodilation and at 1 year following successful mitral valve surgery. The data indicate that, although total left-ventricular stroke work is performed efficiently by coupling of the left ventricle to both the arterial and atrial systems, left-ventricular pump efficiency for performing forward left-ventricular stroke work is impaired, despite preservation of normal forward left-ventricular stroke work. This impairment occurs in patients with long-term mitral regurgitation and a normal left-ventricular ejection fraction as the result of impairment of contractility coupled with a normal total arterial elastance. Acute vasodilation did not significantly improve this impairment in left-ventriculoarterial coupling and pump efficiency. In contrast, successful mitral valve surgery improves forward left-ventricular stroke work and pump efficiency because of recovery of left-ventricular contractility. One potential mechanism for this reversible impairment of left-ventricular contractility is proposed. In conclusion, left-ventricular pump efficiency for performing forward stroke work is impaired in patients with long-term mitral regurgitation, despite the outward appearance of normal left-ventricular ejection performance manifest by a normal ejection fraction. This probably results from coupling of the left ventricle to both the arterial and atrial systems. Consequently, the clinical implication of these data is that mitral valve surgery, particularly mitral valve repair, should be considered earlier in these patients when left-ventricular contractility is reversible and pump efficiency can be returned to normal.

Key words. Left ventricle—Arterial elastance—Ventriculoarterial coupling—Left-ventricular function—Contractility—Mitral regurgitation

Division of Cardiology, University of Michigan Medical Center, Ann Arbor, MI 48105, USA.

Introduction

A complete understanding of the pathophysiology and hemodynamic consequences of long-term mitral regurgitation on left-ventricular pump function and contractility has been elusive. This lack is related to the frequent observation of normal end-systolic stress and the uncoupling of end-systole from end-ejection by ejection into the low-impedance left atrium [1-3]. Thus, conventional indices of left-ventricular pump function, e.g., ejection fraction and fractional shortening, remain normal, so that an outward appearance of an effective and efficient pump is maintained for prolonged periods of time. However, recent highly sophisticated studies in an animal model of mitral regurgitation and in humans with long-term mitral regurgitation have demonstrated that left-ventricular contractility and pump efficiency may undergo an early deterioration, despite the outward appearance of normal left-ventricular pump function [2,4-6]. These data also demonstrate that, following successful mitral valve surgery, left-ventricular contractility and the effectiveness and efficiency of the pump may improve despite an early fall in left-ventricular ejection fraction [2,4,5,7,8]. Thus, these data suggest that left-ventricular pump function would be manifestly impaired if it were not for the low-impedance, left-atrial contribution to the maintenance of ejection performance.

The left-ventriculoarterial coupling relationship provides a unique theoretical framework within which left-ventricular pump function can be more completely assessed in this pathophysiological condition [9,10]. Mitral regurgitation is also unique in that it is the only pathophysiological condition in which the left ventricle is not totally dependent on an interaction with the arterial system; it also interacts with the low-impedance left atrium. The left ventricle in mitral regurgitation ejects, therefore, only a portion of the total output into the arterial system. Using radionuclide angiography to fractionate total left-ventricular output into its forward output and regurgitant output components, we have been able to apply the left-ventriculoarterial coupling concept to evaluate control patients and patients with long-term mitral regurgitation. By this method, we can test the hypothesis that the low-impedance left atrium, which maintains conventional indices of left-ventricular ejection within the normal range, obscures an impairment in left-ventriculoarterial coupling and forward left-ventricular pump efficiency that is caused by early, albeit reversible, left-ventricular contractile dysfunction.

Materials and Methods

Population

Two populations of patients were studied. First, 29 control patients, who were referred for cardiac catheterization to evaluate an atypical chest pain syndrome and who had a normal physical examination, electrocardiogram, and coronary arteriograms at cardiac catheterization, were evaluated [11]. These patients also had normal left-ventricular pressures, volumes, ejection fractions, and masses at cardiac catheterization. The initial 22 control patients were evaluated to assess left-ventriculoarterial coupling relationships during basal hemodynamic conditions, and the remaining 7 control patients were evaluated to assess the effects of enhanced inotropy on these relationships. Control patients included 23 men and 6 women with an age range of 33 to 71 years.

Second, 41 patients with long-term mitral regurgitation, who were referred for cardiac catheterization to study the hemodynamics severity of their valvular heart disease, were also studied [6,8]. All patients with long-term mitral regurgitation were in normal sinus rhythm and had normal coronary arteriograms at cardiac catheterization. The initial 26 patients with long-term mitral regurgitation were evaluated to assess the left-ventriculoarterial coupling relationship both under basal hemodynamic conditions and during vasodilation with nitroprusside before mitral valve surgery. The subsequent group of 15 'patients with long-term mitral regurgitation were evaluated to assess the ventriculoarterial coupling relationship both before and after successful mitral valve surgery. The patients with mitral regurgitation included 39 men and 2 women with an age range of 32 to 74 years.

Protocol

All patients provided written informed consent for this investigation on forms approved by the Human Studies Committees at the University of Michigan or Veterans Affairs Medical Centers in Ann Arbor, MI, USA. All medications were withheld for 24 h before cardiac catheterization. After a diagnostic right and left heart catheterization was completed, a bipolar pacing catheter was placed in the right-atrial appendage to maintain constant heart rate throughout the protocol. A micromanometer left-ventricular catheter (Millar Instruments, Houston, TX, USA) was positioned to measure high-fidelity aortic and left-ventricular pressures. Red blood cells were then tagged with technetium-99m for radionuclide angiography. Both micromanometer left-ventricular pressures and gated equilibrium radionuclide angiograms were acquired in duplicate during multiple left-ventricular loading conditions produced by steady-state infusions of either methoxamine or nitroprusside. All patients completed the protocol without sequelae.

Hemodynamics

Micromanometer aortic and left-ventricular pressures were recorded for 10–20 cardiac cycles at the beginning, middle, and end of each radionuclide acquisition. Left-ventricular pressure waveforms were averaged, digitized at a varying sampling frequency, and interpolated to correspond with each radionuclide frame throughout the cardiac cycle [12].

Radionuclide Angiography

Gated equilibrium radionuclide angiograms were acquired into 30-ms frames throughout the cardiac cycle for 250 cardiac cycles. During the midportion of each radionuclide acquisition, a 2-ml blood sample was drawn and later counted for 2 min. The time delay from acquiring to counting each blood sample was recorded for radioisotope decay correction. At the completion of the protocol, a distance measurement was obtained for attenuation correction. Frame-by-frame absolute radionuclide left-ventricular volumes were then calculated from background-subtracted, hand-drawn, region-of-interest left-ventricular count data that were standardized for frame duration, cardiac cycles acquired, decay-corrected blood sample counts, and attenuation [13,14].

Data Analysis

Corresponding micromanometer left-ventricular pressures and radionuclide left-ventricular volumes for each loading condition were plotted to obtain multiple pressure–volume loops for each patient [2]. The maximum pressure–volume ratio from each pressure–volume loop was then subjected to linear regression analysis to determine the slope (E_{es}), left-ventricular chamber elastance, a relatively load-independent index of left-ventricular contractility [15,16]. To obtain effective arterial elastance (E_a), the end-systolic pressure (P_{es}) was divided by left-ventricular stroke volume (SV) in the control patients and by forward left-ventricular stroke volume (SV_f) in the patients with long-term mitral regurgitation. End-systolic pressure was defined as the left-ventricular pressure at the maximum pressure–volume ratio during control data acquisition.

To obtain forward left-ventricular stroke volume (SV_f) in the mitral regurgitation patients before mitral valve surgery, total left-ventricular stroke output was divided by regurgitant index to partition it into forward and regurgitant stroke volume components. Left-ventricular regurgitant index was obtained by dividing left-ventricular stroke counts by right-ventricular stroke counts; right-ventricular stroke counts were obtained by a modification of the method of Maddahi and colleagues [17,18]. In the 15 patients who were also studied following successful mitral valve surgery, no clinical evidence for residual mitral regurgitation existed by either physical examination or doppler echocardiography. Total left-ventricular stroke output at 1 year was, therefore, considered to be synonymous with forward left-ventricular stroke volume during the preoperative study in these patients. Coupling of the left ventricle to the arterial system was then expressed as the ratio of E_{es} to E_a.

To assess left-ventricular pump efficiency, left-ventricular stroke work was obtained by calibrated planimetry of the control pressure–volume loop. The result was then multiplied by 0.0136 to convert from milligrams Hg per milliliter (mmHg·ml) to gram-meters (g-m). Left-ventricular pressure–volume area was obtained by calibrated planimetry of the area enclosed by the end-systolic and diastolic curves and the systolic portion of each pressure–volume loop [19–21]. The ratio of external work to the corresponding pressure–volume area was considered reflective of left-ventricular pump efficiency, that is, the efficiency of converting the total energy available to the left ventricle into external work. To fully examine the energetics of the left ventricle in patients with long-term mitral regurgitation, a modification of the theoretical concepts of the ventriculoarterial coupling relationship must be introduced [9,10,19–21]. This theoretical construct characterizes a two-step process whereby the conversion of myocardial oxygen consumption into the pressure–volume area, reflecting contractile element efficiency, is followed by the conversion of the pressure–volume area into external work, reflecting left-ventricular pump efficiency. Then, the relationship between myocardial oxygen consumption and external work represents myocardial efficiency, which can be affected (1) by the efficiency of the contractile machinery or (2) by the conversion of total mechanical energy into external work, which is dependent on the coupling of the left ventricle to the arterial system.

In patients with long-term mitral regurgitation, conversion of the left-ventricular pressure–volume area into external work is reflective of not only forward left-ventricular stroke work but also regurgitant stroke work. Therefore, the left ventricle may perform total left-ventricular stroke work efficiently and adapt appropriately to changes in total load, i.e., total arterial and atrial load; however, this does not provide insight into the efficacy of performing forward left-ventricular stroke work which, in the case of long-term mitral regurgitation, may not be performed in an efficient manner. Moreover, control of forward left-ventricular stroke work would be more consistent with the definition used for left-ventricular pump efficiency [9,10]. Accordingly, to obtain forward left-ventricular stroke work (SW_f) in patients with long-term mitral regurgitation before mitral valve surgery, total left-ventricular stroke work was divided by regurgitant index to eliminate regurgitant stroke work. Thus, left-ventricular pump efficiency for performing forward stroke work (SW_f) was more specifically characterized and compared to the corresponding pressure–volume area.

Statistical Analysis

All data are expressed as a mean ± 1 standard deviation (SD). The data were analyzed using either nonpaired or paired *t*-tests or an analysis of variance where indicated. If a significant *F*-statistic was obtained during an analysis of variance. Dunnett's *t*-tests were used to identify specific differences. A difference was considered present when a probability value of 0.05 or less was obtained.

Results

Baseline Studies in Control Patients

Twenty-two control patients were studied only under baseline left-ventricular contractile conditions, while 7 control patients were studied both under control contractile conditions and enhanced contractile conditions during a steady-state infusion of dobutamine at $5\,\mu g \cdot kg^{-1}\,min^{-1}$ [11]. The average left-ventricular chamber elastance (E_{es}) was 3.51 ± 1.26 mmHg/ml with an average volume–axis intercept (V_0) of 1 ± 23 ml. The effects of steady-state left-ventricular load alterations on left-ventriculoarterial coupling are shown in Table 1. An increase in left-ventricular load with methoxamine produced an increase in end-systolic pressure (P_{es}, $P < .0001$) and left-ventricular end-diastolic volume (EDV, $P < .01$). However, there was no change in left-ventricular stroke volume (SV), which resulted in a significant increase in total arterial elastance (E_a, $P < .01$) and a concomitant decrease in left-ventriculoarterial coupling (E_{es}/E_a, $P < .01$). In contrast, decreasing left-ventricular load with nitroprusside produced a significant decrease in P_{es} ($P < .0001$) and left-ventricular EDV ($P < .001$) without changing left-ventricular SV. Consequently, E_a decreased ($P < .001$) and E_{es}/E_a improved ($P < .001$).

The relationship between left-ventricular pump efficiency (SW/PVA) and left-ventriculoarterial coupling is shown in Fig. 1. Over a range of left-ventricular loads and contractile states, the left ventricle operates over a wide range of left-ventriculoarterial coupling ratios during basal contractile conditions, while there is a reduction in the efficiency of converting the pressure–volume area into stroke work as the left-ventriculoarterial coupling ratio declines below 1.0, consistent with the detrimental effects of increasing load and contractile impairment on mechanical efficiency. In contrast, as the left-ventriculoarterial coupling ratio declines toward 1.0, normalized left-ventricular stroke work increases (Fig. 2).

Left-ventricular stroke work was normalized to account for changes in left-ventricular end-diastolic volume produced by variations in left-ventricular load and contractile state. This was done by establishing the left-ventricular stroke work

TABLE 1. Effects of altered load in hemodynamics and left-ventriculoarterial coupling.

	E_{es} (mmHg/ml)	V_0 (ml)	P_{es} (mmHg)	EDV (ml)	SV (ml)	E_a (mmHg/ml)	E_{es}/E_a	SW_{100} (g-m)	SW/PVA
Control	3.51 ± 1.29	1 ± 23	126 ± 18	99 ± 39	59 ± 20	2.32 ± 0.61	1.62 ± 0.80	76 ± 31	0.65 ± 0.10
Methoxamine			166 ± 23††	112 ± 44**	62 ± 26	3.03 ± 1.21**	1.31 ± 0.68**	93 ± 42**	0.60 ± 0.11**
Nitroprusside			92 ± 16††	86 ± 37†	57 ± 21	1.85 ± 0.70†	2.26 ± 1.51†	58 ± 25†	0.72 ± 0.13**

E_{es}, left-ventricular chamber elastance; V_0, unstressed volume; P_{es}, left-ventricular pressure at end-systole; E_a, effective vascular elastance; EDV, left-ventricular end-diastole volume; SV, left-ventricular stroke volume; $SW/_{100}$, left-ventricular stroke work standardized to a common EDV of 100 ml; SW/PVA, left-ventricular mechanical efficiency.
*, $P < .05$; **, $P < .01$; †, $P < .001$; ††, $P < .0001$ vs. control.

TABLE 2. Effects of altered inotropy on hemodynamics and left-ventriculoarterial coupling.

	E_{es} (mmHg/ml)	V_0 (ml)	P_{es} (mmHg)	V_{125} (ml)	EDV (ml)	SV (ml)	E_a (mmHg/ml)	E_{es}/E_a	SW_{100} (g-m)	SW/PVA
Control	2.71 ± 1.07	−9 ± 14	127 ± 34	43 ± 16	114 ± 41	68 ± 26	2.02 ± 0.70	1.48 ± 0.66	78 ± 18	0.68 ± 0.11
Dobutamine	4.93 ± 3.45	−7 ± 17	158 ± 27*	28 ± 11**	97 ± 27	61 ± 22	2.92 ± 1.23*	2.06 ± 1.65	114 ± 27†	0.69 ± 0.18

*, $P \leq .05$; **, $P < .01$; †, $P < .001$ vs. control.

at a common end-diastolic volume of 100 ml. Then, the normalized left-ventricular stroke work (SW_{100}) was plotted assuming that the maximal left-ventricular stroke work would occur at a left-ventriculoarterial coupling ratio of 1.0 [7,9,22]. Although normalized left-ventricular stroke work increases as the coupling ratio approaches 1.0, the left ventricle under basal contractile conditions operates over a narrow range of SW_{100} values. Thus, in control patients, the left ventricle operates under basal contractile conditions over a broad range of left-ventricular coupling ratios within 10% of maximal pump efficiency and maximal normalized left-ventricular stroke work.

Enhanced Inotropy in Control Patients

As shown in Table 2, a steady-state infusion of dobutamine at 5 $\mu g\,kg^{-1}\,min^{-1}$ increased E_{es} without changing V_0. This steepening of the E_{es} slope was also manifested by a smaller left-ventricular volume at a common end-systolic pressure of 125 mmHg (V_{125}, $P < .01$). Because total arterial elastance increased ($P < .05$), there was only a modest increase in the left-ventriculoarterial coupling ratio. This resulted in an insignificant change in left-ventricular pump efficiency. In contrast, there was a substantial increase in standardized left-ventricular stroke work ($P < .001$; see Fig. 2).

Baseline Studies in Long-Term Mitral Regurgitation

The 26 patients with long-term mitral regurgitation were divided into two groups [6]. Group one had normal left-ventricular contractility as manifest by an E_{es} value of 1.00 mmHg/ml or more. This was defined as the lower limit of normal, as it represented 2 SD below the mean E_{es} value for the control patients. The second group of patients with long-term mitral regurgitation had an E_{es} value of less than 1.00 mmHg/

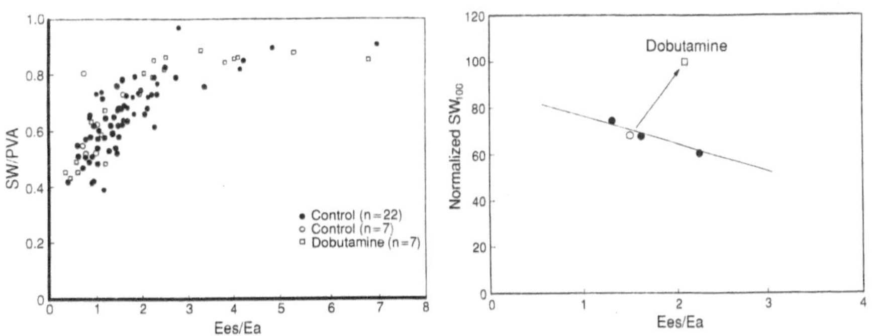

FIG. 1. Relationship between the efficiency of converting the pressure–volume area to stroke work (SW/PVA) compared with the left-ventriculoarterial coupling relationship (E_{es}/E_a). (From [11], with permission)

FIG. 2. Left-ventricular stroke work normalized to a left-ventricular diastolic volume of 100 ml (SW_{100}) plotted for each loading condition and inotropic state relative to their corresponding E_{es}/E_a ratios. It is apparent that the normal left ventricle operates over a broad range of stroke-work values under basal contractile conditions. *Solid circles*, patients who had E_{es}/E_a ratios calculated over a range of loading conditions; *open circles*, patients who had E_{es}/E_a ratios calculated during control hemodynamic state and steady-state dobutamine infusion to alter inotropy. (From [11], with permission)

ml and were considered to have left-ventricular contractile dysfunction. These group II patients with long-term mitral regurgitation were further divided into group IIa and IIb; patients in group IIa had a normal left-ventricular ejection fraction while those in group IIb had an abnormal left-ventricular ejection fraction. As indicated in Fig. 3, there was a progressive deterioration in left-ventricular contractility in the long-term mitral regurgitation patient from group I to groups IIa and IIb in comparison to the control patients.

The patients with long-term mitral regurgitation in all subgroups had P_{es} values, left-ventricular SV_f values, and consequently total arterial elastance values comparable to those of the control patients. However, the left-ventriculoarterial coupling relationship progressively deteriorated in long-term mitral regurgitation patients in comparison to the control patients because of progressive deterioration in left-ventricular contractility.

As expected, total left-ventricular stroke work (SW_t) was significantly increased in all subgroups of patients with long-term mitral regurgitation in comparison to the control patients ($P < .01$ or more for all comparisons; Fig. 4). However, when regurgitant work was taken into account, forward left-ventricular stroke work (SW_f) was similar in all patient groups. Interestingly, in patients with long-term mitral regurgitation and normal left-ventricular contractility or mildly impaired contractility and a normal left-ventricular ejection fraction, total left-ventricular pump efficiency (SW/PVA_t) was preserved. It was impaired only in those patients with long-term mitral regurgitation who had impaired left-ventricular contractility and an abnormal ejection fraction. In contrast, there was a progressive deterioration in left-ventricular pump efficiency for performing forward left-ventricular stroke work (SW/PVA_f) in each subgroup of patients with long-term mitral regurgitation in comparison to the control patients, despite preservation of SW_f.

When left-ventricular pump efficiency for performing either total or forward stoke work was plotted against the left-ventriculoarterial coupling relationship in patients with long-term mitral regurgitation and control patients (Fig. 5), a consistent curvilinear relationship was preserved. Nevertheless, it was apparent that the pump efficiency for performing forward left-ventricular stroke work was much lower and to the

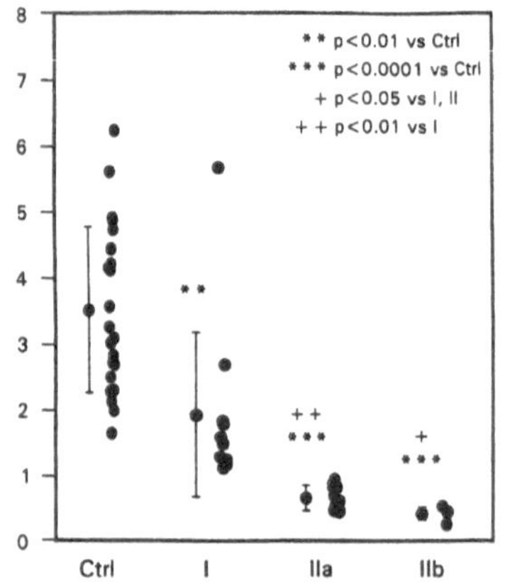

FIG. 3. Individual and mean E_{es} values for control subjects (*Ctrl*) and patients with long-term mitral regurgitation and a normal E_{es} (*I*), those with impaired E_{es} but a normal left-ventricular ejection fraction (*IIa*), and those with impaired E_{es} and ejection fraction (*IIb*). (From [6], with permission)

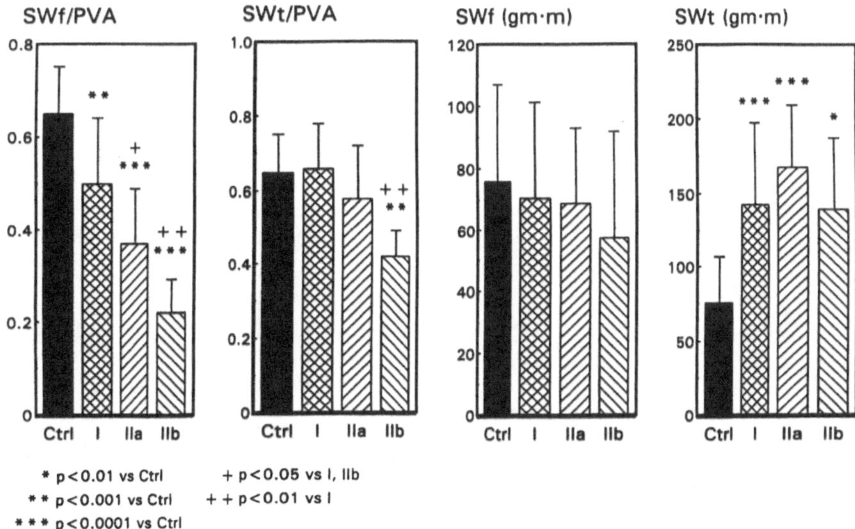

FIG. 4. Average values for forward left-ventricular stroke work (SW_f), total left-ventricular stroke work (SW_t), forward left-ventricular pump efficiency (SW/PVA_f), and total left-ventricular pump efficiency (SW/PVA_t) for control patients and the three subgroups of patients with long-term mitral regurgitation. (From [6], with permission)

left on the downslope of this relationship. Thus, in general, total left-ventricular pump efficiency is well preserved in patients with long-term mitral regurgitation because of the low-impedance left atrium, while the efficiency with which forward left-ventricular stoke work is performed progressively deteriorates in comparison to control patients.

Vasodilator Therapy in Mitral Regurgitation

Acute vasodilator therapy with nitroprusside was used to evaluate the left-ventriculoarterial coupling relationship in patients with long-term mitral regurgitation [11]. A steady-state infusion of nitroprusside decreased P_{es} from 112 ± 19 to 86 ± 12 mmHg ($P < .01$). There was also a decrease in left-ventricular end-diastolic volume from 216 ± 72 to 177 ± 77 ($P < .001$). Because of the decrease in left-ventricular end-diastolic volume, there was a mild but insignificant fall in forward left-ventricular stroke volume (SV_f). Thus, because of the fall in both P_{es} and SV_f, total arterial elastance did not change, nor did left-ventriculoarterial coupling. Further, in the absence of changes in left-ventriculoarterial coupling, left-ventricular pump efficiency for performing forward left-ventricular stroke work also did not change.

Impact of Mitral Valve Surgery in Long-Term Mitral Regurgitation

In 15 patients studied before and 1 year following successful mitral valve surgery, the left-ventriculoarterial coupling relationship demonstrated a significant improvement [8]. Left-ventricular P_{es} increased from 108 ± 17 to 124 ± 19 mmHg ($P < .05$), and forward left-ventricular stroke volume improved from 57 ± 20 to 84 ± 29 ml ($P < .001$).

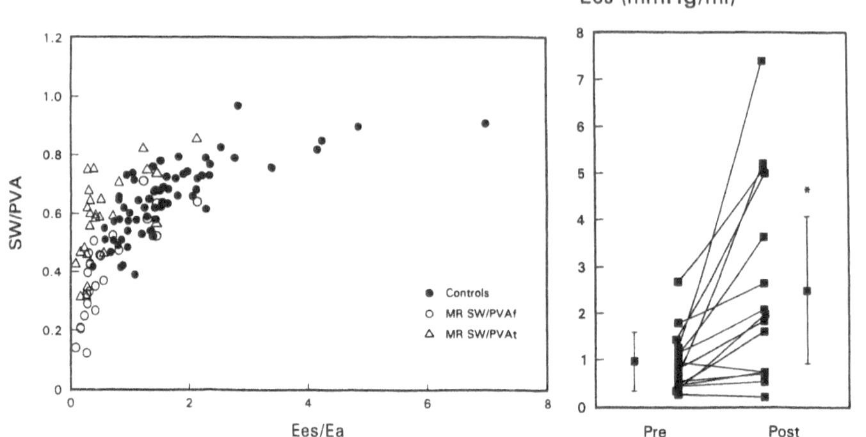

FIG. 5. Individual values for forward left-ventricular pump efficiency (*SW/PVA$_f$*) and total left-ventricular pump efficiency (*SW/PVA$_t$*) plotted against their corresponding left-ventriculoarterial coupling ratios (E_{es}/E_a) for patients with long-term mitral regurgitation during basal hemodynamic conditions and for the control patients over a wide range of loading contions. (From [6], with permission)

FIG. 6. Individual and mean left-ventricular chamber elastance (E_{es}) values in patients with long-term mitral regurgitation both before (*pre*) and after (*post*) successful mitral valve surgery. *, $P < .05$. (From [8], with permission)

Because both P_{es} and forward left-ventricular stroke volume increased, total arterial elastance (E_a) was unchanged. Nevertheless (Fig. 6), because E_{es} increased significantly following mitral valve surgery in these patients with long-term mitral regurgitation (0.95 ± 0.66 to 2.62 ± 2.16 mmHg/ml, $P < .01$), there was a significant improvement in left-ventriculoarterial coupling (0.47 ± 0.39 to 1.81 ± 1.63, $P < .01$). Because left-ventriculoarterial coupling improved, left-ventricular pump efficiency for performing forward left-ventricular stroke work also improved from 0.23 ± 0.10 to 0.55 ± 0.22 ($P < .0001$).

Discussion

The purpose of this investigation was to examine left-ventricular pump function in patients with long-term mitral regurgitation within the context of the left-ventriculoarterial coupling relationship. Understanding left-ventricular pump function in a patient with long-term mitral regurgitation is confounded by the influence of the low-impedance left atrium and uncoupling of the end-systole from end-ejection [3]. Left-ventricular pump function appears normal for a substantial length of time, despite left-ventricular dilation and the recent documentation of occult left-ventricular contractile dysfunction [2,4–8], until late in the hemodynamic course when irreversible myocardial dysfunction is clinically manifest by a decline in left-ventricular ejection fraction [2,23,24].

There are, however, several recent observations which suggest that left-ventricular pump function may not be adequate in patients with long-term mitral regurgitation.

There is evidence that left-ventricular hypertrophy may not be adequate for the degree of left-ventricular dilation manifested by an increase in the left-ventricular radius to thickness ratio. Left-ventricular dilation causes a change in geometry to a more global configuration that may place sarcomeres at a shortening disadvantage [25]. Sophisticated studies in an animal model of mitral regurgitation [4,5,7] and in patients with long-term mitral regurgitation [2,6,8] have suggested that, despite the outward appearance of normal left-ventricular pump function manifest by a normal left-ventricular ejection fraction, left-ventricular contractile function may be deteriorating. The complexities of assessing left-ventricular pump function in patients with long-term mitral regurgitation have previously been impeded by not having a suitable theoretical framework within which to assess left-ventricular pump performance and having a suitable mechanism for separating total left-ventricular stroke work into its forward and regurgitant components.

The left-ventriculoarterial coupling relationship [9,10,19–21] provides a reasonable theoretical framework within which left-ventricular pump function can be evaluated through the interaction of the left ventricle and arterial system. The original concept suggested a two-step process whereby myocardial oxygen consumption is transferred to the pressure–volume area and the total energy in the pressure–volume area is then partially converted into external work, depending on left-ventriculoarterial coupling. Although these two steps hold in most patient populations, a third step must be proposed in patients with long-term mitral regurgitation. Total left-ventricular stroke work must be divided into its components of forward left-ventricular stroke work, which reflects useful work performed for tissue perfusion, and regurgitation work that is wasted into the low-impedance left atrium. Once the ineffective work performed into the low-impedance left atrium is partitioned out of total left-ventricular stroke work, an appreciation of forward left-ventricular pump efficiency in patients with long-term mitral regurgitation can be examined and compared to that in control subjects without evidence of cardiac pathology.

The hypothesis tested in this investigation was that the contribution of the low-impedance left atrium to left-ventricular ejection performance obscures left-ventricular pump dysfunction in patients with long-term mitral regurgitation. The data in this investigation demonstrate that (1) in patients with long-term mitral regurgitation and a normal left-ventricular ejection fraction, left-ventricular contractile dysfunction is present; (2) a decrease in left-ventricular contractility causes left-ventricular coupling to the arterial and atrial systems to decrease toward 1.0, maximizing and maintaining the efficiency of performing total left-ventricular stroke work; (3) in contrast, left-ventricular coupling to the arterial system significantly decreases such that left-ventricular pump efficiency for performing forward left-ventricular stroke work is suboptimal; (4) vasodilation does not significantly affect left-ventricular pump efficiency for performing forward left-ventricular stroke work; and (5) successful mitral valve surgery improves left-ventriculoarterial coupling and pump efficiency as the result of recovery of contractile function.

These data confirm the hypothesis that the low-impedance left atrium maintains left-ventricular coupling to both the arterial and atrial systems, preserving pump efficiency for performing total left-ventricular stroke work. In the face of left-ventricular contractile dysfunction, however, the efficiency of performing forward left-ventricular stroke work is impaired even in the early hemodynamic stages when left-ventricular pump function appears normal, at least by the clinical measurement of ejection fraction. Acute vasodilation does not appear to affect significantly impaired left-ventricular pump efficiency, but definitive surgical correction of mitral regurgitation can return left ventriculoarterial coupling and pump efficiency to normal as the result of recovery of contractile function.

It is important to recognize that studies in an animal model of mitral regurgitation also confirm the presence of occult contractile dysfunction in the face of outward preservation of left-ventricular pump function measured by ejection fraction [4,5,7]. In this animal model of mitral regurgitation, creation of mitral regurgitation causes left-ventricular dilation, an increase in left-ventricular mass, and an increase in left-ventricular ejection fraction. Over time, there is impairment of left-ventricular contractility measured by several sophisticated indices, including both the elastance and systolic stiffness concepts. If, after 3 months of mitral regurgitation, the mitral regurgitation is surgically corrected, left-ventricular ejection fraction decreases and contractile dysfunction recovers. The extent of left-ventricular contractile recovery is dependent on the type of operation [4,5,7]. Similar data in patients with long-term mitral regurgitation indicate that, despite a normal left-ventricular ejection fraction, left-ventricular contractility may indeed begin a slow and occult process of deterioration [2,6]. The reason for the dissociation of left-ventricular contractility and ejection fraction may be related to the uncoupling of end-systole from end-ejection [3]. Although end-ejection is well preserved by the low-impedance left atrium, left-ventricular contractile indices reach their maximum more quickly than in normal patients or patient with long-term aortic regurgitation. As a consequence, conventional measures of left-ventricular pump function are dissociated from more sophisticated contractile indices.

It is this occult impairment of left-ventricular contractility, despite normal total arterial elastance, that leads to impaired pump efficiency for performing forward left-ventricular stroke work. Because of the low-impedance left atrium, total elastance, which includes both the total arterial and atrial elastances, is reduced commensurate with the reduction in left-ventricular chamber elastance such that the coupling ratio approaches 1.0. Consequently, with a E_{es}/E_a ratio approaching 1.0, total left-ventricular stroke work is maximized and performed efficiently in this valve lesion. Nevertheless, if only effective forward work into the arterial system is examined, impaired left-ventricular contractility is coupled only to a normal total arterial elastance, and consequently because of an abnormal left-ventriculoarterial coupling relationship, impaired pump efficiency results.

Although the mechanism for left-ventricular contractile dysfunction in patients with long-term mitral regurgitation is unclear, several important observations have lead to a unifying hypothesis. In an animal model of mitral regurgitation [26], it has been demonstrated that, following the creation of mitral regurgitation, left-ventricular ejection fraction improves, left-ventricular contractility is impaired, and plasma norepinephrine levels increase. If animals with mitral regurgitation are then randomized to placebo or β-adrenergic blockade, those given β-adrenergic blocking agents demonstrate an improvement in left-ventricular contractility. These data suggest that mitral regurgitation may activate neurohumoral systems, particularly the sympathetic and renin–angiotension systems and, through an impairment of β-adrenergic receptor coupling to adenylyl cyclase, lead to altered left-ventricular contractility.

Preliminary observations in our laboratory suggest that this hypothesis may indeed be true in patients with long-term mitral regurgitation. We have demonstrated a relationship between impaired isoproterenol-stimulated adenylyl cyclase production in endomyocardial biopsy specimens from patients with long-term mitral regurgita-

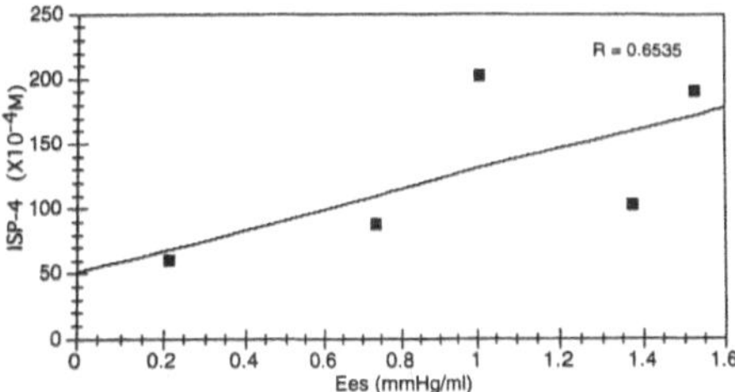

FIG. 7. Relationship between isoproterenol-stimulated adenylyl cyclase production (ISP × 10^{-4} *M*) in endomyocardial biopsy specimens from patients with long-term mitral regurgitation and left-ventricular contractility (E_{es})

tion and left-ventricular contractility (E_{es}) (Fig. 7). Similar relationships have been observed between plasma norepinephrine levels ($r = .91$) and the extent of mitral regurgitation measured as regurgitant volume ($r = .75$) and E_{es}. Furthermore, an observation by Mann and colleagues [27] suggests that catecholamine-mediated myocyte toxicity may also be β-adrenergic receptor-mediated through an increase in cytosolic calcium concentration, because this phenomenon can be ameliorated by β-adrenergic blockade. Thus, early, reversible left-ventricular contractile dysfunction may represent a manifestation of primary neurohumoral-mediated myocardial dysfunction through uncoupling of β-adrenergic receptors from their secondary messenger system, while long-term activation may lead to cardiocyte toxicity and irreversible myocardial dysfunction in patients with long-term mitral regurgitation. The mechanism by which this occurs is under intense investigation in patients with long-term mitral regurgitation and control patients in our laboratory.

Conclusions

These data suggest that left-ventricular contractile dysfunction can occur early in patients with long-term mitral regurgitation when left-ventricular ejection fraction is well maintained. As a consequence, left-ventriculoarterial coupling is impaired and the efficiency for performing forward left-ventricular stroke work is also impaired. In contrast, because of the low-impedance left atrium, left-ventricular ejection performance is maintained and pump efficiency for performing total left-ventricular stroke work is also maintained. Consequently, because patients with long-term mitral regurgitation, a normal left-ventricular ejection fraction, and left-ventricular contractile dysfunction have a predictable response to mitral valve surgery [2] and mitral valve surgery can reverse contractile dysfunction in many of these patients [8], mitral valve surgery should be considered earlier in patients with long-term mitral regurgitation. Moreover, one possible mechanism for this early left-ventricular contractile dysfunction may be activation of neurohumoral systems that eventually impair β-adrenergic receptor system modulation of contractility.

References

1. Wisenbaugh T, Spann JF, Carabello BA (1984) Differences in myocardial performance and load between patients with similar amounts of chronic aortic versus chronic mitral regurgitation. J Am Coll Cardiol 3:916–923

2. Starling MR, Kirsh MM, Montgomery DG, Gross MD (1993) Impaired left ventricular contractile function in patients with long-term mitral regurgitation and normal ejection fraction. J Am Coll Cardiol 22:239–250
3. Brickner ME, Starling MR (1990) Dissociation of end-systole from end-ejection in patients with long-term mitral regurgitation. Circulation 81:1277–1286
4. Nakano K, Swindle MM, Spinale F, Ishihara K, Kanazawa S, Smith A, Biederman RWW, Clamp L, Hamada Y, Zile MR, Carabello BA (1991) Depressed contractile function due to canine mitral regurgitation improves after correction of the volume overload. J Clin Invest 87:2077–2086
5. Urabe Y, Mann DL, Kent RL, Nakano N, Tomanek RJ, Carabello BA, Cooper G (1992) Cellular and ventricular contractile dysfunction in experimental canine mitral regurgitation. Circ Res 70:131–147
6. Starling MR (1994) Left ventricular pump efficiency in long-term mitral regurgitation assessed by means of left ventricular-arterial coupling relations. Am Heart J 127:1324–1335
7. Ishihara K, Zile MR, Kanazawa S, Tsutsui H, Urabe Y, DeFreyte G, Carabello BA (1992) Left ventricular mechanics and myocyte function after correction of experimental chronic mitral regurgitation by combined mitral valve replacement and preservation of the native mitral valve apparatus. Circulation 86:II-16–II-25
8. Starling MR (1995) Effects of valve surgery on left ventricular contractile function in patients with long-term mitral regurgitation. Circulation 92:811–818
9. Sunagawa K, Maughan WL, Burkhoff D, Sagawa K (1983) Left ventricular interaction with arterial load studied in isolated canine ventricle. Am J Physiol 245:H773–H780
10. Sunagawa K, Maughn WL, Sagawa K (1985) Optimal arterial resistance for the maximal stroke work studied in isolated canine left ventricle. Circ Res 56:586–595
11. Starling MR (1993) Left ventricular-arterial coupling relations in the normal human heart. Am Heart J 125:1659–1666
12. Starling MR, Montgomery DG, Mancini GBJ, Walsh RA (1987) Load independence of the rate of isovolumic relaxation in man. Circulation 76:1274–1281
13. Starling MR, Dell'Italia LJ, Walsh RA, Little WC, Benedetto AR, Nusynowitz ML (1984) Accurate estimates of absolute left ventricular volumes from equilibrium radionuclide angiographic count data using a simple geometric attenuation correction. J Am Coll Cardiol 3:789–798
14. Starling MR, Dell'Italia LJ, Nusynowitz ML, Walsh RA, Little WC, Benedetto AR (1984) Estimates of left ventricular volumes by equilibrium radionuclide angiography: importance of attenuation correction. J Nucl Med 25:14–20
15. Suga H, Sagawa K (1974) Instantaneous pressure-volume relationships and their ratio in the excised, supported canine left ventricle. Circ Res 35:117–126
16. Sagawa K (1978) The ventricular pressure-volume diagram revisited. Circ Res 43:677–687
17. Maddahi J, Berman DS, Matsuoka DT, Waxman Ad, Stankus KE, Forrester JS, Swan HJC (1979) A new technique for assessing right ventricular ejection fraction using rapid multiple-gated equilibrium cardiac blood pool scintigraphy. Circulation 60:581–589
18. Sorensen SG, O'Rourke RA, Chaudhuri TK (1980) Noninvasive quantitation of valvular regurgitation by gated equilibrium radionuclide angiography. Circulation 62:1089–1098
19. Suga H, Hayashi T, Shirahata M (1981) Ventricular systolic pressure-volume area as predictor of cardiac oxygen consumption. Am J Physiol 240:H30–H44
20. Suga H, Hayashi T, Shirahata M, Suehire S, Hisano R (1981) Regression of cardiac oxygen consumption on ventricular pressure-volume area in dog. Am J Physiol 240:H320–H325
21. Suga H (1979) Total mechanical energy of a ventricle model and cardiac oxygen consumption. Am J Physiol 236:H498–H505
22. Elzinga G, Westerhof N (1991) Matching between ventricle and arterial load. Circ Res 68:1495–1500

23. Phillips HR, Levine FH, Carter JE, Boucher CA, Osbakken MD, Okada RD, Akins CW, Dagett WM, Buckley MJ, Pohost GM (1981) Mitral valve replacement for isolated mitral regurgitation: analysis of clinical course and late postoperative left ventricular ejection fraction. Am J Cardiol 48:647–654
24. Schuler G, Peterson K, Johnson A, Francis G, Dennis G, Utley J, Daily PO, Ashburn W, Ross J Jr (1979) Temporal response of left ventricular performance to mitral valve surgery. Circulation 59:1218–1231
25. Gould KL, Lipscomb K, Hamilton GW, Kennedy JW (1974) Relation of left ventricular shape, function and wall stress in man. Am J Cardiol 34:627–634
26. Tsutsui H, Spinale FG, Nagatsu M, Schmid PG, Ishihara K, DeFreyte G, Cooper G, Carabello BA (1994) Effects of chronic beta-adrenergic blockade on the left ventricular and cardiocyte abnormalities of chronic canine mitral regurgitation. J Clin Invest 93:2639–2648
27. Mann DL, Kent RL, Parsons B, Cooper G (1991) Adrenergic effects on the biology of the adult mammalian cardiocyte. Circulation 85:790–804

Aging Changes in the Systemic Circulation and Ventriculovascular Coupling

RAYMOND P. KELLY

Summary. With normal aging in humans, loss of distensibility of the large arteries results from loss of the orderly lamellar structure of the arterial media. This loss results in stiffening of vessels, with increased pulse wave velocity and wave reflection. Characteristic changes occur in pulse contour at central and peripheral sites. Changes are also seen in the pattern of arterial impedance. Analysis of pulse wave contour and impedance patterns shows an age-related disturbance in ventriculovascular coupling. Direct studies of left-ventricular function have confirmed such changes with arterial stiffening. These age-related changes in ventriculovascular coupling have implications for effective treatment of hypertension and cardiac failure.

Key words. Aging—Arterial distensibility—Pulse contour analysis—Wave reflection—Impedance

Introduction

The study of "aging" in the past two decades has increased in intensity and relevance because of demographic changes in industrialized societies with an increasing proportion of persons in the older decades of life who require medical care, many as a result of cardiovascular diseases. There is a complex interaction between age-related physiological changes in cardiovascular function and pathological states found in the older population. For example, age-related changes in arterial structure and function variably lead to a rise in blood pressure. These changes interact with other pathological mechanisms that may further increase blood pressure and also interact with other risk factors such as hyperlipidemia to lead to atherosclerosis and target organ damage. The latter are specific disease processes and not part of the normal aging process.

This chapter briefly considers aging studies in general, reviews the arterial and cardiac structural changes of aging, and considers in more detail the resultant functional changes in vascular load and ventriculovascular interaction.

Aging Studies

Weisfeldt [1,2], has highlighted four important factors that need to be considered in

St. Vincent's Clinic, Darlinghurst, NSW, Australia 2010.

studies on aging. These are (1) the need to separate age effects from age-associated disease processes; (2) the study design in assessing age effects; (3) the need for a clear definition of the age range under study in any assessment of "aging"; and (4) the need to assess the physiological relevance of any age-associated change.

The first concern in all aging studies is differentiating between normal aging and changes associated with specific pathological processes that may increase in frequency and severity with age. This may be done by comparative interspecies studies or longitudinal studies of a large population. A clear definition of the portion of the lifespan under study in any investigation is necessary because age-associated changes are often found to greater degree in the latter part of the life span than in the earlier portion. Finally, while there is an observed decline in complex and multisystem integrated functions with age (e.g., maximum exercise ability), one needs to identify which aspects of individual organ structure and function change and which show no change. An observed change in one parameter may not be a physiologically important change to organ function.

Cardiovascular Structural Changes with Aging

Before discussing functional changes of the arterial system and ventriculoarterial coupling associated with aging, structural changes in the heart and large arteries are briefly reviewed. Much has been published on the effects of age on the aorta and large arteries [3,4]. The tunica media forms the largest part of the arterial wall, is the principal determinant of the vessel's mechanical properties, and undergoes important age-related changes. In the "normal" (young) artery, the fibrous elements (collagen and elastin) are arranged in tight helical layers. Wolinsky and Glagov [5] described the orderly lamellar units of the media in which, at physiological distending pressures, elastin and collagen fibers and smooth muscle cells are precisely oriented and form well-defined layers.

With aging, characteristic histological changes in arteries occur independently of changes attributable to atherosclerosis. These include irregularity in size and shape of endothelial cells, a thickened subendothelial layer, and an increase in collagenous fibers and in ground substance, often with calcium deposition in degenerate elastic material. The media experiences loss of the orderly arrangement of elastic fibers and lamellae with thinning, splitting, fraying, and fragmentation. Human aortas show age-related increases in caliber, wall thickness, and stiffness, and a widening of interlamellar distance together with elastin fiber degeneration [6]. Smaller vessels also undergo degenerative changes with age, with hyaline degeneration in arteriolar walls and a rarefaction in density of smaller vessels having been reported [7]. Detailed consideration of all structural changes in systemic vessels is beyond the scope of this chapter, and the reader is referred to other souces for further information [1–7].

With age, there are concomitant changes in cardiac structure. The concept that the heart undergoes atrophy in advanced age as the result of a diminished demand for performance was advocated by some for several decades [8,9]. Data accumulated in many studies now indicate that the human heart hypertrophies with advancing age. Linzbach and Akuamoa-Boateng [10] reviewed autopsy data spanning 91 years in 7112 human hearts (both normal and diseased), and found that between 30 and 90 years of age the heart increases in mass on an average of 1–1.5 g per year. Several echocardiographic studies [11,12] have shown an age-related increase in left-ventricular wall thickness and mass. Changes occur in both muscle cells and interstitial connective tissue that may be age dependent. Cardiac muscle cells show basophilic degeneration and an increase in myocardial cell diameter [13] together with an age-related increase in fibrous tissue and amyloid deposition. The endocardium shows a nonspecific thickening with age, possibly in response to fluctuating tension applied to the heart wall.

Functional Effects of Arterial Aging in Man

The microscopic changes in the arterial wall associated with observable dilation, lengthening, and stiffening of arteries result in important functional effects. The most obvious functional effect is elevation of systolic blood pressure. The increase in systolic blood pressure with age has been documented since early in this century [14]. Findings of the Framingham study have clearly shown the importance of systolic arterial pressure as a risk factor for stroke [15], for coronary events [16], and for cardiac failure [17]; more recently, studies on the efficacy of treatment of isolated systolic hypertension [18–20] have shown the benefits of treatment that reduces systolic pressure. Consideration of peak systolic pressure alone, however, underestimates the real changes in arterial stiffness and increase in cardiac load that occur with age.

The effects of arterial stiffening may be more completely assessed by considering not only changes in systolic or diastolic pressure but also the contour of the pressure wave as it fluctuates between these two extremes of pressure. Furthermore, this analysis in the time domain is extended and further explained by analysis in the "frequency domain." Analysis in the time domain is the most familiar approach, and describes the pressure wave shape and absolute pressure at various times in the cardiac cycle. In youth, "ideal" arterial function maintains minimal pulsation about the mean in the ascending aorta, with a low systolic pressure. The pressure wave in the ascending aorta has a low amplitude, a prominent incisura, and a diastolic secondary wave because of wave reflection (Fig. 1). In children and young adults, mean systolic pressure is only 5–10 mmHg greater than mean pressure throughout the cycle; mean diastolic pressure is only 5–10 mmHg less than mean pressure. With aging, there are changes in the degree of pulsation about the mean, producing not only a rise in systolic pressure level but also other changes in the shape of the pressure wave as are detailed later.

The second form of analysis is to break physiological waveforms down into their component harmonics, thereby obtaining amplitudes at different frequencies. This method gives rise to an impedance plot of harmonic amplitude versus frequency [21,22] (see Fig. 1). This approach measures total vascular load by including both mean and pulsatile components of the load. For many years, the function of the systemic vasculature in terms of cardiac load and the interaction between the heart and the vascular system were considered simplistically in terms of mean resistance at the arteriolar level. However, peripheral arteriolar resistance is only the steady component of the load borne by the left ventricle. The pulsatile component of arterial load arises as a consequence of the intermittent nature of cardiac contraction. The pulsatile flow so generated is cushioned by the arterial bed upstream from the arterioles so that constant flow is available to perfuse distal vascular beds. Large artery function has little effect on peripheral vascular resistance, contributing little to total load in youth, but becomes a significant component of vascular load with age. The total opposition

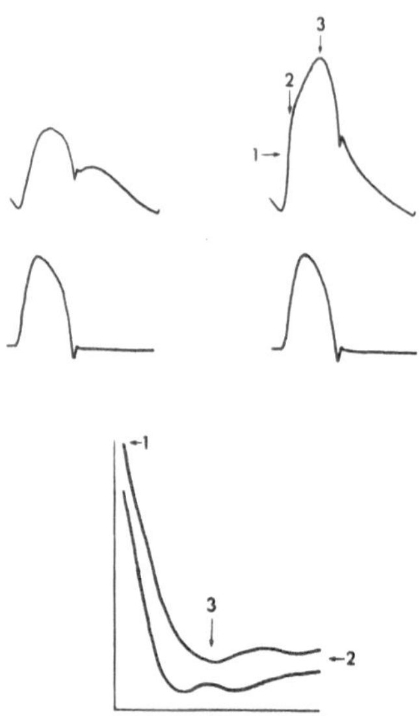

FIG. 1. Diagrammatic illustration of aging (or hypertensive) effects on the ascending aorta pressure wave with the same ascending aortic flow wave (*top*) and on modulus of ascending aortic impedance (*bottom*). Increased peripheral resistance explains increased mean pressure (*top*) and increased impedance modulus at zero frequency (*bottom*). Increased proximal aortic stiffness explains more rapid rate of rise of pressure to a high anacrotic "shoulder" (*1, top*) and increased characteristic impedance (*2, bottom*), while the earlier return of wave reflection explains the late systolic peak of pressure (*2–3, top*) and shift of impedance curves to the right (*bottom*). (From [21], with permission.) The pressure rise from (*2*) to (*3*) divided by the pulse pressure is the basis of the calculated "augmentation index". (From [33], with permission) This pulse contour with a late rise in systolic pressure from point (*2*) to point (*3*) is an example of a Murgo type "A" waveform. (From [35], with permission)

to flow from the heart includes the effects of properties of the large arteries and also wave reflections in the vascular system as well as peripheral resistance.

"Aortic impedance" describes all these components of load on the left ventricle [21–24]. "Impedance" is the total opposition to flow including oscillatory flow, while "resistance" is the opposition to nonoscillatory or steady motion. In the arterial system, the total opposition to blood flow includes both frequency-dependent dynamic (pulsatile) and frequency-independent static (nonpulsatile) components. A plot of aortic impedance (see Fig. 1) is a plot of the ratio of pressure–flow harmonics at different frequencies (analagous to resistance = pressure/flow); a plot of impedance gives information about the peripheral resistance (impedance at 0 frequency, i.e., the steady component), the stiffness of the arterial system excluding the effect of wave reflection (known as "characteristic impedance"), and the effect of wave reflection (degree of oscillation about the value of characteristic impedance and the value of the first impedance minimum). A plot of ascending aortic impedance in a young individual shows that impedance falls from a high value at 0 frequency to low values at frequencies less than 2–3 Hz (Fig. 1). These frequencies correspond to the first three harmonics (those with the most energy) of the left-ventricular flow wave. Thus, in the ideal state, the greatest flow from the heart occurs when the vascular impedance is least.

Consideration of aortic impedance shows clearly how arterial stiffening adversely affects cardiac load in adults. Increased stiffness produces an increase in characteristic impedance [23,25,26]. It also causes an increase in pulse wave velocity [26,27] and thus earlier wave reflection, which is manifest on an impedance plot as an increase in the frequency of the first minimum of the impedance. As a result of these two changes, the impedance curve is shifted. Through increase in characteristic impedance, all values of impedance amplitude are raised and, more importantly, early wave reflection shifts the whole impedance curve to the right. This markedly increases impedance amplitude at 1–2 Hz (i.e., at 60–120 cps, which corresponds to heart rate frequency). The correspondence in youth of aortic impedance minimum with heart rate frequency (and so with the first or largest harmonic component of the ventricular ejection wave) is "ideal" ventricular-vascular coupling [28–31], that is, the greatest pulsatility of flow from the heart occurs at frequencies where the opposition to pulsatile flow is least. The arterial system and the heart are "tuned" to each other. With human aging, the favorable relationship between heartrate and aortic impedance is diminished. The major harmonic component of ventricular ejection now faces a much higher arterial impedance, as seen from the shift in the impedance curve. There develops with age a mismatch between the heart and the arterial system.

These impedance changes give much insight into the changes in arterial function and ventriculovascular coupling with age, but they are complex and not readily assimilated into the clinical domain. Much information can also be gained from consideration of arterial pressure waveforms in the time domain and their associated age-related changes. With aging, there is virtually no alteration in the pattern of ejection from the left ventricle of the heart, just a slight reduction in peak flow. There is however a marked change in the contour of the ascending aortic pressure wave [32,33,34]. Such changes in the arterial pressure pulse were first described in the nineteenth century [35] and were likened to those seen with hypertension, being composed of the appearance of a late systolic peak and the disappearance of the diastolic wave (Fig. 1).

This change in the systolic part of the aortic pressure wave is accompanied by a similar change in left-ventricular systolic pressure. These changes are attributable to arterial stiffening, with earlier return of wave reflection from peripheral sites as a consequence of increased aortic pulse wave velocity as delineated from impedance studies. Ventricular-vascular coupling deviates from its optimal state in youth as (1) arteriolar resistance rises (increasing mean pressure) and (2) arterial properties change, so that mean aortic systolic pressure is increased and mean diastolic pressure is reduced in relation to mean pressure throughout the cardiac cycle. In youth, the pulsatile component of left-ventricular load is much less than the resistive component. This low pulsatile load is the result of the combined effects of high distensibility of the proximal aorta, the arterial branching pattern, peripheral wave reflection, and spatial localization of reflecting sites such that timing of wave reflection is optimized in relation to ventricular ejection [22].

The effects of aging on the pressure and flow wave contours in the ascending aorta have been elegantly detailed by Murgo and colleagues [34] in 18 human subjects without cardiac or vascular disease, using micromanometer multisensor catheters. Murgo et al. described three different patterns of ascending aortic pressure. In the first, designated "A", the pressure wave rose from an early systolic shoulder to a later systolic peak (see Fig. 1). In the second, designated "C", the first systolic peak was not followed by a later pressure rise, and a diastolic pressure wave was apparent. The third pattern, designated "B", was intermediate between types A and C. Murgo noted that the first pattern ("A") was most commonly seen in older subjects and the third pattern ("C") in younger subjects.

These patterns compare with the brachiocephalic flow patterns designated I, II, and III in an earlier study by Mills et al. [36], and have been shown to correspond: type A pressure to type I flow, type B pressure to II flow, and type C pressure to III flow. These brachiocephalic flow patterns and aortic pressure wave patterns may be explained on the basis of timing of wave reflection from the lower body: the type A pressure wave and type I flow wave wave reflection return earlier in the elderly, and other waveforms show later timed reflection in younger subjects [7,37,38]. Further evaluation of human aging changes was performed by Nichols et al. [39], who combined data from different studies to describe changes in ascending aortic pressure wave contour, ascending aortic impedance, and vascular-ventricular interaction with age. A progression of change in ascending aortic pressure wave contour from the C pattern of Murgo, through B to A with advancing years with little change in the ascending aortic flow pattern, was described.

Noninvasive recording of the arterial pulse contour by applanation tonometry has also shown a gradual and progressive change with age. Changes in contour of the radial, carotid, and femoral pressure pulse in 1005 normal subjects aged 2–91 years have been described by Kelly et al. [33].

The changes in carotid, radial, and femoral pressure waveforms with age are summarized in Fig. 2. The femoral pressure waveform shows a smooth contour with a single systolic and single diastolic wave. The diastolic wave becomes less prominent with age, and by the eighth decade, pressure falls smoothly and almost exponentially during diastole. The amplitude of the systolic peak rises by 50% from the first to the eighth decade. The radial pressure wave contour in children shows two prominent fluctuations after the early systolic peak, the first in late systole and the second in diastole. With aging, the diastolic wave become less prominent and the late systolic peak more prominent. The diastolic wave appears to move earlier into systole with age. By the eighth decade, there is a late systolic shoulder just after the systolic peak, and an almost exponential fall in pressure thereafter during diastole. The amplitude of the radial pressure wave increases by 67.5% between the first and eighth decade of life, and this is caused entirely by increase in amplitude of the first systolic wave.

The changes in the carotid pressure waveforms contrast with those in the femoral and radial arteries (Fig. 2). A late systolic pressure shoulder is apparent in the first decade, and this becomes progressively higher with increasing age, equaling the first wave in amplitude by the fourth decade and increasing progressively thereafter. A diastolic wave is apparent in the first four decades, but this decreases progressively with age thereafter. By the eighth decade, the wave shows an almost exponential decay during diastole. The amplitude of the carotid pressure wave increases to a greater degree with aging than that of either the femoral or radial wave. The amplitude of the initial wave increases by 53% from the first to the eighth decade, which is intermediate between the femoral (50%) and radial (67.5%). It is followed however by a further increase of 38% (to a total of 93%) in consequence of the late systolic augmentation.

These carotid pressure wave patterns are similar to those reported in the ascending aorta by Murgo et al. [34] and by Nichols et al. [39] at different ages. Indeed, the

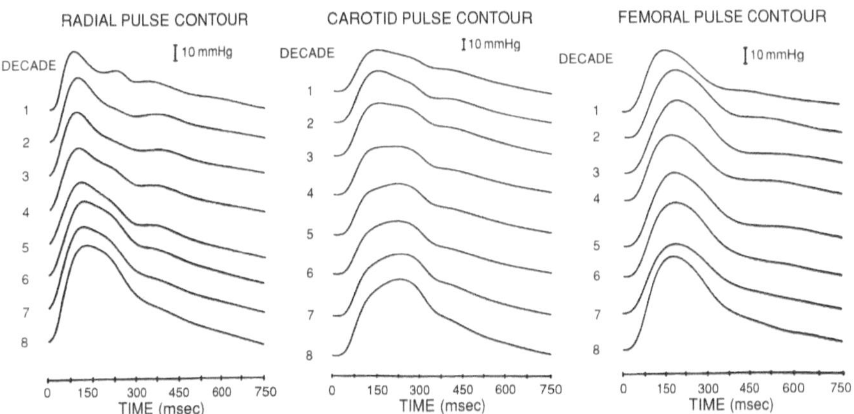

FIG. 2. Radial (*left*), carotid (*center*), and femoral (*right*) waves recorded by applanation tonometry. Each waveform is an averaged wave derived from subjects in each decade of life. (From [33], with permission)

carotid pulse contour has been shown in several studies to be a good indicator of changes in the ascending aortic pressure waves at different ages [40,41]. Carotid pressure waves in the first two decades have contours corresponding to the type C aortic pattern, those in the third and fourth decades to the type B pattern, and those in the fifth to eighth decade to the type C pattern.

The amplitude of the secondary late systolic wave following the shoulder on the upstroke of the carotid or aortic wave can be compared to pulse pressure as an index of pulsatile arterial load. This "augmentation index" [33,35,42,43] (see Fig. 1) may be made from visual inspection of the shoulder of the pressure waves or from timing of peak flow in the ascending aorta or carotid artery. The changes in augmentation index of the carotid pressure wave with age are shown in Fig. 3. Augmentation is less than zero until the fourth decade of life, whereafter it increases progressively. By the eighth decade, augmentation averages 24%, indicating that augmentation of the pressure pulse from wave reflection accounts for 24% of carotid pulse pressure. Although similar, the carotid pressure augmentation appears to underestimate augmentation in the ascending aorta. The combined aortic wave data of Murgo et al. [34] and Takazawa [42] (Fig. 4) show similar results to the carotid data, with augmentation becoming apparent by age 30, but with the degree of augmentation thereafter being almost twice that in the carotid artery. Data from simultaneous recordings of indirect carotid pressure waves and direct ascending aortic pressure waves support this [40,41], indicating that carotid augmentation in mature human adults is a good index but tends to underestimate that in the ascending aorta.

The effects of arterial degeneration on ventriculovascular coupling in humans have been discussed here with respect to functional interaction between ventricular flow and aortic impedance. It should be also be noted that the late augmentation in carotid and aortic pressure waveforms occurs from about the third decade of life, the same time as the progressive age-related increase in left-ventricular mass occurs [10–12].

These detailed studies of the effects of age on pulse contour have not only shown a difference in central (aortic and carotid) pulse contour compared to the peripheral

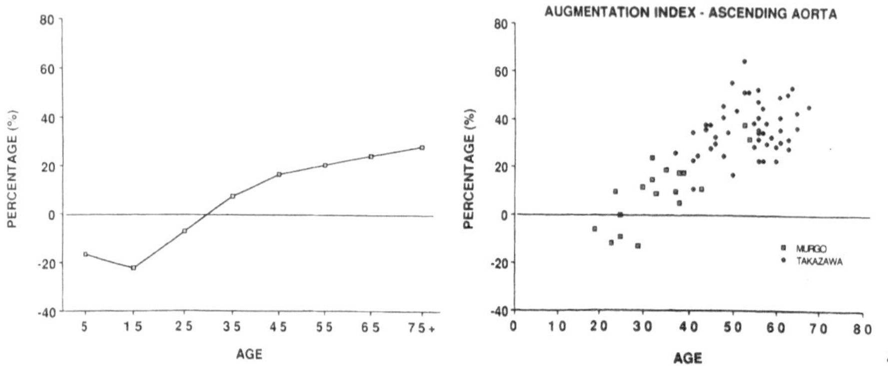

FIG. 3. Change of augmentation index (amplitude of the shoulder to peak divided by pulse pressure) with age in the carotid artery. (From [33], with permission)

FIG. 4. Change of augmentation index (amplitude of the shoulder to peak divided by pulse pressure) with age in the ascending aorta. Cumulative data from Murgo [34] (*squares*) and Takazawa [43] (diamonds)

pulses but also show an age-related decrease in amplification of the pulse between central and peripheral arteries [32,33,44]. In adolescents and young adults the brachial, radial, and femoral arterial pulse pressures are some 50% higher than in the ascending aorta, whereas in elderly subjects they are almost identical. Hence, it is clear that aortic systolic pressure increases more with age, and diastolic pressure less than recorded values in the brachial artery. Such aging changes are usually overlooked when assessing the effects of age on systolic and diastolic blood pressure. Furthermore, with pharmacological challenge such as occurs with the arterial dilator effects of nitroglycerin, it is possible to induce changes in the central pulse contour and systolic pressure with no recordable change in brachial pressure [45]. This remarkable dissociation of drug effects on central and peripheral waveforms may explain many disparities in effects of afterload-lowering drugs in treatment of hypertension and heart failure in elderly subjects.

Of note, while these aging changes in the arterial pressure wave contour are very obvious in humans, they are less apparent in experimental animals. Yin et al. [46,47] showed little or no change in arterial hemodynamics or pressure wave contour in canine studies over a span of 10 years. The explanation appears to be related to concepts of relative and absolute age. The period of 10 years accounts for the full life span of a dog, but this is a small fraction of the human life span. In man, age-related changes are attributable to increased pulse wave velocity caused by arterial stiffening as a result of arterial degeneration consequent on the fatiguing effects of cyclic stress over many decades. In the life span of experimental animals, there is insufficient time for the type of degeneration that is observed almost invariably in humans by middle age [37,47]. This is a cogent example of the need for clear understanding of the need to clearly understand the portion of the life span and absolute time periods under scrutiny in any aging study. Young adults appear to have similar pressure wave patterns in the ascending aorta and peripheral arteries to those observed in dogs [33,34,39]; older adult humans show effects of arterial degeneration not seen in many animal studies.

Arterial Stiffening and Effects on Left-Ventricular Function

With a distensible vasculature as found in young subjects, appropriately timed wave reflection maintains aortic pressure during early diastole without boosting pressure during systole. Thus wave reflection provides a type of "diastolic augmentation" as does intraaortic balloon counterpulsation [37,48]. On the other hand, arterial stiffening and early wave reflection has the effect of an incorrectly timed intraaortic balloon counterpulsation, increasing systolic pressure and decreasing pressure during diastole [28].

The ill effects of arterial stiffening and early wave reflection can readily be understood by comparing pressure waves in the ascending aorta of young and old human subjects at the same mean arterial pressure. There is increase in peak systolic pressure, in mean systolic pressure, and in pressure at end systole, all in association with decrease in pressure throughout diastole. Increased systolic pressure increases myocardial oxygen demands [49,50], while reduced diastolic pressure tends to decrease myocardial blood supply [37,47,48], both predisposing to myocardial ischemia [47,51,52]. In addition, the rise in late systolic pressure predisposes to impaired diastolic relaxation, which may intensify ischemia [53,54], suppresses ventricular ejection, and generates myocardial hypertrophy [54–59].

The effects of aortic stiffening on left-ventricular systolic function have been studied by modeling the arterial pressure and flow wave contours seen in adult human subjects in a canine preparation that assesses ventricular function by measurement of pressure–volume loops [60]. Bypassing the normally compliant canine aorta with a stiff conduit alters the contour of both the pressure–volume loops and the arterial pressure and flow waves. In human studies, pressure–volume loops are characterized by a progressive rise in pressure throughout the period of systolic ejection. In the canine model, the increased pulsatile load that produces the late-peaking loops lowered cardiac efficiency (defined as the ratio of stroke work to oxygen consumption) on average by 16%. More strikingly, for the heart to maintain the same systemic blood flow (stroke volume), myocardial oxygen consumption increases by between 30% and 50% when arterial load is associated with late-peaking loops.

These findings are consistent with previous animal studies [23,61,62] which showed that stiffening of the aorta increases pulsatile energy losses in the circulation and reduces load-dependent indices of contractile state. Stiffening of the aorta in the canine model has also been shown to significantly prolong the time constant of diastolic relaxation [63]. Of note, all these studies have assessed altered loading conditions on normal hearts. They are therefore likely to underestimate the effect of systolic wave reflections and pulsatile load when there is concomitant ventricular hypertrophy or coronary ischaemia.

The typical pressure–volume loop contour found in adult humans with arterial stiffening is not found in young subjects or in animal or computer modeling studies in which effects of pulsatile load are ignored or incompletely assessed. As shown in canine studies [60,63], the late peaking pressure–volume loops are caused by the effects of arterial stiffening and wave reflections and are associated with diminished cardiac efficiency. Analysis of pressure–volume loops emphasizes that pulsatile load causes left-ventricular pressure to peak at end-systole when ventricular volume is minimal, thus increasing ventricular wall stress and oxygen consumption. The late systolic peak also means that the loop reaches its systolic functional limit (the end-systolic pressure–volume relation line) sooner than if it were to continue with a flat contour throughout systolic ejection, thus potentially limiting stroke volume [48].

Conclusion

In humans, aging is associated with changes in the arterial wall structure that cause stiffening and loss of distensibility. These cause a rise in systolic pressure and an increased pulsatile load on the heart. A rise in brachial systolic blood pressure level is the most obvious clinical measure of age-related decrease in arterial distensibility, but much more information may be obtained from consideration of the contour of the central and peripheral pulses and from vascular impedance studies. Concomitant with these arterial changes is an increase in left-ventricular mass and reduced efficiency of myocardial function with increased myocardial oxygen demands. Amelerioration of ventriculovascular interaction based on knowledge of underlying age-related changes provides a basis for improved treatment approaches to cardiovascular disease.

References

1. Weisfeldt ML (1980) Research on aging. In: Weisfeldt ML (ed) The aging heart: Its function and response to stress. Raven, New York, pp 1–14
2. Weisfeldt ML (1980) Aging of the cardiovascular system. N Engl J Med 303:1172–1173
3. Milch RA (1965) Matrix properties of the aging arterial wall. Monogr Surg Sci 2:261–341
4. Yin FCP (1980) The aging vasculature and its effects on the heart. In: Weisfeldt ML (ed) The aging heart: Its function and response to stress. Raven, New York, pp 137–217
5. Wolinsky H, Glagov S (1964) Structural basis for the static mechanical properties of the aortic media. Circ Res 14:400–413
6. Jiang GF (1994) A quantitative investigation of age-related changes and species differences in the structure of the elastin network of the aortic wall. PhD thesis, University of New South Wales, Sydney
7. O'Rourke MF (1982) Arterial function in health and disease. Churchill Livingstone, Edinburgh
8. Batsakis JG (1968) Gould Se. In: Arteriosclerosis (ed) Pathology of the heart and blood vessels, 3rd edn. Thomals, Springfield, pp 519–526
9. Harris R (1975) Cardiovascular aging. In: Goldman R, Rockstem M, Sussman ML (eds) The physiology and pathology of human aging. Academic, New York, pp 109–122
10. Linzbach AJ, Akuamoa-Boateng E (1973) Klin Wochenschr 51:156–163
11. Gerstenblith G, Frederiksen J, Yin FYC, Fortuin NJ, Lakatta EG, Weisfeldt ML (1977) Echocardiographic assessment of a normal adult aging population. Circulation 56:273-278
12. Gardin JM, Henry WL, Savage DP, Epstein SE (1977) Echocardiographic evaluation of an older population without clinically apparent heart disease. Am J Cardiol 39:277
13. Unverferth DV, Fellers JK, Unverforth BJ, Leiser CV, Magorion RD, Arn AR, Baker PB (1983) Human myocardial histologic characteristics in congestive heart failure. Circulation 68:1194–1200
14. Pickering G (1968) High blood pressure. Churchill, London
15. Kannel WB, Wolfe PA, McGee DI, Dawbe TR, McNamara P, Castelli WP (1981) Systolic blood pressure, arterial rigidity and risk of stroke. The Framingham Study. JAMA 12:1225-1229
16. Kannel WB, Gordon T, Schwarte MJ (1971) Systolic versus diastolic blood pressure and risk of coronary heart disease. Am J Cardiol 27:335–345

17. Kannel WB, Castelli WP, McNamara PM, McKee PA, Feinleib M (1972) Role of blood pressure in the development of congestive heart failure. The Framingham Study. N Engl J Med 287:781–787
18. SHEP Cooperative Research Group (1991) Prevention of stroke by antihypertensive drug treatment in older persons with isolated systolic hypertension. JAMA 265:3255–3264
19. Dahlof B, Lindholm LH, Hansson L, Scherston B, Ekbom T, Webster TO (1991) Morbidity and mortality in the Swedish trial in old patients with hypertension (STOP—Hypertension). Lancet 338:1281
20. MRC Working Party (1992) Medical Research Council trial of treatment of hypertension in older adults: principal results. Br Med J 304:405–412
21. O'Rourke MF (1976) Pulsatile arterial haemodynamics in hypertension. Aust N Z J Med 6(suppl 2):40–48
22. O'Rourke MF, Taylor MG (1967) Input impedance of the systemic circulation. Circ Res 20:365–380
23. O'Rourke MF (1967) Steady and pulsatile energy losses in the systemic circulation under conditions and in simulated arterial disease. Cardiovasc Res 1:313–326
24. Milnor WR (1975) Arterial impedance as ventricular afterload. Circ Res 36:565–570
25. O'Rourke MF, Taylor MG (1966) Vascular impedance of the femoral bed. Circ Res 18:126–139
26. O'Rourke MF (1970) Arterial hemodynamics in hypertension. Circ Res 26/27:123–133
27. Avolio AP, Chen S-G, Wang R-P, Zhang C-L, Li M-F, O'Rourke MF (1983) Effects of ageing on changing arterial compliance and left ventricular load in a northern Chinese urban community. Circulation 68:50–58
28. O'Rourke MF, Avolio AP, Nichols WW (1985) Left ventricular systemic arterial coupling in man and strategies to improve coupling in disease states. In: Yin FCP (ed) Vascular ventricular coupling. Springer, Berlin Heidelberg New York, pp 1–19
29. Yaginuma T, Noda T, Tsuchiya M, Takazawa K, Tanaka H, Kotoda K, Hosoda S (1985) Interaction of left ventricular contraction and aortic input impedance in experimental and clinical studies. Jpn Circ J 49:206–214
30. Merillon JP, Fontenier G, Chastre J, Lerallut JF, Jaffrin MY, Gourgon R (1980) Etude du spectre d'impedance chez l'homme normal et hypertendie. Effects de l'accroissement de frequence cardiaque et des croques vasomotrics. Arch Mal Coeur Vaiss 73:83–90
31. Merillon JP, Motte G, Masquet C, Azancot I, Guiomard A, Gourgon R (1982) Relationship between physical properties of the arterial system and left ventricular performance in the course of aging and arterial hypertension. Eur Heart J 3(suppl A):95–101
32. O'Rourke MF, Blazek JV, Morreels CL, Krovetz LJ (1968) Pressure wave transmission along the human aorta; changes with age and in arterial degenerative disease. Circ Res 23:567–579
33. Kelly RP, Hayward CS, Avolio AP, O'Rourke MF (1989) Non-invasive determination of age-related changes in the human arterial pulse. Circulation 80:1652–1659
34. Murgo JP, Westerhof N, Giolma JP, Altobelli SA (1980) Aortic input impedance in normal man: relationship to pressure wave shapes. Circulation 62:105–116
35. Mahomed FA (1872) The physiology and clinical use of the sphygmograph. Medical Times and Gazette 1:62–64, 128–130, 220–222
36. Mills CJ, Gabe IT, Gault JH, Mason DT, Ross J Jr, Braunwald E, Shillingford JP (1970) Pressure-flow relations and vascular impedance in man. Cardiovasc Res 4:405–417
37. Nichols WW, O'Rourke MF (1990) McDonald's blood flow in arteries. Arnold, London
38. O'Rourke MF, Avolio AP (1980) Pulsatile flow and pressure in human systemic arteries; studies in man and in a multi-branched model of the systemic arterial tree. Circ Res 44:363–372
39. Nichols WW, O'Rourke MF, Avolio AP, Yaginuma T, Murgo JP, Pepine CJ, Conti CR (1985) Effects of age on ventricular-vascular coupling. Am J Cardiol 55:1179–1184
40. Kelly RP, Karamanoglu M, Gibbs HH, Avolio AP, O'Rourke MF (1989) Non-invasive carotid pressure wave registration as an indicator of ascending aortic pressure. J Vasc Med Biol 1:241–247

41. Chen C-H, Ting C-T, Nussbacher A, Nevo E, Kass DA, Pak P, Wang S-P, Chang M-S, Yin FCP (1996) Validation of carotid tonometry as a means of estimating augmentation index of ascending aortic pressure. Hypertension (Dallas) 27:168–175
42. Takazawa K (1987) A clinical study of the second component of left ventricular systolic pressure. J Tokyo Med Coll 45:256–270
43. Fujii M, Yaginuma T, Takazawa K, Noda T, Komatsu H, Katsuki T, Watabiki H, Kawada Y, Hosuda S (1989) Non-invasive detection for wave reflection in the arterial system by using the carotid pulse wave and its clinical application. J Jpn Coll Cardiol 29:545–551
44. Rowell LB, Brengelman GL, Blackman J, Bruce RA, Murray JA (1968) Disparities between aortic and peripheral pulse pressures induced by upright exercise and vasomotor changes in man. Circulation 37:954–964
45. Kelly R, Gibbs H, O'Rourke MF, Daley JE, Mang K, Morgan JJ, Avolio AP (1990) Nitroglycerin has more favorable effects on the left ventricular afterload than apparent from measurement of pressure in a peripheral artery. Eur Heart J 11:138–144
46. Yin FCP, Weisfeldt ML, Milnor WR (1981) Role of aortic impedance in the decreased cardiovascular response to exercise with aging in dogs. J Clin Invest 68:28–38
47. Yin FCP (1986) Aging and vascular impedance. In: Yin FCP (ed) Ventricular/vascular coupling. Springer, Berlin Heidelberg New York
48. O'Rourke MF, Kelly RP (1993) Wave reflection in the systemic circulation and its implications in ventricular function. J Hypertens 11:327–337
49. Sarnoff SJ, Braunwald E, Welch GH, Case RB, Stainsby WN, Macruz R (1958) Hemodynamic determinants of oxygen consumption of the heart with special reference to the tension-time index. Am J Physiol 192:148–156
50. Katz LN, Feinberg H (1958) The relation of cardiac effort to myocardial oxygen consumption and coronary flow. Circ Res 6:656–669
51. Buckberg GD, Fisher DE, Archie JP, Hoffman JIE (1972) Experimental subendocardial ischemia in dogs with normal coronary arteries. Circ Res 30:67–81
52. Buckberg GD, Towers B, Paglia DE, Mulder DG, Maloney JV (1972) Subendocardial ischemia after cardiopulmonary bypass. J Thorac Cardiovasc Surg 64:669–685
53. Hori M, Inoue M, Kitakaze M, Abe H (1985) Altered loading sequence as an underlying mechanism of afterload dependency of ventricular relaxation on the heart in situ. Jpn Circ J 49:245–254
54. Yabe R, Takazawa K, Maeda K, Fujita M, Sara T, Ibukiyama C (1991) The influence of ascending aortic augmentation index (reflection coefficient) to left ventricular relaxation. Circulation 84:565
55. Ross J Jr, Covell JW, Sonnenblick EH (1966) Contractile state of the heart chacterised by force-velocity relations in variably afterloaded and isovolumic beats. Circ Res 18:149–163
56. Nichols WW, O'Rourke MF, Avolio AP, Yaginuma T, Murgo JP, Pepine CJ, Conti CR (1987) Age-related changes in left ventricular/arterial coupling. In: Yin FCP (ed) Ventricular vascular coupling. Springer, Berlin Heidelberg New York
57. Lakatta E (1990) Changes in cardiovascular function with aging. Eur Heart J 11(suppl C):22–29
58. Lakatta E (1990) Similar myocardial effects of aging and hypertension. Eur Heart J 11(suppl C):29–38
59. Weisfeldt ML, Lakatta EG, Gerstenblith G (1992) Aging and the heart. In: Braunwald E (ed) Heart disease. Saunders, Philadelphia
60. Kelly R, Tunin R, Kass D (1992) Effect of reduced aortic compliance on left ventricular contractile function and energetics in vivo. Circ Res 71:490–502
61. Salisbury PF, Cross CE, Reiben A (1962) Ventricular performance modified by elastic properties of outflow system. Circ Res 11:319–328
62. Urschel CW, Covell JW, Sonnenblick EH, Ross J Jr, Braunwald E (1968) Effects of decreased aortic compliance on performance of the left ventricle. Am J Physiol 214:298–304
63. Kelly R, Kass D (1992) Effect of pulsatile vascular load on diastolic relaxation. Circulation 86(4):I-554

The Veins and Ventricular Preload

John V. Tyberg, Steven Y. Wang, Nairne W. Scott-Douglas, Vincent J.B. Robinson, Eiichi Chihara, Lisa M. Semeniuk, Debra L. Isaac, Israel Belenkie, and Dante E. Manyari

Summary. A simple model of the systemic circulation is presented in which the heart pumps blood out of a venous reservoir into an arterial reservoir. The arterial and venous reservoirs, which have different compliances, are separated by the systemic vascular resistance. In our view, changes in venous capacitance are represented by shifting pressure-volume relations. We have used this conceptual model to study the effects of various vasoactive drugs (nitroglycerin, enalaprilat, hydralazine, and amlodipine). Effects of interventions (heart failure, vasodilators, etc.) are best defined by changes in capacitance and conductance.

Key words. Venous capacitance—Preload—Vasodilation—Heart failure—Conductance—Systemic vascular resistance

Introduction

Both under normal physiological conditions and in congestive heart failure, the role of the veins in the modulation of cardiac output has been poorly defined and therefore poorly appreciated. The contributions of Starling, Wiggers, Katz, and Sarnoff and their respective collaborators to our understanding of cardiac physiology continue to be properly regarded as seminal. However, in retrospect it seems appropriate to acknowledge that their studies were limited in that the heart was studied in isolation from the normal circulation; it was easier, and for many purposes more satisfactory, to study the performance of the left ventricle when its venous and arterial loads were artificial and controlled. Therefore, it is easy to applaud Guyton's intentions as expressed in his 1955 presentation to the American Physiological Society:

> "Though it is not entirely true that cardiac output is independent of the function of the heart, nevertheless, it is hoped to emphasize in the present paper the special importance of the peripheral circulatory factors which affect venous return." [1]

To best understand the function of the heart, one must understand it within an intact circulation.

Most helpfully, Guyton [2] revived the concept, proposed by Weber [3] and studied by Starling [4], Krogh [5], and Starr [6,7], of mean circulatory filling pressure—"a

Departments of Medicine and Medical Physiology, The University of Calgary, Calgary, Alberta T2N 4N1, Canada.

measure of the fullness of the circulation" [8]. He also suggested that a decrease in venous tone or capacitance could be represented as a leftward shift in the venous pressure–volume relation (see [9], Fig. 14–11).[1] Although Guyton did represent capacitance changes as shifts in pressure–volume relations, he alternatively emphasized the "venous return curve" in which decreases in venous capacitance are represented by upward shifts in the venous return–right-atrial pressure relation (i.e., a change in the pressure–*volume* characteristics of the venous bed is represented by a change in a *flow*–pressure relation) [11]. However, representing a decrease in venous capacitance as an upward shift in the venous-return curve also implies that a decrease in venous capacitance "increases venous return." This may be viewed as a minor semantic inconsistency, but it does lead to an important internally inconsistent conclusion: heart failure simultaneously involves a decrease in cardiac output and an increase in venous return (which must equal steady-state cardiac output) [12,13]. For these and other reasons, we have chosen to represent changes in venous capacitance as changes in venous pressure–volume relations [14].

A Simple Model of the Circulation

If the pulmonary circulation is ignored, the relation of the heart to the systemic circulation can be represented conveniently by the simple hydraulic model shown in Fig. 1a [15]. The heart is represented as a pump that moves blood out of a low-pressure, high-compliance venous reservoir into a high-pressure, low-compliance arterial reservoir. Blood flow back to the venous reservoir is then passive; it is directly proportional to the arteriovenous pressure difference and inversely proportional to systemic vascular resistance, which is composed of each organ's resistance, arranged in parallel. The equilibrium value, mean circulatory pressure, is represented by the dashed line: when cardiac output is zero, pressure in the circulation equals 7 mmHg. Thus, as Levy [16] and others have done, we represent the systemic circulation as arterial and venous capacitances separated by systemic vascular resistance. As Fig. 1 suggests, the faster the heart pumps, the lower the pressure in the venous reservoir falls, reaching zero at a cardiac output of 7 l/min.

Figure 1b is a pressure–volume diagram that is equivalent to Fig. 1a [14]. The arterial pressure–volume relation is 19-fold steeper than that of the veins, reflecting the ratio of compliances [16]. The volume axis is arbitrary and is chosen such that, at a normal resting cardiac output (5 l/min). 70% of the blood is in the venous circulation [17]. Again, mean circulatory pressure (P_{mc}) is represented by the dashed line at 7 mmHg. If it is imagined that the heart is initially arrested and then starts pumping at a cardiac output of 1 l/min, we see that the heart removes an increment of blood volume (ΔV_1) from the venous circulation and adds it to that in the arterial circulation. Because of the disparate compliances, this translocation of blood has the effect of lowering venous pressure by 1 mmHg and raising arterial pressure by 19 mmHg. The system would then achieve a new equilibrium because the arteriovenous pressure difference (26 – 6 = 20 mmHg) is just sufficient to drive a flow of 1 l/min back through the systemic vascular resistance (20 mmHg l^{-1} min^{-1}). The assumption of a normal resting cardiac output (5 l/min) simply involves a greater vein-to-artery translocation

[1] That figure also represents significant changes in arterial capacitance; the physiological significance of such changes remains to be identified [10].

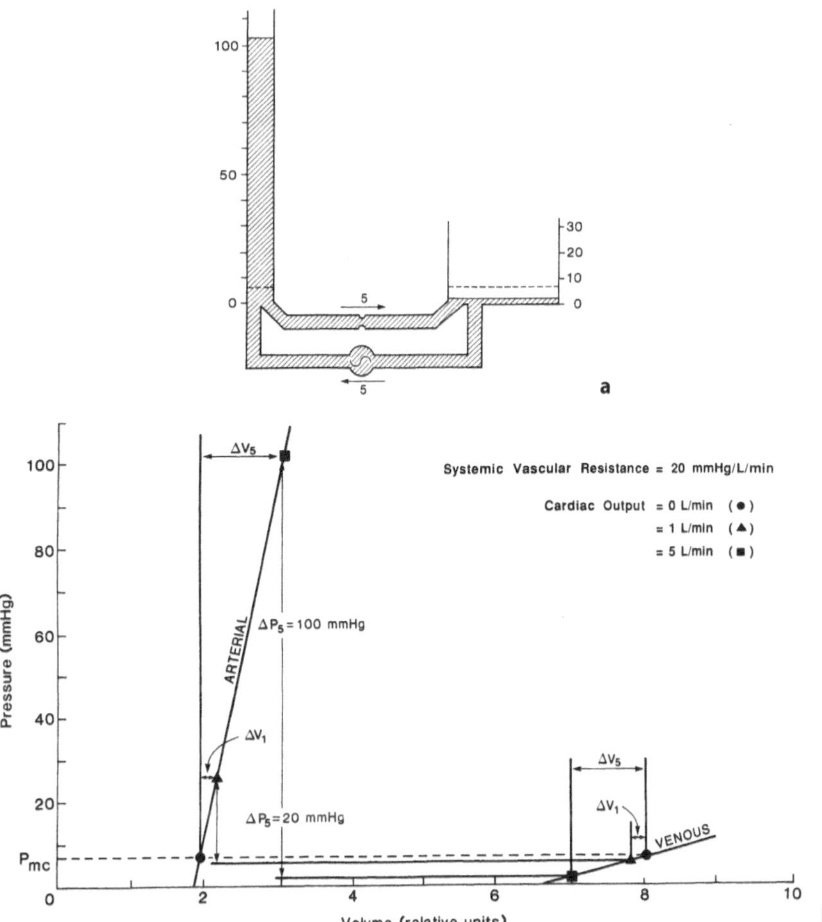

FIG. 1. **a** A simple hydraulic model of the peripheral circulation. For simplicity, the pulmonary circulation is neglected and the heart is represented by a pump that translocates volume from the high-capacity, high-compliance venous reservoir (*right*) into the low-capacity, low-compliance arterial reservoir (*left*). *Vertical scales* represent pressures in mmHg; normal values (i.e., 5 l/min) of cardiac output and flow through the lumped systemic vascular resistance are indicated. *Dashed line* represents the value (7 mmHg) of equilibrium pressure (i.e., mean circulatory pressure) when cardiac output is zero. (From [15] with permission.) **b** Arterial and venous pressure–volume relations that correspond to the hydraulic model (**a**). When cardiac output increases, venous pressure falls because the venous reservoir is depleted (see text). P_{mc}, Mean circulatory pressure. (From [14] with permission)

of blood (ΔV_5). Venous pressure falls to 2 mmHg and arterial pressure rises to a normal mean level of 102 mmHg. Given the constant value of systemic vascular resistance, equilibrium is again achieved with systemic flow also being equal to 5 l/min. This model supports the conclusion that venous pressure falls *because* cardiac output increases (increasing the cardiac output depletes the venous reservoir) and opposes the interpretation that cardiac output (i.e., venous return) increases because venous pressure (i.e., right-atrial pressure, "back pressure") decreases.

While the simplified model just discussed is useful for the circulation at rest, it is clear that additional, compensatory mechanisms must be operative to enable the circulation to respond to stresses that require substantial increases in cardiac output. In principle, the normal circulation has two such mechanisms. In Fig. 2a, these are

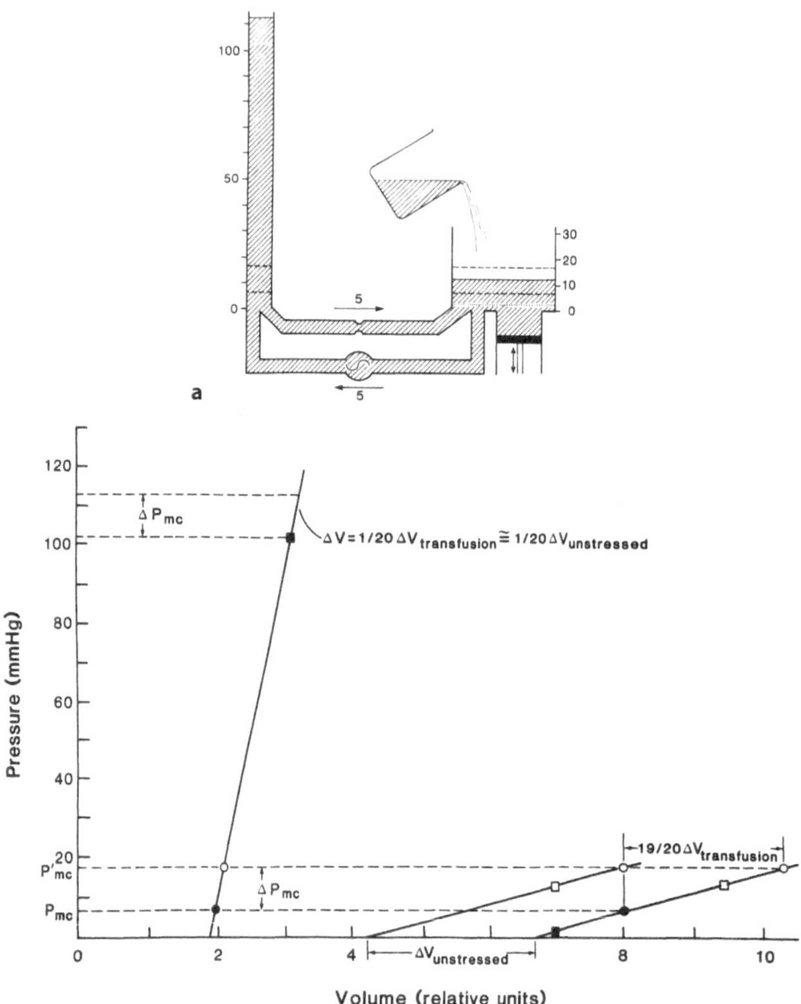

FIG. 2. **a** The hydraulic model shown in Fig. 1a modified to show the equivalence of a blood transfusion (or reduced fluid excretion) and a decrease in venous capacitance (represented by an upward movement of the piston shown in the bottom of the venous reservoir). An increase in mean circulatory pressure of 10 mmHg (from 7 to 17 mmHg) is shown. (From [15] with permission.) **b** Figure 1b has been modified to show the equivalent effects of volume loading and decreases in venous capacitance (i.e., to correspond to Fig. 2a). The total volume of the transfusion is distributed according to the respective venous and arterial compliances (i.e., 19:1). Thus, a 200-ml transfusion is equivalent to a 190-ml decrease in unstressed volume in that both produce the same increase in P_{mc} (see text). *V*, Volume. (From [15] with permission)

represented as a direct addition of blood volume and a mechanism whereby blood that was "stored" within the system can be mobilized [15]. While, of course, a blood transfusion is the prototype of a direct addition of blood volume, water intake, the kidney-mediated mechanisms that conserve sodium and water, and a decrease in insensible water loss all also serve to add volume to the circulation. These mechanisms have slow time-courses that range from several minutes to days and, because we are interested in the rapidly responding mechanism, they are not discussed further in this chapter. The second, rapidly responsive mechanism is that represented by the movable cylinder that can, like a transfusion, increase the "circulating" blood volume, raise P_{mc}, and at any level of cardiac output, increase the arterial and venous pressures.

Figure 2b is a modified pressure–volume diagram that demonstrates the hemodynamic equivalence of the two performance-enhancing mechanisms just presented. The direct addition of blood volume is shown as movement along the original venous pressure–volume relation (Fig. 1b). If a volume of blood sufficient to raise P_{mc} from 7 to 17 mmHg is added, venous and arterial pressures will rise equally (to 12 and 112 mmHg, respectively) at a cardiac output of 5 l/min. Now, cardiac output might increase further (to approximately 17 l/min) before venous pressure would decrease to zero. When no blood is added to the system but venous capacitance is decreased, circulatory changes are the same. The decrease in venous capacitance is represented by a leftward shift in the venous pressure–volume relation; venous *unstressed* volume (the theoretical volume that the veins contain at zero transmural pressure) is decreased, which has the effect of raising P_{mc} (i.e., the diminished circulatory volume is "fuller" [8]). Thus, also in this example, venous pressure is 12 mmHg and arterial pressure is 112 mmHg, even though the volume contained in the venous circulation may be the same. A leftward shift in the venous pressure–volume relation raises venous pressure with no change in volume. Like the direct addition of volume, a decrease in venous capacitance can allow the heart to increase its output further without venous pressure falling to zero.

In Fig. 2, a decrease in venous capacitance is represented as a leftward parallel shift in the pressure–volume relation. A priori considerations suggest that the slope should increase with a leftward shift. Indeed, in well-controlled experiments in which venules were observed directly, small slope changes of that type were demonstrated [18]. However, using measurements of regional blood volume [19] or total blood volume [20], slope changes were not statistically significant. Although small slope changes do occur, it is important to recognize the dominance of the changes in unstressed volume; theoretical models which assume that venous capacitance (like electrical capacitance) can be described by a linear relation that passes through the origin are not satisfactory.

Using pressure-volume diagrams, Fig. 3 shows how the systemic veins are coupled to the right ventricle and therefore how changes in venous capacitance modulate preload, stroke volume or work, and cardiac output. Under control conditions, the pressure in a representative venous capacitance circulation (e.g., the intestinal circulation) might be approximately 10 mmHg. Blood flows from the veins back to the right ventricle under the influence of a pressure difference (ΔP), which is a function of the resistance of the conducting veins and flow. (There is preliminary evidence to show that profound sympathetic stimulation [i.e., the cold-pressor test] may *increase* the caliber of the conducting veins, in spite of an assumed decrease in venous capacitance

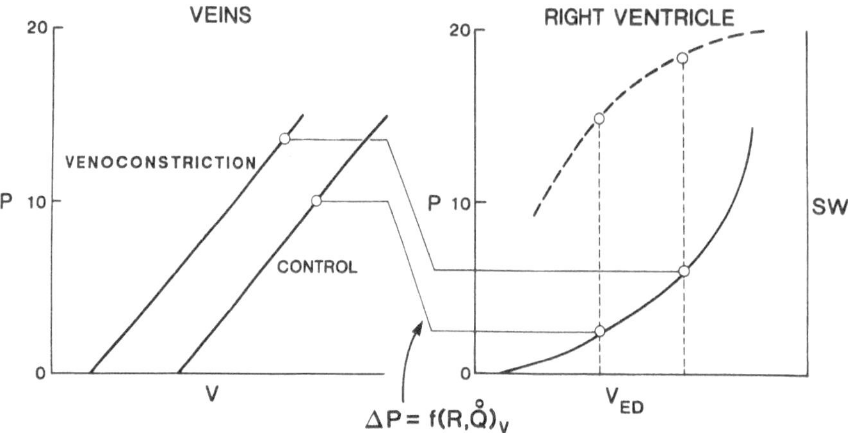

FIG. 3. Schematic diagram depicts the coupling of the venous capacitance bed and the right ventricle, showing how a decrease in venous capacitance (i.e., a decrease in venous unstressed volume) increases cardiac preload and therefore cardiac performance. The difference between the pressure of the venous capacitance bed and right-ventricular end-diastolic pressure is ΔP and is a function of venous resistance, R, and flow, Q. Right-ventricular end-diastolic pressure determines volume, V_{ED} (see *solid line*), which according to the Frank–Starling Law determines stroke work, *SW* (*dashed line*). (From [14] with permission)

[21]). Right-ventricular end-diastolic pressure is approximately 2 mmHg, which determines end-diastolic volume (pericardial effects are ignored), which in turn defines right-ventricular performance (i.e., stroke work) according to the Frank-Starling relation. When venoconstriction occurs (i.e., venous capacitance decreases), the venous pressure-volume relation shifts leftward. This raises P_{mc} (not shown) and, if cardiac output and ΔP are assumed to remain unchanged initially, venous and end-diastolic pressures equally. The resultant increase in end-diastolic volume increases right-ventricular performance. Changes in performance of the right ventricle would then modulate the performance of the left ventricle similarly. This mechanism whereby changes in venous capacitance modulate stroke work and cardiac output defines how the venous capacitance circulation (in addition to systemic vascular resistance, heart rate, and contractility) functions as an effector in baroreceptor-mediated control of the circulation [22,23].

Capacitance Changes in Congestive Heart Failure

The best estimation of the magnitude of heart failure-induced changes in venous capacitance is that of Ogilvie and Zborowska-Sluis [20]. Venous capacitance was measured by classical techniques: they plotted P_{mc} (estimated during acetylcholine-induced cardiac arrest) versus total blood volume (measured using indicator dilution). They produced chronic heart failure in dogs by pacing and found that venous capacitance decreased by approximately 50%. This implies that, compared at a given venous pressure, the systemic veins contained only 50% of the blood that they normally contained—a reflection of the severity of the stress response associated with heart failure.

Effects of "Vasodilators" in Heart Failure

Using an acute coronary embolization model of heart failure [24] and a radionuclide blood-pool scintintigraphic method that we developed to determine relative changes in regional venous capacitance [25], we have shown that acute congestive heart failure decreases intestinal venous capacitance by about 15% [19]. We also compared the venous capacitance effects of three common vasodilators: nitroglycerin, enalaprilat, and hydralazine [19]. After heart failure was induced (i.e., until left-ventricular end-diastolic pressure rose to 20 mmHg and cardiac output decreased by about 50%), each drug was given intravenously in a dosage sufficient to reduce mean aortic pressure by approximately 20%. We found that the effects on venous capacitance were varied. In relation to the heart failure baseline, nitroglycerin increased venous capacitance by 31%, enalaprilat by 13% ($P < .001$ for both), and hydralazine by 2% (NS).

In a more recent study [26], we measured the hemodynamic effects of amlodipine, a second-generation calcium channel blocker that has been reported to increase exercise tolerance in heart failure [27]. Because amlodipine has favorable clinical effects in chronic failure (i.e., increased exercise tolerance and decreased filling pressure), we studied its effects in the venous system. Despite those favorable clinical results in chronic heart failure, we found that amlodipine actually decreased venous capacitance further in our anesthetized dog model of heart failure [26]. The effect of that drug on venous capacitance in chronic failure has not yet been determined; the possibility remains that it has deleterious venous capacitance effects in conscious patients with chronic failure.

Capacitance Versus Conductance Changes

It is assumed that agents that produce vasodilation do so by causing vascular smooth muscle to relax, to assume a longer length when exposed to a given stress. With respect to the arterial or arteriolar circulation, a greater length corresponds to an increased caliber and, therefore, conductance (conductance is the reciprocal of resistance). Increased caliber of the venous capacitance circulation increases capacitance. Therefore, we have begun to characterize the effects of vasodilators in terms of changes in capacitance versus changes in conductance [28]. Not surprisingly, nitroglycerin at low doses produces a substantial increase in capacitance but has trivial effects on conductance. At higher doses, however, conductance increases considerably, with only a trivial further increase in capacitance [28].

The capacitance-conductance plot is also a convenient means to summarize the effects of various vasodilators in our experimental model of heart failure [19,26] (Fig. 4). Heart failure, as produced by acute microsphere embolization of the left coronary arteries, decreases capacitance by about 15% and conductance by about 40%. Nitroglycerin increases capacitance to values greater than that observed during the control period and increases conductance minimally. Enalaprilat, an angiotensin-converting enzyme inhibitor, increases both capacitance and conductance to near-control values. Hydralazine, as expected, does not have a significant effect on capacitance but greatly increases conductance. Amlodipine unexpectedly produced no increase in capacitance or conductance in this experimental model.

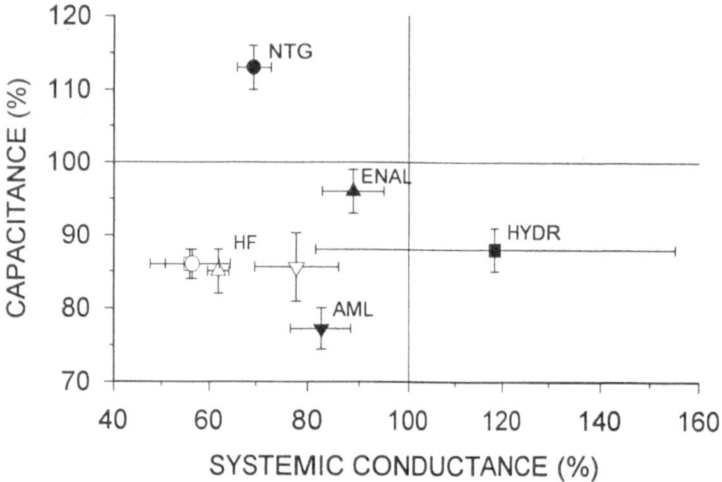

Fig. 4. Changes in capacitance (%) versus changes in systemic conductance (%) expressed in relation to the control observations in anesthetized dogs. *Open symbols* show the changes produced by heart failure (*HF*, large-microsphere embolization of the left coronary arteries) and *filled symbols* indicate the effects of vasodilators (nitroglycerin, *NTG*; enalaprilat, *ENAL*; hydralazine, *HYDR*; amlodipine, *AML*). Systemic conductance data were available for only 2 dogs in the hydralazine study. Data are means ± SEM [19,26]

Future Studies

We are convinced that it is important for cardiologists to consider effects on venous capacitance as well as on arteriolar conductance (resistance) when assessing the utility of vasodilators in the treatment of patients with heart failure. We will continue to use this model of heart failure to explore the comparative capacitance–conductance effects of promising new vasoactive agents. We also look forward to the results of clinical studies, using plethysmographic methods to measure venous capacitance [29,30] in addition to conventional hemodynamic methods to measure arteriolar conductance. We are hopeful that such a combination of experimental and clinical studies will provide a more rational basis for the management of heart failure patients using vasodilators.

References

1. Guyton AC (1955) Determination of cardiac output by equating venous return curves with cardiac response curves. Physiol Rev 35:123–129
2. Guyton AC, Lindsey AW, Kaufmann BN (1955) Effect of mean circulatory filling pressure and other peripheral circulatory factors on cardiac output. Am J Physiol 180:463–468
3. Weber E (1850) On the application of wave theory to the understanding of the circulation (in German). In: Report of the transactions of the Saxon Academy of Science of Leipzig. Weidman, Leipzig

4. Starling EH (1912) Principles of human physiology. Lea and Febiger, Philadelphia, pp 880–884
5. Krogh A (1912) The regulation of the supply of blood to the right heart. (With a description of a new circulation model.) Skand Arch Physiol 27:227–248
6. Starr I, Rawson AJ (1940) Role of the "static blood pressure" in abnormal increments of venous pressure, especially in heart failure. I. Theoretical studies on an improved circulation schema whose pumps obey Starling's law of the heart. Am J Med Sci 199:27–39
7. Starr I (1940) Role of the "static blood pressure" in abnormal increments of venous pressure, especially in heart failure. II. Clinical and experimental studies. Am J Med Sci 199:40–55
8. Rothe CF (1993) Mean circulatory pressure: its meaning and measurement. J Appl Physiol 74:499–509
9. Guyton AC (1977) Basic human physiology: normal function and mechanisms of disease. Saunders, Philadelphia, pp 192–193
10. Hayoz D, Tardy Y, Rutschmann B, Mignot JP, Achakri H, Feihl F, Meister JJ, Waeber B, Brunner HR (1993) Spontaneous diameter oscillations of the radial artery in humans. Am J Physiol (Heart Circ Physiol) 264:H2080–H2084
11. Guyton AC, Lindsey AW, Abernathy B, Richardson T (1957) Venous return at various right atrial pressures and the normal venous return curve. Am J Physiol 189:609–615
12. Packer M (1985) Mechanisms of nitrate action in patients with severe left ventricular failure: conceptual problems with the theory of venosequestration. Am Heart J 110:259–264
13. Packer M (1990) Editorial: Abnormalities of diastolic function as a potential cause of exercise intolerance in chronic heart failure. Circulation 81(suppl III):III-78–III-86
14. Tyberg JV (1992) Venous modulation of ventricular preload. Am Heart J 123:1098–1104
15. ter Keurs HEDJ, Tyberg JV (1996) Control of the circulation: an integrated view. In: Greger R, Windhorst U (eds) Comprehensive human physiology, vol 2. Springer, Berlin Heidelberg, pp 1995–2013
16. Levy MN (1979) The cardiac and vascular factors that determine systemic blood flow. Circ Res 44:739–747
17. Greenway CV (1983) Role of splanchnic venous system in overall cardiovascular homeostasis. Fed Proc 42:1678–1684
18. Shoukas AA, Bohlen HG (1990) Rat venular pressure-diameter relationships are regulated by sympathetic activity. Am J Physiol (Heart Circ Physiol) 259:H674–H680
19. Wang SY, Manyari DE, Scott-Douglas NW, Smiseth OA, Smith ER, Tyberg JV (1995) Splanchnic venous pressure-volume relation during experimental acute ischemic heart failure: differential effects of hydralazine, enalaprilat, and nitroglycerin. Circulation 91:1205–1212
20. Ogilvie RI, Zborowska-Sluis D (1992) Effect of chronic rapid ventricular pacing on total vascular capacitance. Circulation 85:1524–1530
21. Robinson VJB, Yemen GW, Fick GH, Smith ER, Tyberg JV (1988) Paradoxical central venous dilation in response to cold pressor stimulation (abstract). Circulation 78:I-329
22. Haase EB, Shoukas AA (1991) Carotid sinus baroreceptor reflex control venular pressure-diameter relations in rat intestine. Am J Physiol (Heart Circ Physiol) 260:H752–H758
23. Brunner MJ, Greene AS, Frankle AE, Shoukas AE (1988) Carotid baroreceptor control of splanchnic resistance and capacity. Am J Physiol (Heart Circ Physiol) 255:H1305–H1310
24. Smiseth OA, Mjos OD (1982) A reproducible and stable model of acute ischaemic left ventricular failure in dogs. Clin Physiol 2:225–239
25. Scott-Douglas NW, Manyari DE, Smiseth OA, Robinson VJB, Wang SY, Smith ER, Tyberg JV (1995) Measurement of intestinal vascular capacitance in dogs: an application of blood-pool scintigraphy. J Appl Physiol 78:232–238

26. Isaac DL, Belenkie I, Manyari DE, Tyberg JV (1994) Effects of amlodipine on venous capacitance and contractility in acute heart failure (abstract). J Am Coll Cardiol (special issue):383A
27. Packer M, Nicod P, Khanderia BR, Costello DL, Wasserman AG, Konstam MA, Weiss RJ, Moyer RR, Pinsky DJ, Abittan MH, Souhrada JF (1991) Randomized multicenter, double-blind, placebo controlled evaluation of amlodipine in patients with mild-to-moderate heart failure (abstract). J Am Coll Cardiol 17:274A
28. Chihara E, Manyari DE, Tyberg JV (1994) Comparative effects of nitroglycerin on intestinal vascular conductance and capacitance in dogs (abstract). FASEB J 8:A869
29. Manyari DE, Malkinson TJ, Robinson VJ, Smith ER, Cooper KE (1988) Acute changes in forearm venous volume and tone using radionuclide plethysmography. Am J Physiol (Heart Circ Physiol) 255:H947–H952
30. Manyari DE, Wang Z, Cohen J, Tyberg JV (1993) Assessment of the human splanchnic venous volume-pressure relation using radionuclide plethysmography: effect of nitroglycerin. Circulation 87:1142–1151

Role of Endothelium in Control of Regional Blood Flow in Humans

Yoshitaka Hirooka[1], Tsutomu Imaizumi[2], and Akira Takeshita[1]

Summary. We have studied the role of endothelium in control of forearm blood flow in patients with heart failure as well as in normal subjects. First, endothelium-dependent forearm vasodilation in response to acetylcholine (ACh), substance P, and endothelium-independent forearm vasodilation to sodium nitroprusside (SNP) were examined in patients with heart failure and in normal subjects. Endothelium-dependent forearm vasodilation in response to ACh but not to substance P was impaired in patients with heart failure. Endothelium-independent forearm vasodilation to SNP was also preserved in patients with heart failure. Second, the role of nitric oxide (NO) in reactive hyperemia and exercise-induced vasodilation were examined using N^G-monomethyl-L-arginine (L-NMMA), a blocker of NO synthesis, in normal subjects. Results suggested that NO plays a minimal role in reactive hyperemia and exercise-induced vasodilation in normal human forearm vessels. Finally, we determined if L-arginine, a precursor of NO, improves impaired endothelium-dependent vasodilation to ACh and reactive hyperemia in patients with heart failure. L-Arginine augmented impaired ACh-induced vasodilation as well as reactive hyperemia in patients with heart failure. Our results suggested that defective endothelial function may contribute to abnormal control of regional blood flow in patients with heart failure.

Key words. Acetylcholine—Heart failure—Endothelium—Nitric oxide—Forearm blood flow—Human—L-Arginine

Introduction

It has been shown that the endothelium plays an important role in control of regional blood flow in humans [1]. Nitric oxide (NO) is produced during the conversion of L-arginine to L-citrulline by NO synthase in endothelial cells [1,2]. Now, NO is considered to be a major endothelium-derived relaxing factor (EDRF) and a potent vasodilator [1,2]. It has been suggested that a defective NO pathway may play an important role in pathophysiological states such as hypertension, heart failure,

[1] Research Institute of Angiocardiology and Cardiovascular Clinic, Kyushu University School of Medicine, Fukuoka, 812-82 Japan.
[2] Third Department of Internal Medicine, Kurume University School of Medicine, Kurume, 830 Japan.

hypercholesterolemia, and diabetes [1–3]. In humans, changes in the peripheral circulation are often studied in the forearm by measuring forearm blood flow using a plethysmograph with a venous occlusion technique [4,5]. Intraarterial infusion of acetylcholine (ACh) allows us to determine the stimulated endothelium-dependent relaxation in this model [4,5].

Control of regional blood flow is markedly altered in patients with heart failure [3,6]. These patients are characterized by heightened vasoconstriction and reduced vasodilator responses to exercise and ischemia [3,6]. It is likely that fatigue or decreased exercise tolerance in patients with heart failure is caused by not only impaired pump function of the heart but also by inadequate increases in nutritional muscle blood flow resulting from impaired vasodilating response during exercise [6]. We and others have shown that endothelium-dependent peripheral vasodilation evoked by ACh or methacholine is attenuated in patients with heart failure [7–10]. Thus, it is possible that endothelial dysfunction may contribute to impaired vasodilation during exercise or ischemia in patients with heart failure.

The purpose of our studies was to examine the role of endothelium in control of forearm blood flow in patients with heart failure as well as in normal subjects. The studies we discuss were done to answer the following questions:

1. Is only muscarinic receptor-mediated forearm vasodilation impaired, or does generalized endothelial dysfunction exist in patients with heart failure?
2. Does NO plays a role in reactive hyperemia or exercise-induced vasodilation in the forearm resistance vessels in normal subjects? This study utilized N^{G}-monomethyl-L-arginine (L-NMMA), a blocker of NO synthesis.
3. Does L-arginine, a precursor of NO, improve impaired ACh-induced vasodilation and reactive hyperemia in patients with heart failure?

Forearm Blood Flow in Response to Acetylcholine, Sodium Nitroprusside, or Substance P in Patients with Heart Failure and in Normal Subjects

Methods and Results

The study [11] was done with the participants in a quiet and air-conditioned room. Under local anesthesia, the left brachial artery was cannulated for drug infusion with a 20-gauge cannula and was connected to a pressure transducer for direct measurement of arterial pressure. The protocol was explained, and informed written consent was obtained from each subject. The study was approved by the ethical committee for human investigation at our institution. Forearm blood flow was measured by using a mercury in silastic strain gauge plethysmograph with a venous occlusion technique as previously described [12,13].

Forearm blood flow (ml per min per 100 ml of forearm tissue) was calculated from the rate of increase in forearm volume while venous return from the forearm was prevented by inflating the cuff on the upper arm. The pressure in the venous occlusion cuff was 40 mmHg. Circulation to the hand was arrested by a cuff inflated around the wrist. An average of four flow measurements a 15-s interval was used for later analysis. Forearm vascular resistance was calculated by dividing the mean arterial pressure by the forearm blood flow. These values are expressed as units throughout this report. We examined forearm vasodilator responses to intraarterial infusion of ACh, substance P, or sodium nitro prusside (SNP) at graded doses. The drug was infused intraarterially for 2 min at each dose.

We studied 24 patients (16 men and 8 women) with heart failure. Basic disorders of patients with heart failure included dilated cardiomyopathy in 10, ischemic cardiomyopathy in 2, and valvular heart disease in 12 patients. The ages of the patients ranged from 38 to 74 years; they were in New York Heart Association functional class III or IV. The left-ventricular ejection fraction was 30% ± 3% in 12 patients with dilated or ischemic cardiomyopathy and 62% ± 4% in the 12 patients with valvular disease. Eighteen patients were on digoxin and furosemide at the time of study. These drugs were withheld on the day of the study. We also studied 12 normal subjects (10 men and 2 women) ranging in age from 45 to 67 years. The ages did not differ between patients with heart failure and normal subjects.

Mean arterial pressure did not differ between the two groups. Heart rate was faster, forearm blood flow was lower, and forearm vascular resistance was higher ($P < .01$ for each) in patients with heart failure than in normal subjects. The plasma renin activity (7.0 ± 1.5 vs. 1.7 ± 0.6 $ng\,ml^{-1}\,h^{-1}$), norepinephrine concentration (443 ± 63 vs. 149 ± 16 pg/ml), and atrial natriuretic peptide concentration (191 ± 33 vs. 38 ± 3 pg/ml) in patients with heart failure were significantly higher than the corresponding values in normal subjects ($P < .01$ for each). Intraarterial infusions of graded doses of ACh, SNP, and substance P caused progressive increases in forearm blood flow and progressive decreases in forearm vascular resistance in patients with heart failure as well as in normal subjects (Fig. 1).

Forearm blood flow values at rest and in response to three drugs were significantly lower in patients with heart failure than those in normal subjects (Fig. 1a; $P < .01$). The magnitudes of increases in forearm blood flow in response to ACh were less in patients with heart failure than in normal subjects ($P < .01$). However, the magnitudes of increases in forearm blood flow in response to SNP or substance P did not differ between the two groups. Figure 1b shows the percent decreases in forearm vascular resistance during infusions of ACh, SNP, and substance P at the graded doses in the two groups. In patients with heart failure, the percent decreases in forearm vascular resistance during graded infusions of ACh were attenuated ($P < .01$) compared with those in normal subjects. However, the percent decreases in forearm vascular resistance in response to SNP or substance P did not differ between the two groups. Difference in basic disorder of heart failure did not affect the results of the attenuated responses to ACh.

Discussion

We demonstrated that forearm vasodilation evoked by substance P at graded doses did not differ between patients with heart failure and normal subjects although forearm vasodilation evoked by ACh was less in patients with heart failure than in normal subjects. It has been suggested that substance P causes endothelium-dependent vasodilation in coronary artery and forearm vasculatures in humans [14–16]. Our results suggested that endothelium-dependent forearm vasodilation mediated by muscarinic receptor stimulation was attenuated but that endothelium-dependent vasodilation by substance P was preserved in patients with heart failure.

There are several potential mechanisms for endothelial dysfunction in patients with heart failure: (1) decrease of stimulated release of NO from the endothelium, either by chronically reduced blood flow, which causes decreased expression of endothelial NO

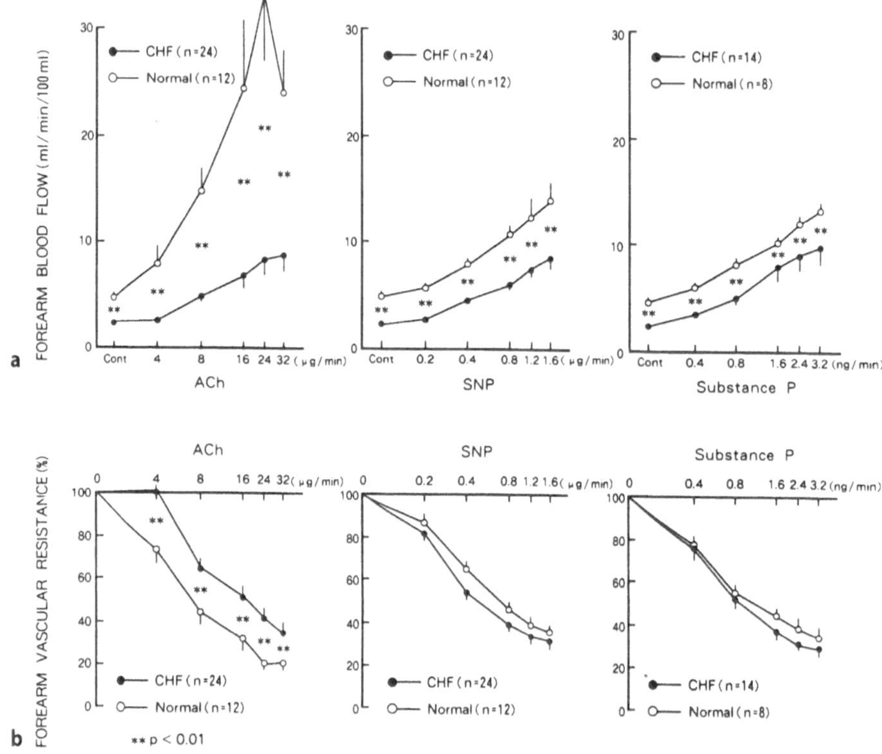

FIG. 1. **a** Forearm blood flow in response to acetylcholine (*ACh*), sodium nitroprusside (*SNP*), or substance P in patients with heart failure (*solid circles*) and in normal subjects (*open circles*). **b** Forearm vascular resistance in response to ACh, SNP, or substance P in patients with heart failure (*solid circles*) and in normal subjects (*open circles*). The values at control are shown as 100%. **, $P < .01$. (Adapted and modified from [11])

synthase [17], or by impaired endothelial receptor signal transduction pathways, which may be muscarinic receptors [18]; (2) inactivation of released NO, which may be caused by increased levels of oxygen free radicals [19]; (3) increased angiotensin-converting enzyme activity, which increases breakdown of bradykinin [20]; or (4) increased levels of cytokines, which impair the synthesis of endothelial NO synthase [21]. It appears likely that release of NO from the endothelium in response to muscarinic receptor stimulation is reduced in heart failure. Because forearm vasodilation evoked by substance P was not attenuated, we speculate the possibility that muscarinic receptors or postreceptor mechanisms coupled with muscarinic receptors in the endothelium are altered in patients with heart failure.

Another possibility is that endothelium-derived hyperpolarizing factor (EDHF) evoked by ACh might be attenuated in patients with heart failure. In supporting this hypothesis, our preliminary study [22] showed that forearm vasodilation evoked by substance P is not blocked by L-NMMA although forearm vasodilation evoked by ACh is partially blocked in normal subjects as in others [23,24]. Also, supplementation of L-arginine does not augment forearm vasodilation evoked by substance P in patients with heart failure as well as normal subjects (Hirooka Y et al., unpublished observation), which is different from the effect of L-arginine on ACh-induced vasodilation [23–25]. We do not know why we failed to see the blocking effect of L-NMMA on substance P-induced forearm vasodilation, which differs from the results of other studies [15,16]. Further studies are required to clarify the difference and the role of EDHF in humans. The reduced response of vascular smooth muscles to NO is unlikely because forearm vasodilating response to SNP was similar in patients with heart failure and normal subjects.

Role of Nitric Oxide in Reactive Hyperemia or Exercise-Induced Vasodilation in Normal Subjects

Methods and Results

Subjects in this study [26,27] were healthy male volunteers (19–28 years old). Forearm blood flow was measured at rest and during reactive hyperemia for 3 min after 10 min of arterial occlusion. After recovery of reactive hyperemia when forearm blood flow had returned to the baseline level, L-NMMA was infused intraarterially at a dose of 4 μmol/min for 5 min while forearm blood flow and arterial pressure were recorded. Immediately after infusion of L-NMMA was stopped, blood flow to the forearm was prevented for 10 min. After release of arterial occlusion, forearm blood flow and arterial pressure were recorded for another 3 min.

Immediately after release of 10 min of arterial occlusion, forearm blood flow increased (4.1 ± 0.6 to 38.2 ± 3.1 ml min^{-1} 100 ml^{-1}, $P < .01$) and forearm vascular resistance decreased (27.9 ± 5.6 to 2.1 ± 0.2 units; $P < .01$). Intraarterial infusion of L-NMMA decreased baseline forearm blood flow (4.1 ± 0.6 to 2.8 ± 0.4 ml min^{-1} 100 ml^{-1}; $P < .01$) and increased forearm vascular resistance (27.9 ± 5.6 to 36.3 ± 5.5 units; $P < .01$). L-NMMA did not alter arterial pressure and heart rate, and did not affect peak forearm blood flow and minimal forearm vascular resistance during reactive hyperemia after release of 10 min of arterial occlusion (Fig. 2). However, L-

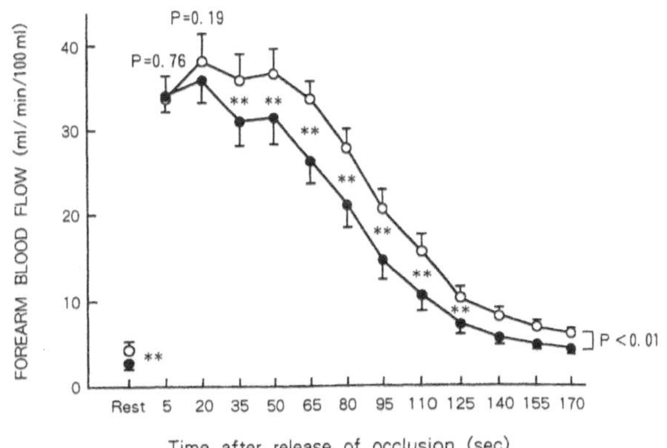

FIG. 2. Forearm blood flow at rest and during reactive hyperemia after 10 min of arterial occlusion before (*open circles*) and after (*solid circles*) ($n = 13$) N^G-monomethyl-L-arginine (L-NMMA). **, $P < .01$. (Adapted and modified from [26])

NMMA reduced blood flow during the mid-to-late phase of reactive hyperemia ($P < .01$) after 10 min of arterial occlusion (Fig. 2).

In another series of this study, after stable forearm blood flow and arterial pressure were obtained, subjects performed static handgrip exercise by pulling a weight (4–5 kg) for 3 min. Forearm blood flow was measured immediately after the cessation of exercise and for 3 min thereafter. After returning to the baseline values, L-NMMA was infused intraarterially at a dose of 4 μmol/min for 5 min and forearm blood flow was measured. Exercise was repeated and forearm blood flow was measured again.

Intraarterial infusion of L-NMMA decreased resting forearm blood flow (4.1 ± 0.7 to 2.8 ± 0.4 ml min^{-1} 100 ml; $P < .01$) and increased resting forearm vascular resistance (28.3 ± 5.6 to 37.6 ± 7.0 units; $P < .01$). Forearm blood flow increased to the peak value immediately after exercise and gradually declined thereafter (Fig. 3). L-NMMA attenuated increases in forearm blood flow after exercise (Fig. 3a). However, percent increases in forearm blood flow after exercise were similar before and after L-NMMA (Fig. 3b).

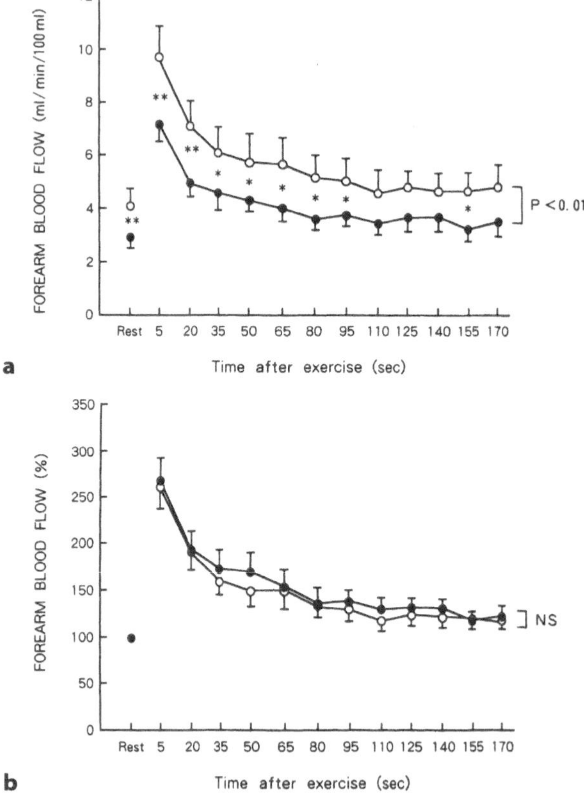

FIG. 3. **a** Blood flow at rest and after static exercise for 3 minutes before (*open circles*) and after (*solid circles*) intraarterial infusion of L-NMMA ($n = 11$). **b** Percent changes in forearm blood flow at rest and after static exercise for 3 min before (*open circles*) and after (*solid circles*) intraarterial infusion of L-NMMA ($n = 11$)., $P < .05$; **, $P < .01$. (Adapted and modified from [27])

Discussion

We studied the role of NO on reactive hyperemia and exercise-induced vasodilation of the forearm in normal subjects. Because these vasodilating responses are impaired in patients with heart failure [6] and endothelial dysfunction might contribute to the impairment, we needed to know the role of NO on such vasodilator capacity in normal subjects. Our results suggest that NO plays a minimal role on reactive hyperemia and exercise-induced vasodilation in normal subjects.

Forearm blood flow and vascular resistance at peak reactive hyperemia after 10 min of arterial occlusion were not different before and after intraarterial infusion of L-NMMA. Thus, our findings are consistent with those in animals [28–32] and suggest that NO plays a minimal role on vasodilation at peak reactive hyperemia in the human forearm. To interpret these results, it is important to consider the efficacy of L-NMMA at the dose used in this study in blocking the synthesis of NO in the forearm vessels. In the previous study [33], we examined the inhibitory effect of L-NMMA at this dose on endothelium-dependent forearm vasodilation evoked by ACh at 4 and 12 μg/min. L-NMMA at this dose almost completely inhibited modest vasodilation induced by intraarterial infusion of ACh at a low dose and inhibited by 50% vasodilation induced by ACh at a high dose. Thus, we consider that L-NMMA at this dose should have altered the peak reactive hyperemia if it had an effect. Although we did not examine the duration of action of intraarterially infused L-NMMA, a previous study by Vallance et al. [23] demonstrated that increases in forearm vascular resistance induced by intraarterial infusion of L-NMMA lasted for at least 30 min after infusion of L-NMMA was stopped. Thus, we consider that the effects of L-NMMA lasted long enough to block the formation of NO during reactive hyperemia.

It has been suggested that NO plays a significant role during the late phase of reactive hyperemia in the coronary circulation of animals [29]. In this study, L-NMMA reduced blood flow during the mid-to-late phase of reactive hyperemia and decreased flow debt repayment. These results suggest that NO plays a significant role in the mid-to-late phase of reactive hyperemia in human forearm vessels. It has been suggested that flow-dependent vasodilation is caused by the production of NO. Thus, the increase in flow causes the increase in shear stress that releases NO from the endothelium. We assume that the marked increases in forearm blood flow during the early phase of reactive hyperemia might have released NO, which contributed to vasodilation during the mid-to-late phase of reactive hyperemia.

Several potential candidates, including prostaglandins, adenosine, and ATP-sensitive potassium channels, are responsible for reactive hyperemia. It has been shown in humans that indomethacin decreased the peak level of reactive hyperemia as well as the duration of the hyperemia and that total reactive hyperemia was limited to about 50% by blockers of prostaglandin synthesis [34,35]. These results suggest that prostaglandins play a significant role not only in peak vasodilation but also during the mid-to-late phase of reactive hyperemia in human vessels. It has also been shown that theophylline reduced the total reactive hyperemia by about 35% in human forearm vessels [35]. These results sugggest that adenosine plays a significant role in reactive hyperemia in humans. In addition, recent studies have shown that ATP-sensitive potassium channels play a role in reactive hyperemia in the canine coronary circulation [36,37]. Thus, it is also possible that ATP-sensitive potassium channels play a role in reactive hyperemia in human forearm vessels.

To determine the role of NO in metabolic vasodilation after exercise, we infused L-NMMA and exercise was repeated. L-NMMA attenuated increases in forearm blood flow after exercise. However, because L-NMMA decreased resting forearm blood flow, the interpretation of these attenuated increases in forearm blood flow after exercise needs great caution. A recent study by Gilligan et al. [38], who performed experiments similar to ours, showed that nitric oxide plays a role in exercise-induced vasodilation. We expressed forearm blood flow after exercise as the percent changes and performed another experiment in which angiotensin II was infused instead of L-NMMA to decrease resting forearm blood flow to the same level as L-NMMA [27]. The percent increases in forearm blood flow after exercise were similar before and after L-NMMA when forearm blood flow after exercise was normalized by resting forearm blood flow. Moreover, the effects of angiotensin II on forearm blood flow after exercise were similar to those of L-NMMA [27]. These results suggest that NO does not play a major role in metabolic arteriolar vasodilation induced by exercise as well as reactive hyperemia in normal subjects.

Effects of L-Arginine on Impaired ACh-Induced Vasodilation and Reactive Hyperemia in Patients with Heart Failure

Methods and Results

We studied 20 patients with heart failure (14 men and 6 women) and 24 normal subjects (20 men and 4 women) [39]. The patients with heart failure were 37–74 years old, and the healthy subjects were 36–71 years old (NS). The underlying heart disease was dilated cardiomyopathy in 11 patients and valvular heart disease in 9 patients. New York Heart Association functional classification was class II to IV. All patients were receiving digoxin and diuretics at the time of the study, and all other medications were discontinued at least 2 days before the study.

Forearm blood flow, arterial pressure, and heart rate were measured at rest and during graded doses of ACh or SNP. After recovery, forearm blood flow measurements during reactive hyperemia following 10 min of arterial occlusion were obtained in 12 normal subjects and 12 patients with heart failure. After forearm blood flow had returned to the baseline level, L-arginine was infused into the brachial artery at 10 mg/min for 5 min, and then ACh or SNP was infused in the same way as before L-arginine while L-arginine was infused simultaneously and continuously. After release of occlusion, reactive hyperemic flows were recorded in the same way as before infusion of L-arginine.

Mean arterial pressure at rest did not significantly differ between patients with heart failure and normal subjects. Heart rate at rest was significantly faster ($P < .05$) in patients with heart failure than in normal subjects. Forearm blood flow at rest was lower ($P < .01$) and forearm vascular resistance at rest was greater ($P < .01$) in patients with heart failure than in normal subjects. Plasma renin activity (5.5 ± 1.1 vs. 1.7 ± 0.6 ng ml^{-1} h^{-1}), norepinephrine concentration (510 ± 110 vs. 150 ± 20 pg/ml), and atrial natriuretic peptide concentration (142 ± 30 vs. 38 ± 3 pg/ml) in patients with heart failure ($n = 9$) were significantly higher than in normal subjects ($n = 8$) ($P < .01$ for each). The magnitude of increase in forearm blood flow in response to ACh was less in patients with heart failure than in normal subjects ($P < .01$). However, the magnitude of increase in forearm blood flow in response to SNP did not differ between the two groups. These results were similar to the first study.

Time-courses of reactive hyperemic flow in normal subjects and patients with heart failure are shown in Fig. 4a. Maximal forearm blood flow during peak reactive hyperemia following release of 10 min of arterial occlusion was significantly lower in patients with heart failure ($n = 12$) than in normal subjects ($n = 12$) (30 ± 3 vs. 43 ± 3 ml min^{-1} 100 ml; $P < .01$). Minimal forearm vascular resistance during peak reactive hyperemia was significantly higher in patients with heart failure than in normal subjects (3.2 ± 0.4 vs. 2.1 ± 0.1 units; $P < .05$). Infusion of L-arginine at this dose did not

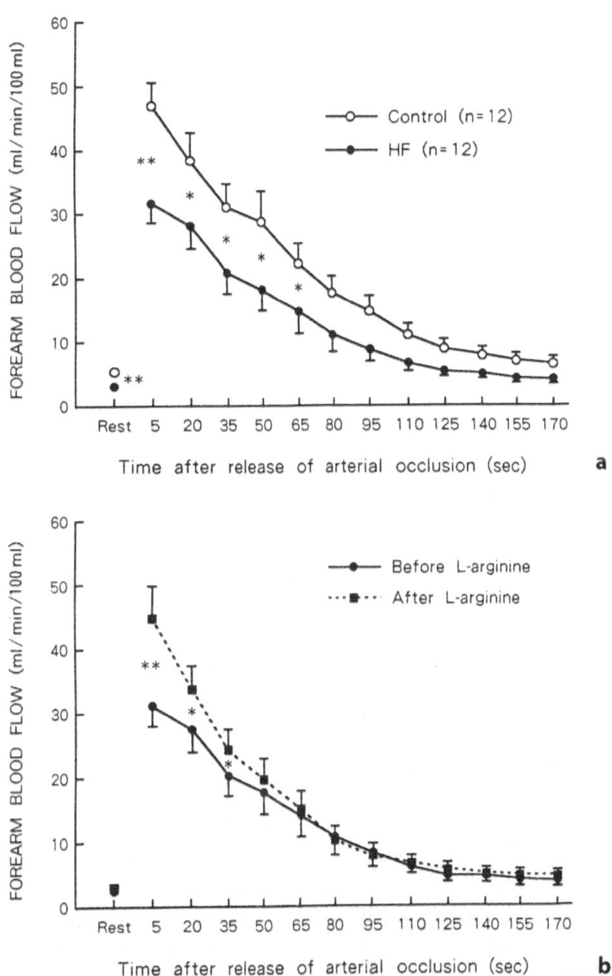

FIG. 4. **a** Time-courses of reactive hyperemic flows in normal subjects (*open circles*) (Control; $n = 12$) and patients with heart failure (*solid circles*) (HF; $n = 12$). *, $P < 0.05$, **, $P < .01$. **b** Pooled data of the time-course of forearm blood flow during reactive hyperemia in patients with heart failure before (*circles*) and after (*squares*) L-arginine. $P < .05$; **, $P < .01$. (Adapted and modified from [39])

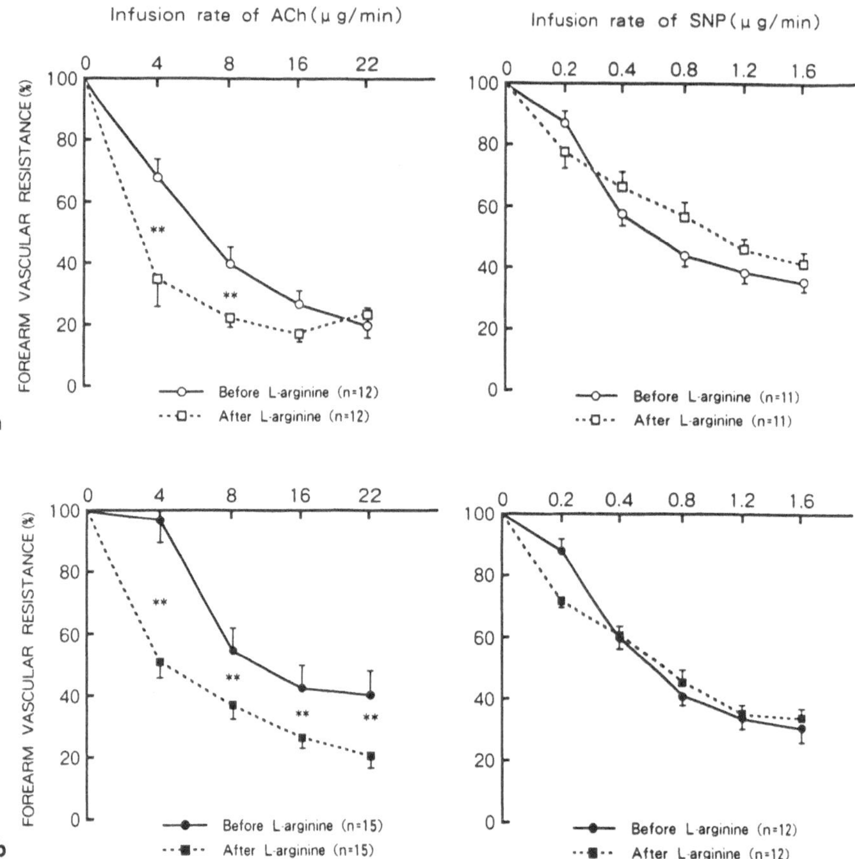

FIG. 5. **a** Responses of forearm vascular resistance in response to infusion of ACh and SNP in normal subjects before (*circles*) and after (*squares*) L-arginine. **b** Responses of forearm vascular resistance in response to ACh and SNP in patients with heart failure before (*circles*) and after (*squares*) L-arginine. The values of forearm vascular resistance before drug infusion are shown as 100%. **, $P < .01$. (Adapted and modified from [39])

alter forearm blood flow, arterial pressure, or heart rate in normal subjects or patients with heart failure as in our previous study [25].

Figure 5 shows forearm vascular responses to infusion of ACh and SNP in normal subjects and patients with heart failure before and after L-arginine. In normal subjects, pretreatment with L-arginine significantly augmented ACh-induced vasodilation at the lower doses but not at the higher doses; in patients with heart failure, however, pretreatment with L-arginine augmented ACh-induced vasodilation not only at the lower doses but also at the higher doses. Thus, pretreatment with L-arginine augmented ACh-induced maximal vasodilation only in patients with heart failure. Pretreatment with L-arginine did not alter responses of forearm vascular resistance to SNP in normal subjects or in patients with heart failure.

Preteatment with L-arginine did not affect the maximal reactive hyperemic flow or minimal forearm vascular resistance in normal subjects. Pretreatment with L-arginine augmented maximal reactive hyperemic flow (Fig. 4b) and decreased minimal forearm vascular resistance ($P < .01$) in patients with heart failure.

Discussion

Our results showed L-arginine augmented ACh-induced vasodilation and reactive hyperemia in patients with heart failure. These findings suggest that decreased maximal vasodilation to ischemic stimuli in patients with heart failure may be caused in part by a defect in the release of NO from the endothelium.

L-Arginine is a precursor of endothelium-derived NO, and supplementation of L-arginine augments coronary and forearm circulation in patients with hypercholesterolemia [40,41]. We and Panza et al. [25,42] have shown that pretreatment with L-arginine augmented ACh-induced vasodilation but did not alter SNP-induced vasodilation in the forearm of healthy humans. Thus, it appears that supplementation of L-arginine facilitates production of NO in the forearm of healthy humans. This study further demonstrated that supplementation of L-arginine augmented ACh-induced vasodilation in patients with heart failure as well as in normal subjects. Of note is that supplementation of L-arginine augmented vasodilation evoked by ACh at the higher doses in patients with heart failure but not in normal subjects. We consider that maximal endothelium-dependent vasodilation that could be evoked by ACh had been achieved in normal subjects so that supplementation of L-arginine did not further augment vasodilation to ACh at the higher doses. Supplementation of L-arginine augmented vasodilation to ACh at the higher doses in patients with heart failure. These results suggest that the decreased vasodilation to ACh at the higher doses in patients with heart failure was most likely caused by defective release of NO from the endothelium. Because venous pressure is high and there may be organic changes in the arteriole, such as an increase in sodium and water content in vessel walls, in patients with heart failure, these mechanical factors could have been responsible for attenuated forearm vasodilation to ACh in these patients. This possibility is unlikely, however, because the vasodilator responses to SNP were comparable between the two groups, and supplementation of L-arginine caused further ACh-induced vasodilation in patients with heart failure. The effects of L-arginine were not nonspecific, because D-arginine did not affect maximal forearm blood flow in some patients with heart failure (data not shown). ACh may cause vasodilation by other mechanisns such as release of prostacyclin and EDHF or inhibition of norepinephrine release from the nerve terminals. It has been shown in normal subjects and patients with essential hypertension that pretreatment with aspirin or phentolamine did not affect the magnitude of ACh-induced forearm vasodilation [43]. Katz et al. [44] showed that administration of indomethacin augments ACh-induced vasodilation in patients with heart failure, suggesting the contribution of cyclooxygenase-dependent vasoconstricting factor. In the same study, phentolamine did not alter ACh-induced vasodilation in patients with heart failure. Thus, increased sympathetic tone in heart failure is not responsible for the impaired vasodilator response to ACh. The contribution of an EDHF is still unknown.

Previous studies have suggested that arterial occlusion for 10 min is the most potent vasodilator stimulus that can be applied to study in the human forearm [45]. In this study, the peak forearm blood flow during reactive hyperemia following 10 min of arterial occlusion was significantly less and the minimal forearm vascular resistance was significantly higher in patients with heart failure than in normal subjects. Our findings are consistent with those of Zelis et al. [46], who showed that maximal vasodilation during reactive hyperemia was reduced in patients with heart failure and suggested that patients with heart failure have impaired ischemic vasodilator mechanisms.

Several mechanisms are considered to account for the decreased reactive hyperemic response in patients with heart failure. A mechanism other than sodium and water retention in vessel walls is involved, because it has been shown that peak reactive hyperemic flow after diuretic therapy was still 32% below nomal [47]. Activated neurohumoral factors or increased venous pressure are unlikely to contribute to the decreased reactive hyperemic response in patients with heart failure [6,46]. The important and unexpected finding of this study is that pretreatment with L-arginine nearly restored impaired ischemic vasodilation during reactive hyperemia in patients with heart failure who did not have peripheral edema at the time of study. This finding may suggest that impaired reactive hyperemic response in our patients with heart failure resulted largely from a defect in release of NO from the endothelium. In this study, peak reactive hyperemic flow was not affected by pretreatment with L-arginine in normal subjects as in the previous study [26]. Thus, supplementation of L-arginine augmented maximal vasodilation evoked by ACh and during reactive hyperemia only in patients with heart failure but not in normal subjects. Interestingly, it has also been shown that L-arginine does not enhance exercise-induced vasodilation in normal subjects [38]. Thus, the capability of the NO pathway during handgrip exercise is not significantly limited by substrate availability.

Conclusion

Our results suggest that defective endothelial function may contribute to impaired forearm vasodilation in patients with heart failure. Also, our results suggest that in normal subjects NO does not play a major role in reactive hyperemia or exercise-induced vasodilation of the forearm. Thus, in patients with heart failure, we speculate that the NO pathway becomes more important as a compensation mechanism in peripheral circulation.

Because such compensation is not enough to retain normal vasodilator capacity, inadequate stimulated release of NO may lead to fatigue or exercise intolerence in patients with heart failure. In supporting this hypothesis, it has been suggested that basal NO production is preserved but stimulated release of NO is attenuated in patients with heart failure [10,48]. It is also possible that improvement of the endothelium dysfunction in heart failure may lead to the increase in exercise tolerance. It has been shown that angiotensin-converting enzyme inhibitors improve impaired vasodilator response and the increase exercise tolerence in patients with heart failure [49,50]. Thus, the beneficial effects of angiotensin-converting enzyme inhibitors may be secondary to improving endothelial function, which also may be the case with supplementation of L-arginine. Further studies are needed to clarify the mechanism(s) of endothelial dysfunction and to determine if the correction of such dysfunction improves impaired exercise tolerence in patients with heart failure.

Acknowledgment. This work was supported by a Grant-in-Aid for Scientific Research from the Japanese Ministry of Education, Science and Culture. We wish to thank our colleagues at our institute who have contributed to research in this field.

References

1. Félétou M, Canet E, Vanhoutte PM (1994) Endothelial NO and vascular regulation. In: Masaki T (ed) Endothelium-derived factors and vascular functions. Elsevier, Amsterdam, pp 47–59
2. Vanhoutte PM, Boulanger CM (1995) Endothelium-dependent responses in hypertension. Hypertens Res 18:87–98
3. Drexler H (1995) Changes in the peripheral circulation in heart failure. Curr Opin Cardiol 10:268-273
4. Kiowski W (1991) Endothelial function in humans. Studies of forearm resistance vessels. Hypertension 18(suppl II):II-84–II-89
5. Benjamin N, Calver A, Collier J, Robinson B, Vallance P, Webb D (1995) Measuring forearm blood flow and interpreting the responses to drugs and mediators. Hypertension (Dallas) 25:918–923
6. Zelis R, Sinoway L, Musch T, Davis D (1989) Vasoconstrictor mechanisms in congestive heart failure. Part 1. Mod Concepts Cardiovasc Dis 58:7–12
7. Hirooka, Y, Imaizumi T, Takeshita A, Ando S, Harada S (1990) Endothelium-dependent forearm vasodilation to intra-arterial acetylcholine is impaired in patients with heart failure. Circulation 82:III–591
8. Kubo S, Rector TS, Bank AJ, Williams RE, Heifetz SM (1991) Endothelium-dependent vasodilation is attenuated in patients with heart failure. Circulation 84:1589–1596
9. Katz SD, Biasucci L, Sabba C, Strom JA, Jondeau G, Balvao M, Solomon S, Nikolic SD, Forman R, Lejemtel TH (1992) Impaired endothelium-mediated vasodilation in the peripheral vasculature of patients with congestive heart failure. J Am Coll Cardiol 19:918–925
10. Drexler H, Hayoz D, Munzel T, Hornig B, Just H, Brunner HR, Zelis R (1992) Endothelial function in chronic congestive heart failure. Am J Cardiol 69:1596–1601
11. Hirooka Y, Imaizumi T, Harada S, Masaki H, Momohara M, Tagawa T, Takeshita A (1992) Endothelium-dependent forearm vasodilation with acetylcholine but not to substance P is impaired in patients with heart failure. J Cardiovasc Pharmacol 20(suppl 12):S221–S225
12. Hirooka Y, Takeshita A, Imaizumi T, Suzuki S, Yoshida M, Ando S, Nakamura M (1990) Attenuated forearm vasodilative response to intra-arterial atrial natriuretic peptide in patients with heart failure. Circulation 82:147–153
13. Imaizumi T, Takeshita A, Suzuki S, Yoshida M, Ando S, Nakamura M (1990) Age-independent forearm vasodilation by acetylcholine and adenosine 5′-triphosphate in human. Clin Sci 78:89–93
14. Förstermann U, Mügge A, Alheid U, Haverich A, Frölich JC (1988) Selective attenuation of endothelium-mediated vasodilation in atherosclerotic human coronary arteries. Circ Res 62:185–190
15. Panza J, Casino PR, Kilcoyne CM, Quyyumi AA (1994) Impaired endothelium-dependent vasodilation in patients with essential hypertension: evidence that the abnormality is not at the muscarinic receptor level. J Am Coll Cardiol 23:1610–1616
16. Cockcroft JR, Chowienczyk PJ, Brettt SE, Ritter JM (1994) Effects of N^G-monomethyl-L-arginine on kinin-induced vasodilation in the human forearm. Br J Clin Pharmacol 38:307–310
17. Miller VM, Vanhoutte PM (1988) Enhanced release of endothelium-derived factor(s) by chronic increases in blood flow. Am J Physiol 255:H446–H451
18. Ontkean M, Gay R, Greenberg B (1991) Diminished endothelium-derived releasing factor activity in an experimental model of chronic heart failure. Circ Res 69:1088–1096

19. Belch JJF, Bridges AB, Scott N, Chopra M (1991) Oxygen free radicals and congestive heart failure. Br Heart J 65:245–248
20. Hirooka Y, Imaizumi T, Masaki H, Ando S, Harada S, Momohara M, Takeshita A (1992) Captopril improves impaired endothelium-dependent vasodilation in hypertensive patients. Hypertension 20:175–180
21. Katz S, Rao R, Berman JW, Schwarz M, Demopoulos L, Bijou R, LeJemtel TH (1994) Pathophysiological correlates of increased serum tumor necrosis factor in patients with congestive heart failure. Circulation 90:12–16
22. Shiramoto M, Imaizumi T, Namba T, Endo T, Takeshita A (1994) Are vasodilator effects of substance P and ATP mediated by nitric oxide in human forearm vessels? Circulation 90:I–321
23. Vallance P, Collier J, Moncada S (1989) Effects of endothelium-derived nitric oxide on peripheral arteriolar tone in man. Lancet 2:997–1000
24. Panza J, Casino PR, Kilcoyne CM, Quyyumi AA (1993) Role of endothelium-derived nitric oxide in the abnormal endothelium-dependent vascular relaxation of patients with essential hypertension. Circulation 87:1468–1474
25. Imaizumi T, Hirooka Y, Masaki H, Harada S, Momohara M, Tagawa T, Takeshita A (1992) Effects of L-arginine on forearm vessels and responses to acetylcholine. Hypertension 20:511–517
26. Tagawa T, Imaizumi T, Endo T, Shiramoto M, Harasawa Y, Takeshita A (1994) Role of nitric oxide in reactive hyperemia in human forearm vessels. Circulation 90:2285–2290
27. Endo T, Imaizumi T, Tagawa T, Shiramoto M, Ando S, Takeshita A (1994) Role of nitric oxide in exercise-induced vasodilation of the forearm. Circulation 90:2886–2890
28. Yamabe H, Okumura K, Ishizaka H, Tsuchiya T, Yasue H (1992) Role of endothelium-derived nitric oxide in myocardial reactive hyperemia. Am J Physiol 263:H8–H14
29. Kostic MM, Schrader J (1992) Role of nitric oxide in reactive hyperemia of the guinea pig heart. Circ Res 70:208–212
30. Wolin MS, Rodenburg JM, Messina EJ, Kaley G (1990) Similarities in the pharmacological modulation of reactive hyperemia and vasodilation to hydrogen peroxide in rat skeletal muscle arterioles: effects of probes for endothelium-derived mediators. J Pharmacol Exp Ther 253:508–512
31. Koller A, Kaley G (1990) Prostaglandins mediate arteriolar dilation to increased blood flow velocity in skeletal muscle microcirculation. Circ Res 67:529–534
32. Messina EJ, Sun D, Koller A, Wolin MS, Kaley G (1992) Role of endothelium-derived prostaglandins in hypoxia-elicited arteriolar dilation on rat skeletal muscle. Circ Res 71:790–796
33. Tagawa T, Imaizumi T, Endo T, Shiramoto M, Hirooka Y, Ando S, Takeshita A (1993) Vasodilatory effect of arginine vasopressin is mediated by nitric oxide in human forearm vessels. J Clin Invest 92:1483–1490
34. Kilbom Å, Wennmalm Å (1976) Endogenous prostaglandins as local regulators of blood flow in man: effect of indomethacin on reactive and functional hyperemia. J Physiol 257:109–121
35. Carlson I, Sollevi I, Wennmalm Å (1987) The role of myogenic relaxation, adenosin and prostaglandins in human forearm reactive hyperemia. J Physiol (Lond) 389:147–161
36. Aversano T, Ouyang P, Silverman H (1991) Blockade of the ATP-sensitive potassium channel modulates reactive hyperemia in the canine coronary circulation. Circ Res 69:618–622
37. Kanatsuka H, Sekiguchi N, Sato K, Akai K, Wang Y, Komaru T, Ashikawa K, Takishima T (1992) Microvascular sites and mechanisms responsible for reactive hyperemia in the coronary circulation of the beating canine heart. Circ Res 71:912–922
38. Gilligan DM, Panza JA, Kilcoyne CM, Waclawiw MA, Casino PR, Quyyumi AA (1994) Contribution of endothelium-derived nitric oxide to exercise-induced vasodilation. Circulation 90:2853–2858

39. Hirooka Y, Imaizumi T, Tagawa T, Shiramoto M, Endo T, Ando S, Takeshita A (1994) Effects of L-arginine on impaired acetylcholine-induced and ischemic vasodilation of the forearm in patients with heart failure. Circulation 90:658–668
40. Drexler H, Zeiher AM, Meinzer K, Just H (1991) Correction of endothelial dysfunction in coronary microcirculation of hypercholesterolemic patients by L-arginine. Lancet 338:1546–1550
41. Creager MA, Gallagher SJ, Girerd XJ, Coleman SM, Dzau VJ, Cooke JP (1992) L-Arginine improves endothelium-dependent vasodilation in hypercholesterolemic humans. J Clin Invest 90:1248–1253
42. Panza JA, Casino PR, Badar DM, Quyyumi AA (1993) Effect of increased availability of endothelium-derived nitric oxide precursor on endothelium-dependent vascular relaxation in normal subjects and in patients with essential hypertension. Circulation 87:1475–1481
43. Linder L, Kiowski W, Buhler FR, Luscher TF (1990) Indirect evidence for release of endothelium-derived relaxing factors in human forearm circulation in vivo: blunted response in essential hypertension. Circulation 81:1762–1767
44. Katz S, Schwarz M, Yuen J, LeJemetel TH (1993) Impaired acetylcholine-mediated vasodilation in patients with congestive heart failure. Circulation 88:55–61
45. Takeshita A, Imaizumi T, Ashihara T, Yamamoto K, Hoka S, Nakamura M (1982) Limited maximal vasodilator capacity of forearm resistance vessels in normotensive young men with a familial predisposition to hypertension. Circ Res 51:457–464
46. Zelis R, Mason DT, Braunwald E (1968) A comparison of the effects of vasodilator stimuli on peripheral resistance vessels in normal subjects and in patients with congestive heart failure. J Clin Invest 47:960–970
47. Sinoway L, Minotti J, Musch T, Goldner D, Davis D, Leaman D, Zelis R (1987) Enhanced metabolic vasodilation secondary to diuretic therapy in decompensated congestive heart failure secondary to coronary artery disease. Am J Cardiol 60:107–111
48. Kubo S, Rector TS, Bank AJ, Raij L, Kraemer MD, Tadros P, Beardslee M, Garr MD (1994) Lack of contribution of nitric oxide to basal vasomotor tone in heart failure. Am J Cardiol 74:1133–1136
49. Mancini D, Davis L, Wexler JP, Chadwick B, Lejemtel TH (1987) Dependence of enhanced maximal exercise performance on increased peak skeletal muscle perfusion during long-term captopril therapy in heart failure. J Am Coll Cardiol 10:845–850
50. Imaizumi T, Takeshita A, Nakamura N, Sakai K, Hirooka Y, Suzuki S, Yoshida M, Nakamura M (1990) Effects of captopril on forearm oxygen consumption during dynamic handgrip exercise in patients with congestive heart failure. Jpn Heart J 31:817–828

Dynamic Characteristics of the Pulmonary Circulation

JOE ALEXANDER, JR.[1], OSAMU KAWAGUCHI[2], and KENJI SUNAGAWA[2]

Summary. The investigations described herein use the white-noise approach together with multiple-input multiple-output systems identification techniques to obtain high-resolution pulmonary vascular admittance spectra. The advantage of such an approach is that it enables characterization of pulmonary vascular properties as a load both for the ejecting right ventricle and the filling left heart, and at the same time accounts for dynamic cross-talk between arterial and venous ports. Moreover, thanks to the white-noise approach, such a characterization is possible despite inherent difficulties that would impede reliable measurement of the relatively low-amplitude pressure–flow signals that characterize the pulmonary venous system port in particular. In one experimental series of open-chest anesthetized dogs, it was found that the retrograde pulmonary venous port admittance could be determined to a 0.1-Hz frequency resolution over a physiological range of frequencies spanning 0.1 to 10 Hz with a relatively high degree of linear correlation (coherence) between the dynamics of left-atrial pressure input and pulmonary venous flow output. The strength of the linear relationship enabled reliable interpretation of the venous admittance in terms of venous compliance, resistance, and inertance, and indicated that under conditions of high dynamic input power, the retrograde admittance of the pulmonary venous system decreased because of the dominance of an inertial component. Similar results were obtained in a separate series of experiments by co-authors Kawaguchi and Sunagawa. The main difference in terms of the pulmonary venous admittance was merely the frequency at which the inertial term became manifest, which was a lower frequency in their experiments.

Key words. White noise—Spectral analysis—Left-ventricular filling—Two-port analysis—Admittance—Impedance

Introduction

In a two-port consideration of its dynamic pressure–flow properties, the pulmonary vascular system acts as a load both for the ejecting right ventricle at its arterial port and for the filling left heart at the venous port. Although the pulmonary arterial port characteristics have been well studied, at least under the assumption that dynamics at

---

[1] Department of Biomedical Engineering, Vanderbilt University, Nashville, TN 37235, USA.
[2] Department of Cardiovascular Dynamics, National Cardiovascular Center, Research Institute, Osaka, 565 Japan.

the venous port do not meaningfully contribute to the dynamic pressure–flow characteristics at the arterial port, pressure–flow dynamics at the venous port and interaction, or cross-talk, between arterial and venous ports have not been systematically investigated. This has been the case despite the obvious importance of the pulmonary venous system in serving as a load for the left ventricle (LV) during filling. Indeed, the venous port admittance characteristics have never before been described. (Admittance is exactly equal to the reciprocal of impedance in a one-port pressure–flow consideration of a vascular bed.)

Such venous admittance data for the left heart constitute essentially the missing link that would enable extension of the ventricular–vascular coupling concept to the filling side of the heart. To be most effective at determining how ventricular and vascular properties interact, the coupling approach relies on the ability to develop mutually independent characterizations of ventricular and vascular properties [1–7]. The dynamics of the left atrium and ventricle during ventricular diastole have been reasonably well characterized [8–11]. The ultimate achievement of a coupling concept at the venoatrioventricular junction might provide new insights into the role of atrial contraction in contributing to left-ventricular filling. For example, because atrial contribution to filling may be regarded as a direct consequence of the relative magnitudes of the retrograde pulmonary venous admittance and the admittance "looking forward" across the mitral valve, dominance of an inertial term in the pulmonary venous admittance, causing it to diminish with high-frequency pressure–flow events (such as during exercise), would favor forward movement of atrial ejected contents and thus promote filling despite the absence of a venoatrial valve to prevent regurgitation.

One problem has been that the measurement of pulmonary venous admittance is very difficult because of the lack of pronounced pulmonary venous pulsations, the difficulty in measuring pulmonary venous flow accurately, the inability (except at harmonics of the fundamental frequency) to resolve admittance spectra by Fourier series analysis, and the fact that some of the measured pulsations in pulmonary venous flow are not determined solely by left-atrial pressure but result from pulmonary arterial pulsations transmitted across the lungs. The purpose of this investigation therefore was to overcome these limitations to determine the pulmonary venous admittance as a quantitative representation of vascular load at the venoatrioventricular junction. To fulfill this ambition, it was necessary to make use of the so-called white noise approach to system identification, initially applied to the cardiovascular system by Taylor [8,12,13], in conjunction with multiple-input multiple-output linear systems analysis techniques [12,14,15] to obtain high-resolution pulmonary venous admittance spectra not contaminated by pulmonary arterial pulsations. Moreover, because of the relatively short sampling period it allows, the approach enabled repeated measures of pressure and flow data for determination of coherence, a kind of frequency-domain correlation function between left-atrial pressure and pulmonary venous flow which, if low, would serve as an index of nonlinear or time-varying vascular system behavior.

Theoretical Background

Compared to the systemic vascular system, the pulmonary vascular bed is complicated by the facts that (a) there is a high ratio of dynamic to static power of pressure and flow phenomena in this low-pressure system (which may have important physiological implications), making it necessary to obtain high-quality instantaneous pressure and flow data for adequate system identification, and that (b) there may be significant communication of dynamic pulsations between the arterial and venous ports [9,11]. This means that in representing either arterial or venous port admittances of the pulmonary vascular bed, one needs to consider the dynamic behavior at both ports. The two-port analysis admittance equations to describe such a system are as follow [12,14]:

$$PAF(\omega) = Yav(\omega)LAP(\omega) + Yaa(\omega)PAP(\omega) \tag{1}$$

$$PVF(\omega) = Yvv(\omega)LAP(\omega) + Yva(\omega)PAP(\omega) \tag{2}$$

where PAP and PAF are pulmonary arterial pressure and flow, respectively, LAP is left-atrial pressure, PVF is total instantaneous pulmonary venous flow, and ω is frequency. If pressures and flows are measured at both ports, it is possible with certain system constraints [12] to solve the equations for all four pulmonary admittances: $Yvv(\omega)$, the pulmonary venous admittance of interest; $Yaa(\omega)$, the pulmonary arterial admittance, and $Yav(\omega)$ and $Yva(\omega)$, the cross-term admittances. As is evident from Eq. 2, the pulmonary venous admittance, $Yvv(\omega)$, is determined not only by $PVF(\omega)$ and $LAP(\omega)$ but also by pulmonary arterial pulsations (manifest through $PAP(\omega)$) operating through the cross-term admittance, $Yva(\omega)$. As we are interested primarily in determining the pulmonary venous admittance, i.e., $Yvv(\omega)$, our analysis may be simplified by considering only Eq. 2. That is to say, the pulmonary venous system may be represented as a two-input single-output system in which the two inputs are PAP and LAP and the single output is PVF. The only mathematical constraint on such a system other than the linearity assumption is that the two inputs not be perfectly correlated [12]. With these constraints in mind, it is possible to expand Eq. 2 into two equations: one equation is obtained by multiplying Eq. 2 by the conjugate of $LAP(\omega)$ and the other by multiplying it by the conjugate of $PAP(\omega)$. The resulting two equations follow:

$$LAP^*(\omega)PVF(\omega) = Yvv(\omega)|LAP(\omega)|^2 + Yva(\omega)LAP^*(\omega)PAP(\omega) \tag{2a}$$

$$PAP^*(\omega)PVF(\omega) = Yvv(\omega)PAP^*(\omega)LAP(\omega) + Yva(\omega)|PAP(\omega)|^2 \tag{2b}$$

where the asterisk (*) denotes a complex conjugate. Having generated Eqs. 2a and 2b in this way, it is then possible to solve them simultaneously for $Yvv(\omega)$. The quantities that result from multiplicative combinations of PAP, LAP, or PVF in the equations are called spectral density functions. (If a signal is multiplied by its own conjugate, the resulting quantity is referred to as the autospectral density function, e.g., $|LAP(\omega)|^2$; otherwise, the product is called the cross-spectral density function, e.g., $LAP^*(\omega)PVF(\omega)$.) Note that the two-input single-output admittance formulation makes it possible to exactly calculate $Yvv(\omega)$ even without simultaneous measurement of $PAF(\omega)$.

One problem that makes determination of pulmonary venous admittance particularly difficult is that, because of the complex manner in which the six or seven pulmonary veins enter the left atrium, there is no current instrumentation that would afford accurate measurement of total instantaneous pulmonary venous volumetric flow. Therefore, we found it necessary in these experiments to determine the admittances in single pulmonary veins and, for extrapolation to total system behavior, assume that the admittances of all the veins are equivalent and arranged in parallel. If such assumptions are accurate, total pulmonary venous admittance can be estimated simply by multiplying the single pulmonary vein admittance by the number of pulmonary veins. The alternative approach would be to individually cannulate all the pulmonary veins and merge their outputs through a common port from which total PVF could be measured instantaneously. The problem with such an approach is that the cannulas, particularly at their common junction, would artificially introduce a significant admittance term of their own which would greatly obscure any measurement of total pulmonary venous admittance.

Materials and Methods

Surgical and Measurement Procedures

Data were obtained from a total of six adult mongrel dogs of either sex weighing 27 ± 1.1 kg. (All summary statistics are reported as mean ± SD.) Each dog was induced with 20–30 mg/kg i.v. sodium pentathol anesthesia, and was subsequently intubated and respired with a 100% O_2/2% isoflurane mixture at a rate of 4–5 l/min. The thorax was opened via median sternotomy and the dog instrumented for measurement of LAP, PVF, aortic flow (AoF), aortic pressure (AoP), and PAP. LAP was measured using a 5F micromanometer tip Millar catheter introduced through a small pulmonary vein. Although it meant sacrificing one of the minor pulmonary veins, this approach to measuring instantaneous LAP was found to be more reliable than introducing the catheter through a left-atrial stab incision because it reduced dynamic artifacts associated with "pressure catheter whip."

The proximal aortic root was exposed by blunt dissection of the surrounding adipose and connective tissue to allow placement of an electromagnetic flow probe (16 or 20 mm in diameter) for measurement of AoF. In some experiments, AoP was measured by use of a 7F Millar pressure catheter advanced to the level of the aortic arch. For measurement of PVF, the left-lower lobe pulmonary vein was exposed by blunt dissection. This particular pulmonary vein, used previously by Rajagopalan et al. [9–11], was selected primarily because it was 4–6 mm in diameter and could generally be isolated close to its termination in the left-atrial chamber as a single vessel, not one that formed multiple terminal branches as it entered the atrium. A Transonic flow probe (Transonic Systems, Ithaca, NY, USA) of either 6 or 8 mm diameter was then positioned around the vein close to its termination at the left-atrial chamber. Acoustical coupling between the flow probe and pulmonary vein was maintained by commercially available jelly.

The flowmeter (Transonic Systems), using the principle of ultrasonic transit time, measured volume flow directly, irrespective of vessel size and flow velocity profile, and provided automatic zeroing and calibration capabilities. With such a flow measurement system, the probe diameter could be chosen to be larger than that of the vessel, eliminating any concern over constriction of the pulmonary vein that would cause secondary alterations in its local mechanical properties. PAP was measured using a 7F micromanometer tip Millar catheter advanced to the root of the pulmonary artery. Finally, the surgical preparation was completed with the suturing of a bipolar electrode to the right-atrial appendage for normal and pseudorandom pacing protocols.

Pseudorandom pacing was achieved by superimposing a train of pseudorandom pulses delivered from a stimulator (Grass S88 Stimulator Grass Instruments, Quincy, MA, USA) on top of natural sinus activity.

Data were acquired and stored using an online computer digitizing at a rate of 200 Hz. All recordings analyzed from these experiments were made with the respirator clamped at end-expiratory volume. Arterial blood gases were sampled periodically and maintained at normal levels by adjustment of ventilatory parameters or by intravenous administration of sodium bicarbonate.

Experimental Protocol

The experimental protocol, which was quite simple, was as follows. With each dog, our first step was to slowly infuse isotonic saline (Normosol, pH = 7.4) to raise mean PAP, and consequently mean LAP, to control levels (24.2 ± 2.6 and 14.2 ± 3.6 mmHg, respectively). We controlled these pressures at moderately elevated levels as a precautionary measure against potential nonlinearities that could arise from venous collapse, inherent nonlinearity of vascular properties, or a "vascular waterfall"-type mechanism in which alveolar pressure, if higher than LAP, would serve as the downstream driving force for transmission of instantaneous flow across the pulmonary vascular bed [16]. Next, we paced the heart and recorded steady-state periodic pressure and flow data over a range of heart rates to verify that the instantaneous single-vein pulmonary venous flow made physiological sense in relation to other more global hemodynamic measurements such as LAP and LVP. Moreover, because our admittance characterization would span a fairly broad frequency range, we examined steady-state periodic data of heart rates ranging typically from about 70 to 150 beats per minute (bpm).

Once the steady-state periodic protocol was complete in confirming physiologically meaningful pulmonary venous flow signals, pseudorandom pacing was initiated. After 5–10 s of pseudorandom pacing, data acquisition was started and continued for at least 51.20 s. After the acquisition period, pseudorandom pacing was discontinued. The respirator was clamped at end-expiratory volume throughout the acquisition period.

Determination of Admittance and Coherence

The use of random and pseudorandom stimulation, or so-called white-noise, to determine the characteristics of biological systems was reviewed in detail by Marmarelis and Marmarelis [15]. In our particular experiments, the digitized LAP, PVF, and PAP signals obtained during pseudorandom atrial pacing were windowed and Fourier transformed to enable calculation of pulmonary venous admittance according to Eqs. 2a and 2b.

Signal processing was as follows. Pseudorandom LAP, PVF, and PAP signals were sampled at a rate of 200 Hz for a total period of 51.20 s that was subsequently divided into five nonoverlapping 10.24-s bins [2^{10} (1024) points in each bin] of pressure and flow data so that the coherence could be calculated. The 10.24-s bin size was chosen so as to provide a 0.1-Hz frequency-domain resolution of admittance spectra. Each bin was analyzed in the following way. First, LAP, PVF, and PAP were windowed with a Hamming window and then Fourier transformed using a Radix-2 FFT algorithm. Next, using these preconditioned signals, the spectral density functions were calculated for each bin separately and then averaged over the five bins. The averaged spectral density functions were used for simultaneous solution of Eqs. 2a and 2b for the pulmonary venous admittance, $Yvv(\omega)$.

We also calculated the ordinary coherence function (γ), which provided an index of the extent to which the LAP(ω) and PVF(ω) at each value of ω were correlated during the data acquisition period. The method for calculating coherence was the same as that offered previously by Alexander et al. [16–18]. To explain briefly, γ(ω) varies between 0 and 1. In the presence of biological variation, system nonlinearity, or sources of random noise not related to the input, the coherence function becomes less than unity. The closer to unity, the more correlated are LAP and PVF, and consequently the more reliable the estimate of pulmonary venous admittance values at that frequency. The determination of coherence is particularly helpful in our analysis because the signals involved are of relatively small magnitude.

Results

Satisfied that the PVF was reasonable as judged from steady-state periodic recordings, we then calculated the pulmonary venous admittance from pseudorandom data. Time-domain signals of PAP, LAP, and PVF were obtained during pseudorandom pacing. From each animal, such data were recorded for at least 51.20 s at 200 Hz and used to calculate pulmonary venous admittance with a 0.1-Hz frequency resolution, as already described.

Figure 1 shows the single-vein pulmonary venous admittance spectrum, Yvv(ω). This is the admittance that the atrium would "see" when it looks back toward the pulmonary veins. (Note that for simplicity in this two-port system analysis approach, admittance may be thought of as the reciprocal of impedance. However, such a

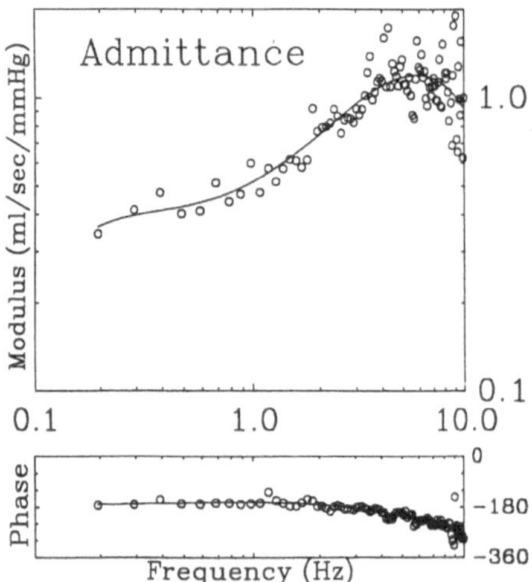

FIG. 1. Example of pulmonary venous admittance spectrum. This is the admittance the atrium "sees" when it "looks back" toward the pulmonary veins. The good frequency resolution is a consequence of pseudorandom pacing. In the case of the pulmonary veins, the admittance modulus reaches a maximum at about 6–7 Hz then starts to decrease. The low-frequency phase angle approximates −180°. As the frequency increases, the phase angle initially becomes less negative because of a capacitative effect. However, as the frequency is further increased to beyond the frequency of maximum admittance modulus, the phase rapidly accelerates toward −270° because of dominance of an inertial term

relationship is exactly true only if the pulmonary vascular bed were considered to be a one-port system.) The calculated spectrum has a good frequency resolution because of the frequency enrichment afforded by the pseudorandom pacing. Nearly 100 admittance values are plotted in Fig. 1, expressed as modulus and phase components.

The admittance modulus is similar to that of the systemic arteries in that in the low-frequency range admittance is a minimum and increases with frequency. However, in the case of the pulmonary veins, the admittance modulus reaches a maximum at around 6–7 Hz then starts to decrease. The low-frequency phase characteristics are consistent with our earlier observation of periodic data: the phase is close to −180°. Inspection of how the phase changes as a function of frequency reveals that as frequency increases the phase becomes less negative because of a capacitative effect. As the frequency is further increased beyond the frequency of maximum admittance modulus, the phase rapidly accelerates toward −270° because of dominance of an inertial term.

The ordinary coherence function associated with the spectrum, $\gamma(\omega)$, was also calculated. Routinely in all animals, the coherence was remarkably good up to almost 10 Hz, except in the vicinity of 1–2 Hz. We suspect that the consistent deterioration in that frequency range was caused by either variability in the diameter of the pulmonary venous ostium with atrial contraction or a heart motion artifact causing intermittent collapse of the pulmonary vein, especially in an open-chest preparation.

The pulmonary venous admittance presented in Fig. 1 is representative of results from all animals in this particular series of experiments in terms of both general characteristics and quantitative findings. Summary results show that the single-vein admittance modulus was at a minimum of $0.39 \pm 0.10\,\mathrm{ml\,s^{-1}\,mmHg^{-1}}$ below 1.0 Hz; that it was at a maximum of $0.97 \pm 0.19\,\mathrm{ml\,s^{-1}\,mmHg^{-1}}$ at 6.9 ± 0.6 Hz; and that the maximum phase angle was $-272° \pm 15.7°$ (Table 1).

Results from a similar series of experiments done by co-authors Kawaguchi and Sunagawa are shown in Fig. 2 for pulmonary arterial and venous port driving point (Fig. 2a) and cross-talk admittances (Fig. 2b). Their venous port driving point admittance, $Yvv(\omega)$, corrected for cross-talk admittance, $Yva(\omega)$, is essentially in agreement with results already presented here, particularly with respect to the importance of inertial characteristics in the venous port admittance compared to the admittance at the arterial port. From their results, however, it appears that the inertial term began to play a significant role even at a lower physiological frequency range.

Discussion

The dynamic characteristics of the pulmonary venous system are of great importance in determining left-ventricular filling. The contracting left atrium, for example, would tend to force blood retrograde into the pulmonary veins were it not for the fact that

TABLE 1. Summary admittance results ($n = 6$ dogs).

Frequency	Admittance ($\mathrm{ml\,s^{-1}\,mmHg^{-1}}$)	Frequency (Hz)
Minimum	0.39 ± 0.10	<1.0
Maximum	0.97 ± 0.19	6.9 ± 0.6

Phase angle at the maximum frequency of 10 Hz = $-272° \pm 15.7°$.

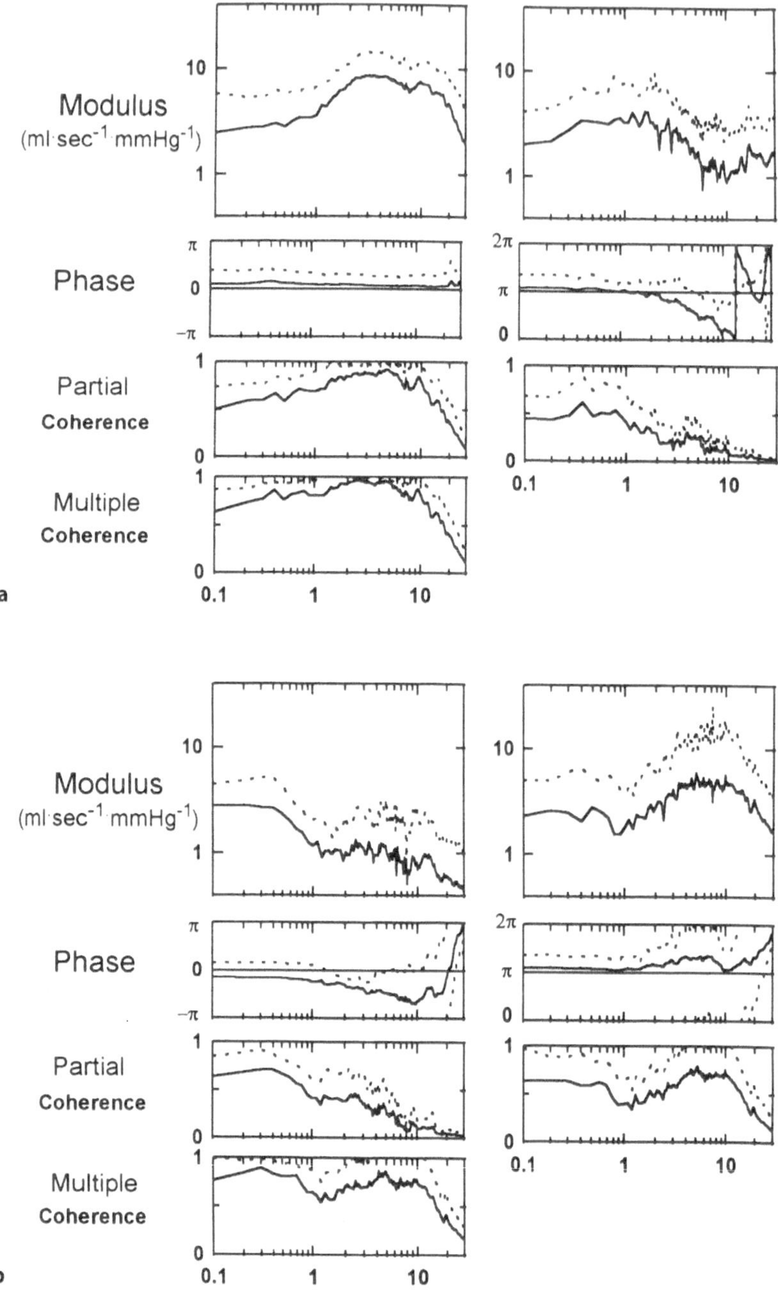

FIG. 2a,b. **a** Summary data from experiments by co-authors Kawaguchi and Sunagawa characterizing the arterial components, Yaa(ω) and Yaa(ω), of the total pulmonary vascular two-port dynamic characteristics in 10 experimental animals. From top to bottom in the *left panel* are the pulmonary arterial driving point admittance modulus, phase, and coherences for Yaa(ω); characteristics of Yav(ω) are shown in the *right panel*. **b** For the same series of animals, for venous components Yvv(ω) and Yva(ω) of the total pulmonary vascular two-port dynamic characteristics: from top to bottom in the *left panel* are the pulmonary venous driving point admittance modulus, phase, and coherences for Yvv(ω); characteristics of Yva(ω) are shown in the *right panel*

the retrograde venous admittance usually is lower than that which the atrium would see looking forward across the mitral valve. Our results indicate that under conditions of high dynamic input power (such as with high heart rates), the retrograde admittance of the pulmonary venous system decreases because of dominance of an inertial component. This mechanism makes it possible for the atrial pump to promote left-ventricular filling during tachycardia when the time available for passive atrial discharge is shortened, even though there is no valve per se to prevent atrial ejected contents from regurgitating into the pulmonary veins.

It is also important to note that exact interpretation of the pulmonary venous admittance spectrum is complicated by the fact that the same admittance characteristics must both promote the forward passage of flow from the pulmonary capillaries through the pulmonary veins on toward the atrium and diastolic ventricle at the same time that it must retard any retrograde flow which might result from atrial contraction. Our results suggest that because pulmonary capillary pressure would represent a relatively low-frequency input to the pulmonary venous system, the forward flow it would induce, because of its operation at frequencies in the spectrum where admittance is relatively high, would travel forward through the pulmonary venous system; however, the higher frequency retrograde left-atrial pressure input would see a decreased admittance looking retrograde into the pulmonary venous system because it is operating at the higher frequency range of the venous admittance spectrum where the inertial effect is dominant, causing a decrease in admittance with increases in frequency.

The white-noise approach implemented through pseudorandom pacing as presented in this study was of tremendous value in enabling us to identify the high-resolution pulmonary venous admittance at the same time that it allowed us to calculate coherence to assess the validity of the linear assumption. However, there are practical limitations in its use in nonacute studies in which the high risk of inducing ventricular fibrillation would constitute an irreconcilable hazard. Overcoming this problem in future studies may prompt investigators to rely on techniques such as autoregressive modeling, a technique shown in some applications to produce accurate, high-resolution system characteristics despite minimal cyclical variations in the input [19,20].

Finally, our analysis showed frequency ranges in the pulmonary venous admittance spectrum at which coherence was diminished, indicating a failure of the linearity assumption. It would be entirely appropriate to extend our analysis to estimation of a nonlinear term as described by Marmarelis and Marmarelis [15] to try to account for this error, the so-called second-order nonlinear kernel. We have not so far undertaken to do so in our investigations of vascular properties, primarily because of the tremendous computational expense of the calculation and also because in a system such as ours in which the input is not truly "white" it is difficult to estimate a nonlinear kernel that does not itself depend on the particular input used for its determination.

Acknowledgments. Joe Alexander Jr. was supported by U.S. Public Health Service, National Institutes of Health, grant R29-HL-46344.

References

1. Latson TW, Hunter WC, Katoh N, Sagawa K (1988) Effect of nitroglycerine on aortic impedance, diameter, and pulse wave velocity. Circ Res 62:884–890

2. McDonald DA (1968) Regional pulse wave velocity in the arterial tree. J Appl Physiol 24:73–78
3. McDonald DA (1974) Blood flow in arteries, 2nd edn. Arnold, London
4. Milnor WR (1975) Arterial impedance as ventricular afterload. Circ Res 36:565–570
5. O'Rourke MF, Taylor MG (1967) Input impedance of the systemic circulation. Circ Res 20:365–380
6. Patel DJ, Defreitas FN, Fry DL (1963) Hydraulic input impedance to the aorta and pulmonary artery in dogs. J Appl Physiol 18:134–140
7. Westerhof N, Sipkema P, Van Den Bos GC, Elzinga G (1972) Forward and backward waves in the arterial system. Cardiovasc Res 6:648–656
8. Alexander J Jr, Sunagawa K, Chang N, Sagawa K (1987) Instantaneous pressure-volume relationship of the ejecting canine left atrium. Circ Res 61:209–219
9. Rajagopalan B, Friend JA, Stallard T, Lee GJ (1979) Blood flow in pulmonary veins: I. Studies in dog and man. Cardiovasc Res 13:667–676
10. Rajagopalan B, Friend JA, Stallard T, Lee GJ (1979) Blood flow in pulmonary veins: II. The influence of events transmitted from the right and left sides of the heart. Cardiovasc Res 13:677–683
11. Rajagopalan B, Bertram CD, Stallard T, Lee GJ (1979) Blood flow in pulmonary veins: III. Simultaneous measurements of their dimensions, intravascular pressure and flow. Cardiovasc Res 13:684–692
12. Bendat JS, Piersol AG (1986) Random data: analysis and measurement procedures, 2nd edn. Wiley, New York
13. Taylor MG (1966) Use of random excitation and spectral analysis in the study of frequency-dependent parameters of the cardiovascular system. Circ Res 18:585–595
14. Javid M, Brenner E (1963) Analysis, transmission, and filtering of signals. McGraw-Hill, New York
15. Marmarelis PZ, Marmarelis VZ (1978) Analysis of physiologic systems: the white-noise approach. Plenum, New York
16. Alexander J Jr, Nikolic S, Fleming GP, Ghai P, Patel P, Yoran C, Frater RWM, Yellin EL (1988) The role of the pulmonary venous impedance and the atrium in determining characteristics of left ventricular filling. In: Proceedings of the IXth international conference of the Cardiovascular Systems Dynamics Society, Halifax, Nova Scotia
17. Alexander J Jr, Burkhoff D, Schipke J, Sagawa K (1989) The influence of mean pressure on aortic impedance and reflections in the systemic arterial system. Am J Physiol 257:H969–H978
18. Burkhoff D, Alexander J Jr, Schipke J (1988) Assessment of Windkessel as a model of aortic input impedance. Am J Physiol 255:H742–H753
19. Kubota T, Itaya R, Alexander J Jr, Todaka K, Sugimachi M, Sunagawa K (1991) Autoregressive analysis of aortic input impedance: comparison with Fourier transform. Am J Physiol (Heart Circ Physiol) 260:H998–H1002
20. Oppenheim AV, Schafer RW (1975) Digital signal processing. Prentice-Hall Englewood Cliffs

Part 4
Coronary Circulation–Ventricular Interaction

Intramyocardial Regulation of Coronary Dynamics

J. Yasha Kresh, H. Frederick Frasch, Igor Izrailtyan, and Stanley K. Brockman

Summary. The intramural blood vessels and fluid-filled interstitial space form a hydraulic continuum enmeshed by myocardial muscle layers and collagen matrix. An underlying theoretical and conceptual construct presented here emphasizes the dynamic interplay between interstitial and intravascular compartments and the resultant microvascular resistance in response to extravascular changes, be these hydraulic, connective tissue (collagen) dependent, or intramyocardial fiber stress in origin. The lumenal patency or narrowing of the coronary resistance vessels depends, in part, on the supporting structure of the myocardium, mediated through collagen matrix attachments (tethering) and interstitial fluid (hydraulic skeleton) matrix. Both these architectural structures provide scaffolding that supports muscle cells and blood vessels. Multiple interactions, including fluid transport, at the vessel-interstitium-muscle interface level are principally responsible for the mechanical regulation of coronary flow dynamics. Analysis of the intramyocardial mechanical milieu in terms of solid elastic theory concepts can be misleading. By exploring the spectrum (beating-isobaric, isovolumic; arrested-systolic, diastolic) of intramyocardial mechanical states, an impetus for a more definitive framework characterizing coronary physiology may emerge.

Key words. Intramyocardial pressure—Interstitial fluid—Collagen matrix—Microcirculation—Dynamic interaction

Introduction

Coronary blood flow regulation has been a much-disputed domain, dominated by a school of thought with a view that myocardial extravascular compression, with each systole, inhibits or shuts off coronary inflow. The resurgence of interest in understanding fully the mechanism(s) regulating coronary flow dynamics (i.e., muscle/interstitial compartment) is in part related to accumulated new evidence pointing to a diminished role of generated chamber pressures to the genesis of phasic attributes of coronary flow [1–3].

The question before us remains: What are the dominant forces contributing to apparent coronary microvessel conductance during contraction and relaxation? Myo-

Departments of Cardiothoracic Surgery and Medicine, Medical College of Pennsylvania and Hahnemann University, Philadelphia, PA 19102-1192, USA.

cardial intrinsic mechanical activity and the associated interstitial muscle-pump mechanism has resurfaced as the major determinant of perfusion [2].

The aim of this writing is to present new experimental data and concepts that advance our knowledge and understanding of the mechanical and hydraulic forces impinging on the coronary microvascular arteriovenous network. Our goal in presenting these concepts (some of which are just emerging) is to stimulate further research into the myocardial mechanics, coronary perfusion, transcapillary fluid transport, and their mutual interactions [3,4].

Myocardial Architecture and Model Description

The myocardium is a composite of muscle (myocytes), coronary microcirculation, and interstitium containing fibrillar collagen. Based on the morphological studies of Caulfield and Borg [5], each point in the myocardial continuum can be represented by a basic structural unit that consists of muscle fibers and microvessels attached to an ensemble of collagen fibers. Cardiac myocytes and capillaries are enmeshed in a network array of extracellular connective matrix [3,6]. This structure is organized into a complex arrangement of intercellular and pericellular fibers and fibrils [7] that serves as a supporting framework (scaffolding) for the contractile cells. Morphologically, the heart's collagen matrix consists of a complex weave with tendinous insertions that surround myocytes, grouping them into myofibers; strands of collagen that connect myofibers; and collagenous struts that join myocytes to other myocytes and capillaries [7,8].

It is important to be reminded that the four-chambered heart develops from a single inflow–outflow tube composed of two layers of muscle (inner endocardium–outer myocardium) separated by a thick extracellular matrix called cardiac jelly [9]. In conceptualizing this complex tissue–fluid milieu, the rheology of the heart may be appropriately described by a continuum of fluid–fiber–collagen stress (σ) tensor [10]:

$$\sigma = \sigma_f + \sigma_c - P_{im} I$$

where σ_f is the muscle fiber stress, σ_c is the collagen fiber stress, P_{im} is the intramyocardial fluid pressure (related to IMP), and I is a unit tensor. This form of stress analysis incorporates a realistic characterization of the myocardium in which intramyocardial pressure has no predetermined value but is dictated by local intramural forces. Alternatively, the resultant intramyocardial body forces can be expressed in terms of an average equivalent (isotropic) stress. In such a framework, intramyocardial tissue pressure will possess both a fluid and a solid phase. A schematic description of the basic constitutive elements in the myocardial tissue that are known to influence and be influenced by the microvascular network is shown in Fig. 1.

The resistance and capacitance of the vascular bed (Fig. 2) are not determined a priori but are functions of the cross-sectional area of the vessel, which is in turn a time-dependent function of intravascular and extravascular pressure gradients [2,11]. It is important to emphasize again that within the myocardial wall the vessels are not free floating but constrained in the interstitial surrounding and tethered by a connective tissue matrix. Accordingly, the equivalent intramyocardial capacitance [1] will be a composite of the interstitial myocardial capacitance (C_m) and vascular capacitance (C_v) such that the total capacitance C is given by $1/C = 1/C_m + 1/C_v$.

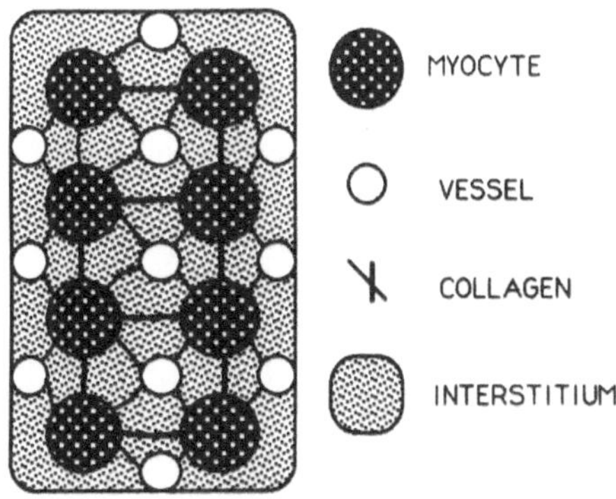

FIG. 1. Idealized myocardial unit. Diagrammatic representation of a structural composite: myocyte–microvessel–collagen (fibrillar connective tissue) spatial organization

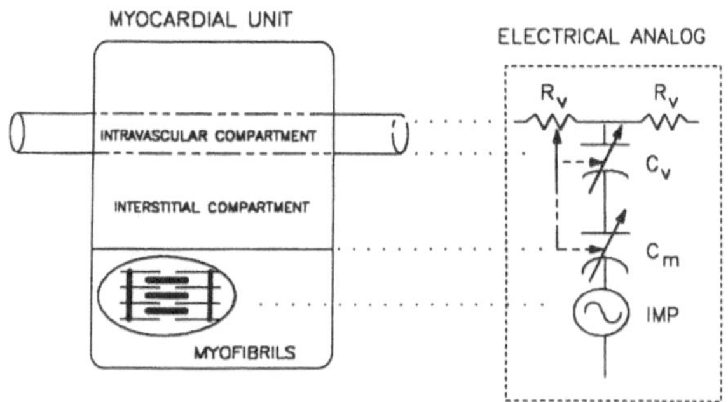

FIG. 2. *Left panel:* Basic myocardial unit consisting of intravascular compartment, interstitial compartment (extravascular fluid–collagen matrix), and extravascular stress generator (myofibrils). *Right panel:* Corresponding electrical analog of the myocardial unit. R_v, Vascular resistance; C_v, vascular capacitance; C_m, extravascular (interstitial) capacitance; *IMP*, intramyocardial pressure source

Moreover, to fully characterize the dynamic relationship between myocardial muscle contraction and resultant coronary flow, one must account for the interstitial fluid mass balance [3,4], by examining a number of functional elements governing the three important interactive phenomena: the net rate of interstitial fluid mass accumulation (dV_{int}/dt) is equal to the net fluid influx across the capillary wall ($S_{cap}J_w$) minus the net lymphatic outflow (F_{lymph}):

$$\frac{dV_{int}}{dt} = S_{cap}J_w - F_{lymph}$$

The fluid influx, J_w, across the capillary wall [4] is given by a modified Starling relation:

$$J_w = L_p\left[\left(P_{cap} - IMP\right) - \nu\Delta\pi\right]$$

where L_p is the hydraulic conductivity, S_{cap} is total capillary wall area, P_{cap} is the capillary pressure, ν is the reflection coefficient, and $\Delta\pi$ is the osmotic pressure gradient. In this context the lymph flow (F_{lymph}) is assumed to be a direct function of IMP [12]:

$$F_{lymph} \propto IMP$$

The details of the constitutive equations including geometrical model constrains and wall kinematics (deformation analysis) integral to the mathematical development are described elsewhere [2,4,13]. The theoretical considerations are presented here to illustrate the close coupling between blood circulation, interstitial fluid mass balance, and muscle mechanics. Clearly, an approach to an integrated model that accounts for the three facets of the myocardial structure (muscle fiber–interstitium–blood vessel) is a prerequisite for greater understanding of the origins and magnitude of intramyocardial tissue stresses and how they are coupled to intravascular pressure, including what constitutes the basis of interaction between systole and diastole in regulating myocardial perfusion. Importantly, a number of characteristic features can be demonstrated by means of this analytical construct [2,4]:

I. The interstitial compartment acts as a hydraulic matrix (skeleton). A rise in intramural fluid volume will increase diastolic intramyocardial pressure (erectile effect), leading to a decreasing diastolic flow.
   i. A selective destruction of the extracellular collagen and partial uncoupling of intramyocardial structures is expected to alter the interstitial hydraulic dynamics.

II. Flow impediment is in part dependent on traction forces imposed on the microcirculation by the surrounding environment (e.g., collagen attachments, interstitial–intracellular fluid transport).
   i. Enhanced muscle contraction alters the extravascular fluid compartment. Thus, active relaxation may enhance diastolic perfusion. (Muscle recoil may generate transiently a negative interstitial fluid pressure.)
   ii. Intramyocardial collagen network supports and tethers intramural vessels, preventing collapse. Disruption of this structure will alter coronary perfusion and interstitial dynamics adversely.

III. Interstitial edema resulting from microvascular displacement of volume (capillary leak) or states in which the interstitial space is not actively pumped will result in an increased effective resistance. Conversely, decrease in tissue fluid pressure (e.g., hyperosmotic perturbation) will reduce the microvascular resistance.

Experimental Findings

Historically, studies of coronary physiology were restricted to the analysis of pressure–flow relationships with considerable overexuberance for the ever-elusive zero-flow pressure intercept point. Much of this emphasis stems from a long-standing dogma which assumes that the intramyocardial structural and functional tissue properties remain invariant to changes in upstream (perfusion) and downstream (sinus/venous) pressures. In fact, coronary perfusion pressure per se influences more than just inflow; it modifies the hydraulic environment (mass/fluid transport) and alters the contractile mechanics of the myocardium. To examine specifically the mechanical coupling function of the interstitial compartment to the embedded microcirculation, coronary blood flow–perfusion pressure (CBF–CPP) relations were studied in an isolated, unloaded [left-ventricular pressure (LVP) ≤ 0], beating, and KCl-arrested hearts. The mean CBF–CPP relation was linear ($r = .91$–$.99$) in both cases.

Most surprisingly, the slope of CBF-CPP did not change ($P > .6$) with arrest. The elevated intramyocardial tissue pressure (IMP) induced by interstitial water swelling in the arrested hearts resulted in a similar mean IMP ($P > .1$), generated in the beating state. Particularly noteworthy is the functional similarity ($P > .1$) between the coupling of coronary perfusion and manifested tissue pressure in the two distinctly different mechanical states (Fig. 3). Equally striking was the observation that the zero-flow (Z_f) pressures (CPP at Z_f and IMP at Z_f) were minimally affected by cardiac arrest ($P > .25$). Under these experimental conditions, the similar flow–pressure responses in beating and arrested hearts reveal the dominant role of extravascular forces, whether they are hydraulic in origin (magnified by the induced interstitial water swelling) or generated by active muscular contraction. Anderson and Jonson [14] showed that tissue fluid volume was greater in the arrested heart when compared to the beating state. In the arrested heart, the increased fluid volume of the interstitial compartment may be responsible for the larger hydraulic interstitial effect and thereby compensating for the absence of active muscle forces on the embedded coronary microvasculature. The finding that the tissue compliance in beating and arrested hearts was shown to be similar or greater in the beating state [14] provides additional support for this interpretation. However, in the beating heart, particularly at low inflow, the distinct phasic nature of the extravascular forces and not merely their average magnitude may affect uniquely coronary flow, including the much "celebrated" zero-flow perfusion pressure values [11].

To heighten the influences of the interstitial compartment on changes in extravascular impedance, an osmotic pressure transient (50 mOsm) was imposed on the arrested (KCl) isolated rabbit heart (Fig. 4). The exaggerated decrease in IMP suggests that this is caused by net fluid shift from the interstitium into the intravascular space

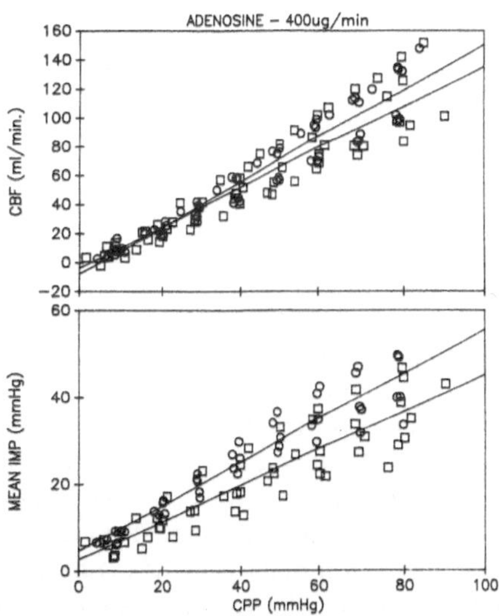

FIG. 3. Pooled data for beating (*circles*) [isobaric left-ventricular pressure (LVP) = 0] and arrested (*squares*) isolated rabbit heart. Similar coronary flow–pressure relations (coronary blood flow–coronary perfusion pressure, *CBF–CPP*) in the beating and arrested heart point to the dominant role of intramyocardial (tissue-fluid) pressure (*IMP*) whether induced as elevated interstitial water swelling or physiologically generated by muscular contraction

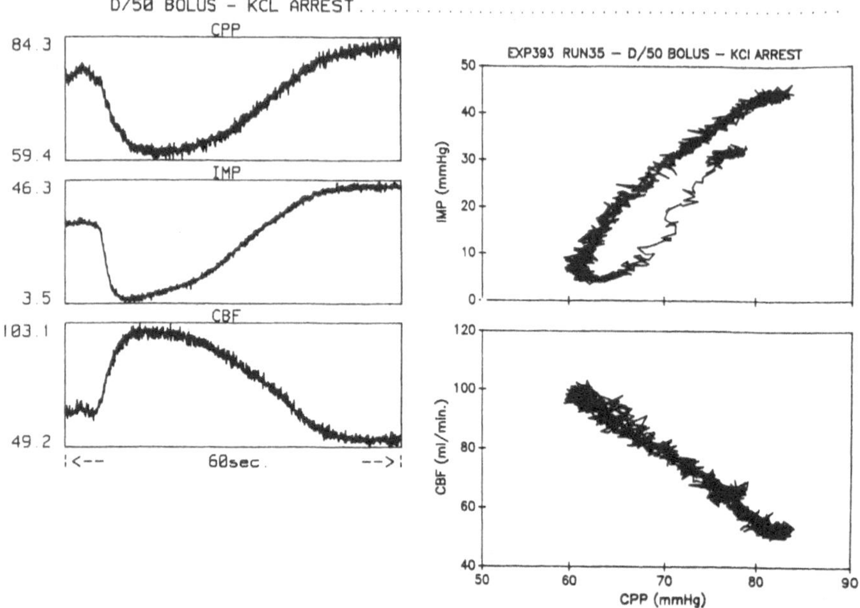

Fig. 4. Effects of osmotic pressure transient (50 mOsm) during KCl-induced arrest. The three frames on *left* show time-course of changes in coronary perfusion pressure (*CPP*), intramyocardial pressure (*IMP*), and coronary blood flow (*CBF*). The decrease in IMP suggests that this is caused by a net fluid flux from interstitium into the intravascular space, i.e., retraction of the vessel lumen into the interstitial space by collagen attachments

or by collagen (re)traction of myocytes, pulling open the vessel lumen. It would seem that the intravascular compartment volume is actively increased (remembering also that resistance = k/vol^2) as evidenced by a drop in CPP, i.e., decrease in microvascular resistance and 100% increase in CBF. Most importantly, when collagen was dissolved, a similarly resulting drop in IMP did not bring about enhancement of CBF.

Hyaluronic acid is the major glycosaminoglycan of the cardiac extracellular matrix (i.e., cardiac jelly). Hyaluronidase, an enzyme that specifically degrades hyaluronic acid, was used to study the influences of extracellular matrix components on structural and functional myocardial changes [15]. Profound changes in coronary vascular resistance were observed (Fig. 5: slope of CBF–CPP relation). It would appear, because of the hydrolysis of the interstitial proteoglycan gels, that the bound water resulting from the iatrogenically induced edema was lost as evidenced by a significant decrease in tissue pressure. The results from these studies are compelling evidence for the way changes in fluid transport and distribution not only alter the functional coronary flow dynamics, but can have a profound influence on the mechanoenergetics of the heart resulting from the "remodeling" of the hydraulic structural organization of the myocardium.

Coronary flow modulation is also a function of passive stresses exerted on the microcirculation by its surrounding environment, which includes a complex collagen network. In particular, the extravascular matrix of collagen struts and weave has been proposed [5,6,16] to maintain vascular patency when extravascular forces exceed the local intravascular pressure. It can be demonstrated [8] that infusion of disulfide reagents [5,5′-dithio-bis(2-nitrobenzoic acid) (DTNB) or oxidized glutathione (GSSG)] activates a collagenolytic system, resulting in near-complete loss of the collagen struts that interconnect myocytes and capillaries to all adjacent myocytes and

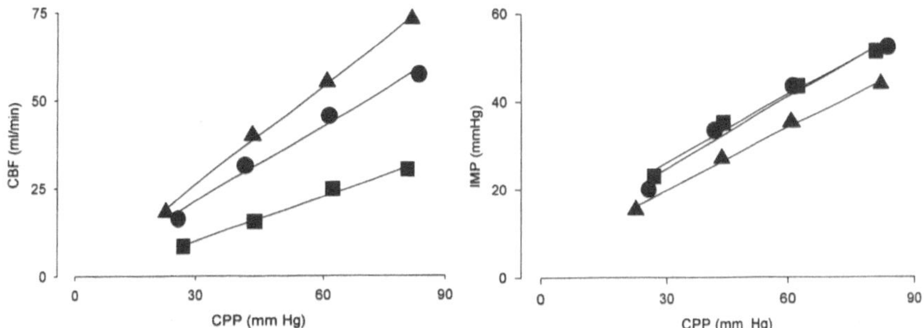

Fig. 5. Myocardial edema (*squares*) and hyaluronidase (*triangles*) effect. Note that profound changes in coronary vascular resistance (slope of CBF–CPP relation) were observed. The iatrogenically induced edema was lost, evidenced by significant decrease in tissue pressure (*IMP*) at all levels of perfusion pressure (*CPP*). *Circles* indicate controls

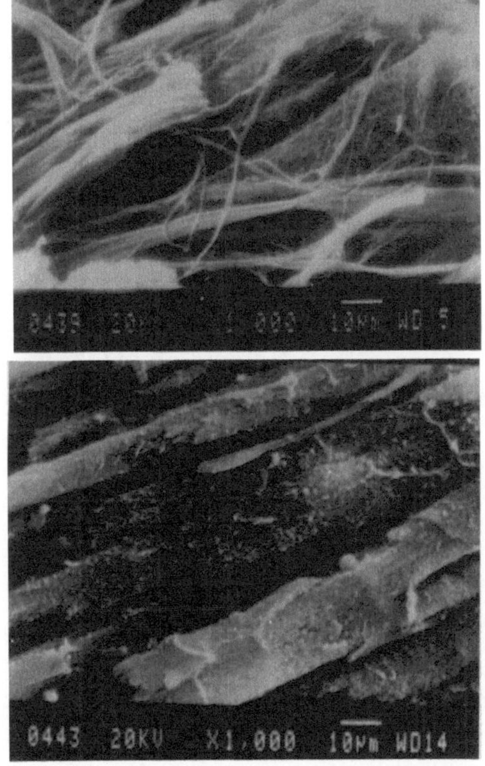

Fig. 6. *Top panel:* Scanning electron micrograph of normal rabbit heart shows complex weave of collagen bundles and large tendon-like structures. *Bottom panel:* Loss of myocyte–connective tissue interface. Usual collagen complex (latticework) is clearly deficient following 2-h 5,5′-dithio-bis-(2-nitrobenzoic acid) (DTNB) perfusion. Most of the collagen bundles on the surface of the myocytes as well as the struts that interconnect the cells are missing, resulting in collagen-free intermuscular space

the weave complex which surrounds groups of myocytes. The acute collagen lysis resulted in a striking reduction in CBF (mean/pulse) and markedly elevated diastolic IMP [16]. Scanning electron microscopy (Fig. 6) revealed degradation of the extracellular collagen matrix and an enlarged interstitial compartment. In view of these findings, it is important to recognize that a tight coupling is present between the extracellular collagen matrix and hydraulic skeleton of the heart.

Indeed, this "remodeling" did not alter the ability of the myocardium to generate similar systolic IMP. The resulting enlarged hydraulic internal load may have compensated for the loss of normal myocardial structural integrity. Mechanically, collagen fibers can support a tensile stress but not a compressive stress. Consequently, increasing size of the extravascular space should have differential effects in the presence or absence of the collagen matrix. In the former case, tensile stress in collagen fibers will be increased and vascular patency will be maintained despite the increase in extravascular pressure accompanying an increased water-containing extravascular space. In the latter case (Fig. 7), i.e., following the acute dissolution of extravascular collagen (2 m*M* DTNB, 2-h infusion), the increased extravascular tissue pressure (diastolic IMP, 15–40 mmHg) resulted in a significant inhibition of coronary flow by compressing the vascular space and thus increasing coronary resistance: change in mean value flow from 82 to 36 ml/min; pulse amplitude excursion (max/min) was drastically altered (131 per 25 ml/min to 70 per −6 ml/min). Compared to 2-h control perfusion, the heart was severely dilated, myocardial edema was present, and water content had increased from 79% to 83%.

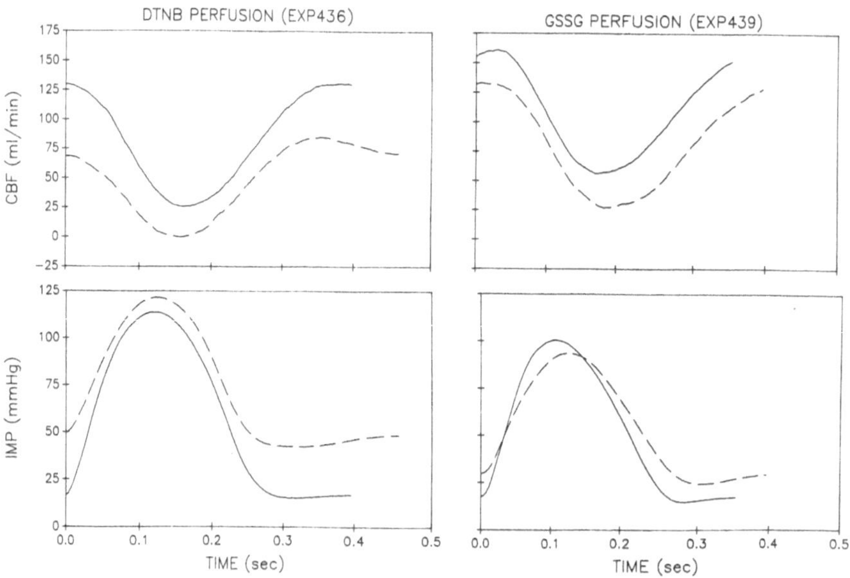

FIG. 7. Acute dissolution of extravascular collagen (2 m*M* DTNB/GSSG, 2-h infusion) resulted in an increased extravascular pressure (diastolic IMP, mmHg) and a significant inhibition of coronary flow (CBF, ml/min). GSSG, Oxidized glutathione. *Solid line*, Control; *dashed line*, treatment

The above demonstrated direct dependence of intramyocardial tissue pressure on coronary arterial perfusion in both beating and arrested states [3,11] attests to the important role of fluid exchange dynamics at the capillary level in the regulation of coronary flow. Coronary venous pressure is also an important determinant of interstitial pressure, and as such, changes in venous pressure may regulate the coronary circulation by a mechanism different from its obvious contribution to the intravascular perfusion pressure gradient. The relatively low vascular resistance to flow between the highly compliant intramyocardial microvessels and coronary veins gives rise to conditions in which intramyocardial volume may be altered significantly by changes in coronary outflow pressure. This suggests that an increase in coronary venous pressure will alter the surrounding environment of the microvasculature embedded in the myocardium and thereby effectively modulate coronary flow. It is assumed that the changes in intramyocardial pressure are directly dependent on the alteration in intramyocardial volume.

In a recent study, Izrailtyan et al. [11] examined the interactions among coronary outflow pressure, intramyocardial tissue pressure, and coronary flow mechanics, including the distribution of venous flow to right and left heart outflow channels. The major new finding was a shift in the intramyocardial pressure-inflow relation to a higher tissue pressure level in response to the rise in right heart pressure. In addition, the modulation of right heart pressure resulted in a functional redistribution and partitioning of myocardial outflow via left heart venous channels (thebesian vessels). Figure 8 shows coronary inflow (Q_{in}) response to a decreasing perfusion pressure (CPP) for a given right heart pressure (RHP: 0, 15, and 25 mmHg). The dependence of right heart venous outflow (Q_{rh}) and left heart outflow (Q_{lh}) on CPP and RHP is also depicted. Note that with an increase in RHP from zero (A) to higher pressures (B) and (C), the left heart outflow (Q_{lh}) remains nearly constant (for a given RHP), irrespective of arterial pressure changes. Relatively small changes (Δ25 mmHg) in RHP affect the left heart outflow to a greater extent than the larger changes (Δ60 mmHg) in CPP.

Volume redistribution in the myocardium is possible principally because of the larger myocardial venous volume and compliance in comparison with the arterial side. The existence of different venous outflow channels could serve a dual design in myocardial volume redistribution. The interaction of these channels can determine the value of the effective venous back-pressure and thereby the global filling state of the coronary bed and myocardium. Furthermore, it is obvious that this "point" of effective venous pressure is not the same throughout the coronary bed. The plurality of different outflow channels leads to a variety of possible local filling patterns of the heart muscle in accordance with specific hemodynamic conditions (pressures in different heart chambers). This mechanism could be especially important in right-left heart mechanical interactions. For example, the enlargement of the left-ventricular myocardial wall volume caused by increasing right heart diastolic pressure brings about the translocation of an external load on the right heart to an internal load of the left heart. This could be a mechanism of hydraulic balance that normalizes the mechanical coupling between different sides of the heart [17].

In addition, interstitial-venous interactions may have a compensatory role in the unloading of specific regions of the heart to regulate their function. With this in mind, the nonhomogeneous distribution of left heart outflow channels acquires a new interpretation. The prevalence of these routes in the inner layers of myocardium can explain the observation that coronary sinus pressure elevation

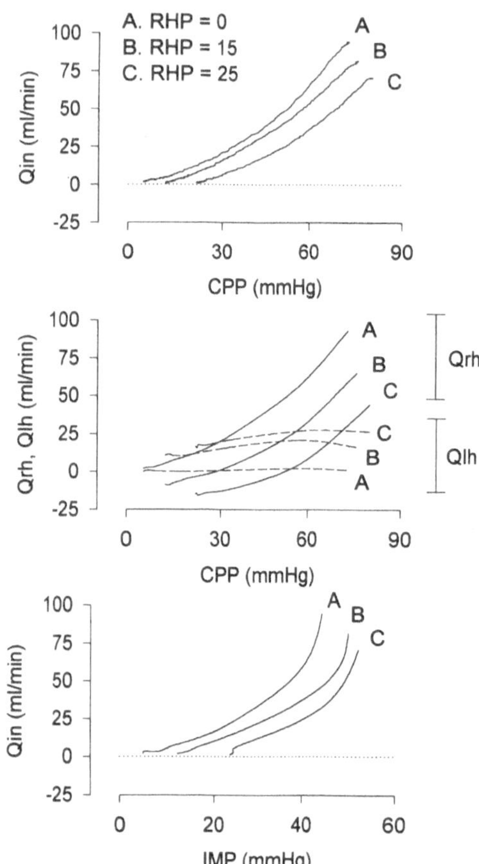

FIG. 8. Coronary pressure-flow relations for different right heart pressures. *Top:* Coronary inflow (Q_{in}) response to a decreasing perfusion pressure (*CPP*) for right heart pressure (*RHP*) of 0, 15, and 25 mmHg. *Center:* Dependence of right heart venous outflow (Q_{rh}) and left heart outflow (Q_{lh}) on CPP and RHP with increase in RHP from zero (A) to higher pressures (B) and (C). *Bottom:* Coronary inflow is displayed as a function of intramyocardial pressure (*IMP*) for a given RHP

shifts the arterial pressure-flow relationship in the epicardial layers but not in the endocardial [18].

Discussion

The prevailing mechanical and fluid transport conditions are interactive and coupled and thus must be analyzed accordingly. This dynamic interdependence is graphically illustrated in Fig. 9, showing that the basic elements which govern muscle mechanics (muscle/collagen fiber stress), fluid transport, and resultant perfusion (flow) can be further elaborated by the neuroendocrine system.

Experimental observations serve to demonstrate that intramyocardial tissue pressure plays a dominant regulatory role. Myocardial contraction inhibits flow by modulating its amplitude, affecting diastolic flow only minimally. Figure 10 is an idealized schematic summarizing and capturing many of the salient experimental features of coronary flow modulation. Shown are the systolic-diastolic envelopes of intramyocardial pressure (IMP) and coronary blood flow (CBF) at constant coronary perfusion pressure (CPP) in the unloaded left ventricle (LVP = 0). As contractility diminishes, peak (systolic) IMP is diminished and the systolic impediment to CBF is

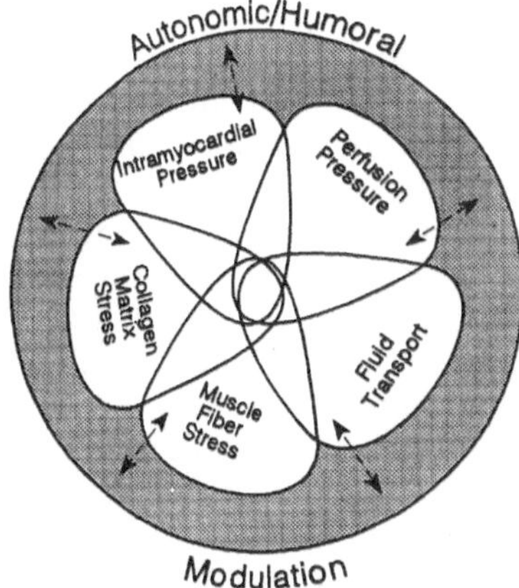

FIG. 9. Interactions between the basic elements that govern muscle mechanics (muscle/collagen fiber stress), fluid transport, and resultant perfusion (flow). Fluid flux and thus interstitial fluid content are interdependent and modifiable by neurohumoral regulation

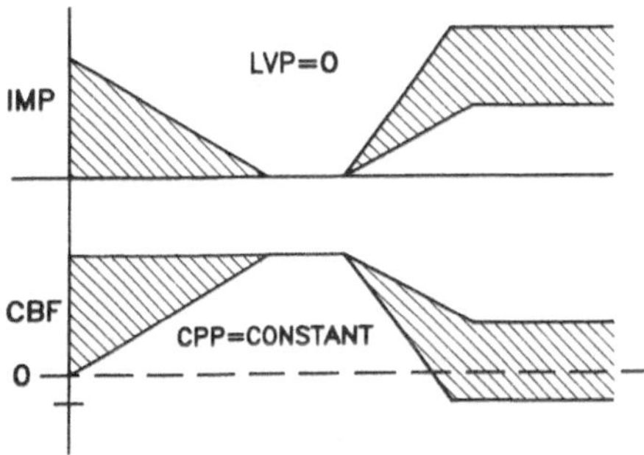

FIG. 10. Idealized schematic of coronary flow modulation showing the systolic–diastolic envelopes of intramyocardial stress field (embodied by tissue/fluid pressure, *IMP*) and coronary blood flow (*CBF*) at a constant coronary perfusion pressure (*CPP*), independent of ventricular load ($LVP = 0$)

removed. As contractility is enhanced (region of increasing IMP), CBF achieves a retrograde component. The actively pumped interstitial compartment will result in a diminished "diastolic" IMP. The degree of inhibition is determined principally by the inotropic state of the myocardium, *not* by the developed left-ventricular pressure. In contrast, as the diastolic portion of IMP is raised, the diastolic component of CBF will be diminished accordingly.

The myocardium has to be viewed as a heterogeneous/anisotropic structure composed of connective tissue and contractile elements (myocytes) vessel-attachments, interposed by an interstitial fluid gel-like matrix. The analysis of intramyocardial mechanical/hydraulic constraint(s) of the coronary microcirculation in terms of classic solid elastic theory is limiting. The dynamic reorganization and fluid shifts that accompany changes in the venous or arterial pressure signify a dominant role of intramyocardial (tissue–fluid) pressure [19] whether manifested as interstitial water swelling or physiologically generated by muscular contraction.

The muscle fiber-microvessel-interstitial interface plays a pivotal role and may prove to be a central link in determining inflow-outflow dynamics in the myocardium. The interaction between coronary outflow channels and the extravascular environment modifies the apparent intravascular pressure gradients. The notion that the "extravascular boundary" conditions remain invariant during changes in arterial or venous pressure must be reexamined. In addition, conditions that affect transcapillary dynamics modify in part the structural and functional organization of the heart. It is tempting to speculate about the functional role of venous channel redistribution and how alteration in the hydraulic structure (matrix) of the heart will moderate the effectual myocardial load.

Equally intriguing is the persistence of phasic inhibition of coronary blood flow in the bypassed left ventricle (i.e., patients undergoing cardiopulmonary bypass) and in isolated heart model, perfused with constant arterial pressure, which serves as indirect evidence for the existence of a residual stress field. Thus, presistence of regional IMP implicates myocardial shortening against an internal "self-imposed" load [20]. Obviously, the left-ventricular pressure must be considered as simply an "extramyocardial boundary" condition to myocardial force generation, which is myocyte in origin. The observed residual mechanical energy generated by the unloaded beating heart [20] is associated with a finite MVO_2, stemming from myofibril cross-bridge cycling and basal metabolism. A number of unresolved related issues remain, including the energetic characterization of myocardial tissue body forces, the coupling function of the collagen matrix, and the hydraulic structure framework of the heart and how they interact in regulating the performance of the heart as an organ system.

References

1. Kresh JY (1990) Myocardial modulation of coronary circulation. Am J Physiol 259:H1934–H1937
2. Kresh JY, Fox M, Brockman SK, Noordergraaf A (1990) Model-based analysis of transmural vessel impedance and myocardial circulation dynamics. Am J Physiol 258:H262–H276
3. Kresh JY (1993) Intramyocardial mechanical states: vessel-interstitium-muscle interface. In: Sideman S, Beyar R (eds) Interactive phenomena in the cardiac system. Plenum, New York, pp 113–123
4. Zinemanas D, Beyar R, Sideman S (1993) Intramyocardial fluid transport effects on coronary flow and left ventricular mechanics. In: Sideman S, Beyar R (eds) Interactive phenomena in the cardiac system. Plenum, New York, pp 219–231
5. Caulfield JB, Borg TK (1979) The collagen network of the heart. Lab Invest 40:364–372
6. Robinson TF, Cohen-Gould L, Factor SM, Eghbali M, Blumenfeld OO (1988) Structure and function of connective tissue in cardiac muscle: collagen types I and III in endomysial struts and pericellular fibers. Scanning Microsc 2(2):1005–1015
7. Weber KT, Janicki JS, Pick R, Abrahams C, Shroff SG, Bashey RI, Chen RM (1987) Collagen in the hypertrophied, pressure-overloaded myocardium. Circulation 75(1 pt 2):140–147
8. Caulfield JB, Norton P, Weaver RD (1992) Cardiac dilation associated with collagen alterations. Mol Cell Biochem 118:171–179
9. Van Mierop LHS (1979) Morphological development of the heart. In: Berne RM, Sperelakis N, Geiger SR (eds) Handbook of physiology, section 2: the cardiovascular system. Waverly Press/American Physiological Society, Baltimore, pp 1–28
10. Ohayon J, Chadwick RS (1988) Effects of collagen microstructure on the mechanics of the left ventricle. Biophys J 54:1077–1088

11. Izrailtyan I, Frasch HF, Kresh JY (1994) Effects of venous pressure on coronary circulation and intramyocardial fluid mechanics. Am J Physiol 36(3):H1002–H1009
12. Laine GA, Granger HJ (1985) Microvascular, interstitial, and lymphatic interactions in normal heart. Am J Physiol (Heart Circ Physiol) 18:H834–H842
13. Arts T, Prinzen FW, Snoeckx LHEH, Rijcken JM, Reneman RS (1994) Adaptation of cardiac structure by mechanical feedback in the environment of the cell, a model study. Biophys J 66:953–961
14. Anderson SE, Jonson JA (1987) Tissue-fluid pressure measured in perfused rabbit hearts during osmotic transients. Am J Physiol (Heart Circ Physiol) 252(21):H1127–H1137
15. Baldwin HS, Lloyd TR, Solursh M (1994) Hyaluronate degradation affects ventricular function of the early postlooped embryonic rat heart in situ. Circ Res 74(2):244–252
16. Kresh JY, McVey M, Frasch HF, Brockman SK (1992) Collagen-matrix dependent modulation of coronary circulation dynamics. Circulation 86(4):I-170
17. Kresh JY, Izrailtyan I, Brockman SK (1996) Microvascular-interstitial interactions in the regulation of coronary flow and cardiac mechanics. In: Kamada T, Shiga T, McCuskey RS (eds) Tissue perfusion and organ function: ischemia/reperfusion injury. Elsevier, Amsterdam, pp 109–121
18. Rouleau JR, White M (1985) Effects of coronary sinus pressure elevation on coronary blood flow distribution in dogs with normal preload. J Physiol Pharmacol 63:787–797
19. Rabbany SY, Kresh JY, Noordergraaf A (1989) Intramyocardial pressure: interaction of myocardial fluid pressure and fiber stress. Am J Physiol 257(26):H357–H364
20. Kresh JY (1991) Myocardial mechanics and energetics revisited. Trans Am Soc Artif Intern Organs 37(4):537–539

Coronary Slosh Phenomenon and Behavior of Intramyocardial Microcirculation

FUMIHIKO KAJIYA, OSAMU HIRAMATSU, and TOYOTAKA YADA

Summary. When the phasic blood flow velocities were evaluated in an intramyocardial artery or a small epicardial artery just before its penetration into the myocardium, the velocity waveform was exclusively diastolic and the systolic retrograde flow was almost always recognized. Since this retrograde flow is related to the wasteful "to-and-fro" movement of the blood between epicardial arteries and intramyocardial arteries, we called it *coronary slosh phenomenon*. The coronary slosh was enhanced by coronary artery stenosis and further increased following the administration of vasodilatory substances. To analyze the mechanism of the slosh phenomenon, we directly observed the behavior of subendocardial microvessels with our CCD videomicroscope. The pulsation of the microvascular diameters during the cardiac cycle was clearly visualized, indicating that the coronary slosh phenomenon is caused by the intramyocardial pulsation of small vessels. The diameter decrease in intramyocardial microvessels during systole also indicated the increase in the vascular resistance during this period. The vascular response following the release of the coronary arterial occlusion for 20 s was greater in subendocardial arterioles than in subepicardial arterioles. This difference may be due, at least partly, to the greater contribution of nitric oxide on the control of blood flow in the subendocardial arterioles, along with the direct mechanical effect of the subendomyocardium on its microvessels.

Key words. Coronary slosh phenomenon—Subendocardial venule—Reactive hyperemic response—Nitric oxide—Coronary artery stenosis—Needle-probe videomicroscope—Subendocardial arterioles

Coronary Inflow into the Myocardium and the Coronary Slosh Phenomenon

For the investigation of phasic myocardial perfusion, it is necessary to evaluate the phasic blood flow velocities in the intramyocardial artery or a small epicardial artery just before penetration into the myocardium, because the proximal coronary arterial flow pattern is influenced by the capacitance effect of large epicardial arteries [1–6].

Department of Medical Engineering and Systems Cardiology, Kawasaki Medical School, 577 Matsushima, Kurashiki, 701-01 Japan.

Accordingly, we measured the phasic flow velocity of an intramyocardial artery in the septal artery by our 20-MHz, 80-channel ultrasound pulsed Doppler method and the velocities in small epicardial arteries by a laser Doppler method with an optical fiber.

Figure 1 shows the phasic blood velocity waveform in the septal artery of the dog under control conditions. Measurement was also obtained during isoproterenol adminstration to enhance the effect of myocardial compressive force on the velocity waveform. The diastolic predominant velocity waveform was observed in this artery, and an early systolic retrograde flow was almost always recognized. Isoproterenol administration enhanced the early systolic retrograde flow and induced the retrograde flow in mid-to-late systole. Thus, the backward movement of the intramyocardial blood by cardiac contraction plays an important role in limiting myocardial inflow in addition to the systolic impeding effect by an increase in intramyocardial resistance. According to the nature of the systolic retrograde flow, i.e., the wasteful "to-and-fro" movement of the blood between epicardial portion of the coronary artery and intramyocardial arteries, we called it the *coronary slosh phenomenon*. This may closely relate to the "intramyocardial steal phenomenon" proposed by Feigl et al. [7].

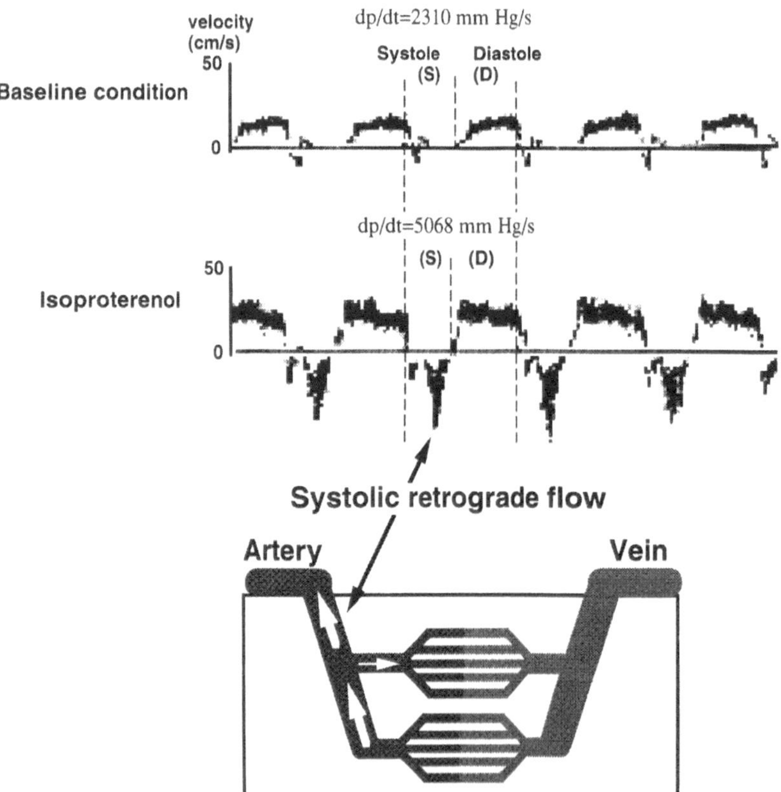

FIG. 1. Typical recordings of septal artery blood flow velocities under control condition (*upper panel*) and during isoproterenol administration (*middle panel*). The *bottom panel* indicates the concept of "coronary slosh phenomenon"

Coronary Slosh Phenomenon During Coronary Artery Stenosis

To evaluate the effect of coronary artery stenosis on the coronary slosh phenomenon, the canine left main coronary artery was perfused from the right carotid artery by a direct shunt circuit using rigid plastic tubing connected to a modified Gregg cannula. An occluder was placed just proximal to the Gregg cannula to realize several degrees of coronary stenosis. The arterial pressure proximal to the stenosis was measured at the tip of the cannula in the left main coronary artery through an external auxiliary tube, and the pressure distal to the stenosis was measured through a stiff cannula with a strain-gauge pressure transducer that was placed in a diagonal branch of the left anterior descending coronary artery. Septal arterial velocity was measured under control conditions with no stenosis (N), in moderate stenosis (S1) with distal perfusion pressure of ~60 mmHg, in severe stenosis (S2) with distal pressure of ~35 mmHg, and in complete coronary atrery occlusion.

Figure 2 shows typical recordings of blood velocity in the septal artery, measured by fast Fourier transform (FFT), with increasing severity of stenosis. The left main coronary artery flows by electromagnetic flowmeter are also displayed in the same figure (right). In the absence of a stenosis, the blood velocity in the septal artery was predominantly diastolic, as shown in Fig. 1. With increasing severity of stenosis (N → S1 → S2), the diastolic antegrade flow decreased while the systolic retrograde flow increased. After coronary occlusion, the blood velocity waveform in the septal artery exhibited a complete "to-and-fro" pattern.

Figure 3 shows the systolic-to-diastolic (S/D) velocity area ratio in the septal artery and the S/D flow ratio in the left main coronary artery under control conditions and in the coronary stenoses. The S/D velocity area ratio in the septal artery decreased as

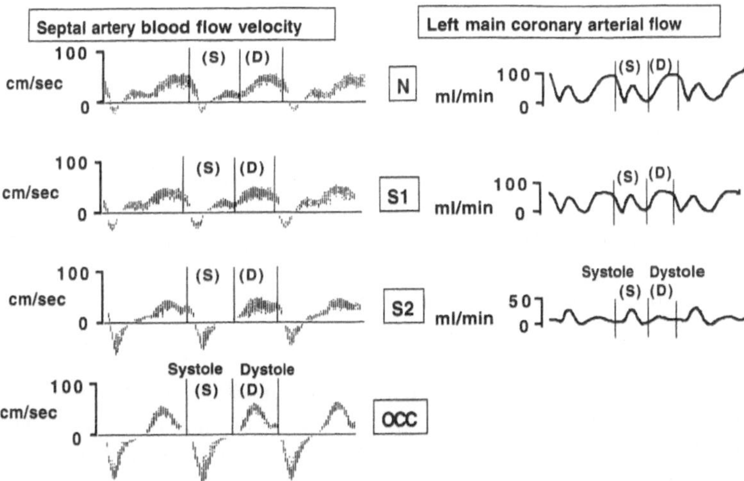

FIG. 2. *Left:* typical recordings of septal artery blood flow velocities (in cm/s) for no stenosis (*N*), moderate stenosis (*S1*), severe stenosis (*S2*), and complete coronary artery occlusion (*OCC*). *Right:* corresponding recordings of left main coronary artery flow except OCC. (From [1], with permission)

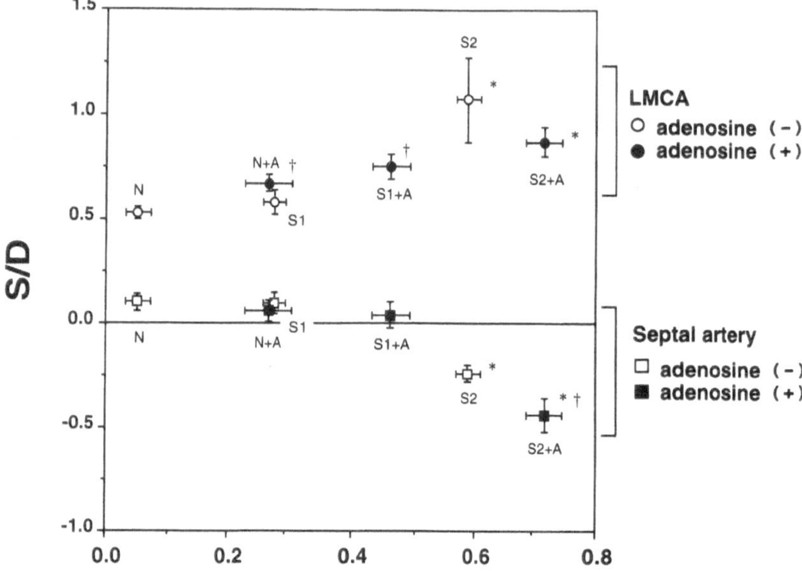

FIG. 3. Relationships between systolic-to-diastolic (*S/D*) velocity area ratio of septal artery and normalized pressure gradient during coronary stenosis and between S/D flow ratio of LMCA and pressure gradient. Normalized pressure gradient = (proximal coronary arterial pressure − distal coronary arterial pressure)/proximal coronary arterial pressure. The relationships following adenosine administration are also superimposed in this figure ($P < .05$). Note that the S/D flow ratio increased with stenosis while S/D velocity area ratio decreased. *LMCA*, Left main coronary artery. (From [1], with permission)

the perfusion pressure decreased. The total inflow into the myocardium through a cardiac cycle decreased progressively with the increase in the severity of stenosis. The decrease in the back-pressure to the retrograde flow during early systole and the increase in the epicardial capacitance distal to the stenosis caused by the decreasing systolic pressure may contribute mainly to the increase in systolic retrograde flow in the septal artery. As for the left main coronary artery flow, in contrast to the velocity area ratio, the S/D flow area ratio increased with the increase in stenosis. This may be explained as follows: although the diastolic flow component decreased with the increase in severity of stenosis as in the septal artery, the systolic component was almost unchanged, unlike the septal arterial flow, because of sustained systole flow through the stenosis resulting from the decrease in poststenotic pressure and the concomitant increase in capacitance (see Fig. 2).

Figure 4 shows typical recordings of blood velocity in the septal artery before and after intracoronary adenosine administration. Adenosine (0.3–0.7 mg/min) was infused continuously into the left main coronary artery through the external lumen of the cannula under three conditions: no stenosis (N + A), moderate stenosis (S1 + A), and severe stenosis (S2 + A). Adenosine increased systolic retrograde blood velocity component significantly, but there was only slight increase (not significant) in the diastolic antegrade blood velocity component. Thus, in the presence of coronary artery stenosis, vasodilation caused by adenosine administration enhanced systolic retrograde movement of the blood in the myocardial arteries and the slight diastolic

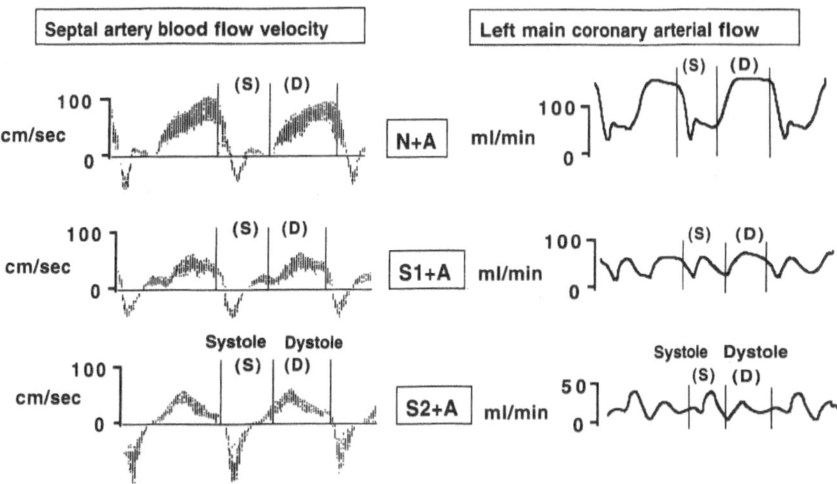

FIG. 4. Typical recordings of septal artery blood flow velocities measured in the same dog as discussed in Fig. 2 for coronary stenosis after adenosine administration (*left*). *Right*, Corresponding LMCA flows. (From [1], with permission)

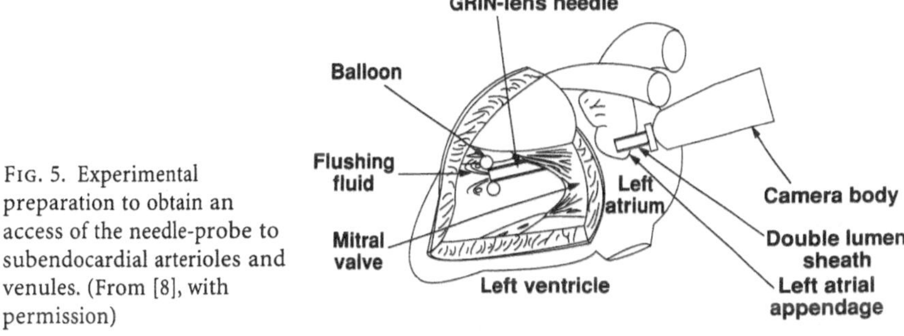

FIG. 5. Experimental preparation to obtain an access of the needle-probe to subendocardial arterioles and venules. (From [8], with permission)

flow increase does not improve the flow reduction caused by stenosis (see also Fig. 3). Therefore, the systolic retrograde flow may be an important factor in explaining the "intramural steal phenomenon" [7].

Direct Observation of Pulsation of Subendocardial Microvessels

We have recently succeeded in the direct observation of the porcine and canine subendocardial microvessels by introducing a novel portable needle-probe video-microscope with a charge-coupled-device (CCD) camera [8,9]. The needle-probe was introduced into the left ventricle through the mitral valve from an incision in the left arterial appendage and was gently placed on the endocardial surface (Fig. 5). The needle-probe was moved slowly and carefully to visualize arteriolar or venular images.

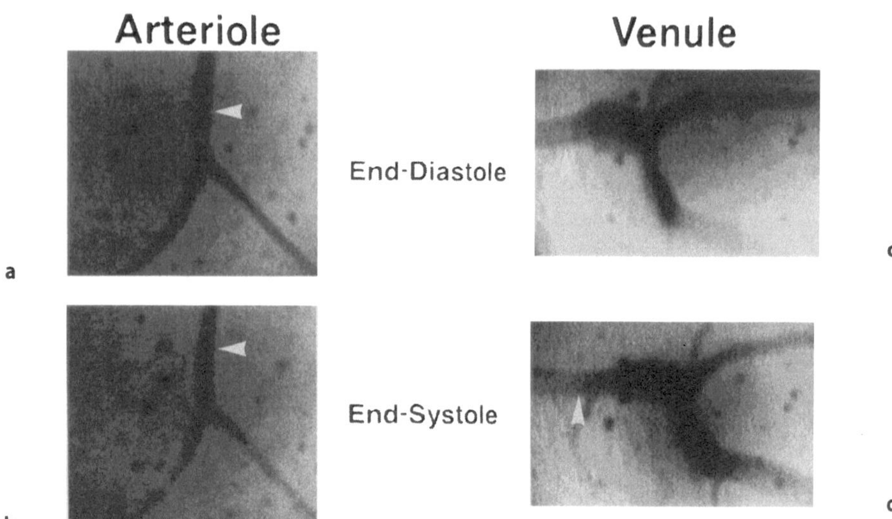

Fig. 6a–d. Typical images of subendocardial arteriole (*left panels*) and venule (*right panels*) in the end-diastole and end-systole. **a** Subendocardial arteriole in end-diastole (*arrowhead*, 90 μm); **b** subendocardial arteriole in end-systole (*arrowhead*, 70 μm); **c** subendocardial venule in end-diastole (*arrowhead*, 157 μm); **d** subendocardial venule in end-systole (*arrowhead*, 120 μm). (From [8], with permission)

Figure 6 shows typical images of a subendocardial arteriole and venule at end-diastole and end-systole. The diameter of subendocardial arterioles and venules decreased from end-diastole to end-systole by cardiac contraction. However, we usually did not observe the collapse of the vessels during systole. Partial pinching and kinking of subendocardial vessels were occasionally observed in both arterioles and venules, although the degree varied individually. Vascular segments with a relatively uniform diameter were used for the observation of phasic diameter change.

Figure 7 shows the diameter changes in subendocardial and subepicardial arterioles and venules between end-diastole and end-systole. Subendocardial arteriolar diameters decreased by 24% and subendocardial venular diameter decreased by 17%. Judd and Levy [10] measured intramyocardial blood volume using systolically and diastolically arrested rat hearts. They found about a 40% decrease in total intramyocardial blood volume in systolically arrested hearts compared to diastolically arrested hearts. Although the vascular diameter or volume of the arrested hearts may be different from that of the beating heart, Judd and Levy's result is close to the phasic change in diameter in the present study; i.e., their 40% volume change is almost equivalent to our 20% diameter change. In contrast, the diameter of subepicardial arterioles was almost unchanged from end-diastole to end-systole (lower panels of Fig. 7).

The coronary microvessels in the deep myocardial layer are pulsated during cardiac cycle, indicating that this is the major origin for the systolic retrograde coronary

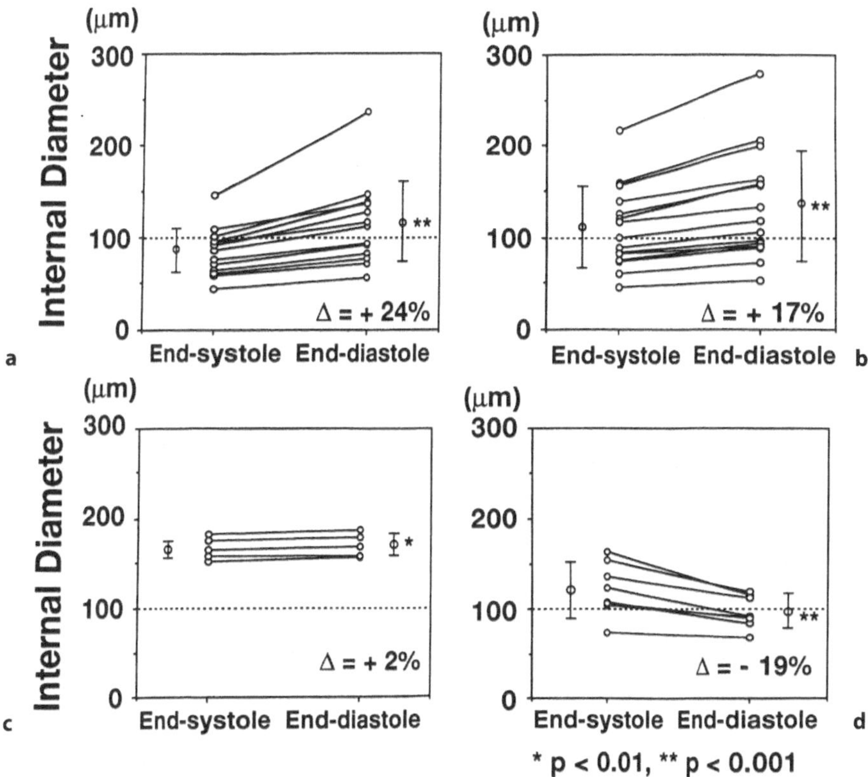

Fig. 7a–d. Effect of cardiac contraction on diameter changes in subendocardial (**a**) and subepicardial (c) arterioles and corresponding venules (**b** and **d**) between end-diastole and end-systole. *, $P < .01$; **, $P < .001$. (From [8], with permission)

blood flow. Thus, the coronary slosh phenomenon may be caused by the intramyocardial pulsation of arteries (small arteries and arterioles). The systolic diameter change in the intramyocardial microvessels can also cause an increase in resistance for systolic myocardial inflow. This systolic narrowing may be also important for explaining the systolic–diastolic interaction of coronary flow [11].

Vasodilatory Capacity of Subendocardial Vessels

It is well accepted that the coronary flow reserve is lower in subendocardium than in subepicardium [12]. When coronary perfusion pressure is reduced stepwise while keeping myocardial oxygen demand constant, autorgeulation works at first and transmural flow remains almost constant. As the pressure is lowered further, subendocardial flow decreases first and then the decrease in subepicardial flow reserve follows [13]. Why is the coronary flow reserve in subendocardium lower than in subepicardium? The possible reason is the difference of vasodilatory response between subendocardial and subepicardial layers. We then evaluated the reactive hyperemic responses of subendocardial arterioles in beating canine hearts directly by our needle-probe intravital microscope. The reactive hyperemic responses of

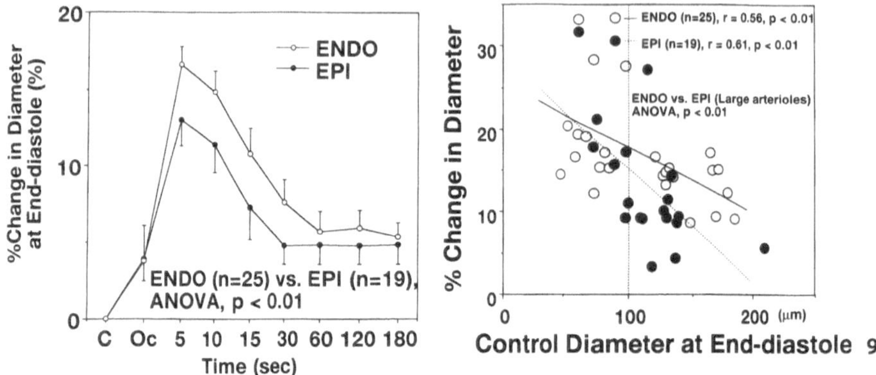

FIG. 8. Time-course of percent diameter changes of subendocardial (*ENDO*) and subepicardial arterioles (*EPI*) during reactive hyperemia. Percent end-diastolic diameter increase of subendocardial arteriole was larger than that of subepicardial arteriole ($P < .01$). *C*, Control; *Oc*, at the end of the occlusion of the left-anterior descending artery (LAD). (From [14], with permission)

FIG. 9. Percent diameter increase in subendocardial (*ENDO*) and subepicardial arterioles (*EPI*) at end-diastole during peak reactive hyperemic flow plotted against the end-diastolic diameter before LAD occlusion. There was a significant difference in the vascular responses between subendocardial and subepicardial arterioles (ANOVA, $P < .01$); the difference was significant for arterioles larger than 100 μm ($P < .01$) but not for the smaller arterioles. The *vertical dotted line* is 100 μm. (From [14], with permission)

subendocardial and subepicardial arterioles were analyzed following 20-s occlusion of the left-anterior descending artery (LAD).

Figure 8 shows the time-course of end-diastolic arteriolar diameter responses in both subendocardium and subepicardium during reactive hyperemia. The time-courses were similar to each other, but the percent end-diastolic diameter increment in subendocardial arterioles during reactive hyperemia was larger than that in subepicardial arterioles ($P < .01$), indicating that the magnitude of the vasodilatory response of subendocardium is even greater than that of the subepicardium [14]. Thus, the hypothesis that decreased subendocardial flow reserve may be caused by decreased vascular reactivity was not shown to be apparently true by this figure. This result, however, should be carefully interpreted by the following reasoning.

Figure 9 shows the percent diameter increase in subendocardial and subepicardial arterioles at end-diastole during peak reactive hyperemic flow plotted against the end-diastolic diameter before LAD occlusion. There is a significant difference in the vascular response between subendocardial and subepicardial arterioles ($P < .01$). The difference was significant for arterioles larger than 100 μm ($P < .01$) but not for the smaller arterioles [14]. Because the larger arterioles (>100 μm) in subendocardium have already dilated at the peak reactive hyperemic flow, the further vasodilatory capacity may be limited compared with larger arterioles in the subepicardium that dilated only a little and therefore may have further vasodilatory capacity.

Another possible mechanical reason for the difference of coronary flow reserve between subendocardium and subepicardium may be the much larger diastolic-to-systolic vascular pulsation amplitude of subendocardial arterioles, which relates to the coronary slosh phenomenon. Figure 10 shows the percent changes in diastolic-to-systolic vascular pulsation amplitude during peak hyperemic flow under control conditions. The pulsation amplitudes of subendocardial arterioles between end-diastole and end-systole were enhanced during reactive hyperemia (25% vs. 13% before LAD occlusion, $P < .01$), while that of subepicardial ones was small even during peak reactive hyperemia (6% vs. 2% before LAD occlusion, NS) [14]. This may result mainly from external forces imposed on vessels by increased vascular volume at reactive hyperemia. As discussed earlier, the increased pulsatility may cause not only systolic impediment of coronary inflow into endomyocardium, but also coronary slosh phenomenon [1].

The transmural difference of coronary flow reserve may also be explained by the difference of the action of vasodilatory substances. For example, Fig. 11 shows the precent end-diastolic diameter changes at the peak flow after inhibitor of nitric oxide synthase (N^G-monomethyl-L-arginine, L-NMMA) for both subendocardial and subepicardial arterioles against the end-diastolic diameters before LAD occlusion. There was a small but significant difference for L-NMMA, i.e., the smaller percent diameter change for subendocardial arterioles [14]. The greater inhibition of nitric oxide (NO) on subendocardial arterioles may relate to the difference of coronary flow reserve between subendocardium and subepicardium. In addition, comparing subendocardial arteriolar response between before (triangle circles in Fig. 11, and after (open circles in Fig. 11) L-NMMA, the large arteriolar response was suppressed more remarkably than the smaller one. The size-dependent difference of NO action was consistent with the result observed by Jones et al. [15]. Our present observations may suggest that the contribution of the endothelium-derived relaxing factor (EDRF) to

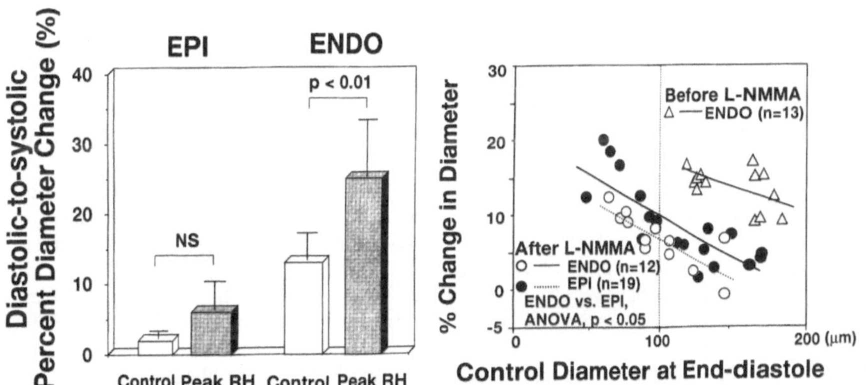

FIG. 10. Percent changes in diastolic-to-systolic vascular pulsation amplitude during peak hyperemic flow under control conditions. The pulsation amplitudes of subendocardial arterioles between end-diastole and end-systole was enhanced during reactive hyperemia (25% vs. 13%: before LAD occlusion, $P < .01$), while that of subepicardial ones was small even during peak reactive phyperemia. *RH*, Reactive hyperemia

FIG. 11. Percent end-diastolic diameter changes at the peak flow after N^G-monomethyl-L-arginine (L-NMMA) for both subendocardial and subepicardial arterioles plotted against the end-diastolic diameters before LAD occlusion. There was a small but significant difference for L-NMMA; i.e., smaller percent diameter change for subendocardial arterioles

the control of the coronary flow in a basal state is greater in the subendocardium where vascular pulsation is larger, leading to the greater inhibition of NO by L-NMMA.

There are several papers supporting the production of NO to be enhanced by the pulsation of the vascular segment. However, the available data about the pulsation effects are mainly from conduit arteries, with only a few from small arteries or arterioles. Lamontagne et al. [16] investigated whether mechanical forces applied to the vascular wall by the myocardial contraction are a stimulus for NO release. Cardiac arrest led to a significant reduction in NO release evaluated by the cyclic guanosine monophosphate (cGMP) content of platelets after passage through the isolated rabbit heart, while the reapperance of heart mechanical activity restored the NO release to the basal level under control beating conditions. This indicates the contribution of the mechanical activity on basal NO release. More recently, Noris et al. [17] examined NO synthesis in cultured endothelial cells by changing flow conditions. They observed that laminar shear stress upregulated the level of NO synthese mRNA from a static control and also that step-change increase of laminar shear increased NO synthesis as compared with steady laminar shear of same magnitude. These results suggested that the phasic change of intramyocardial flow condition affects NO synthesis.

As for other regulatory factors, such as adenosine and the ATP-sensitive potassium channel, we have not observed a significant difference of their transmural effects; this however may be partly the result of the limitation of the number of vessels observed. Further studies are necessary to investigate the transmural effects of various vasoactive substances.

References

1. Kimura A, Hiramatsu O, Yamamoto T, Ogasawara Y, Yada T, Goto M, Tsujioka K, Kajiya F (1992) Effect of coronary stenosis on phasic pattern of septal artery in dogs. Am J Physiol 262:H1690–H1698
2. Carew TE, Covell JW (1976) Effect of intramyocardial pressure on the phasic flow in the intraventricular septal artery. Cardiovasc Res 10:56–64
3. Chilian WM, Marcus ML (1985) Effects of coronary and extravascular pressure on intramyocardial and epicardial blood velocity. Am J Physiol 248:H170–H178
4. Chilian WM, Marcus ML (1982) Phasic coronary blood flow velocity in intramural and epicardial coronary arteries. Circ Res 50:775–781
5. Eckstein RW, Moir TW, Driscol TE (1963) Phasic and mean blood flow in the canine septal artery and an estimate of systolic resistance in deep myocardial vessels. Circ Res 12:203–211
6. Kajiya F, Tomonaga G, Tsujioka K, Ogasawara Y, Nishihara H (1985) Evaluation of local blood flow velocity in proximal and distal coronary arteries by laser Doppler method. J Biomech Eng 107:10–15
7. Feigl EO, Buffington CW, Nathan HJ (1987) Adrenergic coronary vasoconstriction during myocardial under perfusion. Circulation 75(supple I):I-1–I-5
8. Yada T, Hiramatsu O, Kimura A, Goto M, Ogasawara Y, Tsujioka K, Yamamori S, Ohno K, Hosaka H, Kajiya F (1993) In vivo observation of subendocardial microvessels of the beating porcine heart using a needle-probe videomicroscope with a CCD camera. Circ Res 72:939–946
9. Hiramatsu O, Goto M, Yada T, Kimura A, Tachibana H, Ogasawara Y, Tsujioka K, Kajiya F (1994) Diameters of subendocardial arterioles and venules during prolonged diastole in canine left ventricles. Circ Res 75:393–399

10. Judd RM, Levy BI (1991) Effects of barium-induced cardiac contraction on large- and small-vessel intramyocardial blood volume. Circ Res 68:217–225
11. Hoffman JIE, Baer RW, Hanley FL, Messina LM (1985) Regulation of transmural myocardial blood flow. J Biomech Eng 107:2–9
12. Hoffman JIE, Spaan JAE (1990) Pressure-flow relations in coronary circulation. Physiol Rev 70:331–390
13. Guyton RA, McClenathan JH, Newman GE, Michaelis LL (1977) Significance of subendocardial S-T segment elevation caused by coronary stenosis in the dog. Am J Cardiol 40:373–380
14. Yada T, Hiramatsu O, Kimura A, Tachibana H, Chiba Y, Lu S, Goto M, Ogasawara Y, Tsujioka K, Kajiya F (1995) Direct in vivo observation of subendocardial arteriolar response during reactive hyperemia. Circ Res 77:622–631
15. Jones CJ, Kuo L, Davis MJ, Defily DV, Chilian WM (1995) Role of nitric oxide in the coronary microvascular responses to adenosine and increased metabolic demand. Circulation 91:1807–1813
16. Lamontagne D, Pohl U, Busse R (1992) Mechanical deformation of vessel wall and shear stress determine the basal release of endothelium-derived relaxing factor in the intact rabbit coronary vascular bed. Circ Res 70:123–130
17. Noris M, Morigi M, Donadelli R, Aiello S, Foppolo M, Todeschini M, Orisio S, Remuzzi G, Remuzzi A (1995) Nitric oxide synthesis by cultured endothelial cells is modulated by flow conditions. Circ Res 76:536–543

Coronary–Ventricular Interaction: The Gregg Phenomenon

H. FRED DOWNEY

Summary. Despite some negative reports, especially from experiments in pig hearts, there is considerable evidence that changes in coronary perfusion pressure alter myocardial contractile function and oxygen consumption. The mechanism of this Greeg phenomenon has not been delineated; howerver, pressure-induced changes in coronary vascular volume, associated with ineffective coronary pressure–flow autoregulation, are instrumental in producing the Gregg phenomenon. Although changes in coronary vascular volume alter ventricular end-diastolic wall thickness, myocardial fiber length is unchanged. Thus, the Gregg phenomenon does not result from Starling's law. Inotropic factors released from endothelium appear not to be responsible for the Gregg phenomenon. Changes in coronary perfusion pressure alter myocardial stiffness, both in diastole and in systole; this may alter the ratio of internal to external work and modulate myocardial oxygen demand. Because effective coronary autoregulation prevents perfusion pressure-mediated changes in vascular volume, autoregulation protects myocardium from increased oxygen demand when perfusion pressure is elevated. Failure of autoregulation at reduced perfusion pressure reduces myocardial oxygen demand and improves myocardial oxygen utilization efficiency. The Gregg phenomenon may be the initial step in the process of myocardial hibernation.

Key words. Coronary blood flow—Coronary perfusion pressure—Coronary vascular volume—Myocardial contractile function—Myocardial oxygen consumption—Myocardial stiffness—Pressure–flow autoregulation

Introduction

In 1958, Gregg [1] reported data showing that myocardial oxygen consumption increased when coronary perfusion pressure was increased. Since then, this "Gregg phenomenon" [1,2] has interested coronary and cardiac physiologists who have sought to explain how ventricular oxygen demand (as opposed to oxygen consumption) could be directly coupled to coronary circulatory conditions. In 1983 Feigl [3] reviewed possible explanations at that time for the Gregg phenomenon, which included the following points. (1) Hypoperfusion: the increase in myocardial oxygen

---

Department of Integrative Physiology, University of North Texas Health Science Center at Fort Worth, Fort Worth, TX 76107-2699, USA.

consumption could be explained if, under baseline conditions, the myocardium was underperfused, creating an imbalance between myocardial oxygen supply and demand. (2) Nonworking hearts: the more prominent Gregg phenomenon found in non-working heart preparations may suggest its absence under more physiological conditions in in situ working hearts. (3) Cardiac complicance: changes in myocardial compliance would alter the relationships between end-diastolic filling pressure and myocardial fiber length and thus influence myocardial oxygen demand. (4) Coronary distension: pressure-induced distension of intracardiac vessels would stretch myocardial fibers and, by the Frank–Starling mechanism, increase myocardial oxygen demand. (5) Flow and oxygen delivery: changes in coronary flow, rather than pressure, might directly influence myocardial oxygen demand.

Feigl [3] critically analyzed the evidence available at that time supporting or refuting each of these explanations. He concluded that the most coherent explanation for the Gregg phenomenon was pressure-mediated changes in coronary vascular distension and the resulting changes in myocardial fiber length. This review surveys evidence for this explanation and then considers more recent contributions to the understanding of the Gregg phenomenon and related interactions between coronary circulation and ventricular oxygen demand.

Is the Gregg Phenomenon Produced by Changes in Diastolic Fiber Length?

Although coronary perfusion pressure-mediated change in myocardial fiber length is now often cited as the mechanism of the Gregg phenomenon, experimental evidence for this view is limited. An early observation of Poche et al. [4] supported this view with electron microscopic evidence that increased coronary perfusion pressure increased sarcomere length in arrested hearts. However, attempts to measure perfusion pressure-related changes in ventricular diastolic dimensions have not been encouraging. Abel and Reis [5] detected no effect on ventricular end-diastolic pressure when canine coronary flow was elevated by increasing perfusion pressure or by administering nitroglycerine, although ventricular function increased. Following an abrupt reduction in coronary perfusion pressure, Zhao et al. [6] found no immediate shortening of ventricular diastolic segment length. Schouten et al. [7] described perfusion pressure-mediated changes in contractile function of rat papillary muscles and trabeculae with no changes in segment lengths.

In our laboratory, Bai et al. [8] and Iwamoto et al. [9] detected no changes in ventricular end-diastolic or end-systolic segment lengths as coronary perfusion pressure was varied from 60 to 180 mmHg, although myocardial oxygen consumption increased in hearts with poor pressure–flow autoregulation. Ito [10] found that coronary perfusion pressure could be gradually reduced to about 30 mmHg with no significant effect on ventricular end-diastolic segment length. Kitakaze and Marban [11] found that hearts stretched to the optimal end-diastolic length still increased contractile function when perfusion pressure was increased. These pressure-induced increases in contractile function were associated with changes in intracellular calcium transients.

May-Newman et al. [12] performed a careful three-dimensional analysis of the mechanical interaction between the coronary vasculature and passive myocardium. They observed perfusion pressure-induced deformations of the ventricular wall in the radial and longitudinal dimensions but not parallel to the fiber direction. These deformations are consistent with other reported effects of perfusion pressure on coronary blood volume and ventricular wall thickness [8,13–17]. Although wall thickness increases, fiber length may be unaffected or even decreased in the subendocardium, because pressure-induced increases in ventricular wall thickness of dog [16] and rabbit [17] hearts were associated with marked decreases in inner diameter with little or no change in outer diameter. Because myocardial microvessels are oriented primarily parallel to the muscle fibers [18], pressure-induced expansion of the coronary microcirculation would thus expand the ventricular wall but not necessarily distend myocyte sarcomeres. Thus, a change in wall thickness need not reflect a change in sarcomere length. In summary, most experimental evidence does not support the concept of perfusion pressure-mediated changes in diastolic fiber length as the basis of the Gregg phenomenon.

Is the Gregg Phenomenon Produced by Endothelial Factors?

Recent evidence indicates that endothelial cells may modulate the contractile function of nearby cardiac muscle cells [19–21]. Because endothelial release of vasocative substances is flow sensitive, Winegrad's laboratory [22,23] investigated the effect of coronary flow on endothelial modulation of myocardial contractility. Isolated ventricular trabeculae were bathed in reoxygenated venous effluent from an isolated, perfused working rat heart. Contractility of the trabecula was unaffected by coronary flow if the endothelium of the perfused heart was undamaged. After the endothelium was damaged by treatment with 0.5% Triton X-100, contractility of the trabecula declined when flow was reduced, but the responses were quite variable. Oxygen consumption of the trabecula was not measured. Hyperperfusion had no effect on contractility. As this time, it appears that flow-sensitive endocardial release of inotropic factors is not the mechanism of the Gregg phenomenon.

Coronary Pressure–Flow Autoregulation and the Gregg Phenomenon

Although evidence for the Frank-Starling law as the basis of the Gregg phenomenon is contradictory, a consistent association has been found between changes in coronary vascular volume and the Gregg phenomenon [8,13–17]. The dependence of the Gregg phenomenon on changes in coronary vascular volume was recently confirmed by Bai et al. [8] (Fig. 1). However, increases in coronary perfusion pressure did not always produce increases in coronary vascular volume, as might have been expected from previous reports. Bai et al. estimated canine coronary vascular volume from the product of coronary flow and the coronary mean transit time of ^{51}Cr-labeled red blood cells. Myocardial oxygen consumption and contractile function were also measured.

We had noted in many earlier studies an apparent inverse relationship between the potency of coronary pressure–flow autoregulation and the Gregg phenomenon (see [8]), so Bai et al. correlated the degree of autoregulation with changes in coronary volume and with changes in myocardial oxygen consumption following changes in perfusion pressure. Pressure–flow autoregulation was evaluated by calculating the

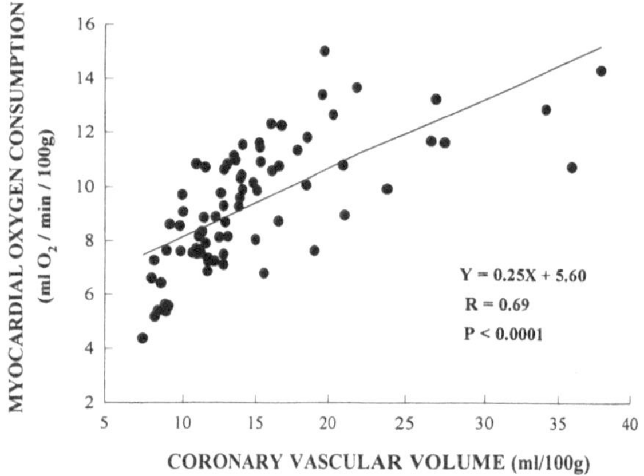

Fig. 1. Relationship between coronary vascular volume and myocardial oxygen consumption for coronary perfusion pressures from 60 to 180 mmHg. (From [8], with permission)

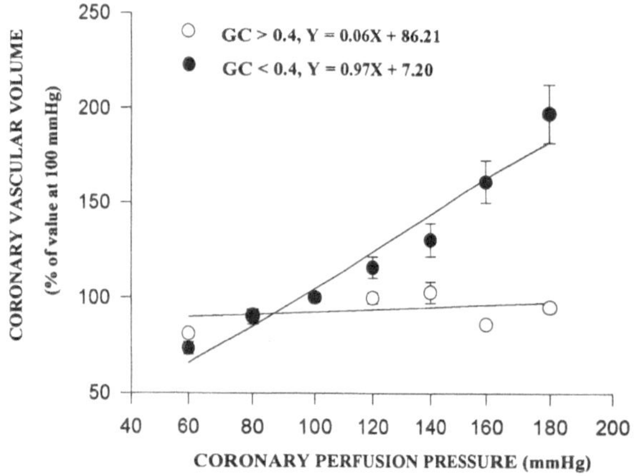

Fig. 2. Relationships between coronary perfusion pressure and mean coronary vascular volume for hearts with effective autoregulation (*open circles*) (Gc > 0.4; $r = .34$, $P < .05$) and for hearts with poor or absent autoregulation (*solid cireles*) (Gc < 0.4; $r = 0.97$, $P < .0001$). Slopes of these relationships differed significantly ($P < .001$). *GC*, Closed-loop gain (Gc). (From [8], with permission)

autoregulatory closed-loop gain (Gc) [24]. This gain reflects the change in coronary blood flow for a corresponding change in coronary perfusion pressure; a gain of 1 reflects perfect autoregulation, i.e., no change in flow for change in pressure, values between 0 and 1 indicate partial coronary autoregulation, and values less than 0 indicate no autoregulation. Data were segregated according to whether the autoregulatory gain exceeded 0.4. In experiments with poor coronary pressure flow autoregulation (Gc < 0.4), pressure-mediated changes in coronary vascular volume were clearly evident; in experiments with intact autoregulation (Gc > 0.4), however, the effect of perfusion pressure on vascular volume was minimal (Fig. 2). Whenever changes in coronary perfusion pressure produced significant changes in vascular volume, there were also coincident changes in myocardial oxygen consumption (Fig. 3). It was apparent from these observations that perfusion pressure-induced changes

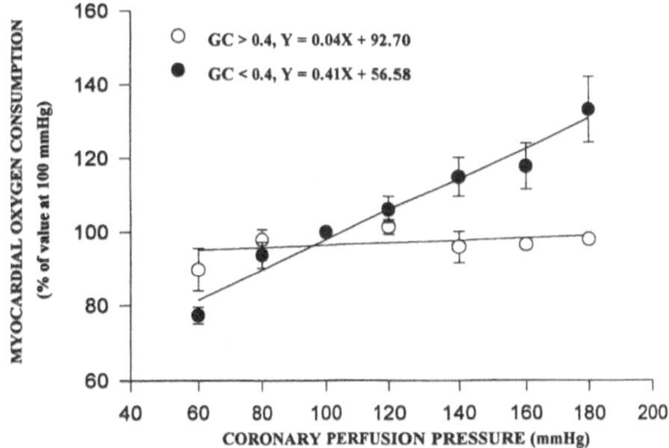

FIG. 3. Relationships between coronary perfusion pressure and mean myocardial oxygen consumption for hearts with effective autoregulation (*open circles*) (Gc > 0.4; $r = .46$, $P < .05$) and for hearts with poor or absent autoregulation (*solid circles*) (Gc > 0.4; $r = .98$, $P < .0001$). Slopes of these relationships differed significantly ($P < .001$). (From [8], with permission)

in coronary vascular volume are prerequisite for expression of the Gregg phenomenon, although these changes in coronary vascular volume do not alter myocardial end-diastolic segment length.

In contrast to the findings of Bai et al., Schulz et al. [25] detected no effect of coronary perfusion pressure on regional myocardial function and myocardial oxygen consumption within the autoregulatory range of pig hearts. Although the perfusion pressures used in this study were within the normal autoregulatory range, Schulz et al. did not evaluate the presence or absence of autoregulation. Their reported mean values of coronary perfusion pressure and flow indicate that autoregulation was effective only between perfusion pressure of 88 and 123 mmHg. From 123 to 186 mmHg, autoregulation was absent (Gc < 0), but there was no significant change in myocardial oxygen consumption or systolic wall thickening. End-diastolic wall thickness however increased about 12%. This small change in wall thickness may indicate that pressure-induced changes in coronary vascular volume are less in the pig heart than in the dog heart, a factor which would minimize the Gregg phenomenon. Accompanying the increase in coronary perfusion pressure and flow was an unexplained but marked increase in myocardial lactate consumption. As noted by Schultz et al., intracellular lactate concentration inhibits glycolysis and free fatty acid utilization and increases lactate consumption. Because the amount of oxygen required to metabolize lactate is less than that required to metabolize fatty acids, this would have to some extent masked a pressure-induced increase in myocardial oxygen demand.

Miller et al. [26] observed a positive effect of coronary hyperperfusion on ventricular function of isovolumic but not in intact ejecting pig hearts. As in the study of Schulz et al. [25], coronary autoregulation of the intact hearts was poor, but increasing perfusion pressure from 100–110 mmHg to 150–160 and 180–190 mmHg did not increase indices of ventricular function or myocardial oxygen consumption. Although ventricular wall thickness increased modestly with perfusion pressure, Miller et al. detected no change in coronary small vessel plasma volume of transmural biopsy samples. These volume measurements may be inaccurate, however, because it was impossible to account for any blood lost from the presumably distended vasculature during the biopsy. Furthermore, no corrections were made for flow-mediated changes in small-vessel hematocrit, which would have occurred during the high-pressure, high-flow condition [27]. In any case, the negative results of these two studies in intact swine do indicate that further work is required to reconcile them with the frequent finding of the Gregg phenomenon in other species. Additional studies of the effects of coronary perfusion pressure on coronary vascular volume in the pig heart would be helpful in this regard.

When autoregulation is abolished by intracoronary infusion of adenosine at constant perfusion pressure, effects on ventricular function and energetics have been variable. Goto et al. [28] and Iwamoto et al. [9] observed improved ventricular function and increased oxygen consumption in rabbit and dog hearts, respectively. However, another negative finding in the pig heart was reported by Schwartz et al. [29]. In their experiments, supranormal coronary flow produced no significant changes in ventricular wall thickening, high-energy phosphates, or oxygen consumption. New evidence has been presented by Wannenburg et al. [30] that adenosine has a negative inotropic action in addition to its antiadrenergic action. If this is indeed the case, effects of altered flow in the absence of autoregulation may have been masked. It would be interesting to extend these studies by examining effects of altered perfusion pressure in the maximally dilated coronary circulation. We predict that the Gregg phenomenon would be enhanced.

For coronary pressure-flow autoregulation to be effective in maintaining flow nearly constant in the fact of an increased perfusion pressure, arteriolar vasoconstriction is required. This vasoconstriction will protect downstream, microvascular elements from increased hydrostatic pressure and thus reduce distending pressure and subsequent increases in vascular volume. On the other hand, poor or absent autoregulation would not dissipate elevated arterial pressure; microvascular distending pressures would increase, and coronary vascular volume would increase. If changes in coronary microvascular volume do mediate the Gregg phenomenon, then it seems reasonable for effective pressure-flow autoregulation to blunt pressure-induced changes in myocardial oxygen consumption. It could be argued, however, that changes in perfusion pressure independently alter myocardial oxygen demand, which is then appropriately met by metabolically influenced changes in coronary blood flow, making autoregulatory capability appear poor. This argument was advanced earlier by our laboratory to explain the poor pressure-flow autoregulatory capability of the right coronary circulation, where changes in perfusion pressure produce marked alterations in myocardial oxygen consumption [31].

The concurrent measurements of coronary vascular volume under conditions of both effective and poor autoregulation provide new information to help resolve this dilemma. Myocardial oxygen consumption always increased with coronary vascular volume (see Fig. 1), but not always with perfusion pressure (see Fig. 3). Although coronary vascular volume and pressure-dependent coronary flow were highly correlated in this study, Crystal et al. [27] found that changes in coronary flow produced by intracoronary infusion of adenosine at constant perfusion pressure had no significant effect on coronary small-vessel blood volume. Thus, it seems most likely that pressure-induced changes in vascular volume, when permitted by ineffective autoregulation, alter myocardial oxygen demand rather than the alternative view that flow alters oxygen consumption directly, but only in some experiments, and then coincidentally alters coronary vascular volume. The data of Bai et al. support the original view of Arnold et al. [13] that myocardial oxygen consumption is modulated by coronary vascular volume, but not directly by coronary perfusion pressure, as proposed originally by Gregg [1,2].

Changes in coronary venous pressure would not be dampened by the precapillary resistance vessels and should readily alter postarteriolar vascular volume independent of the potency of pressure-flow autoregulation. Thus, changes in coronary venous pressure may be a particularly effective and long-overlooked mechanism of the Gregg phenomenon. This intriguing idea is discussed by Dr. Konrad Scheel in another chapter of this volume.

Physiological Benefits of the Gregg Phenomenon

Because autoregulation prevents pressure-mediated changes in vascular volume and associated changes in myocardial oxygen demand, Bai et al. [8] suggested that a previously unrecognized function of coronary pressure–flow autoregulation was to protect the myocardium from pressure-induced increases in oxygen demand. Thus, the myocardial oxygen demand in the hypertensive patient would increase less if coronary autoregulation were intact. On the other hand, pathophysiological conditions or pharmacological interventions that compromise autoregulation would enhance effects of changes in systemic arterial pressure on myocardial oxygen demand. When arterial pressure falls, the autoregulatory response of coronary vessels reduces their resistance to flow, and flow decreases are minimized. Below the left coronary autoregulatory limit of 60–80 mmHg [3], further decreases in coronary perfusion pressure result in nearly proportional decreases in coronary flow. If myocardial oxygen demand is unchanged, a detrimental imbalance between oxygen supply and demand will immediately result. Here, the pressure-related decrease in coronary vascular volume and the resulting decrease in pressure-related myocardial oxygen demand would clearly be beneficial.

The recently described phenomenon of hibernating myocardium [32] may be in part a manifestation of the Gregg phenomenon. Myocardial hibernation refers to the ability of underperfused myocardium to survive for a prolonged period without infarcting. On restoration of normal coronary blood flow, contractile function recovers to near-normal levels. Metabolic mechanisms, perhaps involving adenosine [33], are likely responsible for decreasing myocardial contractile function so that energy supply and demand remain balanced [34,35]. However, at the initiation of hypoperfusion, the Gregg phenomenon would immediately downregulate myocardial oxygen demand. To test this hypothesis, we examined coronary flow and myocardial metabolic responses to moderate reductions in canine coronary perfusion pressure [36,37]. When right coronary perfusion pressure was reduced to 60 mmHg, coronary flow fell (autoregulation was ineffective), while regional contractile function, myocardial high-energy phosphates, and the cytosolic phosphorylation potential of creatine phosphate were unchanged [37]. Myocardial oxygen consumption decreased significantly. In left-ventricular experiments, regional contractile function was also well maintained when left-anterior descending coronary perfusion pressure was reduced to 60 mmHg, although oxygen consumption fell [36]. Therefore, under these moderate hypoperfusion conditions there was an increase in oxygen utilization efficiency. This initial increase in efficiency may be the first step in the hibernating process. Hypoperfused myocardium with minimal contractile dysfunction, normal energy reserves, and increased oxygen utilization efficiency might be called "napping" myocardium to differentiate it from more severely under perfused "hibernating" myocardium in which contractile function is severely depressed. We propose that the Gregg phenomenon is the mechanism of napping myocardium.

A more extreme interaction between the coronary circulation and ventricular function was proposed by Koretsune et al. [38] to explain their observations in isolated ferret hearts. The abrupt fall in contractile force following interruption of coronary flow could not be satisfactorily explained by mechanisms involving impairment of excitation, accumulation of inhibitory metabolites, or depletion of energy-yielding metabolites. Koretsune et al. then studied the role of changes in intravascular pressure on this early ischemic contractile failure. Ischemia was produced by stopping coronary perfusion or by massive embolization of the coronary microcirculation. Functional depression developed more slowly in the microembolized hearts, and this depression more closely paralleled the rate of accumulation of inhibitory metabolites than that produced by stopping perfusion. The investigators concluded that vascular collapse is the mechanism responsible for ischemic contractile failure following coronary artery occlusion. However, the question of how this interaction occurs was unanswered. Assuming the finding of Koretsune et al. [38] was an extreme manifestation of the Gregg phenomenon, it would appear that changes in coronary arterial volume rather than changes in microvascular or venous volumes are important. More research on this question would be helpful. If the Gregg phenomenon permits maintenance of normal function during moderate coronary hypoperfusion, but also causes contractile failure during severe hypoperfusion, then it is important to determine the breakpoint for these opposite interactions between coronary perfusion pressure and myocardial contractile function.

Myocardial Stiffness and the Gregg Phenomenon

If the Gregg phenomenon permits moderately hypoperfused myocardium to function normally, it is important to return to the question of how a decrease in coronary perfusion pressure could decrease myocardial oxygen demand with no decrement in indices of ventricular function. Many investigations have found that coronary perfusion pressure has a direct effect on myocardial stiffness. Salisbury et al. [39] first demonstrated increased diastolic stiffness and speculated that increased coronary perfusion pressure caused engorgement of the coronary vascular bed, resulting in decreased compliance of the ventricular wall. As noted by Feigl [3], this pressure-induced increase in diastolic myocardial stiffness would retard diastolic ventricular filling, lessen sarcomere distension, and act to oppose the Gregg phenomenon. However, when perfusion pressure falls below normal, a decreased in diastolic ventricular stiffness would enhance ventricular filling. This issue is further confused by the recent report of Schipke et al. [17] that perfusion pressure-induced changes in diastolic myocardial stiffness are not necessarily associated with changes in end-diastolic pressure. In any case, the failure of investigators to detect coronary perfusion pressure-related changes in ventricular diastolic segment lengths measured parallel to fiber direction suggests that changes in diastolic stiffness contribute little to the decrease in myocardial oxygen demand at moderately reduced perfusion pressure.

Engorgement of the coronary vasculature should also increase ventricular systolic stiffness. In an early study, Templeton et al. [40] did not detect an effect of coronary perfusion on systolic stiffness when intraventricular pressure was modulated during the cardiac cycle. Livingston et al. [41] evaluated systolic stiffness from small, high-frequency indentations of isolated, tetanized, canine ventricular septa in which coronary autoregulation had been abolished by adenosine. They found a linear dependence of stiffness on coronary perfusion pressure that was much steeper during tetany than during diastole. These findings were interpreted as reflecting a pressure-induced increase in contractility.

Iwamoto et al. [9] assessed perfusion-related changes in myocardial contractile force and systolic ventricular stiffness in working, in situ canine hearts with poor pressure–flow autoregulation. Piezoelectric crystals were used to assess myocardial segment lengths, and a miniature isometric force transducer was implanted nearby to measure developed force. The slope of the force–length relationship during ventricular ejection was used as an index of systolic myocardial stiffness [42]. As coronary perfusion pressure was varied from 60 to 180 mmHg, maximum developed force, stiffness index, and myocardial oxygen consumption increassed with perfusion pressure, while end-diastolic segment length, segment shortening, and other hemodynamic variables stayed constant. When coronary blood flow was increased at constant pressure because of intracoronary infusion of adenosine, maximal developed force, stiffness index, and myocardial oxygen consumption increased, but less steeply than when flow was altered by increased perfusion pressure. Increased flow at constant pressure had no effect on end-diastolic segment length, segment shortening, or other hemodynamic variables.

Systolic stiffness is an index of contractility, and as such was shown to increase with perfusion pressure in the studies of Iwamoto et al. [9] and Livingston et al. [41]. This increase in systolic stiffness may, however, have other mechanical effects relative to the Gregg phenomenon. A less stiff myocardium would be required to perform less internal work, so that for a lesser oxygen consumption, external work might be maintained when perfusion pressure is reduced. This could account for the increase in oxygen utilization efficiency of moderately hypoperfused myocardium [36,37].

The heart muscle performs work on blood within the coronary circulation as blood is displaced from the microcirculation and veins during each contraction [43]. Because ventricular stiffness and microcirculatory blood volume decrease with perfusion pressure, the energy required to drive the intramyocardial blood pump would decrease. Thus, this hydraulic unloading of the coronary circulation during moderate hypoperfusion may also contribute to maintenance of normal external work with reduced myocardial oxygen consumption.

Conclusions

Myocardial contractile function and oxygen consumption are modulated by coronary perfusion pressure if coronary pressure–flow autoregulation is ineffective. This Gregg phenomenon is associated with changes in coronary vascular volume, but not with changes in end-diastolic contractile fiber length. Release of inotropic factors from endothelium is probably not responsible for the Gregg phenomenon. Perfusion pressure alters myocardial stiffness, and changes in systolic stiffness may affect myocardial oxygen demand by changing the ratio of internal to external work. With decreased perfusion pressure, the Gregg phenomenon reduces myocardial oxygen demand and permits an increase in myocardial oxygen utilization efficiency. This pressure-related downregulation of myocardial oxygen demand may be the first step in myocardial hibernation.

Acknowledgment. This work was supported by NIH grant HL 35027, National Institutes of Health, Bethesda, MD.

References

1. Gregg DE (1958) Regulation of the collateral and coronary circulation of the heart. In: McMichael J (ed) Circulation. Proceedings of the Harvey tercentenary congress. Blackwell, Oxford, UK, pp 163–186
2. Gregg DE (1963) Effect of coronary perfusion pressure or coronary flow on oxygen usage of the myocardium. Circ Res 13:497–500
3. Feigl EO (1983) Coronary physiology. Physiol Rev 63:1–205
4. Poche R, Arnold G, Gahlen D (1971) The influence of coronary perfusion pressure on metabolism and ultrastructure of the myocardium of the arrested aerobically perfused isolated guinea-pig heart (in German). Virchows Arch B Cell Pathol 8:252–266
5. Abel RM, Reis RL (1970) Effects of coronary blood flow and perfusion pressure on left ventricular contractility in dogs. Circ Res 27:961–971
6. Zhao M, Franzen D, Eng C (1988) The effect of coronary arterial pressure on myocardial distensibility. Absence of a "garden hose" effect during in vivo conditions. Basic Res Cardiol 83:626–633
7. Schouten VJA, Allaart CP, Westerhof N (1992) Effect of perfusion pressure on force of contraction in thin papillary muscles and trabeculae from rat heart. J Physiol (Lond) 451:585–604
8. Bai X-J, Iwamoto T, Williams AG Jr, Fan W-L, Downey HF (1994) Coronary pressure-flow autoregulation protects myocardium from pressure-induced changes in oxygen consumption. Am J Physiol (Heart Circ Physiol 35)266:H2359–H2368
9. Iwamoto T, Bai X-J, Downey HF (1994) Coronary perfusion-related changes in myocardial contractile force and systolic ventricular stiffness. Cardiovasc Res 28:1331–1336
10. Ito BR (1995) Gradual onset of myocardial ischemia results in reduced myocardial infarction. Association with reduced contractile function and metabolic downregulation. Circulation 91:2058–2070
11. Kitakaze M, Marban E (1989) Cellular mechanism of the modulation of contractile function by coronary perfusion pressure in ferret hearts. J Physiol (Lond) 414:455–472
12. May-Newman K, Omens JH, Pavelec RS, McCulloch AD (1994) Three-dimensional transmural mechanical interaction between the coronary vasculature and passive myocardium in the dog. Circ Res 74:1166–1178
13. Arnold G, Morgenstern C, Lochner W (1970) The autoregulation of the heart work by the coronary perfusion pressure. Pfluegers Arch 321:34–55
14. Arnold G, Kosche F, Miessner A, Neitzert A, Lochner W (1968) The importance of the perfusion pressure in the coronary arteries for the contractility and oxygen consumption of the heart. Pfluegers Arch 299:339–356
15. Scharf SM, Bromberger-Barnes B (1973) Influence of coronary flow and pressure on cardiac function and coronary vascular volume. Am J Physiol 224:918–925
16. Morgenstern C, Holjes U, Arnold G, Lochner W (1973) The influence of coronary pressure and coronary flow on intracoronary blood volume and geometry of the left ventricle. Pfluegers Arch 340:101–111
17. Schipke JD, Stocks I, Sunderdiek U, Arnold G (1993) Effect of changes in aortic pressure and in coronary arterial pressure on left ventricular geometry and function. Anrep vs. gardenhose effect. Basic Res Cardiol 88:621–637
18. Bassingthwaighte JB, Yipintsoi T, Harvey RB (1974) Microvasculare of the dog left ventricular myocardium. Microvasc Res 7:229–249

19. Kramer BK, Nishida M, Kelly RA, Smith TW (1992) Endothelins: myocardial actions of a new class of cytokines. Circulation 85:350–356
20. Mebazaa A, Mayoux E, Maeda K, Martin LD, Lakatta EG, Robotham JL, Shah AM (1993) Paracrine effects of endocardial endothelial cells on myocyte contraction mediated via endothelin. Am J Physiol (Heart Circ Physiol 34) 265:H1841–H1846
21. Winegrad S (1993) Evidence for the existence of endothelial factors regulating contractility in rat heart. Adv Exp Med Biol 332:155–163
22. Ramaciotti C, McClellan G, Sharkey A, Rose D, Weisberg A, Winegrad S (1993) Cardiac endothelial cells modulate contractility of rat heart in response to oxygen tension and coronary flow. Circ Res 72(5):1044–1064
23. Ramaciotti C, Sharkey A, McClellan G, Winegrad S (1992) Endothelial cells regulate cardiac contractility. Proc Natl Acad Sci USA 89:4033–4036
24. Norris CP, Barnes GE, Smith EE, Granger HJ (1979) Autoregulation of superior mesenteric flow in fasted and fed dogs. Am J Physiol 237:H174–H177
25. Schulz R, Guth BD, Heusch G (1991) No effect of coronary perfusion on regional myocardial function within the autoregulatory range in pigs. Circulation 83:1390–1403
26. Miller WP, Nellis SH, Liedtke AJ, Whitesell L, Effron BA (1990) Coronary hyperperfusion and ventricular function in intact and isovolumic pig hearts. Am J Physiol (Heart Circ Physiol 27) 258:H5001–H5007
27. Crystal GJ, Downey HF, Bashour FA (1981) Small vessel and total coronary blood volume during intracoronary adenosine. Am J Physiol 241:H194–H201
28. Goto Y, Slinker BK, LeWinter MM (1991) Effect of coronary hyperemia on E_{max} and oxygen consumption in blood-perfused rabbit hearts. Energetic consequences of Gregg's phenomenon. Circ Res 68:482–492
29. Schwartz GG, Schaefer S, Trocha SD, Garcia J, Steinman S, Massie BM, Weiner MW (1992) Effect of supranormal coronary blood flow on energy metabolism and systolic function of porcine left ventricle. Cardiovasc Res 26:1001–1006
30. Wannenburg T, de Tombe PP, Little WC (1994) Effect of adenosine on contractile state and oxygen consumption in isolated rat hearts. Am J Physiol (Heart Circ Physiol 36) 267:H1429–H1436
31. Yonekura S, Watanabe N, Caffrey JF, Gaugl JF, Downey HF (1987) Mechanism of attenuated pressure-flow autoregulation in right coronary circulation of dogs. Circ Res 60:133–141
32. Rahimtoola SH (1989) The hibernating myocardium. Am Heart J 177:211–221
33. Gao Z-P, Downey HF, Fan W-L, Mallet RT (1995) Does interstitial adenosine mediate acute hibernation of guinea pig myocardium? Cardiovasc Res 29:796–804
34. Marshall RC, Nash WN, Bersohn MN, Wong GA (1987) Myocardial energy production and consumption remain balanced during positive inotropic stimulation when coronary flow is restricted to basal rates in rabbit heart. J Clin Invest 80:1165–1171
35. Arai AE, Pantely GA, Anselone CG, Bristow J, Bristow JD (1991) Active down regulation of myocardial energy requirements during prolonged moderate ischemia in swine. Circ Res 69:1458–1469
36. Lee S-C, Downey HF (1993) Down regulation of oxygen demand in isoprenaline-stimulated canine myocardium. Cardiovasc Res 27:1542–1590
37. Itoya M, Mallet RT, Gao Z-P, Williams AG, Downey HF (1996) Stability of high energy phosphates in right ventricle: myocardial energetics during right coronary hypotension. Am J Physiol (Heart Circ Physiol 40)271:H320–H328
38. Koretsune Y, Kusuoka H, Corretti M, Marban E (1991) Mechanism of early ischemic contractile failure. Inexcitability, metabolic accumulation, or vascular collapse? Circ Res 68:255–262
39. Salisbury PF, Cross CE, Rieben PA (1960) Influence of coronary artery pressure upon myocardial elasticity. Circ Res 8:794–800
40. Templeton GH, Wildenthal K, Mitchell JH (1972) Influence of coronary blood flow on the left ventricular contractility and stiffness. Am J Physiol 223:1216–1220

41. Livingston JZ, Halperin HR, Yin FCP (1994) Accounting for the Gregg effect in tetanised coronary arterial pressure-flow relationships. Cardiovasc Res 28:228–234
42. Kedem J, Lee W, Weiss HR (1994) An experimental technique for quantitative determination of regional myocardial segment work in vivo. Ann Biomed Eng 22:58–65
43. Spaan JAE, Breuls NPW, Laird JD (1981) Diastolic-systolic coronary flow differences are caused by intramyocardial pump action in the anesthetized dog. Circ Res 49:584–593

Influences of Coronary Venous Pressures on Left-Ventricular Function

KONRAD W. SCHEEL[1], DAN MANOR[2], and KRISTIN BRYANT[1]

Summary. In this chapter, we present experimental evidence for the functional influences of the coronary venous system on cardiac performance. The drainage pattern of the canine venous system is described, and casts of the venous circulation establish that venous vascular volume is about twice that of the arterial vascular volume. Experimental evidence is presented demonstrating that coronary arterial flow, both after vasodilation and with intact vasoactivity, is reduced with elevated right heart pressures. Coronary collateral flow (in the presence of arterial occlusion) is also reduced with elevated right heart pressures. Experiments show that increasing left-ventricular volume in a vasodilated heart reduces coronary flow proportionally while there is practically no influence on flow in the vasoactive heart. Pressure interactions between the right heart and left-ventricular pressure are more pronounced in the vasodilated heart compared to the vasoactive heart and more prominent in a stiffer versus a compliant heart. Evidence is presented that elevated right heart pressures reduce the compliance of the left ventricle, which is even more pronounced in the vasodilated state. Finally, evidence for a "venous Gregg effect" is presented. Here we observed that an increase in right heart pressure resulted in an increase in the maximally generated pressure within the left ventricle, increased the maximum rate of change in left-ventricular pressure (index of contractility), and increased myocardial oxygen consumption. The studies suggest a prominent role for the venous circulation on cardiac function, which takes on greater importance in the failing heart when right heart pressures can be as high as 30 mmHg.

Key words. Venous pressure—Coronary flow—Collateral flow—Left-ventricular compliance—Cardiac function—Myocardial oxygen consumption—Left-ventricular volume

Introduction

"Although a great deal of research has been undertaken to elucidate the physiology of the heart, the arterial circulation, and the microcirculation, the physiology of the veins remains much less well understood" [1]. While there exists a large body of literature

[1] Department of Physiology, University of North Texas Health Science Center at Fort Worth, Fort Worth, TX 76107-2699, USA.
[2] Ramez Street, P.O. Box 264, Kadima 60920, Israel.

concerned with the interactions between coronary arterial flow and left ventricular (LV) function and vice versa, far less attention has been given to influences of the coronary venous system on LV function [2,3].

There are several reasons why this area of research has received less attention. (1) The direct effects of coronary venous pressure on LV function are masked by its indirect effects. The indirect effect of coronary venous pressure, as manifested in a rise in right heart pressure (RHP), is to stretch the right-ventricular chambers and elicit an increased contractile response that leads to increased right heart output (the Frank–Starling mechanism). This output is routed through the lungs into the left heart where a similar response produces an increased cardiac output. Therefore, to study the direct affects of RHP on LV function, the pulmonary circulation must be interrupted (uncoupled heart). However, even under this condition, mechanical right-to-left heart interactions remain to be addressed. (2) It is possible that the direct effect of venous pressure on LV function is negligible in the normally functioning heart. However, in a failing heart without pump function reserve (operating in the plateau region of the Frank–Starling curve), the direct effects of elevated coronary venous pressure may have considerable influence on LV function. (3) Under the "waterfall" concept [4], coronary venous or right heart pressure has no influence on coronary arterial flow unless RHP exceeds the coronary "waterfall" pressure. Our investigations into the waterfall phenomenon demonstrated that the waterfall pressure, or pressure at zero coronary arterial flow (PZF), is a function of the coronary collateral circulation and does not exist in the absence of the collateral circulation [5,6]. In additional experiments we observed that pressure at zero flow could indeed become positive and intercept the pressure axis at almost exactly the same value as RHP [7]. Because RHP is often elevated with heart failure, those experiments provided the impetus for further studies into the role of coronary venous pressures on coronary arterial flow and LV function [8].

Anatomy of the Coronary Venous Circulation

Coronary Venous Drainage

Coronary venous drainage has been studied primarily in dog hearts [9–11], although investigations have been conducted on swine [12] and human hearts [13]. In the dog, blood is drained from myocardial tissues via an extensive, richly anastomosed venous network (Fig. 1). Blood to the canine left ventricle is supplied by the left-anterior descending (LAD), circumflex (LCX), and septal arteries. Although venous drainage from the LAD and LCX empties into the right atrium through the coronary sinus (CS) [9,10], the septal artery, which in the dog chiefly supplies the interventricular septum, is drained via the anterior cardiac veins, and a small portion flows to the CS [10]. Most of the venous blood derived from the right heart is channeled through the anterior cardiac veins [9], which open into the right atrium at or below the junction of the atrial wall and the ventral surface of the atrial appendage.

Moir et al. [11], using radioactive dye dilution, found that only 15%–20% of total left common coronary arterial inflow in dogs is drained via noncoronary sinus pathways [11] such as the anterior cardiac veins or thebesian veins, which empty directly into the right and left ventricles [10,11]. In contrast, Hammond and Austen [10] found that 48% of canine left coronary artery flow drains through vessels other than the CS.

FIG. 1. Cast of the canine coronary venous system. Coronary veins parallel the coronary arteries and have numerous anastomotic connections

We used an isolated, vasodilated, canine heart preparation to determine coronary venous drainage distribution. Coronary arterial inflow was occluded in all but one artery, and coronary sinus effluent was collected for 30 s in a graduated cylinder. When all four coronary arteries were perfused simultaneously, CS effluent accounted for 59% of the combined total arterial inflow (LAD, LCX, right and septal artery). When the LCX and LAD arteries were perfused simultaneously while the right and septal artery flow lines were occluded, their combined inflows accounted for 86% of total CS outflow; this means that 14% of left coronary inflow was not accounted for in the CS effluent. The reason for this discrepancy is that the LCX and LAD in the dog heart supply portions of the right heart [14] and probably drain via their regional circulations. The septal artery alone contributed 6% to total CS outflow and the right coronary artery alone contributed only 3% to CS outflow.

Volume of the Coronary Venous Circulation

Although the venous network of the myocardium is quite extensive, the magnitude of coronary venous volume has not been well established. Investigations of coronary vascular volume have centered on determinations of total intramyocardial vascular volume [15–18], arterial volume [17,19], and capillary volume [20,21]. The volume of the coronary venous circulation has been investigated in dog [22,23], rat [24], and pig [25] models. The methods used in these investigations have been either point-counting techniques [23,24] or casting and corrosion techniques [22,25].

In an early morphometric investigation, Grayson et al. [22] indicated that the ratio between coronary venous and arterial volumes in the dog was very large, of the order of 20:1. Although there was no direct measurement of the volumes of the respective coronary circulations, the researchers noted that while 10–12 ml of an injected casting medium was needed to perfuse the entire intramyocardial vasculature of a beating canine heart, 0.5 ml was sufficient to fill the arterial vasculature alone.

Coronary venous volume has also been indirectly estimated using point-counting methods in myocardial tissue samples [23,24]. By this method, venous volumes in papillary muscle regions of the left-ventricular myocardium in dog hearts [23] were studied. The regions were divided into endocardial, intermediate, and epicardial sections and sliced for viewing under light and transmission electron microscopy. The values for venous volume relative to heart tissue volume ranged from 2.3% in the endocardial anterior papillary muscle region to 3.5% in the intermediate anterior papillary muscle region. Arterial volumes were found to range from 0.6% to 0.8%, indicating a ratio of coronary venous to arterial volume of 4:1 or 5:1.

Kassab et al. [25] recently published the results of an extensive investigation of the coronary venous circulation of the pig in which a cast of the coronary vessels was formed by perfusing the coronary circulation with a silicone elastomer. The resulting cast of the vasculature was meticulously measured and assigned an order via a computer imaging process system. Using this methodology, the authors reported that 27% of the blood in the porcine coronary circulation was in the arteries and 37% in the veins. These data indicate that a much narrower ratio exists between the volumes of these two coronary vascular systems; however, in this study vessels less than 200 μm in diameter were considered to be part of the microcirculation and were not distinguished as arterial or venous vessels.

We have recently conducted a study to separately determine the volumes of the coronary venous and arterial systems of the arrested dog heart using the casting and corrosion technique with Batson's No. 17 Plastic Replica Kit (Polysciences Inc., Warrington, PA, USA). The coronary venous vessels were retrogradely perfused with liquid plastic from the right atrium and ventricle and were filled approximately to the level of the capillary (10 μm). We were unable to directly perfuse the coronary veins that drain into the left heart, i.e., the thebesian veins. Because of the extensive anastomoses that exist at all levels of the coronary venous vasculature, however (Fig. 1), minimal portions of the vasculature remained unfilled. After allowing 24 h for the Batson's compound to harden, the heart tissue was macerated with a potassium hydroxide solution. The coronary arterial system was also casted with Batson's via the aorta. The volume of the respective circulations was calculated from the weight of the casts and the specific gravity of the casting medium and normalized for heart weight. We found that venular and arterial volumes were 3.23 and 1.50 ml/100 g heart tissue, respectively, representing a venous-to-arterial volume ratio of 2.2 : 1.

Coronary Venous Pressure and Coronary Arterial Flow

When coronary sinus or venous pressure is increased in a vasodilated heart, the observed results on coronary flow have been controversial. Although some studies [26,27] have reported a venous threshold pressure at about 10 mmHg below which coronary flow remained constant, a phenomenon that was attributed to the waterfall effect, other studies from our and other laboratories found that coronary flow decreased in direct proportion to the elevated venous pressure [7,28]. Several mechanisms have been proposed to explain the reduction in coronary flow with elevated venous pressure. (1) Elevated venous pressure could reduce coronary flow because the pressure gradient between the aorta and the venous system is reduced. (2) Coronary flow could be reduced because filling of the veins increases intramyocardial pressure [29,30], transmitting compressive forces to the coronary vasculature that cause an increase in coronary resistance and thus reduced flow. On the other hand, the existence of circulatory phenomena such as the "waterfall" effect, and the activation of flow regulation mechanisms such as mechanoreceptors in the vascular wall (myogenic response), could oppose the tendency for flow to decrease. Although the response of the vasodilated coronary circulation to elevation in coronary venous pressures is controversial, even less is known regarding the response when coronary vasomotor tone is present (vasoactive).

When RHP is elevated instead of increasing coronary venous pressure alone (via the coronary sinus), another layer of complexity is introduced. Although it would

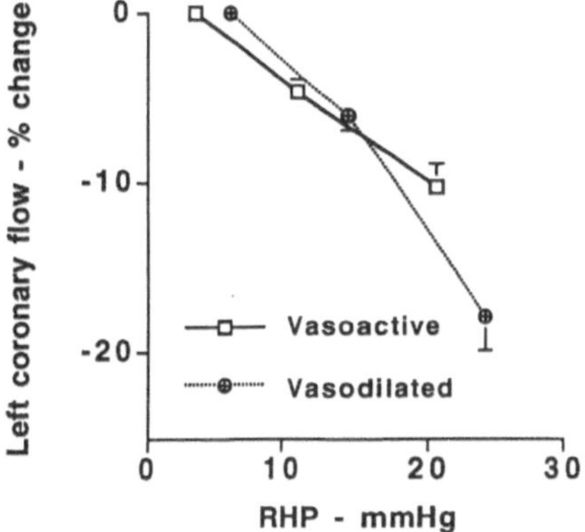

FIG. 2. Left coronary arterial flow as a function of right heart pressure (*RHP*). After vasodilation (*circles*), coronary arterial flow decreased in almost direct proportion to a reduction in perfusion pressure. Even when the vasculature was vasoactive (*squares*), coronary flow decreased with RHP

seem that because the coronary sinus drains into the right heart and all the mechanisms mentioned would be activated, changes in RHP are mechanically coupled to the left ventricle by deflection of the septum separating the two chambers and common fiber bundles at the regions of chamber connection.

To shed light on the questions just raised, we used an isolated, blood-perfused, beating heart preparation in which RHP was increased and coronary arterial flow monitored during maximum vasodilation and under vasoactive conditions. In these studies (Fig. 2), we found that coronary flow decreased with elevated RHP both with vasomotor tone present (vasoactive) and absent (vasodilated). Therefore, the experimental results show that a reduction in coronary arterial flow, induced by elevating RHP, is not compensated for by a change in vascular tone in an effort to maintain constant flow. However, the reduced slope of flow decline in the vasoactive condition suggests some flow regulation.

It is worth exploring whether the mechanisms proposed for regulation and maintenance of coronary arterial flow in response to changes in arterial pressure, namely autoregulation, could also be applied to the venous circulation.

The Myogenic Theory

When examining the applicability of the "myogenic theory" for maintaining coronary arterial flow, an increased venous pressure could be transmitted retrograde via the capillaries to the arterioles, which would respond with vasoconstriction and thus decreased coronary arterial flow. This hypothesis would be consistent with our results.

The Metabolic Theory

When considering the metabolic theory for autoregulation, increased venous pressure would tend to reduce coronary arterial flow because of a decreased perfusion pressure gradient. Also, accumulation of metabolic vasodilator end-products (adenosine) would counteract the decrease in coronary arterial flow. Our results are not consistent with this hypothesis.

The Tissue Pressure Hypothesis

This hypothesis holds that a rise in arterial pressure transiently increases fluid exudation at the capillaries and that the perivascular fluid pressure rises to compress vessels and increase vascular resistance. Johnson [31] has described criteria needed to support this hypothesis. "If a rise in perfusion pressure causes more fluid to leave the capillaries, then tissue pressure should increase and compress the most collapsible structures, probably the small veins. Venous resistance would rise, but resistance in the more proximal vessels would be about normal." In our experiments increased venous pressure would distend coronary veins and lower venous resistance. However, the fact that arteries and veins coexist within the myocardium may cause a rise in intramyocardial pressure [29] and compress the adjacent arterioles, reducing coronary arterial flow.

Flow-Dependent Mechanism

The coronary venular system also actively responds to flow through changes in resistance and permeability [32]. Kuo and colleagues [33] observed, in isolated coronary venules, that the venular microvessels actively dilated on the initiation of flow. This study was accomplished using a specialized microperfusion apparatus that enabled the investigators to discriminate between myogenic and flow-dependent effects. The magnitude of the maximal vasodilatory response was 10%–15%. Although this response is modest when compared to the maximal flow-dependent vasodilatory response to arterioles, which is usually in the range of 30%–35% [34], one must consider that diameter changes are related to resistance by a power of four. Thus, the relatively small change in venular diameter can be translated into a large change in resistance. Moreover, it deserves mention that under certain conditions (e.g., arteriolar vasodilation) the coronary venous system can account for as much as 33% of total coronary resistance [35]. Taken together, flow-dependent vasodilation of coronary venules has the potential to influence total coronary vascular resistance.

Flow-dependent vasodilation of coronary venules appears to be mediated by the endothelium-dependent production of nitric oxide [33]. Interestingly, when the endothelium was mechanically ablated, the vasodilatory response to flow was changed to constriction. The reason(s) underlying constriction to flow are not apparent, but are likely related to mechanostimulation of vascular smooth muscle during flow. Perhaps in the absence of the endothelium, the thin wall of venules and the absence of the internal elastic lamina enable shear forces to be transmitted to venular smooth muscle, enabling constriction to occur.

A flow-dependent vasodilation system is a positive feedback system; that is, as flow increases vascular resistance decreases to further enhance the increase in flow. However, because there is a limited range of effectiveness to this system, flow will not increase without bound.

Effects of Changes in LV Volume on Coronary Arterial Flow

While our experiments established that coronary arterial inflow is changed with changes in RHP (see Fig. 2), these changes could be influenced not only via the coronary venous system but by mechanical coupling between the right and left heart.

The rationale is that mechanical ventricular interaction through the common fiber bundles and septum, shared by the two ventricles, could cause stretching or compression of the left ventricle and the embedded coronary vasculature and thus change coronary resistance and coronary arterial flow [36–39].

To test this hypothesis, we performed experiments to determine how mechanical stretching of the LV (changes in LV volume or pressure) influences left coronary arterial flow during vasodilation and under vasoactive conditions. We used an arrested heart preparation in which cardiac contractions were eliminated. The LV was stretched by increasing the pressure within a balloon located in the LV. Left coronary arterial flow was recorded as the volume within the balloon was increased.

After maximal vasodilation with adenosine, coronary arterial flow decreased linearly with increases in LV volume (Fig. 3). These data established that stretching of the LV myocardium can affect left coronary arterial flow. Interestingly, in the same arrested heart preparation, in the presence of vasomotor tone stretching of the LV resulted in no changes in left coronary arterial flow ("Vasoactive" in Fig. 3); that is, coronary arterial flow was maintained constant. Mechanical compression or stretch of the vasculature and its effects on coronary resistance and flow, discussed previously, holds so long as the coronary vasculature is passively deformed. However, in the presence of coronary vasomotor tone, which is the normal physiological condition of the vasculature, physiological compensatory mechanisms can come into play to maintain a constant coronary arterial flow [8].

Similar results were obtained in isolated rat hearts in which it was found that coronary flow was maintained over a large range in LV volume before a decrease in flow was observed [40]. In that study, however, the average (systole and diastole) of total (right and left) coronary flow was reported, and therefore the affects on diastolic flow remained unclear. Our results suggest that changes in LV volume and coronary arterial flow are either dissociated, or that vascular resistance adaptations occur to maintain a constant left coronary arterial flow. A dissociation between coronary arterial flow and LV volume contradicts the observations seen with vasodilation in which flow changed linearly with changes in LV volume (Fig. 3). Our study, therefore,

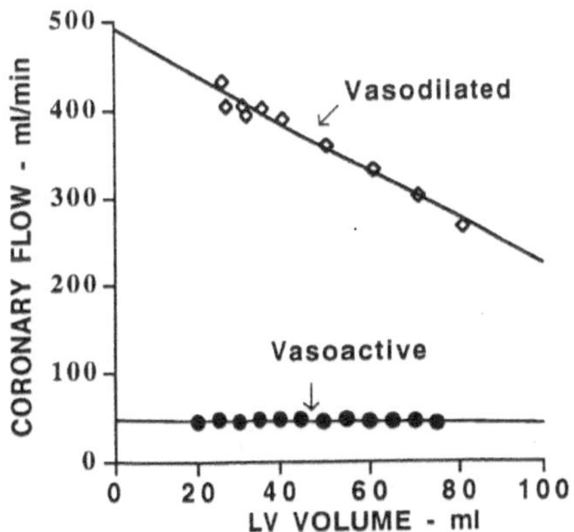

Fig. 3. Relationship of coronary flow as a function of left-ventricular (*LV*) volume. After vasodilation (*squares*) with adenosine, left coronary flow decreased with increasing LV volume; however, when the vasoactive (*circles*) vasculature was intact, coronary flow remained constant with increasing LV volume

suggested that changes in vasomotor tone regulate changes in vascular resistance and thus flow in response to changes in LV volume (or pressure) when the coronary vasculature is vasoactive.

The mechanisms responsible for the unchanged left coronary arterial flow could be attributed to either myogenic or metabolic control [41,42]. Under the myogenic hypothesis, an increase in LV volume results in myocardial stretch [40,43], which changes the stress field within the myocardium [44–46] and activates the mechanoreceptors within the vascular smooth muscle [47] to maintain constant flow. Under the metabolic hypothesis, changes in adenosine or O_2 concentrations may maintain flow regulation. As no changes in O_2 consumption were found over a large range of LV volumes in the beating rat heart [40], it appears that metabolic control is unlikely.

Furthermore, left coronary arterial flow did not decrease with vasomotor tone present, even as LV volumes were increased by three- to sevenfold above control volumes, suggesting a large flow regulation capacity or flow reserve of the coronary vasculature. However, this reserve may be impaired when the LV is dilated as a result of heart failure [48] or volume overload [49,50].

Right-to-Left Heart Interactions

The foregoing experiments established that stretching of the left-ventricular myocardium, whether by changes in LV volume or LV pressure, either have no influence on coronary arterial flow, as in the case of the vasoactive vasculature, or can directly influence flow when the vasculature is dilated. We then tried to determine how changes in RHP are reflected in LV pressures during conditions of vasodilation and when the vasculature was vasoactive.

Ventricular Interaction During Coronary Vasodilation

To determine how RHP is coupled to LV pressures, we conducted experiments in which RHP was changed at various LV volumes (Fig. 4). In this figure, α was defined as $\Delta(\text{RHP})/\Delta(\text{LVP})$ where LVP was the left-ventricular pressure.

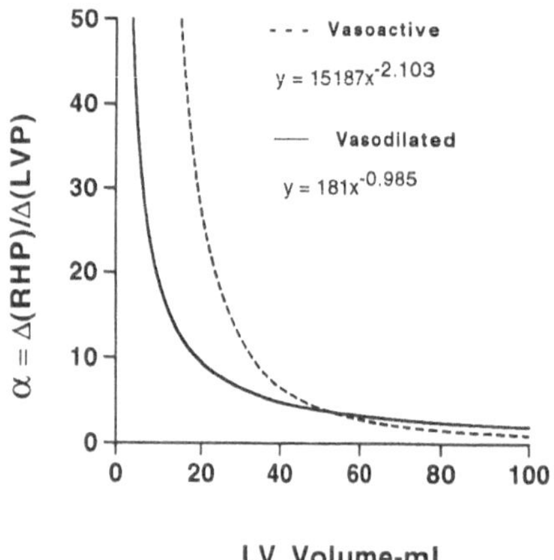

FIG. 4. Interaction of RHP with left-ventricular pressure (*LVP*) as a function of LV volume Note: after vasodilation (*solid line*), RHP had a greater effect on LVP; also, at low LV volume there was practically no effect of RHP on LVP

With coronary vasodilation and at very low LV volumes (~2–5 ml), the value of α was approximately 10 or more, indicating that only one-tenth of the pressure increase in the right heart is reflected as an increase in LV pressure. However, at higher LV volumes when the LV became stiffer (to be shown later), right heart to left heart coupling became tighter and changes in RHP were more closely reflected in LVP. This coupling appears to approach the value of 2 asymptotically; that is, a 2-mmHg change in RHP is reflected as a 1-mmHg change in LVP [45].

When the vasculature was vasoactive (Fig. 4), this relationship was shifted to the right. The results show that there was practically no RHP-to-LVP pressure interaction for LV volumes as great as 20 ml. In subsequent paragraphs we show that LV compliance is reduced with vasodilation. Therefore, the inference is that the more compliant myocardium exhibits less right-to-left heart interactions.

In summary, although elevated RHP can influence left coronary arterial flow by ventricular interactions during vasodilation, it is unlikely that mechanical ventricular interaction plays a major role in the presence of vasomotor tone. Thus, the decrease in coronary arterial flow observed with increased RHP (see Fig. 3) in hearts with vasomotor tone must be explained by mechanisms other than mechanical right-to-left heart interactions. Our experimental evidence suggests that changes in coronary arterial flow are strongly influenced by changes in coronary venous pressures.

Coronary Venous Pressure Reduces Coronary Collateral FLow After Coronary Occlusion

A frequent sequela of congestive heart failure, often precipitated by coronary occlusion, is an elevated RHP that can rise to as high as 20 and 30 mmHg [51,52]. An increase in RHP, in turn, raises coronary venous outflow pressure and reduces coronary perfusion [7,26–28,53–56]. With coronary artery occlusion, the respective myocardial perfusion territory becomes collateral flow dependent. This flow, after traversing the collateral-dependent myocardium, also empties into the coronary venous system. In addition, with coronary occlusion the pressure distal to the occlusion (P_c) drops considerably. Therefore, the pressure gradient across the collateral-dependent myocardium is dependent on (P_c minus the venous outflow pressure). When coronary occlusion is followed by heart failure and an increase in RHP, the pressure gradient for perfusion of the collateral-dependent myocardium becomes P_c minus RHP [7,28]. A reduction in this pressure gradient may in turn lead to a decrease in perfusion of the collateral-dependent myocardium.

In our study, we used an isolated, beating, and vasodilated dog heart preparation [57,58] to study the direct effects of RHP on the coronary collateral circulations [59]. The retrograde flow method [57,60,61] was used to assess collateral flow. An interesting result of this study was the observation that retrograde flow practically doubled as RHP was elevated to 20 mmHg, a result similar to that seen with microembolization of the collateral-dependent myocardium [58]. This flow increase could result from either (a) an increase in collateral flow, caused by an increase in pressure at the stem of the collaterals induced by the elevated RHP or (b) a redistribution of (a portion or all) the antegrade flow in the retrograde direction.

To distinguish between those two possibilities, the vessel on which the retrograde flow was measured was embolized with plastic spheres. The rationale for this protocol was as follows: by embolizing the microcirculation of this vessel, the antegrade component of collateral flow would be eliminated. If retrograde flow still increased by raising RHP after embolization, this increase most likely would be caused by an increase in collateral stem pressure. Alternatively, the increase in retrograde flow seen with increased RHP could be caused by a redistribution of the antegrade flow component added to the retrograde flow. We found that retrograde flow did not change after embolization. This unchanged flow implies that increasing RHP may result in a total cessation or reduction in coronary collateral flow, a condition that would compound myocardial ischemia of the collateral-dependent myocardium. Because coronary occlusion can lead to heart failure and increased RHP, a vicious cycle could occur leading to further ischemia of the collateral-dependent myocardium and deterioration of pump function.

Coronary Venous Pressure Changes the Mechanical Properties of the Left Ventricle

It has been demonstrated that elevating coronary arterial perfusion pressure results in altered LV myocardial compliance [62–64]. The mechanism to explain this change in compliance is thought to be related to changes in LV vascular volume—an "engorgement" of coronary vessels. The rationale is that occupation of a larger percentage of myocardium by an incompressible fluid (blood) would alter the compliance of the myocardium. This same rationale could be applied to the venous vasculature with elevation of pressure. We hypothesized that an increase in RHP, which serves as the effective coronary venous outflow pressure [59], may result in coronary venous distention similar to coronary arterial distension with elevated perfusion pressures.

Studies by others and our observations indicate that venous vascular volume exceeds arterial vascular volume; therefore, it is reasonable to assume that the venous vasculature may play a role in modifying LV properties. Watanabe et al. [3] have shown that elevating RHP results in changes in LV distensibility. A reduction in LV compliance may be an important aspect of diastolic dysfunction in heart failure.

In the beating heart, the myocardium is subjected to cyclic loading and unloading, and its properties must be determined in view of viscoelastic theories [65–67]. Glantz [68] highlighted the difficulties in reliably assessing the diastolic properties of the myocardium by suggesting that the duration of passive late diastolic relaxation may be too short. Therefore, studies utilizing potassium-arrested hearts in which the effects of systolic shortening and active relaxation are eliminated have emerged to determine ventricular properties [69–71]. This model has been recognized by some as a classic preparation by which diastolic ventricular mechanics can be studied [69].

Right Heart Pressure and LV Compliance in the Presence of Vasomotor Tone

We used potassium-arrested dog hearts to assess changes in LV compliance in response to changes in RHP. With vasomotor tone present, the pressure–volume (P–V) relationships were curvilinear (Fig. 5). As RHP was increased, the curve was displaced upward and the slopes of the lines for RHP = 10 and 20 mmHg were increased. Therefore, compliance defined as $\Delta V/\Delta P$ decreased [45].

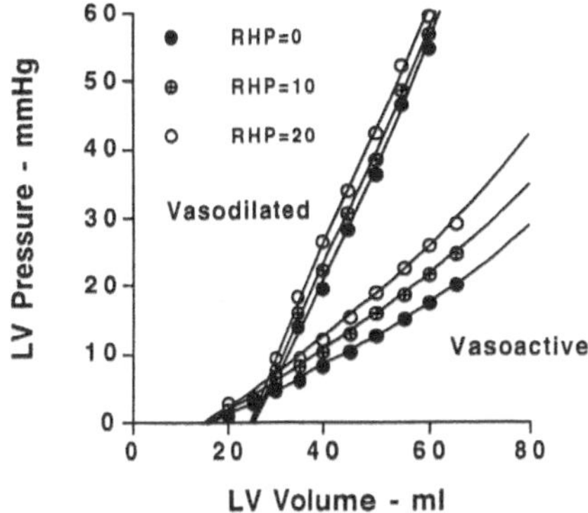

Fig. 5. Effects of RHP on the compliance of the LV. The graph illustrates the LV pressure–volume relationship in which the slope of the lines is defined as LV stiffness. Note: the LV became stiffer with increasing RHP and after vasodilation with adenosine

Right Heart Pressure and LV Compliance with Vasodilation

In the absence of vasomotor tone, the pressure–volume relationship was considerably steeper, indicating that vasodilation results in a decrease in LV compliance (Fig. 5). However, even in this state RHP reduced LV compliance [45]. The mechanism for this observation is thought to be associated with changes in vascular blood volume: before adenosine infusion, end-diastolic blood volume constitutes about 13% of total myocardial mass, but after vasodilation the percentage of blood volume increases to as much as 22% [18,41,72,73].

Changes in vascular blood volume may not only result in changes in myocardial compliance, but increase epicardial segment length [62] and wall thickness [73,74] and decrease LV chamber size [74]. However, in our arrested heart preparation other properties intrinsic to the myocardium such as ventricular relaxation, inertia, or viscous effects [74] were removed. The observed results therefore can be attributed to changes in the passive elastic property of the myocardium.

Coronary Venous Pressure and Its Influence on LV Function

Gregg observed that an increase in arterial perfusion pressure results in a concomitant increase in myocardial oxygen consumption, MVO_2, a phenomena recognized as the "Gregg" effect [75]. This observation was later complemented by Arnold et al. [76], who proposed that the increase in arterial perfusion pressure results in distension of the coronary arteries, stretching of myocardial fibers, and consequently activation of the Frank–Starling mechanism. Activation of this system results in increased cardiac contractile function and increased myocardial oxygen consumption. This proposed mechanism, which was was termed the "gardenhose" effect [76,77], is discussed in greater detail by Dr H.F. Downey in another chapter of this book. We hypothesized that if filling of arterial vessels elicits the Gregg effect, then filling of the venous circulation may result in a similar response, particularly because venous volume exceeds arterial vascular volume.

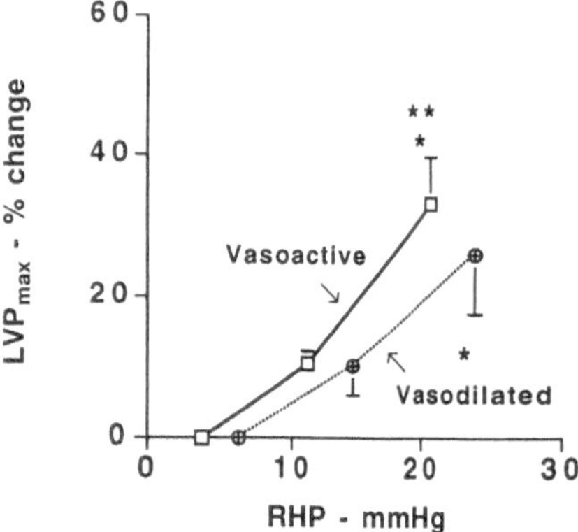

FIG. 6. Relationship between RHP and maximum left-ventricular pressure generation (*LVP max*). LVPmax is an index of contractility; thus, an increase in RHP results in an increase in left-ventricular contractility. *, $P < .05$ relative to control; **, $P < .05$ relative to RHP = 10 mmHg

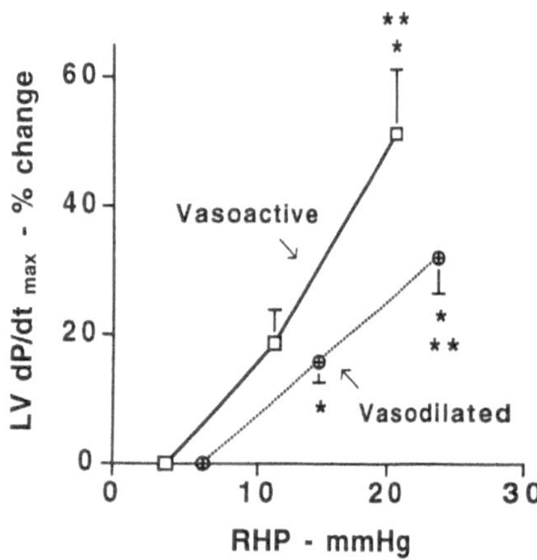

FIG. 7. Relationship between RHP and maximum rate of change of left-ventricular pressure (*LV dP/dt*). LV dP/dt is an index of contractility; thus, an increase in RHP results in an increase in left-ventricular contractility. *, $P < .05$ relative to control; **, $P < .05$ relative to RHP = 10 mmHg

We used an isolated, uncoupled, blood-perfused, beating dog heart preparation. To isolate the effects of coronary venous outflow pressure on myocardial contractile function, the system was designed to maintain LV end-diastolic pressure at a constant level. The experiments were performed both during maximum vasodilation and with intact vasomotor tone. To validate the existence of the gardenhose effect, changes in myocardial contraction were determined. While contractility is difficult to define and measure, changes in myocardial contraction are much simpler to detect [78]. However, because contractile function is dependent to a large extent on LV loading conditions, it is critically important to maintain LV preload constant. In our preparation this was controlled by allowing excess LV blood volume to drain out of the system, thus maintaining a constant LV end-diastolic pressure. Changes in myo-

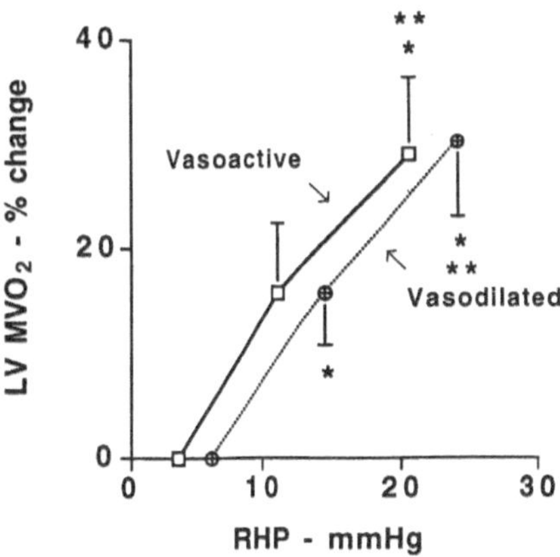

FIG. 8. Relationship between RHP and LV oxygen consumption (MVO_2). *, $P < .05$ relative to control; **, $P < .05$ relative to RHP = 10 mmHg

cardial contractile function were therefore determined by changes in LVP_{max} and dP/dt_{max}, as suggested by Gregg and others (2,27,75,77). Increasing RHP in our experiments resulted in a significant increase in these indices (Figs. 6 and 7), as well as in MVO_2 (Fig. 8), highlighting the causative relationship between these variables.

Proposed Mechanisms

The "Gregg" phenomenon and the "gardenhose effect" have been challenged by two findings. First, it was shown that coronary flow, rather than arterial perfusion pressure, is the dominant factor responsible for the gardenhose effect. The gardenhose effect was not observed in autoregulating hearts where coronary flow was maintained relatively constant over a wide range of arterial perfusion pressures [79–81]. Second, recent studies in isolated papillary muscle [82] and arrested and beating heart preparations [69,83] showed that perfusion pressure-induced changes in myocardial segment length or fiber strain are negligible; this too is in contrast to the proposed gardenhose mechanism. Kitakaze and Marban [83] and Schouten et al. [82] provided an alternative explanation for the increased contractile response with elevated arterial perfusion pressure. Kitakaze and Marban [83] measured intracellular calcium concentrations and segment length in an isovolumically beating ferret heart preparation for different arterial perfusion pressures, and concluded that changes in coronary perfusion modulate intracellular calcium concentrations rather than sarcomere length.

These authors acknowledged, however, that the mechanism of interaction between perfusion pressure and calcium concentration still remains to be determined. This study was later complemented by Schouten et al. [82], who studied the effects of perfusion pressure and calcium in the perfusate solution on the force of contraction in isolated papillary muscle. They too found that increased force of contraction was related to calcium concentration rather than segment length, which was possibly mediated by changes in the volume of the interstitium. They found that the extracellular volume, which includes capillaries and the interstitium, increased with perfusion pressure, and therefore speculated that more calcium would become available to the myocyte, thus enhancing and supporting myocyte force development. Therefore, both of these studies proposed that Gregg's phenomenon is a change in contractility modulated by intracellular calcium rather than changes in myofiber length [82,83].

The common denominator for these proposed mechanisms of myocardial contractile changes with elevated arterial perfusion pressure is increased vascular blood volume. The increased vascular blood volume either stretches the myocardium according to the gardenhose concept [76] or changes the calcium concentrations, possibly secondary to interstitial volume changes, as proposed by the contractility concept [82,83]. The results of our study do not conflict with either of these proposed mechanisms. In fact, because the underlying theories are based on changes in vascular volumes, then they should certainly apply to the coronary venous system, which exceeds coronary arterial blood volume by as much as 2 to 20 fold [22,84]. Thus, elevation of RHP may increase venous blood volume as a result of vessel distension [2,3] and elicit a similar myocardial response, i.e., a "venous gardenhose effect" as increased arterial blood volume.

Schipke et al. [77] have demonstrated that a 20-mmHg increase in arterial perfusion pressure resulted in increased LVP_{max} and dP/dt_{max}. However, the percent changes in their indices were smaller than the relative changes found in our study with a 20-mmHg increase in venous pressure. This is consistent with the observation that venous blood volume is larger than arterial volume, and therefore its relative effect on LV function would be expected to be larger.

Our results with arterial vasomotor tone present and absent further support the volume theory. Recent studies show that the Gregg or gardenhose phenomenon is not found in autoregulating hearts when arterial perfusion pressure is changed [79,80]. These results are consistent with the volume mechanisms just discussed because arterial vascular volume does not change much with perfusion pressure in autoregulating hearts [41,79], and thus the gardenhose or contractility mechanisms are not activated. This condition is different with elevated venous pressure, however, because venous vascular volume is bound to increase regardless of the state of the arterial vasculature (autoregulating or vasodilated). Therefore, an increase in myocardial contractile function would be a direct consequence of changes in vascular volumes with elevated RHP regardless of the vasoactive state of the arterial vasculature (Figs. 6–8), in contrast to changes in arterial pressure that affect myocardial function only in the vasodilated heart.

In summary, elevation of RHP in the absence of the pulmonary circulation has a direct effect on LV function. It increases the maximally generated pressure within the left ventricle, increases the maximum rate of change in LV pressure (index of contractility), and increases myocardial oxygen consumption. While this "venous Gregg" effect or "venous gardenhose" effect seems to be related to changes in coronary venous volume, the exact mechanism by which the response is elicited is as controversial as the effects observed with increased perfusion pressure in the arterial circulation.

Acknowledgments. The authors thank Hong-Wei Wang and Rahmi Heyman for their help in the production of this review. Without the technical assistance of Sue Williams and Robert Ator, the experiments would not have been possible. Kim Chen and John Schofield deserve special recognition for, during their summer work-study, their help in modifying the isolated heart preparation was critical. This work was supported by NIH grant 35030, National Institutes of Health, Bithesda, MD.

References

1. Tyberg JV (1992) Venous modulation of ventricular preload (editorial). Am Heart J 123:1098–1104
2. Scharf SM, Bromberger-Barnea B (1973) Influence of coronary flow and pressure on cardiac function and coronary vascular volume. Am J Physiol 224:918–925
3. Watanabe J, Levine MJ, Bellotto F, Johnson R, Grossman W (1990) Effects of coronary venous pressure on left ventricular diastolic distensibility. Circ Res 67:923–932
4. Downey JM, Kirk ES (1975) Inhibition of coronary blood flow by a vascular waterfall mechanism. Circ Res 36:753–760
5. Scheel KW, Mass H, Williams SE (1989) Collateral influence on pressure flow characteristics of the coronary circulation. Am J Physiol (Heart Circ Physiol 26) 257:H717–H725
6. Scheel KW, Mass H, Gean JT (1993) Interactions of the coronary and collateral circulations. In: Schaper W (ed) Collateral circulation. Kluwer, Norwell, MA, pp 233–260
7. Scheel KW, Williams SE, Parker JB (1990) Coronary sinus pressure has a direct effect on gradient for coronary perfusion. Am J Physiol (Heart Circ Physiol 27) 258:H1739–1744
8. Manor D, Williams S, Ator R, Bryant K, Scheel KW (1995) Modulation of coronary flow by left ventricular volume in the presence and absence of vasomotor tone. Am J Physiol 269:H2010–H2016
9. Gregg DE, Shipley RE (1947) Studies of the venous drainage of the heart. Am J Physiol 151:13–25
10. Hammond GL, Austen WG (1967) Drainage patterns of coronary arterial flow as determined from the isolated heart. Am J Physiol 212:1435–1440
11. Moir TW, Ecxkstein RW, Driscol TE (1963) Thebesian drainage of the septal artery. Circ Res 12:212–219
12. Pantely GA, Bristow JD, Ladley HD, Anselone CG (1988) Effect of coronary sinus occlusion on coronary flow, resistance, and zero flow pressure during maximum vasodilatation in swine. Cardiovasc Res 22:79–86
13. Hood WB (1968) Regional venous drainage of the human heart. Br Heart J 30:105–109
14. Scheel KW, Ingram LA, Gordey RL (1982) Relationship of coronary flow and perfusion territory in dogs. Am J Physiol (Heart Circ Physiol 12) 243:H738–H747
15. Eliasen P, Amtorp O, Tondevold E, Haunso S (1982) Regional blood flow, microvascular blood content and tissue haematocrit in canine myocardium. Cardivasc Res 16:593–598
16. Wu XS, Ewert DL, Lin YH, Ritman EL (1992) In vivo relation of intramyocardial blood volume to myocardial perfusion. Circulation 85:730–737
17. Wusten B, Buss DD, Deist H, Schaper W (1977) Dilatory capacity of the coronary circulation and its correlation to the arterial vasculature in the canine left ventricle. Basic Res Cardiol 72:636–650
18. Crystal GJ, Downey HF, Bashour FA (1981) Small vessel and total coronary blood volume during intracoronary adenosine. Am J Physiol (Heart Circ Physiol 10) 241: H194–H201
19. Douglas JE, Greenfield JC Jr (1970) Epicardial coronary artery compliance in the dog. Circ Res 27:921–929
20. Bassingthwaighte JB, Yipintsoi T, Harvey RB (1974) Microvasculature of the dog left ventricular myocardium. Microvasc Res 7:229–249
21. O'Keefe DD, Hoffman JIE, Cheitlin R, O'Neil MJ, Allard JR, Shapkin E (1978) Coronary blood flow in experimental canine left ventricular hypertrophy. Circ Res 43:43–51

22. Grayson J, Davidson JW, Fitzgerald-Finch A, Scott C (1974) The function morphology of the coronary microcirculation in the dog. Microvasc Res 8:20–43
23. Hyde DM, Buss DD (1986) Morphometry of the coronary microvasculature of the canine left ventricle. Am J Anat 177:415–425
24. Levy BI, Samuel JL, Tedgui A, Kotelianski V, Marotte F, Poitevin P, Chadwick RS (1988) Intramyocardial blood volume measurment in the left ventricle of rat arrested hearts. In: Brun P, Chadwick RS, Levy BI (eds) Cardiovascular dynamics and models. Editions INSERM, Paris, pp 65–71
25. Kassab GS, Lin DH, Fung YC (1994) Morphometry of pig coronary venous system. Am J Physiol (Heart Circ Physiol 36) 267:H2100–H2113
26. Farhi ER, Klocke FJ, Mates RE, Kumar K, Judd RM, Canty JM Jr, Satoh S, Sekovski B (1991) Tone-dependent waterfall behavior during venous pressure elevation in isolated canine hearts. Circ Res 68:392–401
27. Uhlig P, Baer R, Vlahakes G, Hanley F, Messina L, Hoffman J (1984) Arterial and venous coronary pressure-flow relations in anesthetized dogs. Evidence for a vascular waterfall in epicardial coronary veins. Circ Res 55:238–248
28. Bellamy RF, Lowensohn HS, Ehrlich W, Baer RW (1980) Effect of coronary sinus occlusion on coronary pressure-flow relations. Am J Physiol (Heart Circ Physiol 8) 239:H57–H64
29. Izrailtyan I, Frasch HF, Kresh JY (1994) Effects of venous pressure on coronary circulation and intramyocardial fluid mechanics. Am J Physiol (Heart Circ Physiol 36) 267:H1002–H1009
30. Laine GA (1987) Change in $(dP/dt)_{max}$ as an index of myocardial microvascular permeability. Circ Res 61:203–208
31. Johnson PC (1964) Review of previous studies and current theories of autoregulation. Circ Res 15(suppl I):2–9
32. Yuan Y, Granger HJ, Zawieja DC, Chilian WM (1992) Flow modulates coronary venular permeability by a nitric oxide-related mechanism. Am J Physiol (Heart Circ Physiol) 263:H641–H646
33. Kuo L, Arko F, Chilian WM, Davis MJ (1993) Coronary venular responses to flow and presure. Circ Res 72:607–615
34. Kuo L, Davis MJ, Chilian WM (1990) Endothelium-dependent, flow-induced dilation of isolated coronary arterioles. Am J Physiol (Heart Circ Physiol) 259:H1063–H1070
35. Chilian WM, Layne SM, Klausner EC, Eastham CL, Marcus ML (1989) Redistribution of coronary microvascular resistance produced by dipyridamole. Am J Physiol (Heart Circ Physiol) 256:H383–H390
36. Little WC, Badke FR, O'Rourke RA (1984) Effect of right ventricular pressure on the end-diastolic left ventricular pressure-volume relationship before and after chronic right ventricular pressure overload in dogs without pericardia. Circ Res 54:719–730
37. Lorell BH, Palacios I, Daggett WM, Jacobs ML, Fowler BN, Newell JB (1981) Right ventricular distension and left ventricular compliance. Am J Physiol (Heart Circ Physiol 9) 240:H87–H98
38. Maruyama Y, Ashikawa K, Isoyama S, Kanatsuka H, Ino-Oka E, Takishima T (1982) Mechanical Interactions between four heart chambers with and without the pericardium in canine hearts. Circ Res 50:86–100
39. Taylor RR, Covell JW, Sonnenblick EH, Ross JJ (1967) Dependence of ventricular distensibility on filling of the opposite ventricle. Am J Physiol 213:711–718
40. Abe H, Holt W, Watters TA, Wu S, Parmley WW, Schiller N, Higgins C, Wikman-Coffelt J (1988) Mechanics and energetics of overstretch: the relationship of altered left ventricular volume to the Frank-Starling mechanism and phosphorylation potential. Am Heart J 116:447–454
41. Hoffman JIE, Spaan JAE (1990) Pressure-flow relations in coronary circulation. Physiol Rev 70:331–390
42. Klocke FJ, Ellis AK (1980) Control of coronary blood flow. Annu Rev Med 31:489–508
43. Ovize M, Kloner RA, Przyklenk K (1994) Stretch preconditions canine myocardium. Am J Physiol (Heart Circ Physiol 35) 266:H137–H146

44. Holt W, Auffermann W, Wu ST, Parmley WW, Wikman-Coffelt J (1991) Mechanisms for depressed cardiac function in left ventricular volume overload. Am Heart J 121:531-537
45. Manor D, Williams S, Ator R, Bryant K, Scheel KW (1995) Left ventricular mechanics in arrested dog heart: effects of ventricular interaction and vascular volumes. Am J Physiol (Heart Circ Physiol 37) 268:H2125–H2132
46. Resar J, Livingston JZ, Halperin HR, Sipkema P, Krams R, Yin FCP (1990) Effect of wall stretch on coronary hemodynamics in isolated canine interventricular septum. Am J Physiol (Heart Circ Physiol 28) 259:H1869–H1880
47. Rubanyi GM (1993) Mechanoreception by the vascular wall. In: Rubanyi GM (eds) Mechanoreception by the vascular wall. Futura, Mount Kosco, pp xi–xx
48. Shannon RP, Komamura K, Shen Y, Bishop SP, Vatner SF (1993) Impaired regional subendocardial coronary flow reserve in conscious dogs with pacing-induced heart failure. Am J Physiol (Heart Circ Physiol 34) 265:H801-H809
49. Doty DB, Wright CB, Hiratzka LF, Eastham CL, Marcus ML (1984) Coronary reserve in volume-induced right ventricular hypertrophy from atrial septal defect. Am J Cardiol 54:1059-1063
50. Marcus ML, Doty DB, Hiratzka LF, Wright CB, Eastham CL (1982) Decreased coronary reserve—a mechanism for angina pectoris in patients with aortic stenosis and normal coronary arteries. N Engl J Med 307:1362–1367
51. Alcorn JM (1991) Left ventricular diastolic dysfunction presenting as ascites: the importance of clinically assessing central venous pressure. J Clin Gastroenterol 13:83–85
52. Rackley CE, Russell RO (1972) Left ventricular function in acute myocardial infarction and its clinical significance. Circulation 45:231–244
53. Cantin B, Rouleau JR (1992) Myocardial tissue pressure and blood flow during coronary sinue pressure modulation in anesthetized dogs. J Appl Physiol 73:2184–2191
54. Ilbawi MN, Idriss FS, Muster AJ, DeLeon SY, Berry TE, Duffy CE, Paul MH (1986) Effects of elevated coronary sinus pressure on left ventricular function after the Fontan operation. An experimental and clinical correlation. J Thorac Cardiovasc Surg 92:231–237
55. Matsuhashi H, Hasebe N, Kawamura Y (1992) The effect of intermittent coronary sinus occlusion on coronary sinus pressure dynamics and coronary arterial flow. Jpn Circ J 56:272–285
56. Rouleau J, White M (1985) Effects of coronary sinus pressure elevation on coronary blood flow distribution in dogs with normal preload. Can J Physiol Pharmacol 63:787–797
57. Scheel KW, Mass HJ, Williams SE (1989) Pressure-flow characteristics of coronary collaterals in dogs. Am J Physiol (Heart Circ Physiol 25) 256:H441–H445
58. Scheel KW, Daulat G, Williams SE (1990) Functional anatomical site of intramural collaterals in the dog. Am J Physiol (Heart Circ Physiol 28) 259:H707–H711
59. Manor D, Williams S, Ator R, Bryant K, Scheel KW (1994) Reduced collateral perfusion is a direct consequence of elevated right atrial pressure. Am J Physiol (Heart Circ Physiol 36) 267:H1151–H1156
60. Eng C, Kirk ES (1984) Flow into ischemic myocardium and across coronary collateral vessels is modulated by a waterfall mechanism. Circ Res 55:10–17
61. Wyatt D, Lee J, Downey JM (1982) Determination of coronary collateral blood flow by a load line analysis. Circ Res 50:663–670
62. Vogel WM, Apstein CS, Briggs LL, Gaasch WH, Ahn J (1982) Acute alterations in left ventricular diastolic chamber stiffness: role of the "erectile" effect of coronary arterial pressure and flow in normal and damaged hearts. Circ Res 51:465–478
63. Salisbury PF, Cross CE, Reiben PA (1960) Influence of coronary artery pressure upon myocardial elasticity. Circ Res 8:794–800
64. Cross CE, Rieben PA, Salisbury PF (1961) Influence of coronary perfusion and myocardial edema on pressure-volume diagram of left ventricle. Am J Physiol 201:102–108

65. Fung YC (1993) Biomechanics: mechanical properties of living tissue. Springer-Verlag, Berlin Heidelberg New York, pp 41–57
66. Grossman W, McLauria WT (1976) Diastolic properties of the left ventricle. Ann Intern Med 84:316–326
67. Rankin JS, Arentzen CE, McHale PA, Ling D, Anderson RW (1977) Viscoelastic properties of the diastolic left ventricle in the conscious dog. Circ Res 41:37–45
68. Glantz SA (1980) Computing indices of diastolic stiffness has been counterproductive. Fed Proc 39:162–168
69. May-Newman K, Omens JH, Pavelec RS, McCulloch AD (1994) Three-dimensional transmural mechanical interaction between the coronary vasculature and passive myocardium in the dog. Circ Res 74:1166–1178
70. McCulloch AD, Hunter PJ, Smaill BH (1992) Mechanical effects of coronary perfusion in the passive canine left ventricle. Am J Physiol (Heart Circ Physiol 31) 262:H523–H530
71. Olsen CO, Jones RN, Attarian DE, Hill RC, Sink JD, Chitwood Jr. WR, Wechsler AS (1979) Relationship of LV compliance to coronary artery perfusion pressure in potassium-arrested canine heart during cardiopulmonary bypass. Cardiac Surg Forum 30:246–247
72. Chilian WM, Marcus ML (1984) Coronary venous outflow persists after cessation of coronary arterial inflow. Am J Physiol (Heart Circ Physiol 16) 247:H984–H990
73. Farhi ER, Canty JM, Klocke FJ (1991) Dissociation of diastolic pressure-segment length and pressure-wall thickness relations during vasodilation in the conscious dog. J Am Coll Cardiol 18:850–857
74. Verrier ED, Bristow JD, Hoffman JIE (1986) Coronary vasodilation shifts the diastolic pressure-dimension curve of the left ventricle. J Mol Cell Cardiol 18:579–594
75. Gregg DE (1963) Effect of coronary perfusion pressure or coronary flow on oxygen usage of the myocardium. Circ Res 13:497–500
76. Arnold GD, Morgenstern C, Lochner W (1970) The autoregulation of the heart work by the coronary perfusion pressure. Pflügers Arch 321:34–55
77. Schipke JD, Stocks I, Sunderdiek U, Arnold G (1993) Effect of changes in aortic pressure and in coronary arterial pressure on left ventricular geometry and function Anrep vs. gardenhose effect. Basic Res Cardiol 88:621–627
78. Katz AM (1991) Physiology of the heart. Raven, New York, pp 136–137, 371
79. Bai XJ, Iwamoto T, Williams AG, Fan WL, Downey HF (1994) Coronary pressure-flow autoregulation protects myocardium from pressure-induced changes in oxygen consumption. Am J Physiol (Heart Circ Physiol 35) 266:H2359–H2368
80. Schulz R, Guth BD, Heusch G (1991) No effect of coronary perfusion on regional myocardial function within the autoregulatory range in pigs. Evidence against the Gregg phenomenon. Circulation 83:1390–1403
81. Abel RM, Reis RL (1970) Effects of coronary blood flow and perfusion pressure on left ventricular contractility in dogs. Circ Res 27:961–971
82. Schouten VJA, Allaart CP, Westerhof N (1992) Effect of perfusion pressure on force of contraction in thin papillary muscles and trabeculae from rat heart. J Physiol (Camb) 451:585–604
83. Kitakaze M, Marban E (1989) Cellular mechanisms of the modulation of contractile function by coronary perfusion pressure in ferret hearts. J Physiol (Camb) 414:455–472
84. Spaan JAE (1985) Coronary diastolic pressure-flow relation and zero-flow pressure explained on the basis of intramyocardial compliance. Circ Res 56:293–309

Coronary Pressure–Flow Relations

Jos A.E. Spaan

Summary. The hemodynamic properties of the coronary circulation are generally defined by coronary arterial pressure–flow relations. Under normal physiological conditions, such a relationship demonstrates the ability of the coronary system to keep coronary flow at a rather constant level depending on oxygen consumption. We discuss how such a demonstration of flow control may in fact be the result of control of tissue oxygen pressure. The limits of the range of adjustment of coronary resistance are determined by factors such as heartrate, contractility, systolic and diastolic left-ventricular pressure, and the degree of hypertrophy. The influences of these effects on flow impediment become clear from the pressure–flow relations at full coronary vasodilatation. Pressure–flow relations at full vasodilatation are generally curved at low flow rates but rather straight at higher perfusion pressures. The difficulties with the interpretation of these pressure–flow relations are discussed. It is demonstrated on theoretical grounds that the pressure dependence of resistance makes it unjustified to relate the slope of the pressure–flow relation to coronary resistance without taking the pressure dependence of resistance into account. The mechanisms impeding coronary flow and related to heart contraction demonstrate a complicated interaction. On the one hand, contractility is impeding flow, and on the other hand it is protecting the intramural vessels for compression by left-ventricular pressure. One has therefore to be very careful in identifying the cause of flow impediment because it depends on the circumstances.

Key words. Intramyocardial pump—Compliance—Models—Elastance—Left-ventricular pressure

Introduction

A sufficient supply of blood to the myocardium is a prerequisite for its survival and also in stages of remodeling. Numerous factors determine the perfusion of the myocardium, which can be classified as stimulating or inhibiting. The balance between these forces is complex, especially because many physical factors have both stimulating and inhibiting effects. The beating of the heart is central in the analysis of its

Department of Medical Physics, Faculty of Medicine, University of Amsterdam, 1105 AZ Amsterdam, The Netherlands.

perfusion. The heartbeat sets the demands for blood supply, generates the pressure needed for myocardial perfusion, and also forms an impediment for perfusion.

Central to the discussion of myocardial perfusion are the coronary pressure-flow relations. One may wonder why, as coronary pressure is only one of the determinants. Most likely much attention is given to pressure-flow relations because they relate to the most elementary unit of hemodynamics, i.e., the resistance. If the vascular bed were a set of rigid tubes, one would expect a linear relationship between pressure and flow. Deviation from this linearity implicates mechanisms such as flow autoregulation and distention and collapse of vessels with pressure. The interpretation of pressure-flow lines has evolved from a simple resistance analysis to systems analysis, taking into account all these factors. It becomes even more complicated when the dynamics of perfusion are considered. When pressure and flow are changing dynamically, the transient displacements of blood volume in the coronary system must also be taken into account.

The purpose of this chapter is to provide some system analysis of myocardial perfusion, with no intent of giving a full review, let alone a complete picture of the whole system. The focus is on the interaction of the different elements in the balance of supply and demand.

Flow Control

Coronary blood flow is subjected to a control system that matches flow to the oxygen demands of the heart under a wide variety of circumstances. This control system is also multifactorial. At present, many endogeneous factors are known to influence vasomotor tone, but their specific roles in the flow control system as a whole still have to be defined. One may wonder whether a good insight into the functioning of this control system is of importance. The most simple point of view is that only the insight in its malfunctioning is of importance, because when it operates correctly flow is matched to its demand, and what can be more boring than to study an always perfect system? However, one may wonder what happens to the heart if the strength of control is diminished only slightly but chronically. Drugs are given to improve coronary blood flow but in reality manipulate the flow control system, and therefore insight in this system is needed for estimating the efficacy of therapy. In any case, it is important to know the boundaries of the control system because at these boundaries the conditions are created for spiraling down toward heart failure.

The control behavior of the coronary system is demonstrated in Fig. 1. In the study from which the data are taken, the coronary artery was perfused artificially from a pressure reservoir, independent of systemic blood pressure [1]. Such an experimental approach is almost a prerequisite for studying autoregulation [2]. In the case in which the heart generates its own perfusion pressure, changing this pressure will also result in an alteration of cardiac work and thereby oxygen consumption. As is seen later, oxygen consumption is the main determinant of coronary flow. Under control conditions and within a physiological range of pressures, Fig. 1 demonstrates that coronary flow is rather independent of coronary arterial pressure. This phenomenon is denoted as autoregulation. Maintenance of constant flow at changing pressure can only be brought about by changing resistance.

The strong influence of cardiac work on coronary flow is demonstrated by the shift of the autoregulation curve in a downward direction when cardiac work is decreased.

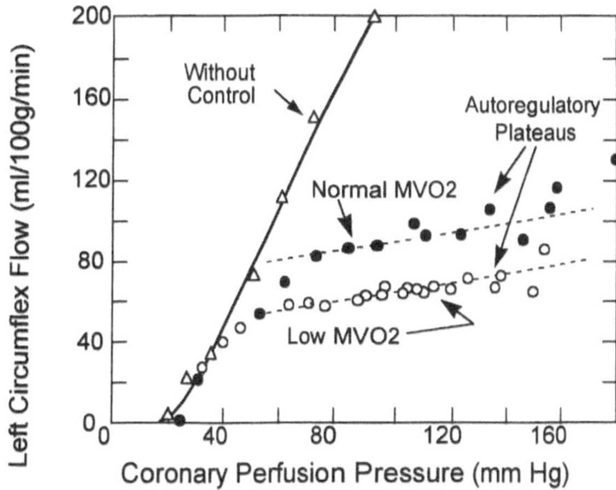

FIG. 1. Pressure-flow relationships with and without local control of coronary blood flow. With local control, coronary flow is rather independent of pressure over a wide range of pressure. Coronary resistance increases with pressure in this range. The level of the autoregulation plateau depends on the oxygen consumption of the perfused myocardium. At full vasodilatation, flow is strongly dependent on pressure. The pressure-flow relation at full vasodilatation is not linear but curved, especially for lower flow rates. (Data from [1])

At the same pressure the flow is lower at a lower level of cardiac work, which implies that resistance has increased. As we discuss, this change in resistance is brought about by decreased metabolism. The adaptation of coronary flow to a change in metabolism is referred to as functional hyperemia, but this author prefers the term metabolic adaptation of flow.

The changes in coronary resistance involved in both autoregualation and metabolic flow adaptation are brought about by changes in resistance of arterioles and small arteries. Because the effectors of both control mechanisms are the same, the two cannot be considered to be independent. A simple model explaining both autoregulation and metabolic flow adaptation is that of control of tissue Po_2. In this model, it is assumed that tissue Po_2 has a feedback to the resistance vessels of the myocardium. Decreasing tissue Po_2, either by increasing the oxygen consumption or decreasing perfusion, results in vasodilatation [3–5]. This oxygen model has been extended to incorporate effects of CO_2 [6].

We have tried several other models of coronary flow control, e.g., one based on adenosine, but the Po_2 model is the only one until now that basically describes the parallel shift of autoregulation curves and metabolic adaptation. It is not yet clear how such a direct Po_2 effect may work because there is no such thing as a Po_2 receptor. The interaction may be indirect, as for example the dissociation of O_2 from hemoglobin, as was recently suggested [7]. In any case, the Po_2 hypothesis is still very much alive.

The assumption of tissue Po_2 being the controlled variable instead of flow results in a different look at the quality of the control loop by which flow is determined. From a control engineering point of view, one of the parameters by which the functioning of the control loop can be quantified is the open- and closed-loop gain. If these gains are calculated on the basis of the oxygen control model, one arrives at values of 1 and 0.5, respectively, for these gains. These values are rather low, and smaller than can be calculated for hormone systems, which have values of 5 and 0.83, respectively [8]. The closed-loop gain provides some insight into how well the control system is influenced by changes in one of its parameters. If the sensitivity to tissue Po_2, for example, is changed by 20% then the tissue Po_2 may change by 10% with the closed-loop gain of 0.5 but only by 2% if this gain was 0.83. Thus, in terms of control system analysis coronary flow control is rather weak.

It is not so illogical to assume that the control system is weak. The purpose of the coronary vasculature is to supply sufficient oxygen to cells, not to maintain a certain fixed Po_2 value. Myocytes can function when a certain minimal Po_2 is maintained, enough to transfer oxygen from the capillaries to the mitochondria in the myocytes. The assumption is then that oxygen consumption of the myocytes is regulated at the mitochondrial level and is not affected or is only slightly affected by tissue Po_2. Such a design of the control system fits with the chaotic nature of the coronary branching patterns [9]. Such a concept does lead to an inhomogeneous distribution of blood flow and makes it impossible to arrive at a homogenous distribution of tissue Po_2. Thus, a tight control of tissue Po_2 would be impossible.

One could argue that the feedback between metabolism and blood supply is controlled at the local level such that the flow could be adapted to the local demand irrespective of the branching nature of the tree. However, control of blood flow is regulated by arteries with diameters smaller than 400 μm [10]. The larger a control vessel, the less it will be under the influence of local metabolism. Moreover, other mechanisms known to affect flow control act on resistance vessels in a diameter-dependent way. There is a need to integrate all these different mechanisms into a single control system. The coordination between all these different mechanisms must allow for some error in the outcome of the whole process. This is possible with a weak feedback loop for the overall system.

As discussed, the idea of a feedback loop for the control of tissue Po_2 is consistent with the high degree of variability found in both the distribution of flow in the myocardium [11,12] and the oxygen saturation of small venules [13].

Pressure–Flow Relations at Vasodilatation

Diastole

When smooth muscle tone in the coronary resistance vessels is completely abolished by the administration of a drug, coronary flow is only determined by mechanical factors. The slope of the pressure–flow relation is rather straight for higher perfusion pressures but becomes curved at the lower flow levels (see Fig. 1).

Often the slope of the pressure–flow relation is interpreted as conductance, G, i.e., being the inverse of resistance. If one aims to relate conductance to the physical properties of blood vessels one should, however, not use the slope of a pressure–flow relation as a measure of conductance. It has been well documented, in isolated small arteries as well as in vivo, that the diameters of resistance vessels when fully dilated are changing with pressure [14,15]. Hence, when one is measuring the slope of the pressure–flow relation by changing pressure over a certain range, the vessel diameters are changing as well. From the simple equation

$$Q = G(P) * P \tag{1}$$

where Q = flow, $G(P)$ = conductance as a function of pressure, and P = pressure, differentiation with respect to P results in

$$\frac{dQ}{dP} = G(P) + \frac{dG}{dP}P \tag{2}$$

Thus, the slope of the pressure–flow relation only relates directly to conductance when the conductance is independent of pressure. When the conductance is pressure dependent, the slope is related in a complex way to conductance as is expressed by the second term on the right-hand side of the equation.

The relation between the slope of the pressure–flow curve and conductance is elucidated in Fig. 2. At a perfusion pressure of P, the flow Q is given by $Q = G*P$. G can be calculated from the geometry of the arterial tree and the rest of the vascular bed at that perfusion pressure. The claim is that if the geometry of the vascular bed were frozen and thereby made indistensible, pressure and flow would relate to each other by a straight line crossing the pressure line at the venous pressure. For Fig. 2, this implies that the pressure–flow relation would be as given by the arrow. Similarly, one can calculate for any given P the flow from the conductance derived from the vascular geometry at that pressure. For Fig. 2, this implies that the slope of the arrow is dependent on P. As demonstrated in this figure, the slope of the pressure–flow relation has no direct relationship to the conductance of the coronary bed because the pressure dependence should be included in its analysis, such as in Eq. 2.

It has been shown that in an arrested heart the pressure–flow relations are strongly affected by left-ventricular pressure [16]. The pressure–flow curves are shifted to the right with increasing pressure [17]; the P_{ZF} values are closely related to the changes in left-ventricular pressure. Coronary veins may collapse in diastole [18], and a functional waterfall behavior has been demonstrated as well [19]. In diastole, the myocardium is very deformable, and left-ventricular pressure is easily transmitted as within

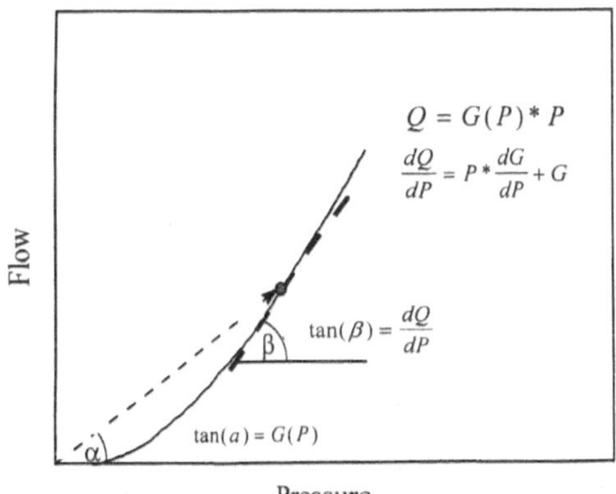

FIG. 2. Demonstration of the relationship between the slope of the pressure–flow relation at full vasodilatation and the pressure dependence of conductance, G. The conductance is the inverse of resistance. The *arrow* indicates the pressure–flow relation if the vascular bed could be fixated at the pressure determined by the coordinates of the *circle*. The slope of the pressure–flow relation at this pressure, dQ/dP, also depends on the sensitivity of conductance from pressure, dG/dP

a fluid. The pericardium, and, in its absence, the epicardium, serves as a constraint for expansion of the ventricle, and the epicardial vessels are compressed by the left-ventricular pressure. Because pressure in the epicardial veins is the outflow pressure of the intramural circulation, it would be correct to interpret the diastolic left-ventricular pressure as the back-pressure for flow. It should be noted, however, that pressure–flow relations also shift under the influence of stretch in an isolated septum [20]. The effect of diastolic pressure therefore needs not only to be the result of pericardium or epicardium. In any case, one has to be careful with the interpretation of the slopes of pressure–flow curves as conductance when curvature is present.

Beating Heart

The beating of the heart provides an additional impeding factor to coronary flow. The effect of contraction on coronary resistance is apparent from a shift of the pressure–flow relation at full vasodilatation. Such a shift was nicely demonstrated by Downey and Kirk [21], who compared the pressure–flow curves between the beating state and cardiac arrest. As is discussed next, contractility of the heart and pressure generated within the left-ventricular cavity are factors that determine the impediment placed upon myocardial perfusion by heart contraction. Before discussing these effects in more detail, let us first consider the pressure–flow relations in the hypertrophied beating heart.

When flow is expressed per unit mass, which generally is per 100g of tissue, the pressure–flow relation will rotate toward the pressure axis at full vasodilatation. This is the expression of the fact that the vascular tree is not generating more branches with the increase of heart muscle mass [22]. Thus it is to be expected that at the same myocardial mass the resistance will be higher. On the basis of vascular changes, one would not predict a shift in P_{ZF} to higher values. The behavior of P_{ZF} and resistance at full vasodilatation has nicely been demonstrated by Duncker et al. [23]. The pressure–flow relations where obtained in the left circumflex artery by occluding this vessel in

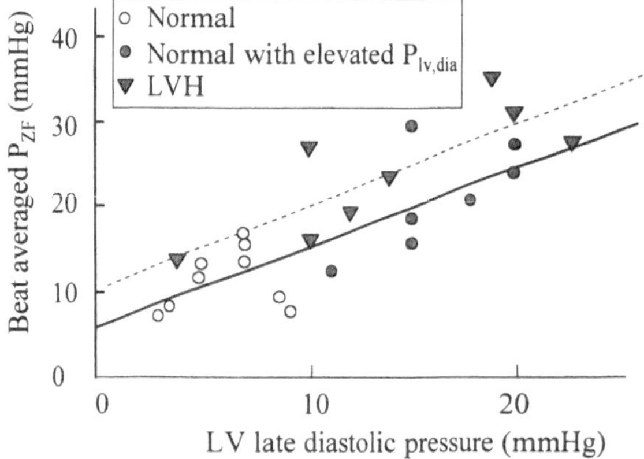

FIG. 3. P_{ZF} as function of end-diastolic left-ventricular (*LV*) pressure as found in control hearts (*open circles*) and hearts with induced hypertrophy (*shaded circles*). Diastolic pressure in the hypertrophic hearts was higher than in the controls. When diastolic left-ventricular pressure was increased to the same level as with hypertrophy (*triangles*), P_{ZF} values became comparable. (Data from [23])

a controlled manner. The heart kept on beating so the effect of heart contraction is taken into account in these relationships. This study demonstrated indeed the expected increase in coronary resistance but also an increase in P_{ZF}. Duncker et al. recognized the effect of diastolic left-ventricular pressure on P_{ZF} and varied this pressure in a set of control dogs, demonstrating the expected relationship. Their results are redrawn in Fig. 3. As the authors concluded, the increased P_{ZF} values occurring in the animals with hypertrophy can be well explained by the higher left diastolic pressures.

The slightly higher values found with hypertrophy compared to normal hearts at equal diastolic pressures suggest that hypertrophy causes some direct increase in P_{ZF}. However, one has to be very careful with such a conclusion. The data of Duncker et al. may be biased by some dynamic effects. Each point on the pressure–flow relation was obtained by partial occlusion of the artery for about 15 s. The rate of change of pressure may become clear after a total coronary occlusion. In that case, pressure decays at a decreasing rate for more than 15 s [24]. Hence, it will take longer for flow and pressure to equilibrate when the flow is brought close to zero.

Contraction of the Heart

It becomes more and more clear that the effect of contraction on the heart is the result of the interplay of contractility and generated left-ventricular systolic pressure. It is also clear that diastole and systole cannot be considered as independent events.

In systole, blood volume that is squeezed out of the myocardium is replaced in diastole in a steadily beating heart. The effect of contraction on intramural blood volume becomes apparent when the heart is suddenly brought into cardiac arrest [25] and after resumption of beating [26]. Intramural volume may change on the order of 4 ml/100 g with these interventions. The change in intramural volume is a good explanation for the increased coronary resistance when the heart is beating compared to cardiac arrest. It is obvious to assume that a vascular bed with less blood volume has a higher resistance. The problem is that these resistance changes cannot be deduced from the coronary pressure and flow signals.

It is convenient to study the coronary flow signals at constant perfusion pressure because then compression of the intramural vessels is visible only in one signal, the flow. In a beating heart at constant perfusion pressure, the flow variations are not only caused by systolic–diastolic resistance changes but also by volume variations in the intramural vessels [27]. These variations are referred to as capacitive flow.

Applying constant perfusion pressure, Katz and Feigl [28] demonstrated by the technique of long diastoles that diastolic flow arrived at a near steady state rather quickly after systole, independent of heart rate. This finding seems at odds with the intramyocardial pump model, which suggests that the flow should decay to a mean value equal to the mean in the beating heart. This discrepancy exists, however, only with the linear intramyocardial pump model [27] in which resistances and compliances are constant and the pressure for generating the pulsatile flow is related to left-ventricular pressure. The linear intramyocardial pump model was used to explain the relation between intramyocardial volume variations and pulsatile coronary arterial and venous flow as well as retrograde coronary arterial flow [29] and, additionally, the effect of arteriolar resistance on coronary flow pulsations. The original model did not take into account the volume dependency of resistances, which was later done in other

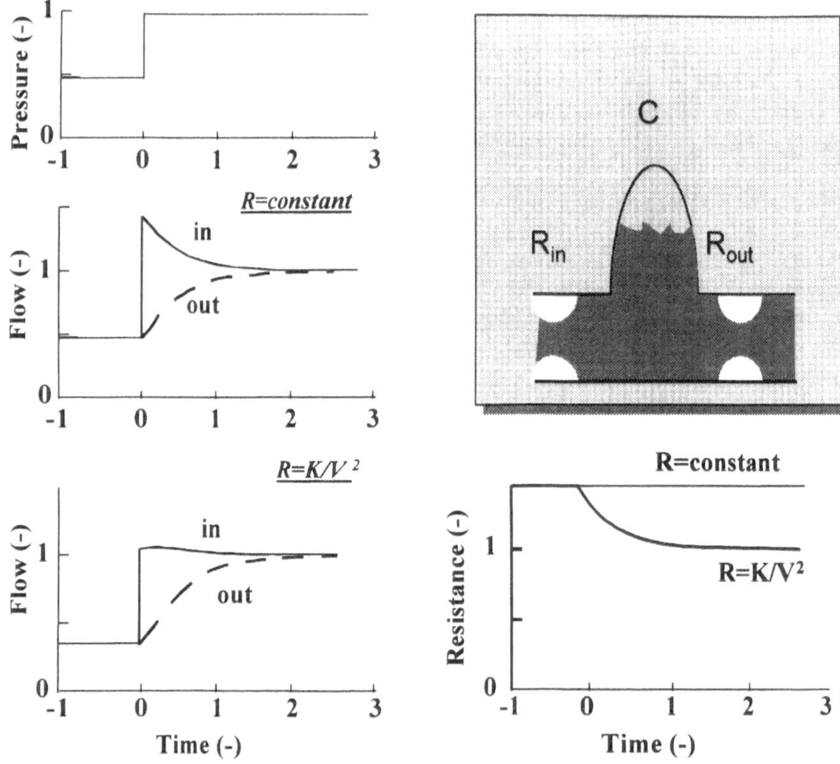

FIG. 4. The effect of pressure volume dependence of resistances (R) on the dynamics of coronary arterial flow to a step change in arterial pressure. The model used to calculate the signals is diagrammed in the *top right panel*. The compliance, C, is filled by R_{in} and emptied by R_{out}. Both resistances are assumed to be equal. The step change in pressure is elucidated in the *top left panel*. The *left middle panel* demonstrates the response of inflow and outflow when the two resistances would be constant. The *left bottom panel* demonstrates the flow signals when it is assumed that resistances are dependent on volume. The *right bottom panel* demonstrates that resistances are changing in a rate comparable to the outflow. The concomitant changes of resistance and volume makes it appear as if the inflow reaches a steady state quickly

studies [30,31]. The volume dependency of resistance complicates insight into the dynamic responses of the coronary circulation.

A compliant volume to be filled via a resistance which value is decreasing when the compliance is filled provides completely different dynamics of inflow. This is illustrated in Fig. 4. The top right panel shows a compliant compartment with an inlet and outlet resistance. In this example, inflow and outflow resistances are taken as equal. If the resistances are constant and a step change in pressure is imposed (as depicted in the top left panel), the inflow and outflow signals behave as shown in the middle left panel of the figure. Note that the time-courses for inflow and outflow are essentially the same. Note also that the area between the inflow and outflow curves equals the amount of filling volume of the compliance induced by the pressure step. If the resistances are dependent on the filling of the compliance, then the time-course of especially the inflow changes. It arrives at the final steady state much faster with only a small overshoot. This diminished overshoot is the result of the increased level of flow in steady state compared to the case with constant resistances, resulting from the decrease in resistance values. This decrease in resistance is shown in the lower right panel of Fig. 4. Note that the time-course of this resistance follows that of the outflow from the compartment.

Figure 4 demonstrates the difficulty with the interpretation of signals from a nonlinear system. It has been argued that the coronary circulation can be approached as a linear system because it seems to follow linear rules. For example, sinusoidal pressure changes on the coronary arterial pressure in long diastoles do not induce higher harmonics in flow responses [32]. Although this is true, it is not a reason to assume that the system parameters, i.e., resistances and compliances, are constant. Most nonlinear systems can be linearized around a working point if the perturbations are small enough. A linear system approach thus provides only a general answer. The question is, however, what the parameters from the linearized model mean physiologically. The discussion is similar to the one with the slope of the pressure–flow curve. The inverse of the slope has the dimension of a resistance but does not represent one. In a nice set of experiments it was recently shown that the diameters of small arteries and veins at the endocardium and epicardium are highly variable indeed, both during the heart cycle and by transition from beating to cardiac arrest [33–35].

Mechanisms of Contraction Effects on Myocardial Perfusion

There are many possible ways that contraction may influence coronary blood flow [2]. Two important factors are the direct effect of muscle contraction on intramural vessels and the generated left-ventricular pressure. Both factors result in compression of intramural vessels, expulsion of intramural blood volume, and thereby pulsatility of coronary arterial and venous flow. The reduction of volume in both instances results in resistance increases and impeding effects of coronary flow. One can best see these mechanisms in relation to the nonlinear intramyocardial pump model. The difference in the results of this model for coronary arterial and venous flow pulsations, having the two mechanisms incorporated, were recently reviewed elsewhere [36,37]. The two different mechanisms and their interaction are briefly discussed next.

The direct effect of contracting myocytes on the intramural vessels is called the elastance effect [38]. The intramural vessels are considered to be influenced by the surrounding muscle, similarly as the left-ventricular cavity [39]. As left-ventricular function can be described by pressure–volume loops and by preload and afterload, so also can the intramural vascular volume be described by pressure–volume loops. Hence, the increasing elastance during systole gives rise to an increased intramural blood pressure that expels blood from the intramural vessels [40]. The amount of pressure increase and volume expelled depends on the intramural blood volume at the onset of contraction on the one hand and the resistances for inflow and outflow of the intramural vessels on the other hand. The expelled intramural volume is then subsequently replaced in diastole, again depending on the diastolic compliance and pressures inside and outside the intramural vessels.

The left-ventricular pressure effect on intramural vessels is similar to the classical tissue pressure concept as applied in the waterfall model [21]. The idea is that because of mechanical equilibrium a balance of forces must exist at the endocardium. This balance requires equilibrium between the radial stress in the ventricular wall and the pressure in the left-ventricular cavity. The same reasoning applied to the epicardial surface should result in a radial stress equal to the pressure in the pericardium. The radial stress is considered as a pressure decreasing from left-ventricular pressure at the endocardium to near atmospheric at the epicardium. In these models, however, it is generally assumed that muscle tension is directed into fiber directions being tangential to a curved plane parallel to endo- and epicardium. However, the myocardial fibers form a network, and at the local level not only tangential stresses but also radial ones are generated. These radial stresses cancel each other out over the myocardial wall but do result in an increased material stiffness. This systolic stiffness increase of the ventricular wall is obvious to anyone who has studied the heart.

These is strong evidence that left-ventricular pressure effects and wall stiffness interact. When the myocardium is relaxed, pressure is easily transmitted through it to the intramural vessels. However, in systole the myocardial wall is stiff and intramural vessels are shielded from the left-ventricular pressure. This results in a complicated interaction of these two factors, changing throughout the cardiac cycle. These interactions have clearly been demonstrated to contribute to the coronary arterial flow pattern and lymph pressure signal [41–43]. Strangely enough, the elastance effect is dominant in the intramural veins [44].

Interaction of Left-Ventricular Pressure and Contractility Effects on Myocardial Perfusion

As outlined, the intramural blood vessels are more easily compressed by the left-ventricular pressure when the contractility of the heart is low. That raises the question of what happens during local ischemia in the myocardium. In the ischemic part of the muscle, contractility will be decreased, while left-ventricular pressure is maintained at normal levels by increased strength of contraction by the healthy part of the muscle. This interaction has been studied in recent years by Hoffman's group. Their studies demonstrate the complex interactions that can occur between several mechanisms. In one study, contractility changes induced by dobutamine only resulted in a 8% reduction of subendocardial flow while the first derivative of left-ventricular pressure doubled. This observation would result in the conclusion that the contractility has hardly any effect on myocardial perfusion [45]. However, a main determinant of coronary perfusion, the diastolic time fraction, was not kept constant and was reduced by 20% because of the increased contractility. It is therefore possible that the increased compression effect has been compensated by its shorter duration.

In a similar study, a major epicardial artery was infused with lidocaine to reduce the contractility [46]. With reduced contractility, a decrease of left-ventricular pressure from 90 to 50 mmHg resulted in a 60% increase of subendocardial flow while such an effect was absent at normal contractility. This demonstrates that indeed with reduced contractility left-ventricular pressure has a strong effect on subendocardial perfusion. However, subendocardial flow increased in the region where contractility was reduced, which would suggest that contractility has a stronger impeding effect than left-ventricular pressure. This may be so, but in the experiments the artery was perfused at 80 mmHg. For the fully dilated heart, this must result in a rather high capillary pressure compared to a coronary bed that is dilated because of a reduced arterial pressure, e.g., as the result of a coronary stenosis. It may well be possible that at a reduced pressure subendocardial flow would have decreased. Subendocardial flow can remain at a sufficient level at a perfusion pressure of 38 mmHg [47]. It would be interesting to know what the left-ventricular pressure would do at this perfusion pressure.

Pressure–Flow Relations and Coronary Flow Reserve

For the heart to keep functioning, its demands for oxygen must be met. This requires a minimum amount of perfusion per unit mass throughout the myocardium. In a healthy heart, this is not a problem. The coronary system is overdesigned, and all cells can be supplied sufficiently. Overperfusion is prevented by the coronary resistance vessels. The range for regulation is so wide that if a factor inhibiting blood flow is occurring locally, vasodilatation can compensate for it. If the range of regulation has become too limited, the possibility for compensation is exhausted and local ischemia is the result.

It is clear from micronecrosis often found in the myocardium that underperfusion can occur locally. Also, an inhomogeneous distribution in coronary reserve is found in healthy canine hearts [48]. Assuming that demand for oxygen per unit mass is rather homogeneous, some parts of the myocardium are bound to be underperfused earlier than others. Flow and oxygen demand are studied predominantly at the global level and therefore it is easier to consider what can go wrong on the basis of these global parameters than to analyze directly the local events.

Figure 5 is a schematic representation of two levels of autoregulation and four conditions of maximum vasodilatation. The solid lines are taken from Fig. 1. For this

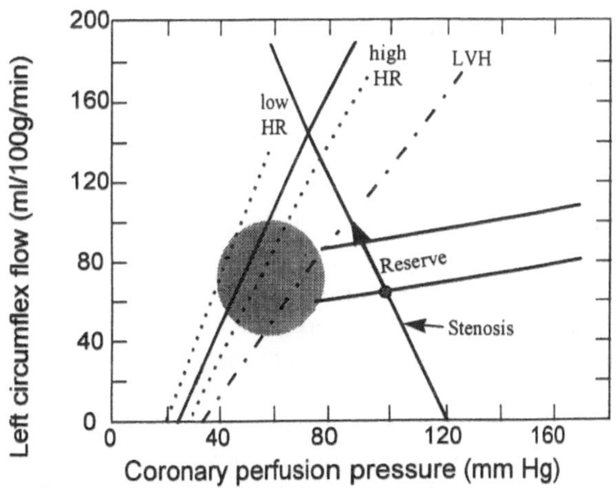

Fig. 5. Schematic representation of the effects of oxygen demand, extravascular factors, and stenosis on coronary flow reserve. The *heavy lines* are from Fig. 1. The *steep line* with negative slope represents the pressure–flow relation of a stenosis (at *isolated arrow*). At constant systemic pressure (120 mmHg in this figure), the perfusion pressure for microcirculation will decrease with increasing flow as demanded by increased oxygen consumption. The vertical component of the *arrow* indicates the flow reserve present when the pressure–flow coordinates are in the *small circle*, and the *dot-dash line* is the pressure–flow line of maximal dilatation. The *dot-dash line* represents the case of hypertrophy (*LVH*) and depicts the increased P_{ZF} value as well as increased resistance per unit mass. The other *dotted lines* indicate the possible shifts from control by varying heartrate (*HR*). Similar shifts may occur in case of hypertrophy

case, the perfused myocardium can function properly so long as the perfusion pressure, normally the arterial pressure or pressure distal of a stenosis when present, is high enough for the flow to remain on an autoregulation curve. In other words, there must still be some increase in flow possible as the result of vasodilatation. This increase is referred to as flow reserve. The flow reserve is dependent on the level of flow corresponding to the autoregulation curve and the maximum flow possible. This level of the autoregulation curve depends on the oxygen consumption, which will increase with increased cardiac work, thereby reducing the reserve when still possible.

The maximum flow possible is determined by the pressure–flow relation at maximal vasodilatation and the perfusion pressure. Without a stenosis, the maximum flow is only determined by the pressure–flow relation at maximum vasodilatation. In the presence of a stenosis, the pressure–flow relation of the stenosis comes into play as well. This relation is indicated by the line with negative slope in Fig. 5. This negative slope expresses the fact that the pressure drop over a stenosis will increase with increasing flow, thereby reducing the reserve capacity. The reduction of the reserve by the stenosis is indicated by the arrow on the pressure–flow line of the stenosis. In this case, the heart would function such that pressure and flow would be at the coordinates indicated by the small circle and oxygen demand would increase then flow would increase, thereby reducing the perfusion pressure. The reserve is exhausted as soon as the prevailing line of maximal vasodilatation (as indicated by the dotted lines) is reached.

As is indicated in Fig. 5, the curve of maximal vasodilatation is not fixed but depends on a given heart condition. The curve can be altered by the way the heart is functioning. Increasing heart rate will shift the curve to the right, decreasing heart rate to the left. In general, it seems understood that an increased demand can result in exhaustion of reserve and therefore underperfusion. However, it is less well accepted that a change in heart function may reduce the reserve by changing the extravascular compression, as expressed by changes in the pressure–flow relation at maximal dilatation. If, for example, the impeding effect of increased heart rate on subendocardial maximal perfusion is compared to the increase in flow needed to meet the increase in oxygen demand, both factors are equally important in reducing coronary reserve [49]. It is therefore of particular importance to know all these effects with parameter and variable values indicated by the gray (shaded) circle drawn in Fig. 5.

As discussed earlier on the basis of the data of Duncker et al. [23], hypertrophy changes the slope and P_{ZF} of the pressure–flow line at maximal vasodilatation. In contrast to the rapid alterations in relation to changes in heart function, this is the result of an adaptation process. Less is known about alterations in pressure–flow relations in other forms of heart remodeling.

The foregoing analysis is simple and may help us to understand why at a certain moment the heart is underperfused. There may be more subtle interactions not yet well understood. Some interactions may result in self-propelling deterioration of the condition of the heart: less perfusion, therefore less left-ventricular function, resulting in further loss of perfusion. Such an interaction will usually be concomitant with a reduction in perfusion pressure. Some interactions may result in stabilizing the condition of the heart: less perfusion resulting in less function with less compression so perfusion can be maintained. This will occur predominantly with local underperfusion, so that perfusion pressure can be maintained by the healthy part of the heart muscle.

References

1. Mosher P, Ross J, McFAte PA, Shaw RF (1964) Control of coronary blood flow by an autoregulatory mechanism. Circ Res 14:250–259
2. HOFFMAN JIE, SPAAN JAE (1990) Pressure-flow relations in coronary circulation. Physiol Rev 70:331–389
3. Drake-Holland AJ, Laird JD, Noble MIM, Spaan JAE, Vergroesen I (1984) Oxygen and coronary vascular resistance during autoregulation and metabolic vasodilation in the dog. J Physiol (Lond) 348:285–299
4. Vergroesen I, Noble MIM, Wieringa PA, Spaan JAE (1987) Quantification of O_2 consumption and arterial pressure as independent determinants of coronary flow. Am J Physiol 252:H545–H553
5. Dankelman J, Spaan JAE, Van der Ploeg CPB, Vergroesen I (1989) Dynamic response of the coronary circulation to a rapid change in its perfusion in the anaesthetized goat. J Physiol (Lond) 419:703–715
6. Broten TP, Feigl EO (1992) RoLE of myocardial oxygen and carbon dioxide in coronary autoregulation. Am J Physiol 262:H1231–H1237
7. Jia L, Bonaventura C, Stamler JS (1996) S-Nitrosohaemoglobin: a dynamic activity of blood involved in vascular control. Nature 380:221–226
8. Spaan JAE (1993) Biological control: A system analysis approach. In: Strackee J, Westerhof N (eds) The physics of heart and circulation. Institute of Physics, Bristol, pp 249–262
9. VANBAVEL E, SPAAN JAE (1992) BRAnching patterns in the porcine coronary arterial tree. Estimation of flow heterogeneity. Circ Res 71:1200–1212
10. Chilian WM, Layne SM, Klausner EC, Eastham CL, Marcus ML (1989) Redistribution of coronary microvascular resistance produced by dipyridamole. Am J Physiol 256:H383–H390
11. Bassingthwaighte JB, King RB, Roger SA (1989) Fractal nature of regional myocardial blood flow heterogeneity. Circ Res 65:578–590
12. Austin RE Jr, Smedira NG, Squiers TM, Hoffman JIE (1994) Influence of cardiac contraction and coronary vasomotor tone on regional myocardial blood flow. Am J Physiol (Heart Circ Physiol) 266:H2542–H2553
13. Weiss HR, Sinha AK (1978) Regional oxygen saturation of small arteries and veins in the canine myocardium. Circ Res 42:119–126
14. Giezeman MJMM, VanBavel E, Grimbergen CA, Spaan JAE (1994) Compliance of isolated porcine coronary small arteries and coronary pressure-flow relations. Am J Physiol 267:H1190–H1198
15. Kanatsuka H, Shikawa K, Komaru T, Suzuki T, Takishima T (1990) Diameter change and pressure-red blood cell velocity relations in coronary microvessels during long diastoles in the canine left ventricle. Circ Res 66:503–510
16. Aversano T, Klocke FJ, Mates RE, Canty JMJ (1984) Preload-induced alterations in capacitance-free diastolic pressure-flow relationships. Am J Physiol 246:H410–H417
17. Watanabe J, Maruyama Y, Satoh S, Keitoku M, Takashima T (1987) Effects of the pericardium on the diastolic left coronary pressure-flow relationship in the isolated dog heart. Circulation 75:670–675
18. Klassen GA, Armour JA (1983) Canine coronary venous pressures: responses to positive inotropism and vasodilation. Can J Physiol Pharmacol 61:213–221
19. Uhlig PN, Baer RW, Vlahakes GJ, Hanley FL, Messina LM, Hoffman JIE (1984) Arterial and venous coronary pressure-flow relations in anesthetized dogs. Evidence for a vascular waterfall in epicardial coronary veins. Circ Res 55:238–248
20. Resar J, Livingston JZ, Halperin HR, Sipkema P, Krams R, Yin FCP (1990) Effect of wall stretch on coronary hemodynamics in isolated canine interventricular septum. Am J Physiol 259:H1869–H1880
21. Downey JM, Kirk ES (1975) Inhibition of coronary blood flow by a vascular waterfall mechanism. Circ Res 36:753–760

22. Hoffman JIE (1984) Maximal coronary flow and the concept of coronary vascular reserve. Circulation 70:153–159
23. Duncker DJGM, Zhang J, Bache RJ (1993) Coronary pressure-flow relation in left ventricular hypertrophy. Importance of changes in back pressure versus changes in minimum resistance. Circ Res 72:579–587
24. Spaan JAE (1985) Coronary diastolic pressure-flow relation and zero flow pressure explained on the basis of intramyocardial compliance. Circ Res 56:293–309
25. Vergroesen I, Noble MIM, Spaan JAE (1987) Intramyocardial blood volume change in first moments of cardiac arrest in anesthetized goats. Am J Physiol 253:H307–H316
26. Kajiya F, Tsujioka K, Goto M, Wada Y, Chen XL, Nakai M, Tadaoka S, Hiramatsu O, Ogasawara Y, Mito K, Tomonaga G (1986) Functional characteristics of intramyocardial capacitance vessels during diastole in the dog. Circ Res 58:476–485
27. Spaan JAE, Breuls NPW, Laird JD (1981) Diastolic-systolic coronary flow differences are caused by intramyocardial pump action in the anesthetized dog. Circ Res 49:584–593
28. Katz SA, Feigl EO (1988) Systole has little effect on diastolic coronary blood flow. Circ Res 62:443–451
29. Spaan JAE, Breuls NPW, Laird JD (1981) Forward coronary flow normally seen in systole is the result of both forward and concealed back flow. Basic Res Cardiol 76:582–586
30. Bruinsma P, Arts T, Dankelman J, Spaan JAE (1988) Model of the coronary circulation based on pressure dependence of coronary resistance and compliance. Basic Res Cardiol 83:510–524
31. Arts T, Reneman RS (1985) Interaction between intramyocardial pressure (IMP) and myocarcial circulation. J Biomed Eng 107:51–56
32. Canty JMJ, Klocke FJ, Mates RE (1985) Pressure and tone dependence of coronary diastolic input impedance and capacitance. Am J Physiol 248:H700–H711
33. Hiramatsu O, Goto M, Yada T, Kimura A, Tachibana H, Ogasawara Y, Tsujioka K, Kajiya F (1994) Diameters of subendocardial arterioles and venules during prolonged diastole in canine left ventricles. Circ Res 75:393–399
34. Kajiya F, Yada T, Kimura A, Hiramatsu O, Goto M, Ogasawara Y, Tsujioka K (1993) Endocardial coronary microcirculation of the beating heart. Adv Exp Med Biol 346:173–180
35. Yada T, Hiramatsu O, Kimura A, Goto M, Ogasawara Y, Tsujioka K, Yamamori S, Ohno, K, Hosaka H, Kajiya F (1993) In vivo observation of subendocardial microvessels of the beating porcine heart using a needle-probe videomicroscope with a CCD camera. Circ Res 72:939–946
36. Spaan JAE (1995) Mechanical determinants of myocardial perfusion. Basic Res Cardiol 90:89–102
37. Spaan JAE (1995) Nonlinear models of coronary flow mechanics. In: Jaffrin MY, Caro CG (eds) Biological flows. Plenum, New York, pp 267–286
38. Krams R, Sipkema P, Westerhof N (1989) The varying elastance concept may explain coronary systolic flow impediment. Am J Physiol 257:H1471–H1479
39. Suga H, Sagawa K, Shoukas AA (1973) Load independence of the instanteneous pressure-volume ratio of the canine left ventricle and effects of epinephrine and heart rate on the ratio. Circ Res 32:314–322
40. Westerhof N (1990) Physiological hypotheses—intramyocardial pressure. A new concept, suggestions for measurement. Basic Res Cardiol 85:105–119
41. Kouwenhoven E, Vergroesen I, Han Y, Spaan JAE (1992) Retrograde coronary flow is limited by time-varying elastance. Am J Physiol 263:H484–H490
42. Han Y, Vergroesen I, Spaan JAE (1993) Stopped flow epicardial lymph pressure is affected by left ventricular pressure in anesthetized goats. Am J Physiol (Heart Circ Physiol) 264:H1624–H1628
43. Han Y, Vergroesen I, Goto M, Dankelman J, Van der Ploeg CPB, Spaan JAE (1993) Left ventricular pressure transmission to myocardial lymph vessels is different during systole and diastole. Pflügers Arch 423:448–454

44. Vergroesen I, Han Y, Goto M, Spaan JAE (1994) Cardiac contraction and intramyocardial venous pressure generation in the anaesthetized dog. J Physiol (Lond) 480:343–353
45. Iwanaga S, Ewing SG, Husseini WK, Hoffman JIE (1995) Changes in contractility and afterload have only slight effects on subendocardial systolic flow impedediment. Am J Physiol 269:H1202–H1212
46. Doucette JW, Goto M, Flynn AE, Austin RE, Husseini WK, Hoffman JIE (1993) Effect of cardiac contraction and cavity pressure on myocardial blood flow. Am J Physiol 265:H1342–H1352
47. Van der Ploeg CPB, Dankelman J, Spaan JAE (1995) Heart rate affects the dependence of myocardial oxygen consumption on flow in goats. Heart Vessels 10:258–265
48. Austin REJ, Aldea GS, Coggins DL, Flynn AE, Hoffman JIE (1990) Profound spatial heterogeneity of coronary reserve. Discordance between patterns of resting and maximal myocardial blood flow. Circ Res 67:319–331
49. Spaan JAE, Verburg J (1991) Limitation of coronary flow reserve by a stenosis. In: Spaan JAE (ed) Coronary blood flow: Mechanics, distribution, and control. Kluwer Academic, Dordrecht, 131–162

Nitric Oxide and the Heart: Implications in Physiological and Pathological Conditions

SEINOSUKE KAWASHIMA, KEN-ICHI HIRATA, and MITSUHIRO YOKOYAMA

Summary. Nitric oxide (NO) is one of the most important intracellular messenger molecules with multiple biological functions. The heart contains two cellular components that produce NO, one via constitutive-type nitric oxide synthase (cNOS) and the other via inducible-type NOS (iNOS). Coronary vascular endothelial cells and cardiac myocytes contain both cNOS and iNOS, but coronary vascular smooth muscle cells and leukocytes have only iNOS. In coronary vessels, NO operates as an endothelium-derived relaxing factor (EDRF) and participates in the regulation of the vascular tone of the resistance vessels as well as the epicardial artery. NO is involved not only in the regulation of basal coronary vascular tone but also in the metabolic control of coronary vascular resistance. The impairment of the L-arginine–NO pathway by endothelial dysfunction plays an important role in the pathogenesis of coronary artery disease. The reduced vasodilatory response of the epicardial artery to agonist stimulation may be related to the mechanisms of coronary artery spasm. The impaired metabolic vasodilation is likely to result in abnormal myocardial perfusion. NO is also involved in the regulation of cardiac functions. NO produced via the cNOS-mediated pathway in response to parasympathetic stimulation serves to regulate heart rate and cardiac contractility. The cNOS-derived NO may protect against myocardial injury by preserving coronary flow, inhibiting leukocyte adhesion to endothelium and cardiac myocytes, and inhibiting platelet aggregation. On the other hand, a large amount of NO produced via the iNOS-mediated pathway in cardiac myocytes, leukocytes, and vascular cells is likely to depress cardiac function in pathological conditions in which immune and inflammatory processes are involved.

Key words. Nitric oxide—Nitric oxide synthase—Coronary artery—Cardiac myocytes—Cardiac function

Introduction

Nitric oxide (NO), a short-lived radical gas, has been revealed to be one of the most important intracellular messenger molecules [1–8]. NO is produced by the oxidation of guanidinonitrogens in L-arginine to form NO and L-citrulline [3,4,6]. This reaction is catalyzed by a family of catalytic enzymes termed NO synthases (NOS) and uses reduced nicotinamide adenine dinucleotide (NADPH) and O_2 as cosubstrates.

First Department of Medicine, School of Medicine, Kobe University, Chuo-ku, Kobe, 650 Japan.

The synthesis of NO also requires tetrahydrobiopterin, flavin mononucleotide (FMN), flavine adenine dinucleotide (FAD), and calmodulin as cofactors. NO synthesis has been found in a variety of different cell types, in which three distinct isoforms of NOS have been cloned and sequenced [2,9–14]; these are neuronal NOS, endothelial NOS, and inducible-type NOS (iNOS). Neuronal and endothelial NOS are constitutively expressed in each tissue; iNOS is not present at the steady-state level and is expressed on stimulation by such substances as cytokines and lipopolysaccharide [3,8].

Constitutive-type NOS (cNOS) requires Ca^{2+} calmodulin for activation, while activation of iNOS is independent of intracellular Ca^{2+} concentration. Although the cell types expressing cNOS are limited, iNOS is present in such cell types as macrophages, leukocytes, vascular smooth muscle cells, hepatocytes, respiratory epithelial cells, renal tubular epithelial cells, cardiomyocytes, and fibroblasts. NO production by the cNOS-mediated pathway is transient and at a low level. On the other hand, the iNOS-mediated L-arginine–NO pathway requires several hours until NO production, but a large amount of NO is produced over a prolonged time.

The NO produced in various cells diffuses to adjacent target cells in an autocrine or paracrine manner and there exhibits a variety of functions [5–9]. In the central nervous system, NO mediates several functions, including memory. In the peripheral neurons, NO serves as a neurotransmitter excreted from nonadrenergic, noncholinergic (NANC) nerves. In vessels, NO synthesis via constitutive-type vascular endothelial NOS (ECNOS) plays an essential role in the regulation of vascular tone [3,15]. NO also inhibits platelet aggregation and leukocyte adhesion. The presence of a large amount of NO produced by iNOS has been demonstrated to have tumorcidal and bacteriocidal actions. NO is also shown to inhibit proliferation of vascular smooth muscle cells.

L-Arginine analogs such as N^G-monomethyl-L-arginine (L-NMMA), N^G-nitro-L-arginine (L-NNA), and N^G-nitro-L-arginine methylester (L-NAME) compete with L-arginine for NOS and serve as NO synthesis inhibitors [16]. On the other hand, it is widely accepted that nitrovasodilators yield their effects by behaving as NO donors, although the mechanisms by which these compounds release NO are still not well understood [17]. These NO synthesis inhibitors and NO donors have been widely used as tools to investigate the role of NO in the cardiovascular system under physiological as well as pathological conditions.

In this chapter, we review the role of NO in the heart under physiological as well as pathophysiological conditions.

Nitric Oxide and Cardiovascular Function

In the heart, there are two components of the L-arginine-NO pathway. As cellular components producing NO via cNOS, vascular endothelial cells and cardiac myocytes are found; vascular smooth muscle cells, leukocytes, and fibroblasts produce NO induction via iNOS. Endothelial cells and cardiac myocytes also contain iNOS.

Coronary Circulation and NO

In coronary vessels, NO operates as an endothelium-derived relaxing factor (EDRF) to regulate coronary blood flow [18]. In the endothelial cells, NO is synthesized by receptor-mediated agonist stimulation such as acetylcholine, bradykinin, adenosine

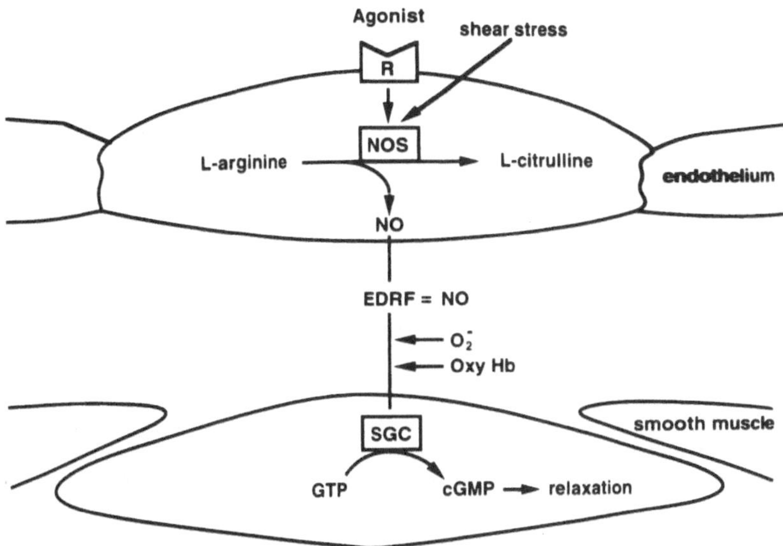

FIG. 1. L-Arginine–NO pathway in vascular endothelium. *NO*, Nitric oxide; *NOS*, NO synthase; *R*, receptor; *SGC*, soluble guanylate cyclase; *EDRF*, endothelium-derived relaxing factor

diphosphate (ADP), and serotonin [19,20]. Another important stimulation of NO synthesis in the endothelium is mechanical stimulation by shear stress, and shear stress is thought to play a key role in the regulation of coronary vascular tone [3]. The endothelium-derived NO reaches vascular smooth muscle cells as EDRF and there activates soluble guanylate cyclase by binding its Fe-heme center. The subsequent increases in cyclic guanosine monophosphate (GMP) decrease the intracellular Ca^{2+} and relax vascular smooth muscle (Fig. 1). In coronary circulation, the vascular tone of the epicardial artery is largely dependent on coronary flow, and recent experimental studies have revealed that flow-mediated vasodilation is caused by the release of endothelium-derived NO [21].

On the other hand, regulation of microvascular tone is influenced by many factors such as local metabolites, adenosine, prostaglandins, and neural stimulation. Recent experimental as well as clinical studies have revealed the participation of endothelium-derived NO in the regulation of microvascular tone. Quyyumi et al. demonstrated that, in patients without overt coronary artery disease, intracoronary administration of L-NMMA resulted in a 16% decrease in basal coronary vascular resistance. They also demonstrated that L-NMMA caused a significant inhibition of pacing-induced metabolic coronary vasodilation [22]. Therefore, NO is involved not only in the regulation of basal coronary vascular tone but also in the metabolic control of coronary vascular resistance [23].

NO is also shown to participate in reactive hyperemic response after transient ischemia. Experimental studies have demonstrated that approximately 30% of the component of reactive hyperemia was attributable to NO [24]. Adenosine and NO are involved independently in hyperemic response. Kitakaze et al. showed in open-chest dogs that L-arginine analogs further decreased coronary flow and aggravated myocardial ischemia when administered into the coronary artery with severe stenosis [25]. From these observations, they suggested that NO served to maintain coronary flow under ischemia. The mechanism of activation of endothelial cNOS (ECNOS) during myocardial ischemia is still unclear and awaits future investigation.

Impairment of the L-arginine-NO pathway in coronary arteries as the result of endothelial dysfunction is involved in the pathogenesis of coronary artery disease. In experiments with isometric tension measurement, reduced endothelium-dependent vasorelaxation and enhanced response to vasoconstrictor stimuli have been shown in atherosclerotic vessels [18,19]. The vasodilatory response to endothelium-independent vasodilatory substances is not impaired in these vessels. In the human coronary artery with atherosclerosis, the impaired vasodilatory response to endothelium-dependent vasodilators such as acetylcholine is demonstrated by coronary angiography [25]. The impaired vasodilatory response is also observed in angiographically normal coronary arteries of patients with coronary risk factors [26]. In addition, acetylcholine and serotonin, both endothelium-dependent vasodilators, cause coronary artery spasms in patients with vasospastic angina pectoris, and the impaired vasodilatory response has been suggested as one of the mechanisms [27]. As mechanisms of this impairment, reduced production, reduced diffusion, and enhanced inactivation of NO may be involved [28,29].

The reduced vasodilatory response is not confined to the epicardial coronary artery. Although they are free of histological evidence of atherosclerotic changes, impaired dilation of small-resistance coronary arteries to endothelium-dependent vasodilatory stimuli has been reported in patients with hypercholesterolemia. Using an intracoronary Doppler catheter, Egashira et al. demonstrated impaired coronary blood flow response to intracoronary acetylcholine in patients with coronary risk factors and proximal atherosclerotic lesions [30]. In addition, cholesterol-lowering therapy is shown to result in the reversal of vasoconstrictor response to vasodilatory response of epicardial coronary artery to intracoronary acetylcholine [31]. Therefore, the pathophysiological manifestations of atherosclerosis may extend to small-resistance coronary arteries. The impaired response is a manifestation of endothelial dysfunction and is likely to result in altered regulation of myocardial perfusion. Because NO is partly involved in the metabolic control of coronary circulation, the disturbed metabolic vasodilation in face of increased cardiac work may be related to the pathophysiology of exercise-induced myocardial ischemia in some patients [23].

Cardiac Function and NO

Another cellular component possessing cNOS in the heart is the cardiac myocyte. Although the existence of both cNOS and iNOS has been reported in cardiac myocytes, cNOS in cardiac myocytes is not as well characterized compared to iNOS. Most of the data indicating the presence of cNOS in cardiac myocytes are based on physiological studies [32]. Although Balligand et al. recently reported that they succeeded in cDNA cloning of cNOS in adult rat cardiac myocytes, which showed high homology to rat ECNOS, further studies for characterization of cNOS in myocytes are needed [33].

There are a number of reports demonstrating the physiological role of NO in regulating cardiac contractility as well as heart rate [34–37]. Most of these studies were performed in isolated or cultured myocytes. Brady et al. observed in isolated, electrically stimulated, contracting cardiac myocytes that endothelium-derived endogenous NO decreased myocyte contraction [34]. This effect was mimicked by sodium nitroprusside, NO gas, and 8-bromo-cGMP. Balligand et al. reported that carbacol, a muscarinic cholinergic agonist, decreased the beating rate in spontaneously beating neonatal rat cardiomyocytes, and that this effect was partially restored by a NO synthase inhibitor [35]. Because the negative chronotropic effect of carbacol was mimicked by cGMP analogs, the effects of NO on cardiomyocytes are likely mediated by cGMP. cGMP is thought to be a second messenger of cholinergic stimulation, and cGMP analogs have been shown to produce a negative chronotropic as well as inotropic effects; thus this study suggested the involvement of NO in parasympathetic nerve-mediated cholinergic control of heart rate and cardiac contractility [35]. In addition, the role of NO produced in cardiac pacemaker cells in the cholinergic regulation of heart rate is further supported by the study of Han et al. Using isolated spontaneously beating cells from the rabbit sinoatrial node, they showed that an acetylcholine analog inhibited L-type calcium currents that had been augmented by β-adrenergic stimulation and that an NO synthase inhibitor abolished this effect [36].

The role of NO in autonomic regulation of cardiac contractility is also shown in the heart in situ. In the heart, parasympathetic stimulation causes little effect on cardiac contractility under basal conditions, but it attenuates sympathetic nerve-mediated contractile response. Hare et al. examined the effect of vagal stimulation on dobutamine-induced increase in cardiac contractility in closed-chest dogs [37]. They found that vagal stimulation attenuated the inotropic response to dobutamine, while L-NMMA, an NO synthase inhibitor, diminished and L-arginine, a substrate for NO, restored the vagal inhibitory effect. On the basis of these observations they suggested that NO plays a role in the normal physiological regulation of myocardial autonomic responses (Fig. 2). The source of the cardiac NO elicited by parasympathetic stimulation, which is likely produced by cNOS, has not been identified. Because a dense network of NOS-contained nerve fibers is shown in the heart [38], the nerve ending is a putative candidate for the source of cardiac NO. Other possible sources of NO production during vagal stimulation are the vascular endothelium and cardiac myocytes, but the mechanisms by which vagal stimulation facilitates cNOS activity in these cells are unclear.

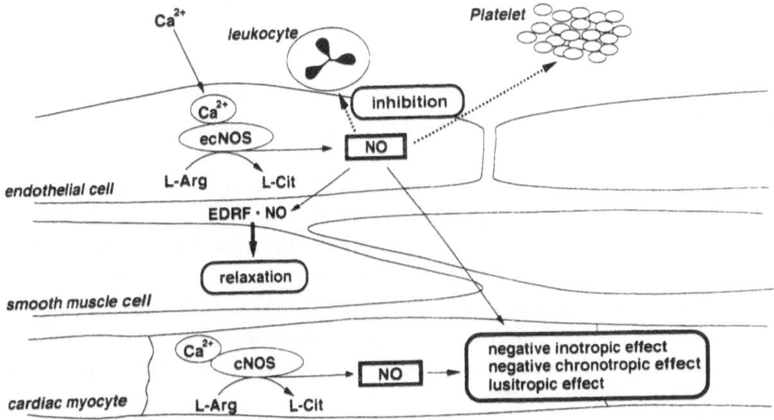

FIG. 2. Networks of cNOS-mediated NO generators and targets in the heart. *L-Arg*, L-arginine; *L-Cit*, L-citrulline

Several mechanisms can be considered to explain NO attenuation of myocardial contractility in vivo. Changes in loading condition or coronary flow caused by the vasodilatory effect of NO on systemic and coronary vessels may indirectly modify cardiac contractile status. On the other hand, NO may directly attenuate myocardial contractility by increasing cGMP production. cGMP may inhibit the L-type slow inward Ca^{2+} channel by stimulation of a cGMP-dependent cAMP phosphodiesterase or via a cGMP-dependent protein kinase [39]. cGMP may also decrease the calcium sensitivity of the contractile element via cGMP-dependent protein kinase.

On the other hand, cNOS is associated with low levels of NO formation. Although the aforementioned studies suggested the possible regulation of cardiac contractility by cNOS-derived NO, it is still unclear whether the NO produced at low levels by cNOS in vivo is actually involved in the regulation of cardiac contractility. The recent report of Weyrich et al. indicated that physiological levels of NO did not exert a major effect on cardiac contractility in isolated papillary muscle [40]. In their study, NO was found to exert a negative inotropic effect only in the presence of β-adrenoceptor stimulation by norepinephrine. Therefore, NO-mediated parasympathetic modulation of cardiac contractility may be confined to conditions with elevated sympathetic activity.

In addition to cNOS-derived NO, NO is induced via the iNOS-mediated pathway in the heart. iNOS is induced in the heart in several cell components: vascular smooth muscle cells, vascular endothelial cells, cardiac myocytes, macrophages, and leukocytes. Among them, cardiac myocytes are an important component of NO production. Tsujino et al. showed that interleukin-1β induced expression of iNOS mRNA and protein in cultured cardiac myocytes [41]. Balligand et al. demonstrated production of NO from a single, isolated adult cardiac myocyte stimulated by interleukin-1β and interferon γ [42]. In cardiac myocytes, induction of iNOS requires a relatively long period after stimulation, and it has been demonstrated in cultured rat cardiac myocytes that induction of iNOS mRNA is fairly consistently detected until 6h after stimulation and then shows a drastic time-dependent increase with marked production of NO [41,42].

In the physiological condition, iNOS is not expressed in the heart, while on stimulation by cytokines and bacterial lipopolysaccaride, iNOS is induced in the aforementioned cells. The excess and prolonged production of NO in the heart is thought to depress the functional status of the heart in pathological conditions [3-5,43]. Balligand et al. demonstrated the depressed contractile response of adult rat ventricular myocytes to β-adrenergic agonists by a 24-h exposure to activated macrophage-conditioned medium [44]. The depression is likely to be mediated by NO produced in cardiac myocytes stimulated by macrophage-derived inflammatory cytokines.

In regard to the mechanisms of myocardial depressant action of NO, Kinugawa et al. showed that iNOS-mediated production of NO in cardiac myocytes decreased contractility in an autocrine manner in association with reduction of the peak value of the intracellular Ca^{2+} concentration [45]. On the other hand, recent studies have demonstrated that NO produced in cytokine-stimulated cardiomyocytes as well as in cytokine-stimulated macrophages caused myocardial death as in an autocrine or paracrine manner [46]. Therefore, a large amount of NO is not only depressive but also cytotoxic to cardiac myocytes. The cytotoxic effect of NO is shown to be mediated by its direct inhibiting action on the mitochondrial respiratory chain reaction and aconitase of the Krebs cycle [47]. NO also inhibits synthesis of DNA by inactivating ribonucleotide reductase [48]. Moreover, NO may combine with oxygen-derived free

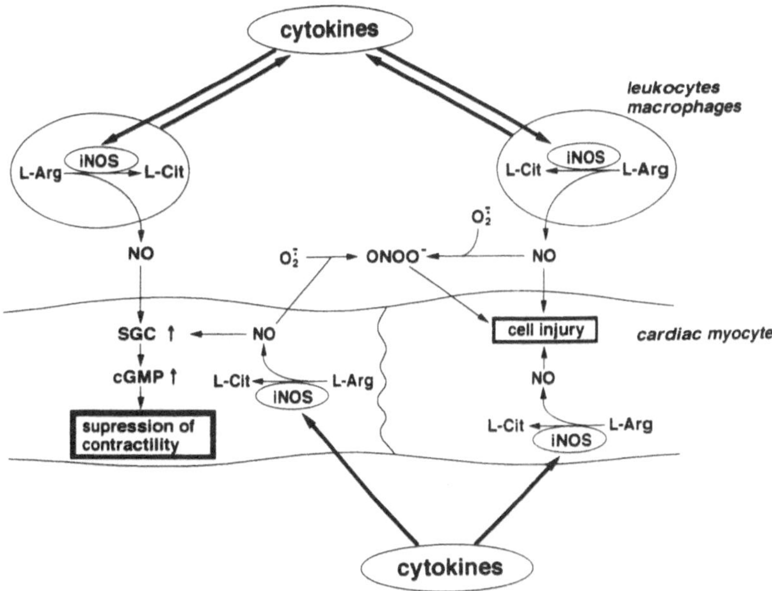

FIG. 3. Inducible NOS-(*iNOS*) mediated L-arginine–NO pathway and cardiac function

radicals to form peroxynitrite and cause further change to hydroxy radicals, which are potent cytotoxic molecules (Fig. 3) [49].

A new clinical entity known as systemic inflammatory response syndrome (SIRS) has attracted attention [50]. It is characterized by shock and multiorgan failure accompanied by hypotension, tachypnea, hypothermia or hyperthermia, and leukocytosis in clinical conditions such as advanced viral or bacterial infection, severe trauma, and burns. SIRS is also characterized by the presence of depressed myocardial contractile function, which is usually reversible and does not accompany overt signs of organic changes in the heart such as inflammation or myocardial necrosis.

Experimental as well as clinical studies have revealed the pathogenic role of iNOS-derived NO in SIRS [51]. Various inflammatory cytokines are produced in SIRS, and circulating or locally released cytokines induce iNOS expression in a variety of tissues including the heart and vessels, where NO exhibits its cardiodepressant and vasodilatory actions by the aforementioned mechanisms. In an experimental model of sepsis, increased iNOS activity and generation of NO have been noted and NOS synthesis inhibitors such as L-NMMA have been shown to improve systemic hemodynamics, cardiac function, and survival ratio [51,52]. The beneficial effects of NO synthesis inhibitor in septic shock have also been noted clinically [53].

A recent report by Yang et al. showed the expression of iNOS mRNA and protein in the rat heart with allograft rejection [54]. This expression was noted in cardiac myocytes, infiltrating inflammatory cells, and microvascular endothelial and smooth muscle cells. NO produced in these cells may be involved in the pathogenic mechanisms of myocardial damage during allograft rejection. Because viral myocarditis is mediated by an immune and inflammatory process [55], we investigated the role of NO in the pathogenesis of murine viral myocarditis induced

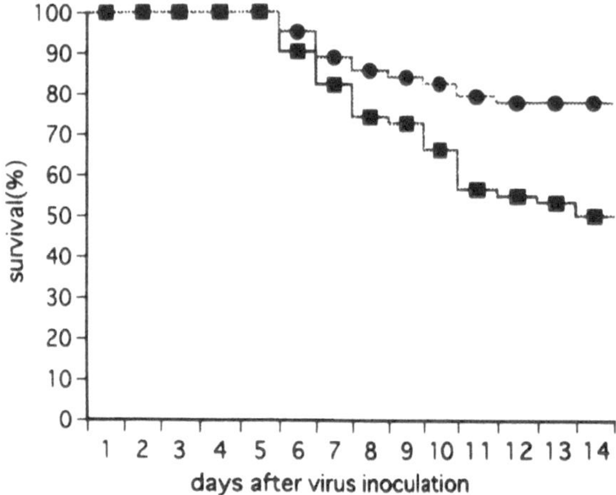

FIG. 4. Effect of N^G-nitro-L-arginine methylester (L-NAME) treatment on survival in coxsackievirus B3-induced myocarditis. Normal drinking water was given to the non-treated mice (*squares*); to the L-NAME-treated mice (*circles*), drinking water containing 0.01% (10 mg/100 ml) L-NAME was given after virus inoculation

by coxsackievirus B3 [56]. Using reverse transcriptase polymerase chain reaction (RT-PCR), we found the expression of iNOS mRNA in the heart to be first detected on day 4 after virus inoculation and to continue for 1 month. Six days after virus inoculation, iNOS activity in the heart measured by the conversion of ^{3}H-L-arginine to ^{3}H-L-citrulline showed an eightfold increase as compared with controls. Positive immunostaining for iNOS protein was observed exclusively in the infiltrating inflammatory cells.

To study whether NO plays a role in the process of myocardial damage in acute viral myocarditis, we examined the effect of L-NAME, a NO synthesis inhibitor, on mortality and cardiac damage in the same model. Treatment with L-NAME significantly decreased the mortality rate and reduced the severity of heart failure, the area of myocardial necrosis, and the degree of inflammatory cell infiltration as compared to those in nontreated mice (Fig. 4). Thus, NO and NO-derived products are an important mediator of myocardial damage in viral myocarditis. Because viral myocarditis is suggested to be closely related to the pathogenesis of idiopathic dilated cardiomyopathy, it is interesting that deBelder et al. reported the predominant activity of iNOS as compared witn cNOS in the homogenate of endomyocardium obtained by biopsy from patients with dilated cardiomyopathy [57]. Therefore, NO can be regarded as a common important modulating factor of myocardial damage in various pathological conditions mediated by immune and inflammatory processes such as SIRS, allograft rejection and viral myocarditis [58].

NO exhibits other actions on the heart. NO inhibits platelet aggregation by the mechanism dependent on cGMP [59]. Experimental data indicated that inhibition of NO production promotes mechanical injury-induced arterial thrombosis, while increasing NO synthesis or administration of exogenous NO donors protects against arterial thrombosis [60]. Yao et al. recently reported that in anesthetized dogs increasing NO production by L-arginine or administration of sodium nitroprusside, a NO donor, inhibited platelet aggregation and delayed intracoronary thrombus formation and reocclusion after thrombolysis [61].

NO is also known to inhibit leukocytes adhesion, although the mechanisms are still unknown [62]. Adhesion of leukocytes to the endothelium has been shown to play an important role in the initial steps of cardiac damage during ischemia reperfusion. Ma et al. showed increased leukocyte adhesion to the endothelium following ischemia reperfusion and implicated the loss of endothelium-derived NO during reperfusion as the mechanism [63]. Endothelium-derived NO is likely to serve to protect myocardial ischemia by inhibiting adhesion of leukocytes and aggregation of platelets in addition to its vasodilatory action [64].

The mechanisms of NO production from the endothelium during ischemia and ischemia reperfusion are unclear, but an increased concentration of intracellular Ca^{2+} with the increased influx of calcium is suggested to play a key role in that process. The importance of inhibition by NO of leukocyte adhesion in protecting against myocardial injury is supported by the report of Lefer et al., who showed that intracoronary administration of a NO donor resulted in a modest restoration of regional function and a reduction in the infarct size in open-chest dogs that underwent myocardial ischemia-reperfusion [65]. This improvement was associated with attenuation of adherence of LTB4-stimulated neutrophils to the coronary artery. Adhesion of leukocytes to cardiac myocytes is also an important factor in the early process of myocardial injury [66]. Experimental studies in vitro demonstrated that the adhesion of leukocytes to cardiac myocytes produced highly compartmentalized oxidative direct cardiotoxic effects [67]. We recently found that NO donors inhibit adhesion of platelet-activating factor-(PAF) activated leukocytes to interleukin-1β-stimulated isolated rat cardiac myocytes. NO inhibited adhesion by acting mainly on leukocytes without affecting the up-regulation of CD11b/CD18 (in manuscript). The inhibition of leukocyte adhesion may operate as an important mechanism in the effects of NO donors on protection against ischemia reperfusion damage (see Fig. 2) [68].

NO may also exert direct myocardial relaxant effects. A recent experimental report using isolated ejecting guinea pig heart demonstrated that administration of sodium nitroprusside, a NO donor, resulted in a premature onset and acceleration of early left-ventricular (LV) relaxation without change in the rate of pressure rise [69]. These effects were independent of changes in cardiac loading and possibly attributable to the reduced myofilament response to Ca^{2+} caused by increases in cGMP. In support of this experimental observation, Paulus et al. showed that in humans the infusion of NO donor sodium nitroprusside into the coronary artery resulted in a significant reduction in left peak and end-systolic pressures, an early onset of left-ventricular isovolumic relaxation, and increased left-ventricular distensibility [70]. These effects likely resulted from activation of guanylate cyclase in the cardiomyocytes or were mediated by other proteins influencing various cellular functions.

Nitric oxide may also serve to protect against myocardial stunning. In preinstrumented conscious dogs, Hasebe et al. reported that inhibition of endogenous NO synthesis by L-NNA enhanced myocardial stunning independent of its effects on coronary blood flow [71]. They thought that the origin of NO was likely to be the vascular endothelium and implied that the inhibitory effect of NO was caused by the supression of Ca^{2+} influx by increasing cGMP. In addition, the action of NO on oxygen radicals and platelet aggregation may participate in the mechanisms of protection against stunning by NO.

Conclusion

Because of its multiple biological actions, NO is likely to play several important roles in the cardiovascular system. NO is involved in such aspects of the physiological regulation of the cardiovascular system as control of coronary vascular tone and cardiac contractility by the cNOS-mediated pathway. In addition, cNOS-derived NO may serve to protect against myocardial injury in pathological conditions such as ischemia reperfusion by preserving coronary blood flow and inhibiting platelet aggregation and leukocyte adhesion. On the other hand, the presence of a large amount of NO generated via the iNOS-mediated pathway in the heart is likely to depress cardiac function in the pathological conditions.

Therefore, NO is regarded as a molecule with two divergent effects, one cytoprotective and the other cytotoxic, and the role of NO may differ in different pathological conditions. Future development combining physiological, biochemical, and molecular approaches will lead to further understanding of the role of NO in physiological and pathological conditions in the heart.

References

1. Furchgott RF, Zawadzki JV (1980) The obligatory role of the endothelial cells in the relaxation of arterial smooth muscle by acetylcholine. Nature 288:373–376
2. Palmer RM, Ferrige AG, Moncada S (1987) Nitric oxide release accounts for the biological activity of endothelium-derived relaxing factor. Nature 327:524–526
3. Moncada S, Palmer RMJ, Higgs EA (1991) Nitric oxide: physiology, pathology, and pharmacology. Pharmacol Rev 43:109–141
4. Nathan C (1992) Nitric oxide as a secretory product of mammalian cells. FASEB J 6:3051–3064
5. Schmidt HHHW, Walter U (1994) NO at work. Cell 78:919–925
6. Moncada S, Higgs A (1993) The L-arginine-nitric oxide pathway. N Engl J Med 329:2002–2012
7. Lowenstein CL, Snyder SH (1992) Nitric oxide, a novel biologic messenger. Cell 70:705–707
8. Nathan C, Xie Q-W (1994) Regulation of biosynthesis of nitric oxide. J Biol Chem 269:13725–13728
9. Dinerman JL, Lowenstein, Snyder SH (1993) Molecular mechanisms of nitric oxide regulation. Potential relevance to cardiovascular disease. Circ Res 73:217–222
10. Nishida K, Harrison DG, Navas JP, Fisher AA, Dockery SP, Uematsu M, Narem RM, Alexander PW, Murphy TJ (1992) Molecular cloning and characterization of the constitutive bovine aortic endothelial nitric oxide synthase. J Clin Invest 90:2092–2096
11. Hirata K, Kuroda R, Sakoda T, Katayama M, Inoue N, Suematsu M, Kawashima S, Yokoyama M (1995) Inhibition of endothelial nitric oxide synthase activity by protein kinase C. Hypertension 25:180–185
12. Lyons CR, Orloff GJ, Cunningham JM (1992) Molecular cloning and functional expression of an inducible nitric oxide synthase from a murine macrophage cell line. J Biol Chem 267:6370–6374
13. Koide M, Kawahara Y, Nakayama I, Tsuda T, Yokoyama M (1993) Cyclic AMP elevating agents induce an inducible type of nitric oxide synthase in cultured vascular smooth muscle cells. J Biol Chem 268:24959–24966
14. Hirata K, Miki N, Sakoda T, Kawashima S, Yokoyama M (1995) Low concentration of oxidized low-density lipoprotein and lysophosphatidylcholine upregulate constitutive nitric oxide synthase mRNA expression in bovine aortic endothelial cells. Circ Res 76:958–962

15. Ueno M, Kawashima S, Morita M, Tsumoto S, Iwasaki T (1994) Depressed endothelium-dependent vasodilatory response in the hindquarter resistance vessels in heart failure dogs. Jpn Circ J 58:778–786
16. Rees DD, Palmer RMJ, Schultz R, Hodson HF, Moncada S (1990) Characterization of three inhibitors of endothelial nitric oxide synthase in vitro and in vivo. Br J Pharmacol 101:746–752
17. Harrison DG, Bates JN (1993) The nitrovasodilators: new ideas about old drugs. Circulation 87:1461–1467
18. Kelm M, Schrader J (1990) Control of coronary vascular tone by nitric oxide. Circ Res 66:1561–1575
19. Vanhouette PM, Shimokawa H (1989) Endothelium-derived relaxing factor and coronary vasospasm. Circulation 80:1–9
20. Flavahan NA (1992) Atherosclerosis or lipoprotein-induced endothelial dysfunction. Potential mechanisms underlying reduction in EDRF/nitric oxide pathway. Circulation 85:1927–1938
21. Rubanyi GM, Romero JC, Vanhoutte PM (1986) Flow-induced release of endothelium-derived relaxing factor. Am J Physiol 250:H1145–H1149
22. Quyyumi AA, Dakak N, Andrews NP, Gilligan DM, Panza JA, Cannon RO (1995) Contribution of nitric oxide to metabolic coronary vasodilation in the human heart. Circulation 92:320–326
23. Parent RM, Al-Obaidi, Lavallee M (1993) Nitric oxide formation contributes to β-adrenergic dilation of resistance coronary vessels in conscious dogs. Circ Res 73:241–251
24. Yamabe H, Okumura K, Ishisaka H, Tsuchiya T, Yasue H (1992) Role of endothelium-induced nitric oxide in myocardial reactive hyperemia. Am J Physiol 70:123–130
25. Node K, Kitakaze M, Kosaka H, Komamura K, Minamino T, Tada M, Inoue M, Hori M, Kamada T (1995) Plasma nitric oxide end products are increased in the ischemic canine heart. Biochem Biophy Res Commun 211:370–374
26. Forstermann U (1986) Properties and mechanisms of production and action of endothelium-derived relaxing factor. J Cardiovasc Pharmacol 8:S45–S51
27. Yasue H, Horio Y, Nakamura N, Fujii H, Imoto N, Sonoda R, Kugiyama K, Obata K, Morikami Y, Kimura T (1986) Induction of coronary artery spasm by acetylcholine in patients with variant angina: possible role of the parasympathetic nervous system in the pathogenesis of coronary artery spasm. Circulation 74:955–963
28. Yokoyama M, Hirata K, Miyake R, Akita H, Fukuzaki H (1990) Lysophosphatidylcholine: essential role in the inhibition of endothelium-dependent vasorelaxation by oxidized low density lipoprotein. Biochem Biophys Res Commun 168:301–308
29. Ohara Y, Peterson TE, Harrison DG (1993) Hypercholesterolemia increases endothelial superoxide anion production. J Clin Invest 91:2546–2551
30. Egashira K, Inou T, Hirooka Y, Yamada A, Maruoka Y, Kai H, Sugimachi M, Suzuki S, Takeshita A (1993) Impaired coronary blood flow response to acetylcholine in patients with coronary risk factors and proximal atherosclerotic lesions. J Clin Invest 91:29–37
31. Leung W-H, Lau C-P, Wong C-K (1993) Beneficial effect of cholesterol-lowering therapy on coronary endothelium-dependent relaxation in hypercholesterolemic patients. Lancet 341:1496–1500
32. Schulz R, Nava E, Moncada S (1992) Induction and potential biologic relevance of a Ca^{2+}-independent nitric oxide synthase in the myocardium. Br J Pharmacol 105:575–580
33. Balligand J-L, Kelley RA, Smith TW, Michel T (1994) Identification of a constitutive isoform of NO synthase in adult rat cardiac myocytes (abstract). Circulation 90:I-193
34. Brady AJB, Warren JB, Poole-Wilson PA, Williams TJ, Harding ASE (1993) Nitric oxide attenuates cardiac myocyte contraction. Am J Physiol 265:H176–H182
35. Balligand J-L, Kelley RA, Marsden PA, Smith TW, Michel T (1993) Control of cardiac muscle cell function by an endogenous nitric oxide signalling system. Proc Natl Acad Sci USA 90:347–351

36. Han X, Shimoni Y, Giles WR (1994) An obligatory role for nitric oxide in autonomic control of mammalian heart rate. J Physiol (Camb) 476:309–314
37. Hare JM, Keaney JF, Balligand J-L, Loscalzo J, Smith TW (1995) Role of nitric oxide in parasympathetic modulation of β-adrenergic myocardial contractility in normal dogs. J Clin Invest 95:360–366
38. Klimaschewski L, Kummer W, Mayer B, Couraud JY, Pressler B, Philippin B, Heym C (1992) Nitric oxide synthase in cardiac nerve fibers and neurons of rat and guinea pig heart. Circ Res 71:1533–1537
39. Mery PF, Pavoine C, Belhassen L, Pecker F, Fisch Meister R (1993) Nitric oxide regulates cardiac Ca^{2+} current. Involvement of cGMP-inhibited and cGMP-stimulated phosphodiester through guanylyl cyclase activation. J Biol Chem 268:26286–26295
40. Weyrich AS, Ma X-l, Buerke M, Murohara T, Armstead VE, Lefer AM, Nicolas JM, Thomas AP, Lefer DJ, Vinten-Johansen J (1994) Physiological concentrations of nitric oxide do not elicit acute negative inotropic effect in unstimulated cardiac muscle. Circ Res 75:692–700
41. Tsujino M, Hirata Y, Imai T, Kanno K, Eguchi S, Ito H, Marumo F (1994) Induction of nitric oxide synthase gene by interleukin-1β in cultured rat cardiomyocytes. Circulation 90:375–383
42. Balligand J-L, Ungureanu D, Simmons WW, Pimental D, Malinski TA, Kapturczak M, Taha Z, Lowenstein CJ, Davidoff AJ, Kelly RA, Smith TW, Michel T (1994) Cytokine-inducible nitric oxide synthase (iNOS) expression in cardiac myocytes. J Biol Chem 269:27580–27588
43. Pagani FD, Baker LS, Hsi C, Knox M, Fink MP, Visner MS (1992) Left ventricular systolic and diastolic dysfunction after infusion of tumor necrosis factor-δ in conscious dogs. J Clin Invest 90:389–398
44. Balligand J-L, Ungureanu D, Kelly RA, Kobzik L, Pimental D, Michel T, Smith TW (1993) Abnormal contractile function due to induction of nitric oxide sunthesis in rat cardiac myocytes follows exposure to activated macrophage-conditioned medium. J Clin Invest 91:2314–2319
45. Kinugawa K-I, Takahashi T, Kohmoto O, Yao A, Aoyagi T, Momomura S-I, Hirata Y, Serizawa T (1994) Nitric oxide-mediated effects of interleukin-6 on $(Ca^{2+})_i$ and cell contraction in cultured chick ventricular myocytes. Circ Res 75:285–295
46. Pinsky DJ, Cai B, Yang X, Rodriguez C, Sciacca RR, Cannon PJ (1995) The lethal effects of cytokine-induced nitric oxide on cardiac myocytes are blocked by nitric synthase antagonism or transforming growth factor β. J Clin Invest 95:677–685
47. Stuehr DJ, Nathan CF (1989) Nitric oxide. A macrophage product responsible for cytostasis and respiratory inhibition in tumor target cells. J Exp Med 169:1543–1555
48. Lepoivre M, Fieschi F, Coves J, Thelander L, Fotecave M (1991) Inactivation of ribonucleotide reductase by nitric oxide. Biochem Biophys Res Commun 179:442–448
49. Beckman JS, Beckman TW, Chien J, Marshall PA, Freeman BA (1990) Apparent hydroxyl radical production by peroxynitrite: implications for endothelial injury from nitric oxide and superoxide. Proc Natl Acad Sci USA 87:1620–1624
50. Bone RC, Balk RA, Cerra FB, Dellinger RP, Feim AM, Knaus WA, Schein RM, Sibbald WJ (1992) Definitions for sepsis and organ failure and guidelines for the use of innovative therapies in sepsis. The ACCP/SCCM Consensus Conference Committee, American College of Chest Physicians/Society of Critical Care Medicine. Chest 101:1644–1655
51. Ungureanu-Longrois D, Balligand J-L, Kelly RA, Smith TW (1995) Myocardial contractile dysfunction in the systemic inflammatory response syndrome: role of a cytokine-inducible nitric oxide synthase in cardiac myocytes. J Mol Cell Cardiol 27:155–167
52. Thiemermann C, Vane J (1990) Inhibition of nitric oxide synthesis reduces the hypotension induced by bacterial lipopolysaccharides in the rat in vivo. Eur J Pharmacol 182:591–595
53. Nava E, Palmer RMJ, Moncada S (1991) Inhibition of nitric oxide synthesis in septic shock. How much is beneficial? Lancet 38:1555–1557

54. Yang X, Chowdhury N, Cai B, Brett J, Marboe C, Sciacca RR, Michler RE, Cannon PJ (1994) Induction of myocardial nitric oxide synthase by cardiac allograft rejection. J Clin Invest 94:714–721
55. Estrin M, Huber SA (1987) Coxsackievirus B3-induced myocarditis. Am J Pathol 127:335–341
56. Mikami S, Kawashima S, Kanazawa K, Hirata K, Katayama Y, Hotta H, Hayashi Y, Ito H, Yokoyama M (1996) Expression of nitric oxide synthase in a murine model of viral myocarditis induced by coxsackievirus B3. Biochem Biophys Res Commun 220:983–989
57. deBelder AJ, Radomski MW, Why HJF, Richardson PJ, Bucknall CA, Salas E, Martin JF, Moncada S (1993) Nitric oxide synthase activity in human myocardium. Lancet 341:84–85
58. Russel ME, Wallace AF, Wyner LR, Newell JB, Karnovsky MJ (1995) Upregulation and modulation of inducible nitric oxide synthase in rat cardiac allografts with chronic rejection and transplant arteriosclerosis. Circulation 92:457–464
59. Radomski MW, Palmer RMJ, Moncada S (1987) The role of nitric oxide and cGMP in platelet adhesion to vascular endothelium. Biochem Biophys Res Commun 148:1482–1489
60. Golino P, Cappelli-Bigazzi M, Ambrosio G, Ragni M, Russolillo E, Condrelli M, Chiariello M (1992) Endothelium-derived relaxing factor modulates platelet aggregation in an in vivo model of recurrent platelet aggregation. Circ Res 71:1447–1456
61. Yao S-K, Akhtar S, Scott-Burden T, Ober JC, Golino PG, Buja M, Casscells W, Willerson JT (1995) Endogenous and exogenous nitric oxide protect against intracoronary thrombosis and reocclusion after thrombolysis. Circulation 92:1005–1010
62. Kubes P, Suzuki M, Granger DN (1991) Nitric oxide: an endogenous modulator of leukocyte adhesion. Proc Natl Acad Sci USA 88:4651–4655
63. Ma XL, Weyrich AS, Lefer DJ, Lefer AM (1993) Diminished basal nitric oxide release after myocardial ischmeia and reperfusion promotes neutrophil adherence to coronary endothelium. Circ Res 72:403–412
64. Weyrich AS, Ma XL, Lefer AM (1992) The role of L-arginine in ameliorating reperfusion injury after myocardial ischemia in the cat. Circulation 86:279–288
65. Siegfried MR, Erhardt J, Rider T, Ma XL, Lefer AM (1991) Cardioprotection and attenuation of endothelial dysfunction by organic nitric oxide donors in myocardial ischemia-reperfusion. J Pharmacol Exp Ther 260:668–675
66. Entman ML, Youker K, Shappell SB, Siegel C, Rothlein R, Dreyer WJ, Schmalstieg FC, Smith CW (1990) Neutrophil adherence to isolated adult canine myocytes: evidence for a CD18-dependent mechanism. J Clin Invest 85:1497–1506
67. Entman, ML, Youker K, Shoji T, Kukielka G, Shappell SB, Taylar AA, Smith CW (1992) Neutrophil-induced oxidative injury of cardiac myocytes a compartmented system requiring CD11b/CD18-ICAM-1 adherence. J Clin Invest 90:1335–1345
68. Lefer DJ, Nakanishi K, Johnston WE, Vinten-Johansen J (1993) Antineutrophil and myocardial-protecting actions of a novel nitric oxide donar after acute myocardial ischemia and reperfusion in dogs. Circulation 88:2337–2350
69. Grocott MR, Fort S, Lewis MJ, Shah AM (1994) Myocardial relaxant effect of nitric oxide in isolated ejecting hearts. Am J Physiol 266:H1699–H1705
70. Paulus WJ, Vantrimpont PJ, Shah AM (1994) Acute effects of nitric oxide on left ventricular relaxation and diastolic distensibility in humans: assessment by bicirinary sodium nitroprusside infusion. Circulation 89:2070–2078
71. Hasebe N, Shen Y-I, Vatner SF (1993) Inhibition of endothelium-derived relaxing factor enhances myocardial stunning in conscious dogs. Circulation 88:2862–2871

Cardiac Mechanoenergetics and Coronary Circulation

Kunihisa Kohno, Miyako Takaki, Takehiko Matsushita, Yasunori Nakayama, Shunsuke Suzuki, Juichiro Shimizu, Junichi Araki, Hiromi Matsubara, and Hiroyuki Suga

Summary. Coronary circulation variously modifies cardiac performance. We have been characterizing the left-ventricular mechanoenergetics modified by interventions related to the coronary circulation by fully using the Vo_2–PVA–E_{max} framework, in which Vo_2 is myocardial oxygen consumption, PVA is the systolic pressure-volume area as a measure of the total mechanical energy generated by ventricular contraction, and E_{max} is the end-systolic maximum elastance as an index of ventricular contractility. This framework has been powerful in analyzing cardiac mechanoenergetics under various physiological and pathophysiological inotropic conditions. Coronary hypoperfusion depresses E_{max}, PVA, and Vo_2 at a given preload and lowers the volume-loading Vo_2–PVA relation in a parallel manner until the coronary reserve exhausts at a high PVA: coronary hyperperfusion does the opposite. These mechanoenergetic responses to changes in coronary perfusion partly resemble those to negative and positive inotropism. Postischemic reperfusion causes myocardial stunning, characterized by disproportionately increased coronary flow and Vo_2 despite depressed E_{max}, resulting in a twice-normal oxygen cost of E_{max}. This type of mechanoenergetic change contrasts with ordinary negative inotropism. We also referred to some other interventions accompanying changes in coronary perfusion. By using E_{max} and PVA, we have thus been able to characterize left-ventricular mechanoenergetics related to primarily and secondarily modified coronary circulation in a manner such as could not have been possible otherwise.

Key words. Cardiac mechanics—Cardiac energetics—E_{max}—PVA—Gregg phenomenon—Oxygen consumption—Oxygen cost of contractility

Introduction

Cardiac contraction consumes ATP produced mainly (95%) by oxidative phosphorylation under aerobic conditions. Cardiac oxygen consumption (Vo_2) amounts to $8\,ml\,min^{-1}\,100\,g^{-1}$ at rest and as much as $50\,ml\,min^{-1}\,100\,g^{-1}$ during heavy exercise. This range of oxygen supply is essential for the full swing of cardiac performance between rest and heavy exercise. Because the coronary oxygen extraction ratio is usually 75–85% and the oxygen capacity of arterial blood is 20 vol%, this range of myocardial Vo_2

Department of Physiology II, Okayama University Medical School, Okayama, 700 Japan.

requires coronary blood flow to be 50–300 $ml\ min^{-1}\ 100\ g^{-1}$. Any limitation of coronary flow limits maximum exercise; examples are ischemic cardiac diseases, including angina pectoris and myocardial infarction. Reperfusion after ischemia causes myocardial stunning instead of immediately restoring normal myocardial function. Frequent brief ischemic insults augment resistance against ischemic injury (perconditioning) rather than cumulate myocardial ischemic injury. Thus, coronary circulation variously modifies cardiac performance [1–3], and much of the underlying mechanoenergetics is still unknown. We briefly review what we have understood so far of this aspect.

Coronary Flow and Heart Preparation

The first investigator to discover the vital importance of coronary circulation in the mammalian heart was Henry Newell Martin, the first professor of physiology at the Johns Hopkins University [4]. Before this discovery, physiologists could only use cold-blooded animal hearts, such as a frog heart as used by Otto Frank [5]. These hearts have no coronary circulation; oxygen diffuses from the epicardium and endocardium directly into the myocardium. Martin recognized that a mammalian heart could not continue to beat in saline after excision, unlike a frog heart. He found a "Eureka" when he noted the necessity of coronary circulation for a continuously beating mammalian heart preparation. He then established a heart-lung preparation (Fig. 1) [4], as early as 12 years before O. Langendorff established the isolated heart preparation now well known as the Langendorff preparation [6]. E. H. Starling used Martin's heart-lung preparation and added a Starling resister in the arterial circulation to find the law of the heart [7].

Suga and Sagawa [8] established a new type of excised, cross-circulated canine heart preparation without any interruption of coronary circulation for the study of

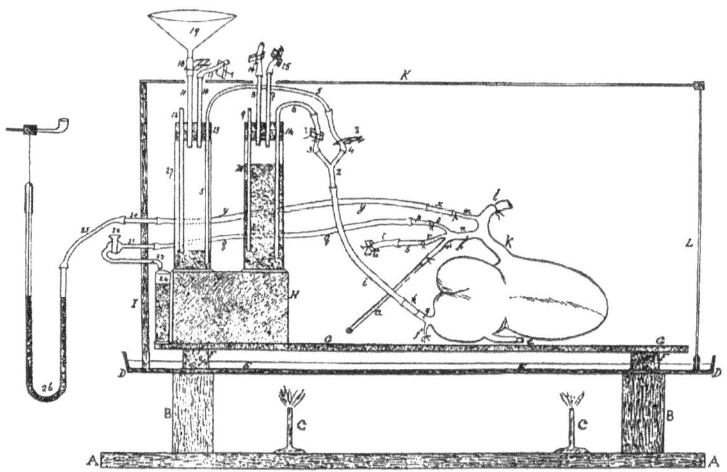

FIG. 1. The world's first mammalian heart-lung preparation made by H.N. Martin in 1883, 12 years before the famous Langendorff heart preparation. For more details, see [4]. (Reproduced from this century-old reference [4], with our acknowledgments to both the publisher and the author)

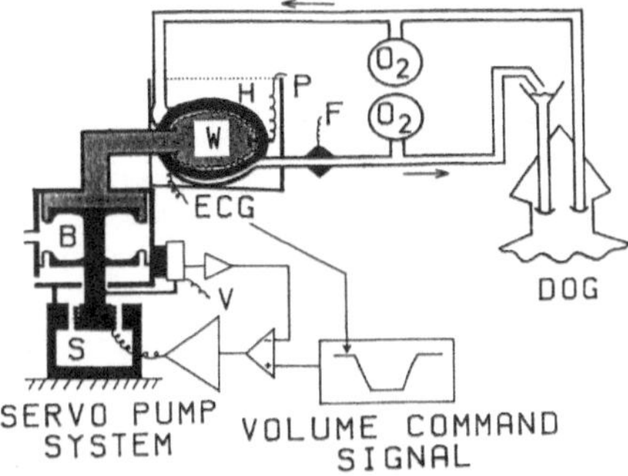

FIG. 2. Our excised, cross-circulated canine heart preparation. *B*, Bellofram pump (Nitto-Shoji, Osaka, Japan); *DOG*, support dog; *ECG*, electrocardiogram; *F*, electromagnetic flowmeter; *H*, excised cross-circulated heart; *P*, ventricular pressure signal; *S*, servo pump; *V*, volume signal; *W*, water in the intraventricular balloon. For more details, see [12]. (From [12], with permission)

left-ventricular (LV) pressure–volume (P–V) relation in 1974. Previous investigators used to prepare excised, cross-circulated hearts by first excising the heart and connecting it to already prepared cross-circulation tubes after a finite period (5–15 min) of myocardial ischemia [9–11].

We have experienced that even a temporary interruption of coronary circulation causes a long-lasting reactive hyperemia and loss of coronary autoregulation. We believe that our heart preparation, with no experience of interruption of coronary circulation, is still the most viable type of blood-perfused heart preparation that enables accurate instantaneous measurements of ventricular volume, coronary flow, and its arteriovenous oxygen content difference [8,12].

Coronary Flow and Arteriovenous Oxygen Content Difference

In our excised, cross-circulated canine heart preparation (Fig. 2), coronary perfusion is essential and we take great care not to stop it at all during the surgical preparation. If coronary flow stops by accident for minutes, the heart becomes pale and weakens in contractility. When the reperfusion starts, a reactive hyperemia occurs and the heart becomes mosaically flushed and pale, gradually recovering contractility. The increased coronary flow accompanies a decreased arteriovenous oxygen content difference (AVO_2D) for 30 min to a few hours, reflecting a sustained failure of the coronary autoregulation [13].

We pooled coronary flow, AVO_2D, and Vo_2 data in many successful experiments on excised cross-circulated canine hearts suffering no accidental ischemia (Fig. 3) [13]. This figure also plots data from in situ right-heart-bypass canine hearts and in situ intact canine hearts. When LV mechanical load increased in terms of PVA (systolic P–V area, a measure of total mechanical energy generated by ventricular contraction; see following), LV Vo_2 also increased with various combinations of coronary flow and AVO_2D changes among the three different types of heart preparations. The excised, cross-circulated hearts increased Vo_2 by increasing AVO_2D more than coronary flow. The other two types of in situ heart preparations increased Vo_2 by increasing coronary

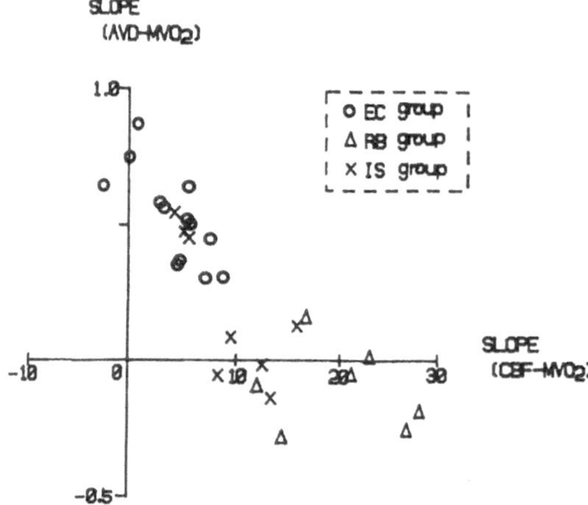

FIG. 3. Different responses of coronary flow and arteriovenous oxygen content difference (AVO_2D) to changes in myocardial oxygen consumption (Vo_2) with systolic pressure–volume area (PVA) by left-ventricular (LV) volume loading. *EC* (*circles*), Excised, cross-circulated canine hearts; *RB* (*triangles*); canine right-heart-bypass hearts; *IS* (*crosses*), in situ intact canine hearts. Slope (*CBF-MVO_2*) on the abscissa indicates the ratio of change in coronary blood flow in ml min^{-1} $100 g^{-1}$ to change in myocardial Vo_2 in ml O_2 min^{-1} $100 g^{-1}$ (*CBF*, coronary blood flow). Slope (*AVD-MVO_2*) on the ordinate indicates the ratio of change in coronary arteriovenous oxygen content difference in ml O_2 $100 g^{-1}$ to change in myocardial Vo_2 in ml O_2 min^{-1} $100 g^{-1}$. For more details, see [13]. (From [13], with permission)

flow more than AVO_2D. Intriguingly, these disproportionate changes in coronary flow and AVO_2D did not significantly affect the Vo_2–PVA relation under LV volume loading in a given contractile state.

Myocardial Vo_2–PVA–E_{max} Framework

Using the excised, cross-circulated canine hearts, we have extensively studied the relation between LV mechanical energy (PVA, systolic pressure–volume area) and Vo_2 in variably loaded contractions in a stable contractile state as well as in variably altered contractile states (Fig. 4) [14–18]. We have shown that the total mechanical energy generated by a ventricular contraction is quantifiable by a specific area in the P–V diagram [19]. The specific area is the area circumscribed by the systolic P–V loop segment, the end-systolic P–V relation (ESPVR), and the diastolic P–V relation (Fig. 4a). We designated this area as systolic P–V area (PVA) [19].

We have found that LV Vo_2 correlates linearly and closely with PVA in a stable contractile state [20]. The Vo_2–PVA relation line divides Vo_2 into the PVA-dependent and PVA-independent components at the Vo_2 axis [intercept (b) in Fig. 4b]. The PVA-dependent component is the Vo_2 primarily for cross-bridge cycling, and the PVA-independent component is the Vo_2 for Ca^{2+}, Na^+, and K^+ handling in the excitation–contraction (E–C) coupling and basal metabolism [14–18]. The slope (a) of this line represents the oxygen cost of PVA and its reciprocal represents the contractile efficiency of the contractile machinery [14–18]. This cost characterizes the efficiency of the energy conversion from Vo_2 via ATP to the mechanical energy generated

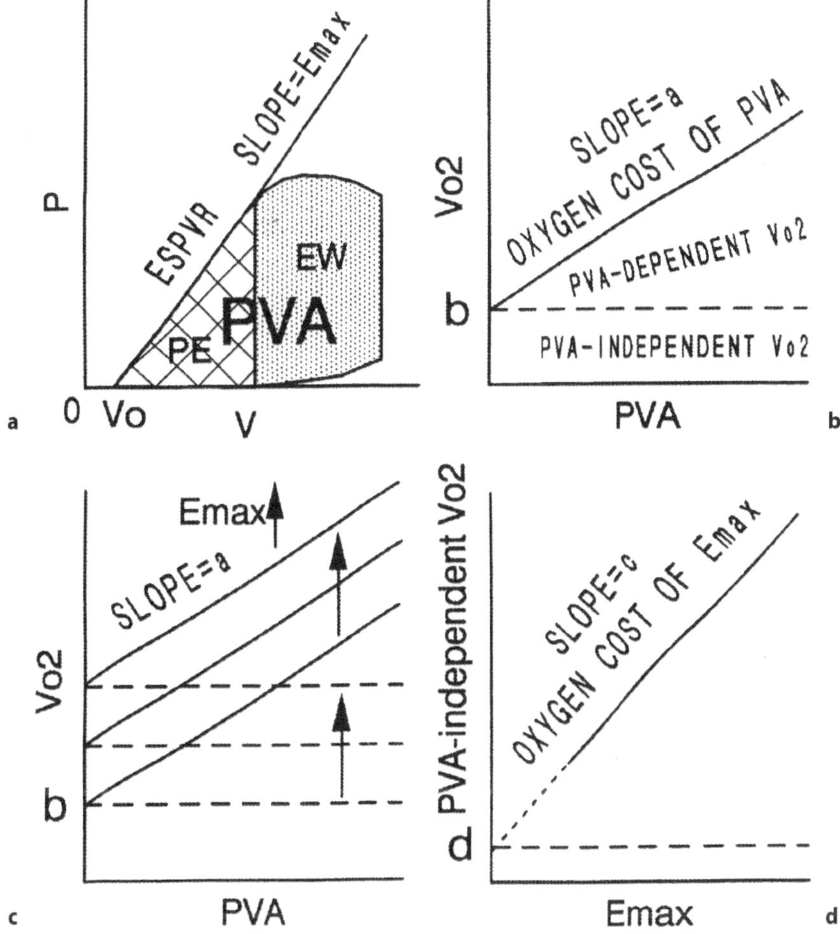

FIG. 4a–d. Schematic illustration of E_{max} and PVA (**a**), the Vo_2–PVA relation in a stable E_{max} (**b**), parallel elevations of the Vo_2–PVA relation with increasing E_{max} (**c**) and the PVA-independent Vo_2–E_{max} relation (**d**). *ESPVR*, end-systolic pressure–volume relation; E_{max}, slope of the ESPVR; *EW*, external work; *PE*, potential energy, *Vo*, unstressed volume. **a** Slope of the Vo_2–PVA relation during LV volume loading in a stable E_{max}. **b** *y*-Axis intercept of the Vo_2–PVA relation, which divides Vo_2 into the PVA-dependent Vo_2 and PVA-independent Vo_2. **c** Slope of the PVA-independent Vo_2–E_{max} relation. **d** Basal metabolism. For more details, see [36]. (From [36], with permission)

by cross-bridge cycling [14–18]. We have assumed that the PVA-independent Vo_2 component is constant at different preloads and thus PVA levels in a stable contractile state [21].

The Vo_2–PVA relation line shifts upward with an increase in end-systolic maximum elasticity (E_{max}) by Ca^{2+}, catecholamines, digitalis, paired pulse pacing, and new cardiotonic agents such as milrinone, amrinone, besnarinone, pimobendan, DPl-201 106, and EMD 53998 [14–17,22]. On the contrary, the Vo_2–PVA relation shifts down with a decrease in E_{max} by β-blockers, calcium antagonists, or pentobarbital sodium [23].

Figure 4c illustrates the parallel shifts of the Vo_2–PVA relations with increases in E_{max}. The slope of the Vo_2–PVA relation usually remains unchanged. Therefore, the oxygen cost of PVA also remains unchanged by most positive and negative inotropic agents [14–18]. On the other hand, the PVA-independent Vo_2 changes in proportion to E_{max} [14–18].

Figure 4d shows the relation between PVA-independent Vo_2 and E_{max}. It is linear and has a *y*-axis intercept (d) that corresponds to basal metabolism. We have

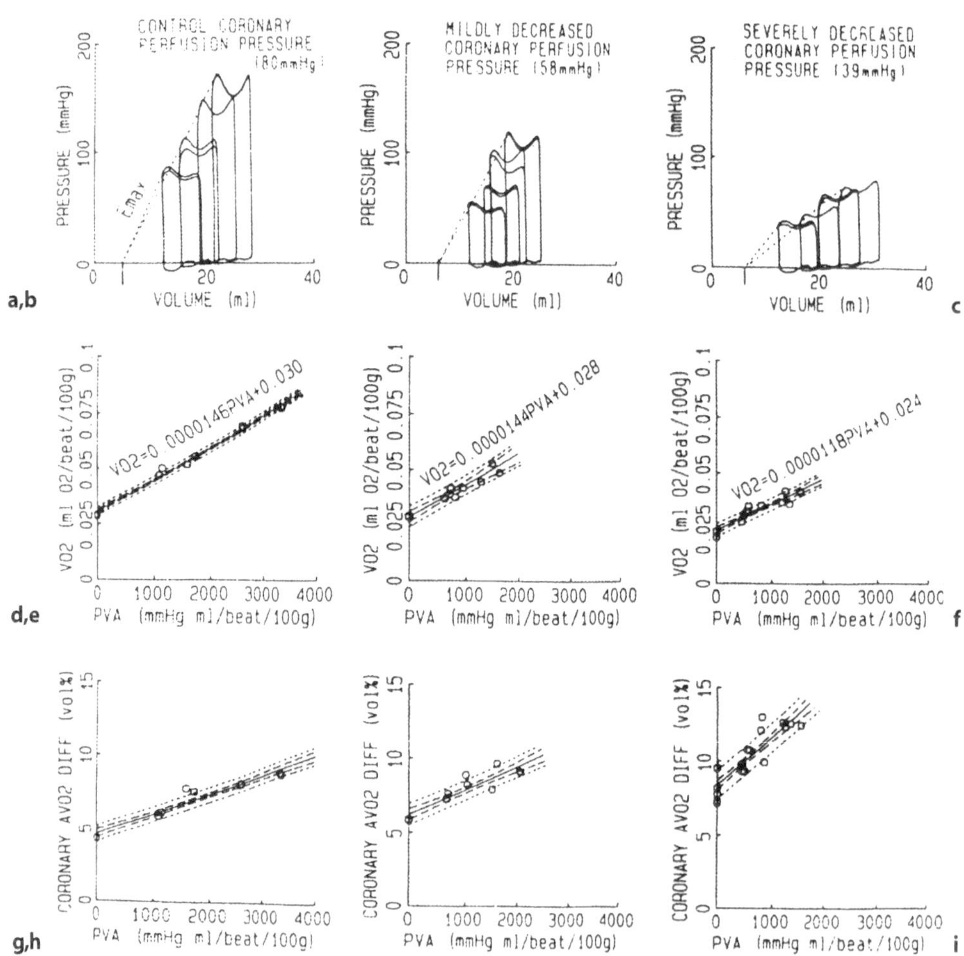

Fig. 5a–i. Mechanoenergetic effects of low coronary perfusion. **a–c** Pressure–volume loops and the E_{max} (*dotted*) lines under control (**a**, 80 mmHg), moderately decreased (**b**, 58 mmHg), and severely decreased coronary pressure (**c**, 39 mmHg), left through right, respectively, in one representative excised, cross-circulated canine left ventricle. **d–f** Corresponding Vo_2–PVA relations. **g–i** Corresponding relations between coronary arteriovenous oxygen content difference (AVO_2) and PVA. For more details, see [25]. (From [25], with permission)

found that most positive and negative inotropic interventions have a linear PVA-independent Vo_2–E_{max} relation [14,16–18,23].

Inotropic interventions that yielded strange responses in the oxygen costs of PVA and E_{max} include cardiac cooling and warming, ventricular pacing, intraventricular block, and ventricular wall vibration [24]. Because these interventions did not involve obvious changes in coronary circulation, we do not discuss on these interventions further in this chapter.

Low Coronary Perfusion Pressure

With these conceptual backgrounds, we decreased coronary perfusion pressure of the excised, cross-circulated canine heart to study directly the effect of coronary perfusion on cardiac mechanoenergetics [25]. When mean coronary pressure decreased from 82 ± 8 to 51 ± 6 mmHg, we found that the Vo_2-PVA relation obtained by LV volume loading shifted down without a change in the slope. Therefore, the oxygen cost of PVA and contractile efficiency remained unchanged. When coronary pressure further decreased from 51 ± 6 to 32 ± 6 mmHg, we found a different response of the Vo_2-PVA relation; the Vo_2-PVA relation further shifted down, but the slope also decreased (Fig. 5).

We measured coronary arteriovenous lactate difference and always found lactate extraction, rather than its production, even at the lowest coronary pressure. However, lactate extraction almost vanished at the largest PVA under the lowest coronary pressure. At the same time, AVO_2D was maximum and E_{max} was minimum. This result suggested to us that the heart may have been suffering from subendocardial ischemia at a large PVA under a limited coronary perfusion. Even no lactate extraction as net may mean lactate production in the subendocardium; if lactate extraction in most layers exceeds the lactate production in a layer, lactate in the coronary venous blood will show a net extraction.

Unlike most conventional positive and negative inotropic agents that did not change the basal metabolic Vo_2 [14], basal metabolism measured as Vo_2 under KCl arrest decreased with decreases in coronary pressure (Fig. 6h). Therefore, the decreases in PVA-independent Vo_2 with an increasing coronary insult consist of not only suppression of the E–C coupling but also depression of basal metabolism.

We could draw a schema of the smaller oxygen cost of PVA observed at a severely decreased coronary pressure (CP) (Fig. 7a). Our point is that the lowest Vo_2-PVA relation under severe coronary hypoperfusion (SD) does not indicate an increased contractile efficiency [25]. It is a composite Vo_2-PVA relation; it passes from the lowest Vo_2-PVA point on the normal Vo_2-PVA relation at a lower E_{max} to a higher Vo_2-PVA point on the most depressed Vo_2-PVA relation at the lowest E_{max}. It crosses the other Vo_2-PVA relations at gradually decreasing E_{max} levels (Fig. 7b).

Exactly the same phenomenon as this may not occur in in situ normal hearts because mean coronary pressure and ventricular ejection pressure change together. Although Sunagawa et al. [26] did not study cardiac energetics, they observed that E_{max} decreased when they simultaneously decreased coronary pressure and ejection pressure (and end-systolic pressure) in the excised, cross-circulated canine LV. As the result, the ESPVR became curled as E_{max} decreased (Fig. 8b). This and our previous observations seem to be different aspects of the same phenomenon that coronary insult downregulates myocardial mechanoenergetics; i.e., depression of ventricular

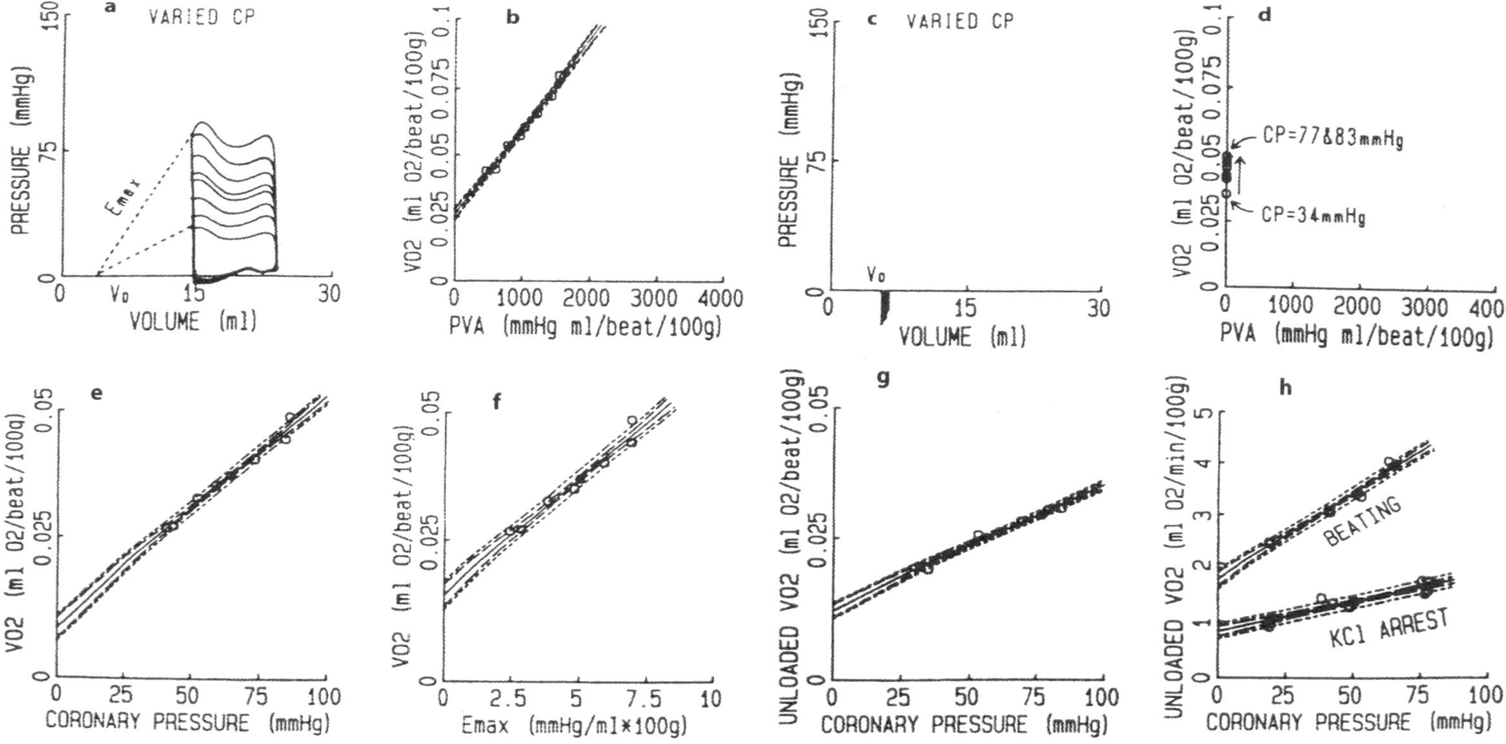

FIG. 6a–h. Coronary dependence of left-ventricular mechanoenergetics. *Top row*, Changes in pressure–volume trajectories of ejecting contractions (**a**) and unloaded contractions (**c**) with decreasing coronary pressure and their V_{O_2}–PVA relations (**b** and **d**). *Bottom row*, Corresponding relations of changes in V_{O_2} with changes in coronary pressure (**e** and **g**) and E_{max} (**f**). **h** The dependence of both beating-heart unloaded V_{O_2} and KCl-arrest V_{O_2} on coronary pressure (*CP*). For more details, see [25]. (From [25], with permission)

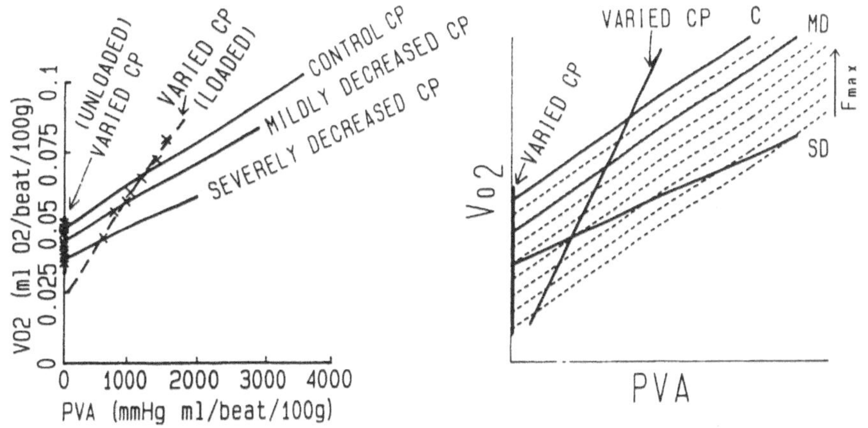

FIG. 7. **a** Vo_2–PVA relations under LV volume loading at three differnet coronary perfusion pressures (*solid lines*) and in unloaded and constantly preloaded contractions during varied coronary pressure (*dashed lines*). **b** Schematic illustration of the concept of the mechanoenergetic effects of low coronary perfusion. *CP*, Coronary pressure; *C*, *MD*, and *SD* (**b**) are control CP, mildly decreased CP, and severely decreased CP, respectively. In **b**, the composite Vo_2–PVA relation for SD starts from one Vo_2–PVA line and traverses lower parallel Vo_2–PVA relations, unlike the two Vo_2–PVA relations for C and MD. The slope of this composite Vo_2–PVA relation is less than the other Vo_2-PVA relations for C and MD. For more details, see [25]. (From [25], with permission)

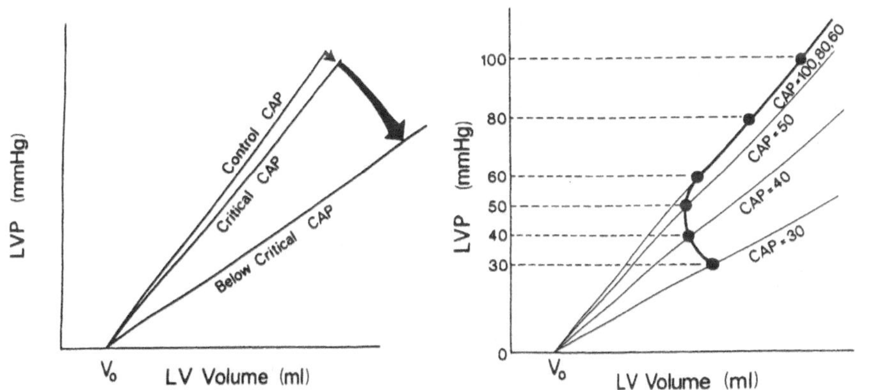

FIG. 8a,b. Effects of three different levels of coronary perfusion pressure on the end-systolic pressure-volume relation (**a**) and of combined decreases in coronary pressure and end-systolic pressure (**b**). *CAP*, Coronary arterial pressure in mmHg, *LV*, left ventricle; *LVP*, LV pressure. *Control CAP*, *Critical CAP*, and *Below Critical CAP* in **a** correspond to CAP = 100, 80, 60, CAP = 50, and CAP = 40, 30, respectively, in **b**. For more details, see [26]. (From [26], with permission)

contractility to match ventricular energy supply and demand. This downregulation may be caused by limited supplies of oxygen and metabolic substrates and limited washout of carbon dioxide and metabolic wastes by the coronary circulation. However, we do not yet know which component of basal metabolism is dependent on coronary pressure.

Gregg Phenomenon

The Gregg phenomenon mentions an augmentation of contractility by coronary hyperperfusion [27]. When we changed mean coronary pressure (= mean arterial pressure) between 50 and 175 mmHg in a specially prepared canine heart preparation, LV E_{max} [as well as peak isovolumic pressure P_{max} and peak rate of rise of isovolumic pressure $(dP/dt)_{max}$] changed from the control level (100%) obtained at 100 mmHg of mean arterial and coronary pressure (Fig. 9) [28]. Figure 9 also shows the E_{max} responses to different combinations of carotid sinus (CS) and aortic arch baroreceptor (AA) input and coronary pressures (COR). These results suggest that the Gregg phenomenon usually hides behind the negative inotropic effect of the intact baroreceptor reflex.

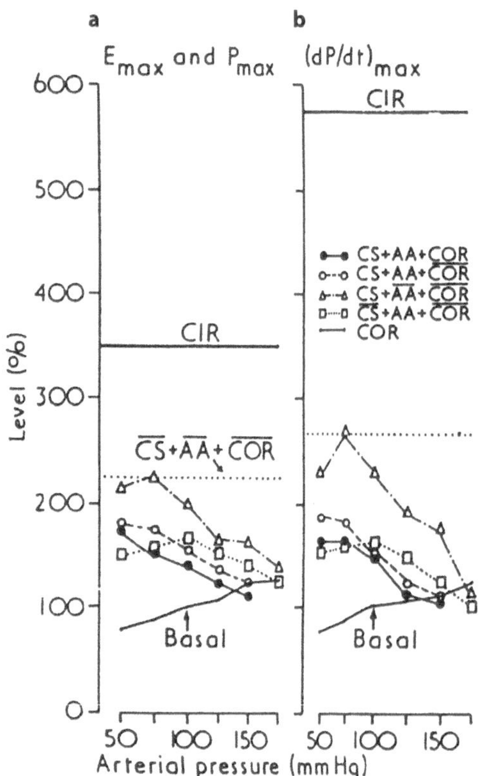

FIG. 9a,b. Effects of mean arterial pressure on E_{max}, *Pmax* (peak isovolumic pressure at a fixed volume), and (dP/dt)max of the left ventricle under graded denervation and hemodynamic isolation. Mean arterial pressure served not only as the coronary pressure but also as the carotid sinus and aortic arch baroreceptor input pressures. *CS*, Carotid sinus baroreceptor nerves intact; $\overline{CS}$, carotid sinus baroreceptor nerves denervated; *AA*, aortic arch baroreceptor and vagus nerves intact; $\overline{AA}$, aortic arch baroreceptor and vagus nerves denervated; *COR*, coronary pressure changing simultaneously with mean arterial pressure; $\overline{COR}$, fixed coronary pressure at 100 mmHg; *CIR*, cerebral ischemic response; *basal*, both sympathetically and vagally denervated, and only coronary pressure changes (COR alone), indicating the pure effect of coronary pressure changes on left-ventricular contractility. For more details, see [28]. (From [28], with permission)

The Gregg phenomenon seems to depend on cardiac loading conditions. Although we observed it explicitly, as just mentioned, the Sunagawa et al. data do not show when peak ventricular pressure increased with increasing ventricular volume in proportion to increases in coronary pressure (>80 mmHg) [26]. A possible explanation would be as follows: as soon as the Gregg phenomenon starts to appear when coronary pressure rises, ventricular load simultaneously increases and the energy and coronary demands also increase, thereby leaving no room for this phenomenon to appear.

Goto et al. [29] put adenosine into the coronary circulation of the excised, cross-circulated rabbit heart and increased coronary blood flow to restore coronary pressure. The Vo_2–PVA relation ascended in a parallel manner by this procedure (Fig. 10). This result indicates that coronary flow rather than coronary pressure is the determinant of the Gregg phenomenon. Coronary hyperperfusion distends the coronary vasculature and myocardium, in turn augmenting myocardial E–C coupling. Kitakaze et al. [30] showed that Ca^{2+} handling in the E–C coupling increases in the Gregg phenomenon. This seems consistent with the parallel elevation of the Vo_2–PVA relation.

We compared the behavior of the Vo_2–PVA relation between high Ca^{2+} and high coronary flow in the newly instituted, excised, cross-circulated guinea pig heart preparation [30a]. We imposed the Ca^{2+} and coronary perfusion interventions on the isovolumically contracting LV at the same volume after determining the reference Vo_2–PVA relation by changing intraventricular volume. The two Vo_2–PVA relations under increased coronary Ca^{2+} infusion rate and perfusion pressure were steeper than the reference Vo_2–PVA relation, and the two were mutually superimposable. Moreover, the relations between PVA-independent Vo_2 and E_{max} were also superimposable on each other between the Ca^{2+} and coronary interventions. This indicates that Ca^{2+} and coronary interventions have the same oxygen cost of E_{max}. This result is consistent with more Ca^{2+} involvement in the E–C coupling during the coronary hyperperfusion, but not so with the gardenhose effect (or coronary and myocardial distension) by coronary hyperperfusion. Although the length-dependent Ca^{2+} recruitment exists in the law of the heart, the mechanoenergetic

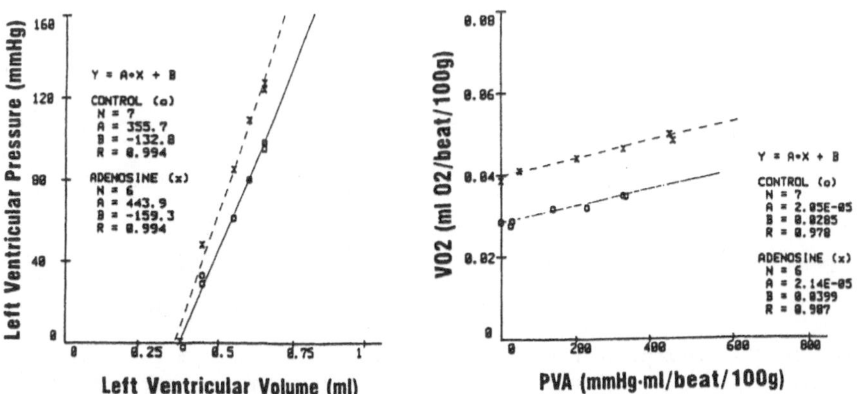

FIG. 10a,b. Mechanoenergetic effects of coronary hyperperfusion by adenosine (*crosses and dashed line*) under a fixed coronary pressure (93 ± 11 mmHg); *open circles and solid line*, control data. For more details, see [29]. (From [29], with permission)

result suggests more Ca^{2+} recruitment in the Gregg phenomenon than in this "law."

Constant Pressure Versus Constant Flow of Coronary Perfusion

Saeki et al. [31] compared the volume-loading V_{O_2}–PVA relation under a constant coronary perfusion pressure and a constant coronary flow in excised, cross-circulated canine hearts. They set constant coronary flow and pressure during volume loading at the values observed at zero PVA. The volume-loading V_{O_2}–PVA relation obtained under the constant flow condition was linear, as is the relation obtained under the constant pressure condition that is the usual condition in our studies. However, the slope of the volume-loading V_{O_2}–PVA relation under the constant flow was less by about 20% on average than that under the constant pressure. During volume loading under constant flow, coronary pressure gradually decreased with increases in PVA. On the contrary, during volume loading under constant pressure, coronary flow gradually increased with increases in PVA, as usually observed in our studies. These changes in coronary pressure and flow are caused by normal coronary autoregulation [1–3]. The Gregg pheomenon seems responsive to the different slopes of the V_{O_2}–PVA relations observed under constant pressure and flow conditions. We would interpret the smaller slope of the V_{O_2}–PVA relation as indicating the slope of a composite V_{O_2}–PVA relation in the same way as under the severely decreased coronary pressure referred to previously [25].

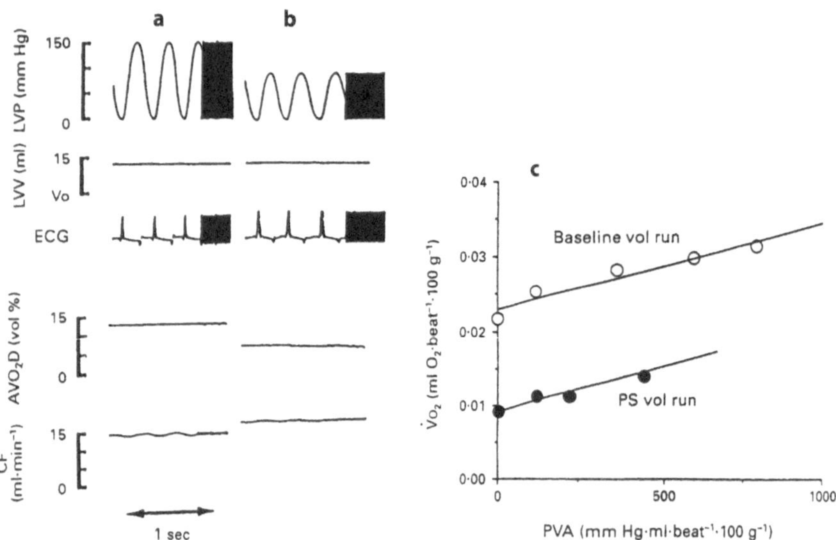

FIG. 11a–c. Effects of pentobarbital sodium given intracoronarily on E_{max} and coronary flow of an excised, cross-circulated canine heart. **a** and **b** Data compared before and during pentobarbital administration (6.7 mg/min into coronary circulation, corresponding to 40 mg/l coronary blood). **c** Corresponding V_{O_2}–PVA relations compared. *PS*, Pentobarbital sodium; *vol run*, LV volume-loading V_{O_2}–PVA relation at a stable E_{max}. For more details, see [23]. (From [23] with permission)

Pentobarbital

We have been using pentobarbital sodium as an anesthetic to prepare the excised, cross-circulated heart; it is a cardiovascular depressant and a coronary dilator. We studied its effect on the Vo_2–PVA relation because we expected that coronary dilation and thus coronary hyperperfusion followed by the Gregg phenomenon would compensate the depression of E_{max} and the descent of the Vo_2–PVA relation.

Figure 11 shows a representative example of decreased E_{max} associated with increased coronary flow [23]. This contrasts with a combination of decreased E_{max} and coronary flow in response to propranolol, for example [25]. When the additional pentobarbital dose was very small or less than 1 mg/min into the coronary flow above the basal pentobarbital anesthesia (25 mg/kg iv), both coronary flow and E_{max} increased, probably by the Gregg phenomenon, in 5 of the 12 hearts. Otherwise, both coronary flow and E_{max} rather gradually decreased with the increased pentobarbital dose. The Vo_2–PVA relation descended in proportion to E_{max}, and both oxygen costs

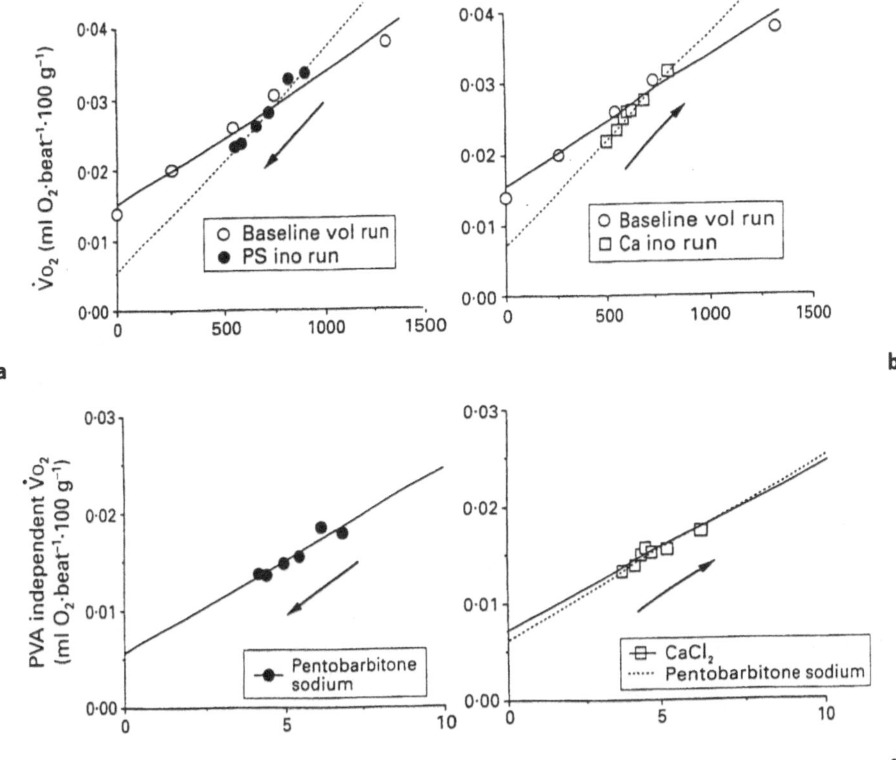

FIG. 12a–d. Comparisons of the volume-loading Vo_2–PVA relation (**a** and **b**) and the PVA-independent Vo_2–E_{max} relation (**c** and **d**) before and during either pentobarbital (**a** and **c**) or $CaCl_2$ administration of graded doses (**b** and **d**). *Baseline vol run* (*open circles*), LV volume-loading Vo_2–PVA relation before pentobarbital; *PS ino run* (*solid circles*), Vo_2–PVA relation during graded doses (to 40 mg/l coronary blood) of pentobarbital; *$CaCl_2$* (*squares*) graded doses (to 0.7 mmol/l coronary blood) of $CaCl_2$ into the coronary flow. *Arrows* indicate the direction of the changes. For more details, see [23]. (From [23], with permission)

of PVA and E_{max} remained unchanged (Fig. 12). Pentobarbital of the subanesthetic dose may dilate coronary circulation, and the Gregg phenomenon may compensate the depressions of E_{max} and the Vo_2–PVA relation.

Nipradilol

Nipradilol is a relatively new agent to possess β-blocking and vasodilating actions. We found that E_{max} decreased and PVA at a constant isovolumic volume also decreased with increasing doses of nipradilol (<1.3 mg/l blood in coronary circulation); the Vo_2–PVA relation descreased [32]. The mechanoenergetic responses were similar to those of propranolol. Because coronary flow did not increase, the Gregg phenomenon seems not to be involved. No increase in coronary flow is consistent with the vasodilating effect of nipradilol, primarily on large coronary vessels.

Capsaicin

Capsaicin, a component of hot pepper, selectively acts on sensory nerve endings in cardiac muscles and coronary arterial smooth muscles. Capsaicin-sensitive C fibers seem to release calcitonin gene-related peptide (CGRP), which causes vasodilation and has positive inotropic and chronotropic effects on the heart. It has, however, nonselective effects including inhibition of cardiac muscle excitability and enhancement of vascular smooth muscle tone at high doses. We studied whether and how intracoronary infusion of capsaicin affects mechanoenergetics of the excised blood-perfused canine heart and coronary vascular resistance.

We found that capsaicin at low concentrations (0.4–2.8 μmol/l coronary blood) increased E_{max} by 8–30% and Vo_2 in only three of ten hearts [33]. Coronary flow did not change, but the AVO_2D increased, indicating no contribution of the Gregg phenomenon in this phase. This effect is possibly a specific action of capsaicin on capsaicin-sensitive sensory nerves in the LV. This result coincides with the reported histochemical observations that the distribution of capsaicin-sensitive sensory nerves is not dense in the canine LV. Capsaicin at high doses (0.9–14 μmol/l), however, decreased E_{max} and coronary flow dose dependently [33]. It also lowered the linear Vo_2–PVA relation without a change in the slope, decreasing unloaded Vo_2.

These effects of high-dose capsaicin seem to be a direct negative inotropic action on cardiac muscles associated with enhancement of coronary arterial smooth muscle tone, as these effects were not desensitized [33]. The negative inotropic action was not the toxic effect of capsaicin, because there were no morphological changes of myocardial cells and mitochondria [33]. The Vo_2–PVA relation descended in proportion to decreases in E_{max}. Both oxygen costs of PVA and E_{max} remained unchanged.

Ischemic Arrest and Postmortem Rigor

Rigor is the state of muscle in which tight cross-bridges are formed because of lack of ATP. When we completely stopped coronary perfusion and observed the LV P–V relation, the P–V relation became flat when anoxic arrest occurred. Then, the relation steepened gradually over 30 min to 1 h until it became vertical (Fig. 13) [34]. The rigor means that ATP reached zero.

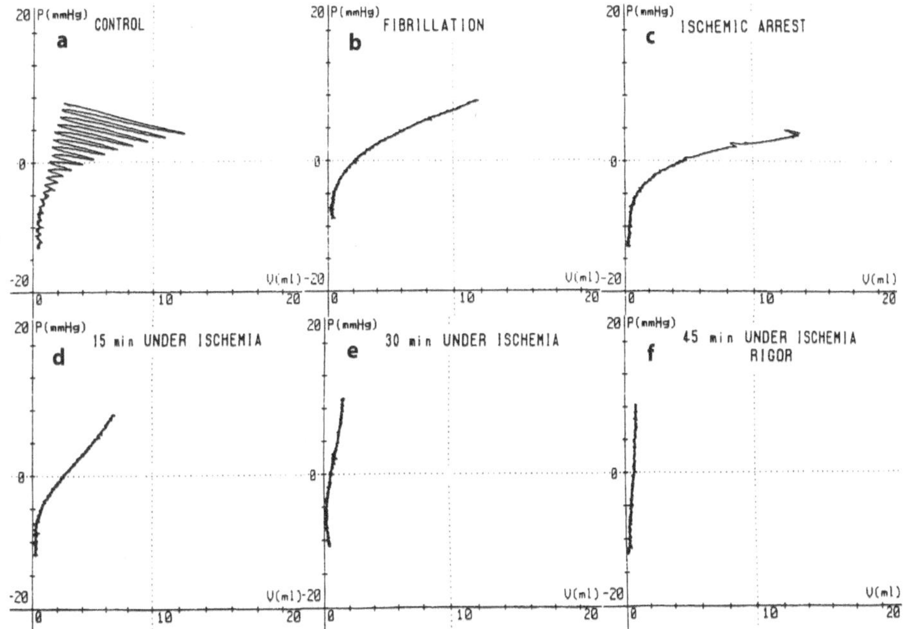

FIG. 13a-f. Left-ventricular pressure-volume (P-V) relations gradually changing from control (**a**) to fibrillation (**b**), ischemic arrest (**c**), and 15, 30, and 45 min of ischemia (**d**, **e**, and **f**, respectively). The final state is the rigor. For more details, see [34]. (From [34], with permission)

Postischemic Reperfusion Stunning

We produced global ischemia for 15 min at 36°C and started reperfusion, watching the consequent cardiac mechanoenergetics in the excised, cross-circulated canine heart [35]. Although coronary perfusion became twice normal, LV E_{max} remained depressed over at least 2 h. This condition is myocardial stunning. The end-diastolic P–V curve was normal. Figure 14 compares the Vo_2–PVA relations obtained by changing ventricular loads before and after stunning. Although E_{max} decreased, the Vo_2 axis intercept was comparable between control and stunning. In addition, the slope of the Vo_2–PVA relation was smaller after stunning, indicating a smaller oxygen cost of PVA or a greater contractile efficiency. We do not yet know the mechanism of this change.

LV contractility increased before and after stunning with coronary infusion of Ca^{2+} at the same isovolumic volume. The obtained Vo_2–PVA relation was steeper than the control Vo_2–PVA relation obtained by LV volume loading in either case. After plotting PVA-independent Vo_2 against E_{max}, we found that the oxygen cost of E_{max} was nearly twice as steep and the PVA-independent Vo_2 at any E_{max} was greater in the stunned heart than in the control. The basal metabolic Vo_2 under KCl arrest was however comparable between control and stunning.

Various mechanisms could explain the increased oxygen cost of E_{max}. One of them may be a decreased Ca^{2+} sensitivity (or responsiveness) of the contractile proteins. In this case, a greater amount of Ca^{2+} must be released to produce the same contractility

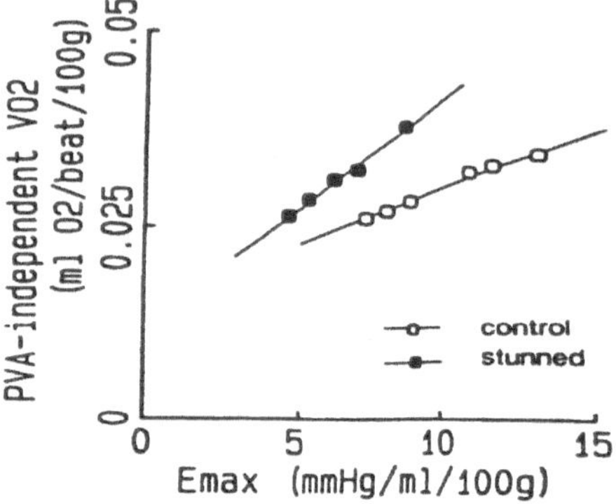

FIG. 14. Increased oxygen cost of E_{max} in stunned myocardium (*solid symbols*; control, *open symbols*) of the excised, cross-circulated canine heart. E_{max} was increased with Ca^{2+} given intracoronarily. For more details, see [35]. (From [35], with permission)

and a greater amount of ATP be consumed for removal of the greater amount of released Ca^{2+}. Another mechanism would be a leaky sarcoplasmic reticulum (SR), which allows futile Ca^{2+} cycling. In this case, the Ca^{2+} pump normally removes sarcoplasmic Ca^{2+} by consuming ATP in the $2Ca^{2+}$:1ATP stoichiometry, but part of the sequestered Ca^{2+} would leak out and additional ATP would be necessary to remove the leaking Ca^{2+}. A third mechanism would be the contribution of the Na^+–Ca^{2+} exchanger. In this case, if Ca^{2+} removal by the SR fails, Ca^{2+} removal by the Na^+–Ca^{2+} exchanger ($3Na^+$: $1Ca^{2+}$) would increase. Although this exchanger per se does not consume ATP, the Na^+–K^+ pump consumes ATP in the $3Na^+$:1ATP stoichiometry when this pump removes the Na^+ influx via the exchanger. As a whole, 1 ATP is consumed for removal of 1 Ca^{2+}. As the result, twice as much ATP would be consumed for removal of the same amount of Ca^{2+} if the SR Ca^{2+} pump does not remove it. We have not directly identified any of these mechanisms in our canine stunned hearts.

Hypercapnic Acidosis, Postacidotic Stunning, and Its Prevention

Coronary hypercapnic acidosis occurred after loading carbon dioxide in the coronary arterial blood with a gas exchanger inserted in the arterial cross-circulation tubing. Coronary arterial Pco_2 was 90 mmHg and pH was 7.0 without hypoxia. E_{max} and PVA at a fixed LV volume were nearly halved. Coronary flow increased at the smallest LV volume but slightly decreased at the maximum LV volume, resulting in decreased coronary autoregulation [36]. The Vo_2–PVA relation decreased with a decreased PVA-independent Vo_2. However, when E_{max} recovered to control with intracoronary Ca^{2+} infusion, the PVA-independent Vo_2 increased by 20% from control. Coronary flow did not significantly change. The oxygen cost of E_{max} during acidosis was 53% greater than control.

When the acidosis was rapidly corrected, E_{max} and the Vo_2–PVA relation did not recover to the preacidosis levels [37]. The oxygen cost of contractility with Ca^{2+} further increased to a level comparable to the postischemic stunning level. However, when a selective Na^+–H^+ exchanger, 5-(*N*,*N*-dimethy)-amiloride (DMA), was administered before the correction, the mechanoenergetics returned to the preacidosis control.

Assuming the expected effect of DMA, the result suggests that the stunning involves Ca^{2+} overloading via the combination of the Na^{+}–H^{+} and Na^{+}–Ca^{2+} exchangers [37].

Ca^{2+}-Free, High-Ca^{2+} Perfusion

We attempted to produce stunned myocardium by Ca^{2+} overload without hypoxia or ischemia. To this end, we interrupted the blood circulation of the excised, cross-circulated canine heart with oxygenated (Po_2 > 600 mmHg) Ca^{2+}-free, high-Ca^{2+} (16 mmol/l) Tyrode perfusions for 10 min (5 min each) and investigated the mechanoenergetics thereafter [38]. One hour after the restart of blood perfusion, coronary flow almost recovered but E_{max} remained decreased by 40%. The PVA-independent Vo_2 also decreased, unlike the stunned myocardium. These conditions continued for 1–4 h. However, oxygen costs of both PVA and E_{max} remained unchanged in this type of failing heart. This result suggests that this cardiac failure primarily involved a suppressed Ca^{2+} handling in the E–C coupling. A 10-min Ca^{2+}-free Tyrode perfusion followed by blood perfusion resulted in an enhanced E_{max}, against an expected Ca^{2+} paradox. A 10-min, high-Ca^{2+} Tyrode perfusion followed by blood perfusion resulted in no change in E_{max}. These results with three different Ca^{2+} interventions indicate that Ca^{2+}-overloading episodes do not necessarily elicit myocardial stunning [38].

Conclusion

From all these experimental findings together, we have confirmed that coronary circulation is a vitally important factor to modify cardiac mechanoenergetics. This review could not cover more details of the individual studies, not only those of other investigators but also ours. For more details of individual studies, we advise the reader to see the corresponding references.

The sensitivies of cardiac mechanoenergetic parameters (such as E_{max}, PVA, and their oxygen costs) to changes in coronary perfusion pressure and flow depend on the normality or abnormality (or physiology or pathophysiology) of the myocardium and the existing mechanoenergetic level. In general, coronary hyperperfusion upregulates myocardial mechanoenergetics by augmenting both Ca^{2+} handling and cross-bridge cycling, and the coronary hypoperfusion down-regulates them. These effects of coronary perfusion variably modify, whether by compensating or aggravating, the simultaneous changes in mechanoenergetics as the result of a positive or negative inotropic intervention. Consequently, observed changes in LV mechanoenergetics induced by an inotropic intervention are the integrated responses of its (positive or negative) inotropic action and any accompanying action on the coronary circulation (dilating or constricting). Therefore, clear decomposition of the integrated mechanoenergetic responses into the myocardial and coronary factors may not always be easy, because an imposed intervention affects coronary circulation not only directly but also indirectly via the resultant myocardial metabolism.

Acknowledgment. This work was partly supported by Grants-in-Aid for Scientific Research (04454267, 04557041, 04237219, 04557041, 05305007, 06213226 06770494, 07508003, 08670052) from the Ministry of Education, Science, Sports and Culture, Research Grants for Cardiovascular Diseases (4C-4, 7C-2) and a Research Grant on Aging and Health from the Ministry of Health and Welfare, 1994 and 1995 Joint

Research Grants Utilizing Scientific and Technological Potential in Region from the Science and Technology Agency, a grant from the Terumo Life Science Foundation, and a grant from the Murata Science Foundation, all of Japan.

References

1. Marcus ML (1983) The coronary circulation in health and disease. McGraw-Hill, New York
2. Feigl EO (1983) Coronary physiology. Physiol Rev 63:1–205
3. Hoffman JI, Spaan JA (1990) Pressure-flow relations in coronary circulation. Physiol Rev 70:331–390
4. Martin HN (1883) A new method of studying the mammalian heart. In: Studies from the biological laboratory, Johns Hopkins University, vol 2. Johns Hopkins, Baltimore, pp 119–130
5. Frank O (1899) Die Grundform des Arteriellen Pulses. Z Biol 37:483–526. English translation by Sagawa K et al (1990) J Mol Cell Cardiol 22:253–277
6. Langendorff O (1895) Untersuchungen am uberlebenden Saugethierherzen. Pflugers Arch Physiol 61:291–332
7. Patterson SW, Piper H, Starling EH (1914) The regulation of the heart beat. J Physiol (Lond) 48:465–513. [Starling EH (1918) The Linacre lecture on the law of the heart (given at Cambridge, 1915). Longmans Green, London]
8. Suga H, Sagawa K (1974) Instantaneous pressure-volume relationships and their ratio in the excised, supported canine left ventricle. Circ Res 35:117–126
9. Heymans JF, Kochmann M (1905) Une nouvelle methode de circulation artificielle a traverse le coeur isole de mammifere. Arch Int Pharmacodyn Ther 13:379–385
10. Hashimoto K, Shigei T, Imai S, Sato Y, Yago N, Uei I, Clark RE (1960) Oxygen consumption and coronary vascular tone in the isolated fibrillating dog heart. Am J Physiol 198:965–970
11. Monroe RG, French GN (1961) Left ventricular pressure-volume relationships and oxygen consumption in the isolated heart. Circ Res 9:362–374
12. Suga H, Hayashi T, Shirahata M (1983) Ventricular systolic pressure-volume area as predictor of cardiac oxygen consumption. Am J Physiol 240:H39–H44
13. Futaki S, Nozawa T, Yasumura Y, Tanaka N, Uenishi M, Suga H (1988) The different contributions of coronary blood flow to changes in myocardial oxygen consumption between excised and in situ canine hearts. In: Mochizuki M, Honig CR, Koyama T, Goldstick TK, Bruley DF (eds) Oxygen transport to tisssue X. Plenum, New York, pp 437–445 (Advances in experimental medicine and biology, vol 222)
14. Suga H (1990) Ventricular energetics. Physiol Rev 70:247–277
15. Sagawa K, Maughan WL, Suga H, Sunagawa K (1988) Cardiac contraction and the pressure-volume relationship. Oxford University Press, New York
16. Suga H, Goto Y (1991) Cardiac oxygen costs of contractility (E_{max}) and mechanical energy (PVA): new key concepts in cardiac energetics. In: Sasayama S, Suga H (eds) Recent progress in failing heart syndrome. Springer, Tokyo, pp 61–115
17. Suga H (1994) Paul Dudley White international lecture: Cardiac performance as viewed through the pressure-volume window. Jpn Heart J 35:263–280
18. Takaki M, Namba T, Araki J, Ishioka K, Ito H, Akashi T, Zhao LY, Zhao DD, Liu M, Fujii W, Suga H (1993) How to measure cardiac energy expenditure. In: Piper HM, Preusse CJ (eds) Ischemia-reperfusion in cardiac surgery, Kluwer, Dordrecht, pp 403–419
19. Suga H (1979) Total mechanical energy of a ventricle model and cardiac oxygen consumption. Am J Physiol 236:H498–H505
20. Suga H, Hayashi T, Shirahata M (1981) Ventricular systolic pressure-volume area as predictor of cardiac oxygen consumption. Am J Physiol 240:H39–H44
21. Yasumura Y, Nozawa T, Futaki S, Tanaka N, Suga H (1989) Minor preload-dependence of O_2 consumption of unloaded contraction in dog heart. Am J Physiol 256:H1289–H1294

22. Suga H, Hisano R, Goto Y, Yamada O, Igarashi Y (1983) Effect of positive inotropic agents on the relation between oxygen consumption and systolic pressure-volume area in canine left ventricle. Circ Res 53:306–318
23. Namba T, Takaki M, Araki J, Ishioka K, Suga H (1994) Energetics of the negative and positive inotropism of pentobarbitone sodium in the canine left ventricle. Cardiovasc Res 28:557–565
24. Nishioka T, Goto Y, Hata K, Takasago T, Saeki A, Taylor TW, Suga H (1996) Mechanoenergetics of negative inotropism of ventricular wall vibration in dog heart. Am J Physiol 270:H583–H593
25. Suga H, Goto Y, Yasumura Y, Nozawa T, Futaki S, Tanaka N, Uenishi M (1988) O_2 consumption of dog heart under decreased coronary perfusion and propranolol. Am J Physiol 254:H292–H303
26. Sunagawa K, Maughan WL, Friesinger G, Guzman P, Chang M, Sagawa K (1982) Effects of coronary arterial pressure on left ventricular end-systolic pressure-volume relation of isolated canine heart. Circ Res 50:727–734
27. Gregg DE, Fisher LC (1963) Blood supply to the heart. In: Handbook of physiology, section 2: circulation, vol II. American Physiological Society, Washington, DC, pp 1517–1584
28. Suga H, Sagawa K, Kostiuk DP (1976) Controls of ventricular contractility assessed by pressure-volume ratio, E_{max}. Cardiovasc Res 10:582–592
29. Goto Y, Slinker BK, LeWinter MM (1991) Effect of coronary hyperemia on E_{max} and oxygen consumption in blood-perfused rabbit hearts. Energetic consequences of Gregg's phenomenon. Circ Res 68:482–492
30. Kitakaze M, Marban E (1989) Cellular mechanism of the modulation of contractile function by coronary perfusion pressure in ferret hearts. J Physiol (Lond) 414:455–472
30a. Matsushita T, Takaki M, Fujii W, Matsubara H, Suga H (1995) Left ventricular mechanoenergetics under altered coronary perfusion in guinea pig hearts. Jpn J Physiol 45:991–1004
31. Saeki A, Goto Y, Futaki S, Kawaguchi O, Hata K, Takasago T, Taylor TW, Nishioka T, Suga H (1991) Mode of coronary perfusion modifies left ventricular contractility and energetics. Circulation 84(suppl II):11–46
32. Zhao DD, Namba T, Araki J, Ishioka K, Takaki M, Suga H (1993) Nipradilol depresses cardiac contractility and O_2 consumption without decreasing coronary resistance in dogs. Acta Med Okayama 47:29–33
33. Takaki M, Akashi T, Ishioka K, Kikuta A, Matsubara H, Yasuhara S, Fujii W, Suga H (1994) Effects of capsaicin on mechanoenergetics of excised cross-circulated canine left ventricle and coronary perfusion. J Mol Cell Cardiol 26:1227–1239
34. Suga H, Yasumura Y, Nozawa T, Futaki S, Tanaka N (1988) Pressure-volume relation around zero transmural pressure in excised cross-circulated dog left ventricle. Circ Res 63:361–372
35. Ohgoshi Y, Goto Y, Futaki S, Yaku H, Kawaguchi O, Suga H (1991) Increased oxygen cost of contractility in stunned myocardium of dog. Circ Res 69:975–988
36. Hata K, Goto Y, Kawaguchi O, Takasago T, Saeki A, Nishioka T, Suga H (1994) Hypercapnic acidosis increases oxygen cost of contractility in the dog left ventricle. Am J Physiol 266:H730–H740
37. Hata K, Takasago T, Saeki A, Nishioka T, Goto Y (1994) Stunned myocardium after rapid correction of acidosis. Increased oxygen cost of contractility and the role of the Na^+-H^+ exchange system. Circ Res 74:795–805
38. Araki J, Takaki M, Namba T, Mori M, Suga H (1995) Ca^{2+}-free, high-Ca^{2+} coronary perfusion suppresses contractility and excitation-contraction coupling energy. Am J Physiol 268:H1061–H1070
39. Takaki M, Matsubara H, Araki J, Zhao LY, Ito H, Yasuhara S, Fujii W, Suga H (1996) Mechanoenergetics of acute failing hearts characterized by oxygen costs of mechanical energy and contractility. In: Sasayama S (ed) New horizons for failing heart syndrome. Springer, Berlin Heidelberg New York Tokyo, pp 133–164

Index